DISEASES OF FRUITS AND VEGETABLE CROPS

Recent Management Approaches

DISEASES OF FRUITS AND VEGETABLE CROPS

Recent Management Approaches

Edited by
Gireesh Chand, PhD
Md. Nadeem Akhtar, PhD
Santosh Kumar, PhD

Apple Academic Press Inc.
4164 Lakeshore Road
Burlington ON L7L 1A4, Canada

Apple Academic Press Inc.
1265 Goldenrod Circle NE
Palm Bay, Florida 32905, USA

© 2021 by Apple Academic Press, Inc.

Exclusive worldwide distribution by CRC Press, a member of Taylor & Francis Group

No claim to original U.S. Government works
International Standard Book Number-13: 978-1-77188-836-3 (Hardcover)
International Standard Book Number-13: 978-0-42932-218-1 (eBook)

All rights reserved. No part of this work may be reprinted or reproduced or utilized in any form or by any electric, mechanical or other means, now known or hereafter invented, including photocopying and recording, or in any information storage or retrieval system, without permission in writing from the publisher or its distributor, except in the case of brief excerpts or quotations for use in reviews or critical articles.

This book contains information obtained from authentic and highly regarded sources. Reprinted material is quoted with permission and sources are indicated. Copyright for individual articles remains with the authors as indicated. A wide variety of references are listed. Reasonable efforts have been made to publish reliable data and information, but the authors, editors, and the publisher cannot assume responsibility for the validity of all materials or the consequences of their use. The authors, editors, and the publisher have attempted to trace the copyright holders of all material reproduced in this publication and apologize to copyright holders if permission to publish in this form has not been obtained. If any copyright material has not been acknowledged, please write and let us know so we may rectify in any future reprint.

Trademark Notice: Registered trademark of products or corporate names are used only for explanation and identification without intent to infringe.

Library and Archives Canada Cataloguing in Publication

Title: Diseases of fruits and vegetable crops : recent management approaches / edited by Gireesh Chand, Md. Nadeem Akhtar, Santosh Kumar.

Names: Chand, Gireesh, editor. | Akhtar, Md. Nadeem, editor. | Kumar, Santosh (Professor of plant pathology), editor.

Series: Innovations in horticultural science.

Description: Series statement: Innovations in horticultural science | Includes bibliographical references and index.

Identifiers: Canadiana (print) 20200178857 | Canadiana (ebook) 2020017889X | ISBN 9781771888363 (hardcover) | ISBN 9780429322181 (electronic bk.)

Subjects: LCSH: Fruit—Diseases and pests—Control. | LCSH: Vegetables—Diseases and pests—Control.

Classification: LCC SB608.F8 D57 2020 | DDC 634/.049—dc23

Library of Congress Cataloging-in-Publication Data

Names: Chand, Gireesh, editor. | Akhtar, Md. Nadeem, editor. | Kumar, Santosh (Professor of plant pathology), editor.

Title: Diseases of fruits and vegetable crops : recent management approaches / Gireesh Chand, Md. Nadeem Akhtar, Santosh Kumar.

Description: Palm Bay, Florida, USA : Apple Academic Press, [2020] | Includes bibliographical references and index. | Summary: "Diseases of Fruits and Vegetable Crops: Recent Management Approaches covers certain basic aspects of knowledge on diagnostic symptoms, modes of perpetuation and dissemination of pathogens, favorable conditions for disease development, and the latest management strategies for disease prevention and mitigation in vegetable crops, fruit crops, and plantation crops. With chapters written by experts working on specific fruit and vegetables diseases, the volume covers many vegetable and fruit crops, including pineapples, grapes, apples, guava, litchi, potatoes, peas, beans, ginger and turmeric, and many more. Each chapter reviews the specific diseases relevant to the crop and their management and includes recent research findings. The information presented here will be valuable for plant protection officers, district horticulture officers, and other government personnel in the directorates and agencies of agriculture, horticulture and plant protection, as well as plant protection experts, vegetable specialists, and others"-- Provided by publisher.

Identifiers: LCCN 2020006107 (print) | LCCN 2020006108 (ebook) | ISBN 9781771888363 (hardcover) | ISBN 9780429322181 (ebook)

Subjects: LCSH: Fruit--Diseases and pests--Control. | Vegetables--Diseases and pests--Control.

Classification: LCC SB608.F8 D555 2020 (print) | LCC SB608.F8 (ebook) | DDC 634--dc23

LC record available at https://lccn.loc.gov/2020006107

LC ebook record available at https://lccn.loc.gov/2020006108

Apple Academic Press also publishes its books in a variety of electronic formats. Some content that appears in print may not be available in electronic format. For information about Apple Academic Press products, visit our website at **www.appleacademicpress.com** and the CRC Press website at **www.crcpress.com**

About the Editors

Gireesh Chand, PhD
*Associate Professor-cum Senior Scientist,
Department of Plant Pathology, Bihar
Agricultural University, India*

Gireesh Chand, PhD, is currently an Associate Professor-cum Senior Scientist in the Department of Plant Pathology at Bihar Agricultural University, India. He was previously working as an Assistant Professor at N.D. University of Agriculture and Technology, Faizabad, India. His research specialization is in the study of molecular plant pathology of crop diseases and their management. He has a decade-long career dedicated to teaching undergraduate and postgraduate classes, research, and extension activities. He is the author or co-author of 10 books and manuals on plant pathology and has published numerous research papers, book chapters, and popular articles. He has presented many research papers at national and international seminars and symposia. With research specialization in the study of molecular plant pathology of crop diseases and their management, he has led more than 10 research projects. He was recently elected as Zonal President (EZ)–2018 of the Indian Phytopathological Society, Indian Agricultural Research Institute, New Delhi, India. He received his PhD degree in plant pathology from C.S. Azad University of Agriculture and Technology in Kanpur. India. He has received a number of awards and honors, including the P.R. Verma Award, a PhD Research Fellowship, Best KVK Award, SPPS Fellow Award, ISHA Best Student-Guide Award, Prof. M.J. Narsingham Award, Young Scientist Award, Best Paper Presentation Award, and the Excellence in Teaching Award. He visited Beijing, China, through the International Financial Support Scheme by DST, New Delhi.

About the Editors

Md. Nadeem Akhtar, PhD
Scientist (Plant Pathology), Bihar Agricultural University, Sabour, Bhagalpur, India

Md. Nadeem Akhtar, PhD, is presently working as a Scientist (Plant Pathology) at Bihar Agricultural University, Sabour, Bhagalpur, India. He earned his BSc (Agriculture) degree from Uttar Banga Krishi Vishwavidyalaya, Cooch Behar, W. B., and his master's degree in Plant Pathology from G.B. Pant University of Agriculture and Technology, Pantnagar, India. Mr. Akhtar also received the prestigious P. R. Verma Award from the Society of Mycology and Plant Pathology for his MSc thesis and an INSPIRE Fellowship from the Department of Science & Technology, Government of India during his PhD program. He was a recipient of an Excellence in Extension Scientist Award from the Agricultural Technology Development Society. He has published several research and review papers and 30 popular articles in international and national journals as well as several book chapters. He has also published more than 20 extension bulletins on different aspects of integrated disease management of different crops and mushroom production. He has attended several national and international symposia and has presented his research papers, receiving best paper presentation awards several times. He has delivered many radio and TV talks for the upscaling of farming communities. His research interests are in soil-borne fungal diseases of vegetable crops and their alternative management practices, biological control, and mushroom production technology.

Santosh Kumar, PhD
Assistant Professor-cum-Junior Scientist, Department of Plant Pathology, Bihar Agricultural University, Sabour, Bhagalpur, India

Santosh Kumar, PhD, is presently working as an Assistant Professor-cum-Junior Scientist in the Department of Plant Pathology, Bihar Agricultural University, Sabour, Bhagalpur, India. He teaches many undergraduate and postgraduate courses. He received a Young Scientist Award

About the Editors   vii

from the Bioved Research Institute of Agriculture and Technology, Allahabad, U.P.; a Best Teacher Award from Bihar Agricultural University,
Sabour; and a Prof. K.S. Bilgrami best poster presentation commendation
award from the Indian Society of Mycology and Plant Pathology. He has
published many research papers and review articles in international and
national journals as well as several book chapters. He has also published
several books, a practical manual, and two extension bulletins in mushroom
production. He has attended several national and international symposia and
has presented his research papers, receiving several best paper presentation
awards. His research interests include management of diseases of pulses and
rice, mushroom and biocontrol. He earned his PhD degree in Plant Pathology
from G.B. Pant University of Agriculture and Technology, Pantnagar, India.
Dr. Kumar was awarded with a Senior Research Fellowship from the Indian
Council of Agricultural Research, New Delhi; and a Rajiiv Gandhi National
Fellowship through the University Grants Commission, New Delhi, during
his PhD program.

Contents

Contributors .... *xi*

Abbreviations .... *xv*

Preface .... *xvii*

Part I: Fruits 1

1. **Pineapple Diseases and Their Integrated Management** 3
   Sanjeev Kumar and Erayya

2. **Diseases of Apples and Their Management** 19
   J. N. Srivastava, A. K. Singh, and R. K. Sharma

3. **Citrus Diseases and Their Management** 41
   V. Jyothi and M. E. Shilpa

4. **Diseases of Grapes and Their Management** 65
   B. D. Devamani

5. **Diseases of Custard Apples and Their Management** 75
   G. L. Sharma and N. Lakpale

6. **Diseases of Guava and Their Management** 87
   B. D. Devamani

7. **Diseases of Jackfruit Crops and Their Management** 95
   R. C. Shakywar

8. **Diseases of Litchi and Their Management** 107
   Vinod Kumar

9. **Major Diseases of Mangoes and Their Management** 125
   Supriya Gupta, Pankaj Rautela, C. S. Azad, and K. P. Singh

Part II: Vegetables 145

10. **Diseases of Potato Crops and Their Management** 147
    Shailbala

11. **Diseases of Tomato Crops and Their Management** 181
    Pankaj Rautela, Supriya Gupta, C. S. Azad, and R. P. Singh

x

Contents

12. **Diseases of Elephant Foot Yams and Colocasia Crops and Their Management** 211
   Shailbala

13. **Diseases of Garden Peas (*Pisum sativum* L.) and Their Management** ...229
   Ramesh Nath Gupta

14. **Diseases of Ginger and Turmeric and Their Management** 261
   Sunil Kumar and Gireesh Chand

15. **Diseases of Pointed Gourd Crops and Their Management** 273
   R. C. Shakywar, M. Pathak, Mukul Kumar, and R. B. Verma

16. **Diseases of Brinjals (Eggplant) and Their Management** 303
   Narender Kumar, Ajay Kumar, Gireesh Chand, and S. K. Biswas

17. **Diseases of Carrots, Radishes, and Knol Khol (Kholrabi) and Their Management** 323
   Shailbala

18. **Major Diseases of Chili and Their Management** 353
   C. S. Azad, Pankaj Rautela, Supriya Gupta, and R. P. Singh

19. **Diseases of Cucurbits and Their Management: Integrated Approaches** 379
   Gireesh Chand

20. **Diseases of Beans Crop and Their Management** 401
   Shailbala

Part III: Other Food Crops 433

21. **Economically Important Diseases of Tea (*Camellia* sp.) and Their Management** 435
   Kishor Chand Kumhar and Azariah Babu

22. **Symptomatology and Etiology of Alternariose in Root, Fruits, and Leafy Vegetables** 461
   Udit Narain, Alka Kushwaha, Rajendra Prasad, and Ved Ratan

23. **Micronutrients Deficiency in Vegetable Crops and Their Management** 489
   Shweta Shambhavi, Rakesh Kumar, Rajkishore Kumar, and Mahendra Singh

Index 509

Contributors

Md. Nadeem Akhtar
Scientist (Plant Pathology), Bihar Agricultural University, Sabour, Bhagalpur, India
E-mail: nadeemgbpuat@gmail.com

C. S. Azad
Department of Plant Pathology, College of Agriculture G.B. Pant University of Agriculture and Technology, Pantnagar–263145, Uttarakhand, India, E-mail: azadbau81@gmail.com

Azariah Babu
Tea Research Association, North Bengal Regional Research and Development Center, Nagrakata, Jalpaiguri–735225, West Bengal, India

S. K. Biswas
Department of Plant Pathology, C.S. Azad University of Agriculture and Technology, Kanpur, Uttar Pradesh, India

Gireesh Chand
Department of Plant Pathology, Bihar Agricultural University, Sabour, Bhagalpur–813210, India, E-mail: gireesh_76@rediffmail.com

B. D. Devamani
Department of Plant Pathology, UAS, GKVK, Bangalore–560065, Karnataka, India

Erayya
Department of Plant Pathology, Bihar Agricultural University, Sabour, Bhagalpur, Bihar, India

Ramesh Nath Gupta
Department of Plant Pathology, Bihar Agricultural College, BAU, Sabour, Bhagalpur, Bihar, India, E-mail: rameshnathgupta@gmail.com

Supriya Gupta
Department of Plant Pathology, College of Agriculture, G.B. Pant University of Agriculture and Technology, Pantnagar–263145, Uttarakhand, India, E-mail: gupta.supriya15@gmail.com

V. Jyothi
Scholar, Department of Plant Pathology, UAS, GKVK, Bangalore–65, Karnataka, India

Ajay Kumar
Department of Plant Pathology, C.S. Azad University of Agriculture and Technology, Kanpur, Uttar Pradesh, India

Mukul Kumar
Department of Tree Improvement, College of Horticulture and Forestry, Central Agricultural University, Pasighat–791102, Arunachal Pradesh, India

Narender Kumar
Department of Plant Pathology, C.S. Azad University of Agriculture and Technology, Kanpur, Uttar Pradesh, India, E-mail: kumar.narendra6887@gmail.com

Rajkishore Kumar
Department of Soil Science and Agricultural Chemistry, Bihar Agricultural University, Sabour–813210, Bhagalpur, Bihar, India

Rakesh Kumar
Department of Soil Science and Agricultural Chemistry, Bihar Agricultural University, Sabour–813210, Bhagalpur, Bihar, India

Sanjeev Kumar
Department of Plant Pathology, Bihar Agricultural University, Sabour, Bhagalpur, Bihar, India, E-mail: drsanjeevdmr@gmail.com

Santosh Kumar
Assistant Professor-cum-Junior Scientist, Department of Plant Pathology, Bihar Agricultural University, Sabour, Bhagalpur, India E-mail: santosh.kumar@agr.gc

Sunil Kumar
School of Agricultural Sciences and Rural Development, Nagaland University, Medziphema–797106, Nagaland, India, E-mail: drsunilk81@gmail.com

Vinod Kumar
Senior Scientist, National Research Center on Litchi, Mushahari, Muzaffarpur, Bihar–842002, E-mail: vinod3kiari@yahoo.co.in

Kishor Chand Kumhar
Tea Research Association, North Bengal Regional Research and Development Center, Nagrakata, Jalpaiguri–735225, West Bengal, India, E-mail: kishorkumarc786@gmail.com

Alka Kushwaha
Assistant Professor, Department of Botany, D.A-V, College, Kanpur–208001, Uttar Pradesh, India, E-mail: alkakushwaha17march@gmail.com

N. Lakpale
College of Agriculture, Indira Gandhi Krishi Vishwavidyalaya, Raipur, Chhattisgarh, India

Udit Narain
Professor (Retd.), Department of Plant Pathology, C.S. Azad University of Agriculture and Technology, Kanpur–208002, Uttar Pradesh, India

M. Pathak
KVK, East Siang, College of Horticulture and Forestry, Central Agricultural University, Pasighat–791102, Arunachal Pradesh, India

Rajendra Prasad
Professor and Head (Retd.), Department of Plant Pathology, C.S. Azad University of Agriculture and Technology, Kanpur–208002, Uttar Pradesh, India

Ved Ratan
Professor and Head, Department of Plant Pathology, C.S. Azad University of Agriculture and Technology, Kanpur – 208002, Uttar Pradesh, India

Pankaj Rautela
Department of Plant Pathology, College of Agriculture, G.B. Pant University of Agriculture and Technology, Pantnagar–263145, Uttarakhand, India

Contributors

Shailbala
Junior Research Officer, Plant Pathology, Sugarcane Research Center, Bazpur Road, Kashipur–244713, G.B. Pant University of Agriculture and Technology, Uttarakhand, India,
E-mail: shailbalasharma10@gmail.com

R. C. Shakywar
Department of Plant Protection, College of Horticulture and Forestry, Central Agricultural University, Pasighat–791102, Arunachal Pradesh, India, E-mail: rcshakywar@gmail.com

Shweta Shambhavi
Department of Soil Science and Agricultural Chemistry, Bihar Agricultural University, Sabour–813210, Bhagalpur, Bihar, India

G. L. Sharma
Department of Plant Pathology and Fruit Science, Indira Gandhi Krishi Vishwavidyalaya, Raipur, Chhattisgarh, India

R. K. Sharma
Sher-e-Kashmir University of Agriculture Sciences and Technology Jammu, Jammu and Kashmir, India

M. E. Shilpa
Scholar, Department of Agriculture Microbiology, UAS, GKVK, Bangalore–65, Karnataka, India

A. K. Singh
Sher-e-Kashmir University of Agriculture Sciences and Technology Jammu, Jammu and Kashmir, India

K. P. Singh
Department of Plant Pathology, College of Agriculture, G.B. Pant University of Agriculture and Technology, Pantnagar–263145, Uttarakhand, India

Mahendra Singh
Department of Soil Science and Agricultural Chemistry, Bihar Agricultural University, Sabour–813210, Bhagalpur, Bihar, India

R. P. Singh
Department of Plant Pathology, College of Agriculture G.B. Pant University of Agriculture and Technology, Pantnagar–263145, Uttarakhand, India

J. N. Srivastava
Department of Plant Pathology, Bihar Agricultural University, Sabour, Bhagalpur, Bihar, India,
E-mail: j.n.srivastava1971@gmail.com

R. B. Verma
Department of Horticulture (Veg. Sci. and Flr.), Bihar Agricultural University, Sabour, Bhagalpur–813210, Bihar, India

Abbreviations

BABA	beta-amino butyric acid
BCAs	biocontrol agents
BCTV	beet curly top virus
CEVd	citrus exocortis viroid
CFB	Corrugated Fiber Board
CMV	cucumber mosaic virus
CTV	citrus tristeza virus
CYVV	clover yellow vein virus
DSMV	dasheen mosaic virus
EB	endophytic bacteria
FAO	Food and Agriculture Organization
ICM	integrated crop management
ISR	induce systemic resistance
KoMV	konjac mosaic virus
LP	light pruning
MM.106	Malling Merton 106
MP	medium pruning
NBM	neem bark methanol extract
NEPZ	northeastern plain zone
PAL	phenylalanine ammonia-lyase
PGP	plant growth-promoting
PMMV	pepper mild mottle virus
PO	peroxidase
PPC	plant protection code
PPO	polyphenol oxidase
PVX	potato virus X
PVY	potato virus Y
RH	relative humidity
RP	rejuvenation pruning
SOPP	sodium ortho-phenylphenate
SqMV	squash mosaic virus
TBZ	thiabendazole
TmRSV	tomato ringspot virus
TMV	tobacco mosaic virus

ToLCV	tomato leaf curl virus
ToMV	tomato mosaic virus
TPS	true potato seed
TRSV	tobacco ringspot virus
TSWV	tomato spotted wilt virus
TYLCV	tomato yellow leaf curl virus
WMV-1	watermelon mosaic virus 1
WMV-2	watermelon mosaic virus 2
ZYMV	zucchini yellow mosaic virus

Preface

Plant pathology is one of the most vital branches of agricultural sciences. Plant diseases render a huge loss of agricultural produce every year in terms of both quality and quantity. The better diagnosis and detection of the pathogens in the early stages of infection are quite necessary for the effective management of these diseases. There has been a long-felt need by the student and teacher for a comprehensive book on diseases of vegetable crops, fruit crops, and plantation crops. Because of that, it has been my ambition to write a book on plant pathology for the benefit of students of botany and agriculture and which is devoted to covering, in general, certain basic aspects of knowledge on diagnostic symptoms, mode of the perpetuation of the pathogen and dissemination, favorable conditions for disease development, and latest management strategy seem. This volume has been produced to introduce the student to the basic and fundamental aspects of the subject before he attempts to learn about the diseases of vegetable crops, fruit crops, and plantation crops. The selected bibliography given at the end of each chapter should help the student to acquire additional knowledge on the subject. In the last chapter, an attempt is made to critically review the present status of plant pathology in India and to assess the future needs so as to give perspective to young minds desirous of specializing in this branch of science.

It is sincerely hoped that this book would be extremely useful to students, teachers, and researchers. This book will also assist in various competitive examinations, such as those for plant protection officer, district horticulture officer, and other government jobs in agriculture, horticulture, and plant protection. Moreover, plant protection experts, vegetable specialists, horticultural officers, and extension workers may utilize this informative book as a ready reckon.

—Gireesh Chand, PhD
Santosh Kumar, PhD
Nadeem Akhtar, PhD

PART I
Fruits

CHAPTER 1

Pineapple Diseases and Their Integrated Management

SANJEEV KUMAR* and ERAYYA

Department of Plant Pathology, Bihar Agricultural University, Sabour, Bhagalpur, Bihar, India

Corresponding author. E-mail: drsanjeevdmr@gmail.com

1.1 INTRODUCTION

Pineapple, *Ananas comosus*, is a herbaceous biennial or perennial plant, and it belongs to family Bromeliaceae, grown for its edible fruit. The pineapple plant has a short, stout stem and a rosette of sword-shaped leaves with needle-like tips. The leaves are waxy and have upturned spines on the margins. Leaves are solid green or striped with red, white, or cream color stripes. When the plant flowers, the stem begins to elongate and produces a flower head of small purple or red flowers, each with a pointed bract. The stem continues to elongate and sets down a tuft of short leaves called a 'crown.' Individual fruits develop from the flowers and fuse to form one large cylindrical fruit topped by the crown. This fruit, known as pineapple, has a tough rind made up of hexagonal units and a fibrous, juicy flesh that may be yellow to white in color. Pineapple may reach 1.5–1.8 m (5–6 ft) in height, and some varieties can grow for in excess of 20 years. Pineapple is a tropical plant and grows best in temperatures between 23 and 32°C (73.4–89.6°F). The plant can tolerate colder temperatures for short periods but will be killed by frosts. Pineapple will grow optimally in well-draining sandy loam, which is rich in organic matter. The optimum pH for pineapple growth is between 4.5 and 6.5. Established pineapple plants are tolerant of drought but will not tolerate waterlogged soil, which quickly leads to root rot. It is a single plant can produce fruit over many years, with no need for replanting. However, with the prolonged growing of pineapples come the risks of the diseases that might infect them. Since they are grown at ground level, several diseases are the most concerning with pineapples.

1.2 FUNGAL DISEASES

1.2.1 *PHYTOPHTHORA HEART (TOP) ROT*

Causal organism: *Phytophthora parasitica.*

1.2.1.1 SYMPTOMS

Plants of all ages are attacked, but three to four-month-old crown plantings are most susceptible. Fruiting plants or suckers on ratoon plants may be affected. The color of the heart leaves changes to yellow or light coppery brown. Later, the heart leaves wilt (causing the leaf edges to roll under), turn brown, and eventually die. Once symptoms become visible, young leaves are easily pulled from the plant, and the basal white leaf tissue at the base of the leaves becomes water-soaked and rotten with a foul smell due to the invasion of secondary organisms. The growing point of the stem becomes yellowish-brown, with a dark line between healthy and diseased areas.

1.2.1.2 SURVIVAL AND SPREAD

Chlamydospores of the two species are the primary inoculum, and they can survive in the soil or in infected plant debris for several years. They germinate directly to produce hyphae that are able to infect roots and young leaf and stem tissue, or indirectly to produce sporangia. *Phytophthora* pathogens are soil inhabitants and require water for spore production and infection. As free water is required for producing sporangia and releasing motile zoospores, infection, and disease development are exacerbated in soils with restricted drainage.

1.2.1.3 MANAGEMENT

Planting in raised beds helps to drain the soil and reduces the incidence of the disease. Mulch from pineapple debris should be avoided. Pre-planting dips and foliar applications of Fosetyl Al 80% WP are very effective at controlling the disease.

Pineapple Diseases and Their Integrated Management 5

1.2.2 PHYTOPHTHORA ROOT ROT

Causal organism: *Pythium arrhenomanes.*

1.2.2.1 SYMPTOMS

The symptoms above ground are similar to those caused by nematodes, mealybug wilt, and low levels of soil oxygen. Leaves change in color from a healthy green through various shades of red and yellow. Leaf tips and margins eventually become necrotic, the root system is dead, and plants can easily be pulled from the ground. From infected plants, color prematurely becomes small and unmarketable. If symptoms are recognized early and control measures are taken, plants can recover. If roots are killed right back to the stem, they often fail to regenerate.

1.2.2.2 SURVIVAL AND SPREAD

Losses might be severe in poorly drained fields. Plants on even relatively well-drained soils can be affected during prolonged wet weather. Losses from root rot can be serious in high rainfall areas where prolonged rains extend into the winter months. The disease can eliminate the ratoon crop. Rough leaf varieties and some low acid hybrids are more susceptible than smooth cayenne.

1.2.2.3 MANAGEMENT

Drenching or spraying with Fosetyl Al 80% WP reduces root rot. Infected plants can be saved only if treated soon after symptoms appear. Avoid excessively deep planting and prevent soil from entering the hearts during planting. Well-drained soils are essential for minimizing the risk of *Phytophthora* infection. This can be achieved through careful field selection, planting on raised beds at least 20 cm high, constructing drains to intercept run-off before it reaches the plantation, constructing drains within the field so that water is removed rapidly without causing erosion and installing underground drains.

1.2.3 BASE (BUTT) ROT

Causal organism: *Chalara paradoxa.*

1.2.3.1 SYMPTOMS

Symptoms are seen only on crowns, slips, and suckers before or immediately after planting. A grey to black rot of the soft butt tissue develops, leaving stringy fibers and a cavity at the base of the stem. If affected material is planted, partial decay of the butt severely reduces plant growth. When butt decay is severe, plants fail to establish, wilt rapidly, and leaf tissue dies. Unlike *Phytophthora* heart rot, the young leaves remain firmly attached to the top of the stem. Infected plants can easily be broken off at ground level.

1.2.3.2 SURVIVAL AND SPREAD

The fungus is important in the breakdown of pineapple residues after cropping and survives as chlamydospores in soil and decaying pineapple residues. The fungus commonly infects plants through fresh wounds occurring where the planting material has been detached from the parent plant and destroys the soft tissue at the base of the stem. Material removed during showery weather and stored in heaps is particularly prone to infection. Tops (crowns) used for planting are particularly susceptible. Conidia are produced under conditions of high humidity and can be dispersed by wind. Losses of planting material and plantings from diseased material can be severe at times.

1.2.3.3 MANAGEMENT

Seed material should be stored on mother plants during dry weather and with good air circulation. Freshly removed seed material should be dipped in Carbendazim 50WP (0.01%) solution within 12 hours of removal from the mother plant. During harvest, avoiding bruising and wounding of fruit helps to reduce black rot. Harvested fruit should be dipped in Carbendazim 50WP (0.01%) solution within 6–12 hours of harvest to prevent disease development during shipping.

Pineapple Diseases and Their Integrated Management

1.2.4   FRUITLET CORE ROT (GREEN EYE)

Causal organism: *Penicillium funiculosum/Fusarium moniliforme.*

1.2.4.1   SYMPTOMS

This is an internal fruit disease. Smooth Cayenne fruits do not usually show any external symptoms. However, the fruit of the rough-leaf may produce fruitlets that fail to color—a condition often referred to as "green eye." Severely affected fruitlets may become brown and sunken as the fruit ripens. Internal symptoms consist of browning of the center of the fruitlets starting below the floral cavity and sometimes extending to the core. The browning, which remains quite firm, varies in size from a speck to complete discoloration of one or more fruitlets.

1.2.4.2   SURVIVAL AND SPREAD

Penicillium funiculosum infects the developing fruit at some stage between initiation and an open flower. Infection is favored by cool temperatures (16–20°C) during the five weeks after flower initiation, during which time the fungus builds up in leaf hairs damaged by mites. Similar cool temperatures are required for infection from about 10–15 weeks after flower induction. Symptoms of fruitlet core rot on a fruit cylinder in damaged leaf hairs. *Fusarium moniliforme* enters the fruit through open flowers or injury sites. The risk of disease caused by this fungus is higher when flowers are initiated and fruit mature under warm conditions.

1.2.4.3   MANAGEMENT

Fungicides have not been effective except when applied directly into the opening of the terminal leaves that is created by the emerging inflorescence.

1.2.5   FUSARIOSIS

Causal organism: *Fusarium subglutinans* f.sp *ananas.*

1.2.5.1 SYMPTOMS

It is sporadic and affects all parts of the pineapple plant but is most conspicuous and damaging on fruit. Fruits exhibit stem rosetting and curvature of the plant because portions of the stem are girdled or killed. Rough leaf pineapple cultivars are more susceptible than smooth-leaf varieties.

1.2.5.2 SURVIVAL AND SPREAD

Infections of the inflorescence and fruit occur primarily via injuries caused by insects, particularly the pineapple fruit caterpillar (*Thecla basilides*) and by infected planting materials.

1.2.5.3 MANAGEMENT

The sporadic nature of the disease makes chemical control impractical and uneconomic. Fungicide, Carbendazim 12WP+ Mancozeb 63WP 2 g/l, and insecticide, Trizophos 40%EC 1.5 ml/l applications at flower induction and three weeks after forcing can reduce disease.

1.2.6 GREEN FRUIT ROT

Causal organism: *Phytophthora cinnamomi.*

1.2.6.1 SYMPTOMS

Green fruit in contact with the soil is liable to be infected. A water-soaked rot develops internally behind affected fruit lets with no external symptoms. As the disease progresses, a general, water-soaked rot of green fruit with a distinct brown margin develops in green fruit.

1.2.6.2 SURVIVAL AND SPREAD

The pathogen lives in the soil and requires free water for spore production and fruit infection. Ratoon crop fruit lying close to or touching soil are most affected. Spores may be splashed by rain on to fruit near the ground.

Pineapple Diseases and Their Integrated Management

1.2.6.3 MANAGEMENT

Apply Metalaxyl 8%+ Mancozeb 64%WP 2 g/l to control root and heart rot, protecting the inflorescence and young fruit with fungicides.

1.2.7   INTER FRUITLET CORKING

Causal organism: *Penicillium funiculosum.*

1.2.7.1 SYMPTOMS

Fruits affected by inter fruitlet corking often show shiny patches on the shell early in their development, where the trichomes (hairs) have been removed by mite feeding. Externally, corky tissue develops on the skin between the fruitlets, but usually, only "patches" of eyes are affected. Fine, transverse cracks may also develop on the sepals and bracts. In moderate to severe cases, corkiness surrounding fruitlets prevents their development, and one side of the fruit will be malformed.

1.2.7.2 MANAGEMENT

Inter fruit let corking is limited almost exclusively to fruit initiated in early autumn. It is sporadic and often confused with boron deficiency. Fungicides have not been effective except when applied directly into the opening of the terminal leaves that is created by the emerging inflorescence.

1.2.8 LEATHERY POCKET

Causal organism: *Penicillium funiculosum.*

1.2.8.1 SYMPTOMS

Fruits do not usually show any external symptoms. Internally, the formation of corky tissue on the walls of the fruitlets makes them leathery and brown.

1.2.8.2   SURVIVAL AND SPREAD

See fruitlet core rot. Leathery pocket occurs sporadically. At some stage between initiation and an open flower, *Penicillium funiculosum* infects the developing fruit. Infection is favored by cool temperatures (16–20°C) during the five weeks after flower initiation, during which time the fungus builds up in leaf hairs damaged by mites. Similar cool temperatures are required for infection from about 10–15 weeks after flower induction.

1.2.8.3   MANAGEMENT

The sporadic nature of the disease makes chemical control impractical and uneconomic. Miticide, Dicofol 4 ml/l applications at flower induction, and three weeks after forcing can reduce disease.

1.2.9   WATER BLISTER

Causal organism: *Cerratocystis paradoxa.*

1.2.9.1   SYMPTOMS

Symptoms include water blister, which is also referred to as black rot or soft rot. This causes a soft, watery rot of the fruit flesh and makes the over-lying skin glassy, water-soaked, and brittle. The skin, flesh, and core disintegrate, and the fruit leaks through the shell. In advanced cases, this leaves a fruit shell containing only a few black fibers. This shell collapses under the slightest pressure.

1.2.9.2   SURVIVAL AND SPREAD

Infection occurs through shell bruises and growth cracks but mainly through the broken fruit stalks. The disease is most active in warm, wet weather and is most severe from January to April, when the summer crop is harvested. (The correlation between rainfall before harvest and disease after harvest has resulted in the name "water blister"). When fresh fruits are marketed with the crowns left on, this eliminates a major point of entry for the fungus.

Pineapple Diseases and Their Integrated Management 11

1.2.9.3 MANAGEMENT

Handle fruit carefully to avoid bruising and scuffing. Rapid fungal invasion occurs through even minute, weeping fractures. Reject sunburnt and damaged fruit, because these often have minor skin cracks that are readily infected. Dip the base of the fruit in a Carbendazim 50WP (0.01%) solution within five hours of harvesting and store fruit at 9oC. This is most important for fruit harvested during warm, wet weather. Remove pineapple refuse and rejected fruit from in and around the packing shed. Treat the shed with the Formalin 0.03% once a week.

1.2.10 WHITE LEAF SPOT

Causal organism: *Chalara paradoxa.*

1.2.10.1 SYMPTOMS

The first symptom is a small, brown spot on the leaf, usually where the leaf margin has been rubbed by another leaf during strong winds. These spots lengthen rapidly during wet weather. During prolonged wet periods, spots may reach more than 20 cm in length and spread to the leaf tip. Fine weather rapidly dries the affected area leaving cream-colored or almost white, papery pots, hence the name "white leaf spot." The margins of the spot often remain brown.

1.2.10.2 SURVIVAL AND SPREAD

Chalara paradoxa is common in pineapple plantations. The fungus will only invade wounds and is most active in warm, wet weather.

1.2.10.3 MANAGEMENT

Seed material should be stored on mother plants during dry weather and with good air circulation; freshly removed seed material dipped in Carbendazim 50WP (0.01%) solution within 12 hours of removal from the mother plant. Avoiding bruising and wounding of fruit during harvest helps to reduce black rot. Harvested fruit dipped in Carbendazim 50WP (0.01%) solution within 6–12 hours of harvest to prevent disease development during shipping.

12   *Diseases of Fruits and Vegetable Crops*

1.2.11   *FRUIT ROT BY YEAST*

Causal Organism: *Saccharomyces* spp.

1.2.11.1   *SYMPTOMS*

Yeasts ferment sugar solution, producing alcohol, and releasing carbon dioxide. The first symptom is a bubbling exudation of gas and juice through the crack or injury where infection occurred. The shell then turns brown and leathery, and as the juice escapes, the fruit becomes spongy. Internally, the decaying flesh turns bright yellow and develops large gas cavities. Finally, all that remains of the fruit is the shell and spongy tissue.

1.2.11.2   *SURVIVAL AND SPREAD*

In spring, rapid changes in fruit growth, resulting from the shift from cold and dry to warm and wet weather, can result in the pineapple skin cracking between fruit lets. Fruits affected by even minor frost damage are prone to cracking as they ripen in spring. Yeasts immediately invade the juice weeping from those wounds, and these fruits are severely damaged or destroyed as they ripen. The disease may occur before or after harvest.

1.2.11.3   *MANAGEMENT*

Yeasty rot is widespread but occurs mainly during spring in overripe or damaged fruit. Protect fruit that will ripen in spring in frost-prone areas by covering young developing fruit with paper bags. Fruit showing even minor inter fruitlet cracking should not be consigned to the fresh-fruit market. Any fruit showing fractures between fruitlets should be picked at the earliest stages of fruit maturity to minimize losses.

1.3   BACTERIAL DISEASES

1.3.1   *MARBLING*

Causal organism: *Erwinia ananas.*

1.3.1.1 SYMPTOMS

Infected fruits do not show any external symptoms. Internally, the flesh is red-brown and granular and has a woody consistency.

1.3.1.2 SURVIVAL AND SPREAD

The disease occurs when flowers are initiated and when fruit mature under warm, wet conditions. The bacteria enter through the open flower and natural growth cracks on the fruit surface. Infected fruits are usually low in both acid and sugars.

1.3.1.3 MANAGEMENT

There are currently no methods of controlling the disease. The pineapple variety Smooth Cayenne appears to be moderately resistant to the disease.

1.3.2 PINK DISEASE

Causal organism: *Pantoea citrea.*

1.3.2.1 SYMPTOMS

Infected fruits do not show any external symptoms, even when fully ripe. Internally, the flesh may be water-soaked or light pink and have an aromatic odor, although these symptoms may not be obvious immediately. When sterilized by heat during canning, infected tissue darkens to colors ranging from pink to dark brown. In some fruits, only one or a few fruitlets may be infected. In highly translucent, low-Brix fruit, the entire cylinder can be invaded.

1.3.2.2 SURVIVAL AND SPREAD

The bacteria infect through the open flower during cool weather. Disease incidence increases in dry conditions before flowering, followed by rainfall during flowering. The bacteria are thought to be carried by nectar-feeding insects and mites to open flowers from infected, decaying fruit near flowering fields.

1.3.2.3   MANAGEMENT

This disease occurs only sporadically when fruits develop under cool, wet conditions. Since the bacteria are killed by high temperatures, the pink disease occurs mainly in spring (September–October). The incidence of infected fruit is very low. Management is not usually warranted. Smooth Cayenne is relatively resistant.

1.3.3   BACTERIAL HEART ROT

Causal Organism: *Erwinia chrysanthemi.*

1.3.3.1   SYMPTOMS

Water-soaked lesions on the white basal sections of leaves in the central whorl, which may spread to all leaves in the central whorl, mid portions of leaves become olive green in color with a bloated appearance. Infected fruits exude juices, and the shell becomes olive green, cavities form within the fruit.

1.3.3.2   SURVIVAL AND SPREAD

The disease is thought to be spread from the juices of infected fruits; bacteria in the juice can enter leaves through wounds; ants act as vectors for the bacteria.

1.3.3.3   MANAGEMENT

Remove and destroy infected fruits. Avoid the use of infected crowns for seed material to prevent the spread of the disease. Planting to avoid flowering when the adjacent field is fruiting can reduce disease development. Apply Dicofol 4 ml/l to control of ants can significantly reduce disease

Pineapple Diseases and Their Integrated Management

incidence. Foliar application of streptomycin sulfate @ 0.05% along with blitox-50 @ 1–2 g/l is effective to manage bacterial diseases of pineapple.

1.4 VIRAL DISEASES

1.4.1 MEALYBUG WILT DISEASE

Causal organism: *Ampelovirus.*

1.4.1.1 SYMPTOMS

The early symptoms are a slight reddening of leaves about halfway up the plant. The leaf color then changes from red to pink, and leaves lose rigidity, roll downwards at the margin, and the tip of the leaf dies. The root tissue also collapses, and the plant appears wilted. Plants can recover to reduce symptomless leaves and fruit that are markedly smaller than fruit from healthy plants. Symptoms are most obvious in winter when plant growth and vigor are reduced. Disease development and the incidence is affected by plant age at the onset of mealybug infestation, with younger plants displaying symptoms two to three months following feeding, while older plants may take up to 12 months to develop symptoms.

1.4.1.2 SURVIVAL AND SPREAD

The disease is thought to be caused by viruses transmitted by mealybugs with the pink mealybug (*Dysmicoccus brevipes*) being the main vector. The disease is probably introduced in planting material that may not show obvious disease symptoms. Once established, the viruses are transmitted when the mealybugs feed on young leaves. Mealybugs are sedentary insects that are moved from plant to plant by attendant ants or by the wind. Ants actively tend to mealy bugs. The coastal brown ant (*Pheidole megacephala*) is common and active, but many other species can be involved in raising mealy bugs. Mealybugs produce honeydew, which is harvested by ants for food. Ants also protect mealybugs from predators and move them around and between plants. The removal of spiders from fields by ants often allows large populations of mealy bugs to develop, increasing the risk of severe mealybug wilt outbreaks. The incidence is variable and sometimes high. The amount of wilt in a field is related to the number of mealy bugs present, the length of time they feed, and the activity of ants.

1.4.1.3 MANAGEMENT

Use planting material from wilt-free areas or from fields with a low level of wilt disease. If less than 3% of plants show wilt symptoms, remove infected plants by hand, and destroy them. Use recommended insecticides for mealybug and ant control, where more than 3% of plants show wilt symptoms. If more than 10% of plants show wilts symptoms, do not use the field as a source of planting material. Eradicate badly affected areas immediately after harvest. Keep headlands and field boundaries free from weeds and rubbish as these may act as reservoirs for ants and mealy bugs.

1.4.2 YELLOW SPOT

Causal organism: *Tomato spotted wilt virus* (TSWV) (Tospoviruses).

1.4.2.1 SYMPTOMS

Infection occurs on young crowns when they are still on the fruit or during the first few months after planting. Small (2–5 mm), round, yellow spots appear on the upper surface of the leaves of young plants. These spots fuse and form yellow streaks in the leaf tissue, which soon become brown and die. The virus spreads to the leaves in the plant heart, causing the plant to bend sideways. The infection eventually kills the plant so that the virus is not transmitted to subsequent plantings. If the crown is infected while still on the fruit, the fruit dies from the top downwards. Infections can occur through open blossoms, causing the development of large, blackened cavities in the side of the fruit.

1.4.2.2 SURVIVAL AND SPREAD

The viruses are transmitted to pineapple plants by small flying insects (thrips). Infection occurs mostly on plants during early growth, and crowns on developing fruit are occasionally infected. As infection is always fatal, vegetative propagation does not spread the virus to subsequent plantings. Tospoviruses have a wide range of hosts among weed and crop plants. The disease is rarely seen.

1.4.2.3 MANAGEMENT

Keep the plantation free from weeds. Avoid destroying old weedy patches near young crown plantings or fields with developing fruit. If this is impossible, it may be necessary to first spray Acetamirid 20%SP 0.5 g/l of the old infected field to control thrips.

KEYWORDS

- **Ampelovirus**
- **bacterial heart rot**
- ***Erwinia chrysanthemi***
- **pink disease**
- **tomato spotted wilt virus**
- **white leaf spot**

REFERENCES

Anonymous (2014). *AESA Based IPM Package, Pineapple* (p. 64). NIPHM, Ministry of Agriculture, India.

Bartholomew, D. P., Rohrbach, K. G., & Evans, D. O., (2002). *Pineapple Cultivation in Hawaii*. University of Hawaii Cooperative Extension Service. Available at: http://www.ctahr.hawaii.edu/oc/freepub (Accessed on 18 November 2019).

CABI Crop Protection Compendium, (2013). *Ananas Comosus (Pineapple) Datasheet*. Available at: http://www.cabi.org/cpc/datasheet/5392 (Accessed on 18 November 2019).

Crane, J. H., (2013). *Pineapple Growing in the Florida Home Landscape*. Available at: http://edis.ifas.ufl.edu/mg055 (Accessed on 18 November 2019).

Ploetz, R. C., Zentmyer, G. A., Nishijima, W. T., Rohrbach, K. G., & Ohr, H. D., (1994). *Compendium of Tropical Fruit Diseases*. American Phytopathological Society Press. Available at: http://www.apsnet.org/apsstore/shopap (Accessed on 18 November 2019).

CHAPTER 2

Diseases of Apples and Their Management

J. N. SRIVASTAVA,[1*] A. K. SINGH,[2] and R. K. SHARMA[2]

[1]Department of Plant Pathology, Bihar Agricultural University, Sabour, Bhagalpur, Bihar, India

[2]Sher-e-Kashmir University of Agriculture Sciences and Technology Jammu, Jammu and Kashmir, India

[*]Corresponding author. E-mail: j.n.srivastava1971@gmail.com

2.1 INTRODUCTION

Apple scab is the most destructive disease of apple throughout the world. Apple scab is an economically important disease resulting in a direct loss from fruit or pedicel infections. Indirectly, repeated defoliation reduces tree growth and yield. The most adverse effect of the disease is the reduction of fruit quality. The disease was first described by Fries in 1819 from Sweden. In the United States, it was first described in 1834 by Schweinitz, who collected it from cultivated apples in New York and Pennsylvania. The earliest report of the disease in England was by Berkeley in 1855. In India, the disease was first reported in Kashmir valley in 1935 by P. Nath on an indigenous apple cultivar, "Ambri." In 1973 the disease appeared in epidemic form in Kashmir valley, and it severely damaged the commercial "Red Delicious" cultivar, and this ruined the apple crop worth Rs. 54 Lakhs in a single season (Joshi et al., 1975). Apple scab was also detected in Himachal Pradesh in an orchard and in a few localized pockets of Kullu, Mandi, and Chamba district.

2.1.1 FUNGAL DISEASES

2.1.1.1 APPLE SCAB

2.1.1.1.1 Symptoms

The characteristic symptoms of scab are mainly observed on leaves and fruits. Petioles, pedicels, blossoms, and young shoots also exhibit scab symptoms under severe infestation. First, scab lesions are observed on the lower surface of young leaves from the opening bud, later on, both the upper and lower surface of the leaf may be infected (Gupta, 1987).

1. **Leaves:** Young leaf lesions are velvety brown to olive green within distinct margins. Lesions in a more advanced stage become black and slightly raised. As the infected leaf ages, several lesions may coalesce, and tissues adjacent to lesions thicken, causing the leaf to become curled, dwarfed, or distorted. Premature yellowing and defoliation may also take place (Agrios, 1997).
2. **Flowers:** Lesions on flowers resemble those on the leaves. A single lesion on the pedicel or a sepal can make the flower desiccate and fall.
3. **Fruits:** Early infections can lead to abnormal growth (fruit deformation) and fruit drop. The lesions on the fruits are similar to those on the leaves; as infected fruit enlarge, the lesions become brown and corky and but as they age, they will produce cracks. Early-season infection may result in uneven fruit development, causing cracks in the skin and flesh. Late-season infection or just before harvest, black, circular, very small (0.1–4 mm diameter) lesions called pinpoint scabs, where circular lesions are rough and black. The term "storage scab" refers to incipient infections that were too small to see prior to fruit storage or maybe the result of infections during storage that occur as a result of sporulation from older scab lesions (Odile et al., 2006).

2.1.1.1.2 Causal Organism: Venturia Inaequalis (Cre) Wint

Apple scab is caused by a fungus, *namely Venturia inaequalis* (Cre) Wint. Its imperfect stage is *Spilocaeapomi* Fr. Pseudothecia, produced in overwintered leaves or fruit, are separate, dark brown to black in color, and spherical, with a short beak and distinct ostioles with single-celled bristles at the apex. Asci are fasciculate, cylindrical, short stipitate, eight spored, and have thin, bitunicate walls (Gupta, 1987). Ascospores are yellowish-green to tan and unequally two-celled. Conidia are olive, have one or two cells, are ovate to lanceolate, and are produced sequentially by a series of abscission ridges on the conidiophores (Odile et al., 2006).

2.1.1.1.3 Disease Cycle and Epidemiology

The scab pathogen has two stages in its life cycle, the saprophytic stage, which is found on dead leaves, and where the pathogen overwinters. The parasitic stage is the second stage in which the pathogen completes its parasitic life cycle on living parts of the plants, such as on leaves, flowers, twigs, and fruits. Ascospores developed on dead leaves are the primary source of infection. They infect young leaves and buds during spring. Secondary spread of the disease is by conidia developed after the primary infection sets in. After fungal penetration of the cuticle, conidiophores, and conidia are produced in a visible lesion. Conidia are then disseminated by rain and wind to new leaves and fruit, initiating a secondary infection cycle.

Weather conditions are very important for the development and release of ascospores and formation of the conidial cycle. The optimum temperature for ascospore maturation is about 20°C. Rains favor the ejection of ascospores. The ascospores germinate and cause infection at a temperature of 10–22°C and free water or 90% relative humidity (RH).

2.1.1.1.4 Management

1. Spray apple trees with urea (5 kg/100 L water) at pre-leaf fall.
2. Give one chemical spray before initiation of leaf fall with Carbendazim (50 g/100 L water) or Thiophanate methyl (75 g/100 L water).
3. Clean Cultivation: Collect and destroy the fallen leaves and pruned material by burning. Plow the orchards to burry plant debris, harboring the pathogen.
4. Attention to weather forecasts, particularly those of extended wet periods, can assist in following chemical spray schedule:

    - **1st Spray:** Silvertip to green tip stage: (Mid-March to Mid-April): Dodine (100 g/100 L water) or Mancozeb (200 g/100 L water) or Chlorothalonil (400 g/100 L water) or Dithanon (50 g/100 L water) or Captan (300 g/100 L water).
    - **2nd Spray:** Pink bud or partial bloom stage: (1st week of April to end to April): Hexaconazole (300 ml/100 L water) or Mancozeb +Sulphur (200 g+200 g/100 L water) or Carbendazim (50 g/100 L water) or Captan (300 g/100 L water).
    - **3rd Spray:** After petal fall or fruitlet stage: (Last week of April to end to May): Carbendazim (50 g/100 L water) or Thiophanate methyl (50 g/100 L water) or Hexaconazole (300 ml/100 L water).

- **4th Spray:** Fruitlet stage: (Mid-May to Mid-June): Dodine (100 g/100 L water) or Dithanon (50 g/100 L water) or Mancozeb (200 g/100 L water) or Bitertanol (50 g/100 L water).
- **5th Spray:** Fruit development stage: Walnut size fruit: 15–21 days after 4th spray (1st week of June to 1st week of July): Dodine (100 g/100 L water) or Dithanon (50 g/100 L water) or Thiophanate methyl (50 g/100 L water) or mancozeb + Carbendazim (250 + 50 g /100 L water) or Mancozeb (200 g/100 L water) or Captan (300 g/100 L water).
- **6th Spray:** Fruit development stage: 40 days before harvest (Mid-August): Carbendazim (50 g/100 L water) or Thiophanate methyl (50 g/100 L water) or Bitertanol (50 g/100 L water) or Mancozeb (200 g/100 L water) or Captan (300 g/100 L water).
- **7th Spray:** Pre-harvest spray: (20–25 days before harvest): Carbendazim (50 g/100 L water) or Mancozeb (200 g/100 L water.
5. After harvest, spray Copper oxychloride (600 g/100 L water).
6. Add stickers like Sandovit (50–75 ml) or Sel-Wet-E (50–75 ml) or Triton (50–75 ml) in 100-liter water in each fungicidal spray for improving the solubility, retention, and spreading of fungicide (Anonymous, 2001).

2.1.2 POWDERY MILDEW

2.1.2.1 INTRODUCTION

The disease causes severe losses to apple nurseries and orchard trees throughout the apple-growing regions of the world. A persistent disease, the severity of powdery mildew and resulting economic loss varies with environmental conditions, cultivar susceptibility, and management practices. Powdery mildew can be especially damaging in nursery production. The disease was first noticed on apple seedlings in Iowa, USA, in 1871 (Bessey, 1877).

2.1.2.2 SYMPTOMS

A white powder is seen on leaves, young shoots of plants, flowers, and fruits. Affected leaves are narrower and longer than healthy leaves. Later, the leaves curl from margins and become brown colored. Infected buds die; fruits are reduced in size, deformed with a rough skin (Burr, 1980). Infected flower buds

Diseases of Apples and Their Management 23

open five to eight days later than healthy buds and exhibit reduced fruit set; flower petals are distorted and pale yellow or light green. Apples affected during bloom are stunted in growth and covered with a network pattern of cork cells (russet) that may be so closely woven as to appear as a solid patch (Gupta, 1979).

2.1.2.3 CAUSAL ORGANISM

Podosphaera leucotricha (anamorph *Oidium farinosum*) is the causal organism of powdery mildew on apple. Produced in long chains on thin, amphigenous mycelia, conidia are ellipsoidal, truncate, hyaline, and contain fibrosin bodies. Aerial conidiophores arise from the mycelium on leaves and shoots. Each conidiophore develops conidia in the chain, which is 25–30 x 10–12μ in size. Cleistothecia are embedded in the mycelium. Each cleistothecium contains a single ascus which bornsascus (William et al., 2004).

2.1.2.4 DISEASE CYCLE AND EPIDEMIOLOGY

P. leucotricha overwinters as mycelia in dormant buds infected during the previous growing season. The primary source of inoculum is the resting mycelium or encapsulated haustoria in the buds. When such buds sprout in next season, a large number of conidia are formed, which are wind-borne and serve as secondary inoculum. Healthy buds often open earlier than infected buds, thus providing susceptible tissue upon conidia development.

Powdery mildew infections occur when the RH is greater than 70%. Even on days when RH is low, infections may occur during the night or early morning hours when RH usually rises. Infections can occur when the temperature lies between 10 to 25°C. The optimum temperature range for infection is between 19 to 22°C. Unlike other foliar diseases, leaf wetting is NOT a requirement for powdery mildew infection. Conidia will not germinate if immersed in water, although high RH is required for infection. Under optimum conditions, powdery mildew will be visible 48 hours after infections are initiated; new infections produce spores in about 5 days (Agrios, 1997).

2.1.2.5 MANAGEMENT

1. Dormant-season pruning may remove infected buds and reduce the level of primary inoculums and avoid infection the next season.

24 *Diseases of Fruits and Vegetable Crops*

2. During the growing season, severely infected shoots should be pruned and destroyed.
3. Spraying the tree at the green tip, petal fall, 20 days and 40 days after fruitlet stage with Dinocap (50 g/100 L water) or Tridemorph (50 g/100 L water) or wettable sulfur (300 g/100 L water) or Carbendazim (50 g/100 L water) or Karathane (100 ml/100 L water) (Gupta, 1991).

2.1.3 RUST

2.1.3.1 INTRODUCTION

There are several species of rusts that attack apple. Rusts are unusual diseases in that many require two hosts (i.e., two different plant species) to complete their life cycle. Consequently, the geographic distribution and the natural density of the alternate host can have a significant impact on how important a particular rust disease is in any given region. The most commonly encountered and commercially important rusts are caused by fungi in the genus *Gymnosporangium*. Cedar-apple rust, caused by *G. Juniper-virginianae* Schwein, is the most important rust fungi.

The fungus attacks both the fruit and leaves of apples and can cause serious losses through a direct reduction yield or diminished fruit quality.

The alternate host for this fungus is the red cedar (*Juniperus virginiana* L.). Quince rust, caused by *G. clavipes* (Cooke and Peck), is also widely distributed in eastern North America. Unlike cedar apple rust, the fungus attacks only the fruit of apple. The alternate hosts are trees in the genus *Juniperus*. Hawthorne rust, caused by *Gymnosporangium globosum* (Farl.) Farl., is less common than quince rust but, unlike quince rust, Hawthorne rust infects only leaves and not the fruit (Gupta, 1986; Burr, 1995).

The disease was first reported by Schweinitz in 1822 on Junipers and later on wild apple. The fungus attacks both the fruit and leaves of apples and can cause serious losses through a direct reduction yield or diminished fruit quality (Pathak, 1986).

2.1.3.2 SYMPTOMS

Cedar apple rust affects leaves, petioles, and fruit, beginning as small yellow lesions (upper surface of leaves), which may be surrounded by chlorotic

Diseases of Apples and Their Management

halo or red band. Small orange-brown pustules (pycnia) develop within the lesions, producing watery orange drops. Later, yellow-brown lesions form on the undersurface of leaves and produce small, dark, tubular structures (aecia) that fracture to release red-brown spores. Fruit lesions are usually superficial, causing a brown necrosis 1–5 mm into the flesh. Cedar apple rust on alternate host juniper often forms a gelatinous horned gall (Jones et al., 1990).

2.1.3.3 CAUSAL ORGANISM

Gymnosporangium junipri-virginianae Schwin., are the causal organisms of rust on apple. The name comes from the fact that red cedar (*Juniperus virginianae*) is the alternate host.

2.1.3.4 DISEASE CYCLE AND EPIDEMIOLOGY

On native cedars, *Gymnosporangium* species induce a gall from which telial horns emerge under wet conditions. During rains, telia swell and appear jellylike, releasing teliospores, which then germinate to produce basidiospores. Basidiospores are immediately discharged into the air and can travel more than a mile on air currents. Those landing on susceptible apple tissue may germinate and infect the host if a film of water is present for a suitable length of time. Aeciospores are later released from aecia during dry weather and may germinate and infect native cedars. At the optimum temperature of 11–25°C, basidiospores are produced within 4 hours, whereas at 7–11°C, it takes about 5–7 hours (Burr, 1995).

2.1.3.5 MANAGEMENT

1. Removal of infected native cedars within close proximity may reduce infection pressure; however, elimination is unlikely as basidiospores can travel great distances.
2. Cultivars susceptible to cedar-apple rust include Golden Delicious, Rome, Jonathan, Lodi, Idared, Mutsu (=Crispin), Fuji, Braeburn, Gala, Cameo, Ginger Gold, Gold Rush, and Alert. 'Honeycrisp' is moderately susceptible, and McIntosh and Delicious are considered resistant to the disease.

3. Spray twice or thrice with Zineb@0.25% or Mancozeb@0.25% or Plantavax @0.1%, at 15 days interval.

2.1.4 WHITE ROOT ROT/HITE ROOT ROT

2.1.4.1 INTRODUCTION

Disease is highly destructive both in nurseries and orchards, and the affected plants are killed within a short period depending upon the age of host and environmental condition. The fungus has a very wide host range of about 158 species belonging to 45 families, comprising of horticultural, field, and other plants of economic importance. The disease was first observed on apple in 1900 from Norwich and later in 1913 from Canterbury (Salmon and Wormald, 1913). In 1925, *Rosellinia necatrix*, the perfect stage of *the Dematophora necatrix,* was identified as a cause of death of the apple tree at Winscombe-Somerset (Nattras, 1927). In India, the disease was recorded first time in 1929 by Singh from Uttar Pradesh Hills (Bose and Sindhan, 1976) and later from Himachal Pradesh.

2.1.4.2 SYMPTOMS

The disease symptoms are observed on the underground plant parts and also give a reflection on foliage as premature defoliation and premature chlorosis of leaves. Initially, the rotting of fine roots is observed, which extended to secondary and tertiary roots and ultimately to the tree trunk. After a few days of infection or rainy season, the roots are covered with white cottony mycelial growth extending to adjacent soil. The infected trees are given a sickly appearance with bonze colored leaves during August-September. In advance stage, the root is completely devoured, and disease seedlings or trees are easily uprooted from soil (Agarwal and Sharma, 1966).

2.1.4.3 CAUSAL ORGANISM

Rosellinia necatrix is the perfect stage of *Dematophora necatrix,* has not been observed on the rotten root in India. However, the *Dematophora necatrix* is very common and produce septate hyphae with a characteristic pear-shaped swelling near each septum. This character separates the fungus from other root rot organisms (Agrios, 1997).

Diseases of Apples and Their Management   27

2.1.4.4   DISEASE CYCLE AND EPIDEMIOLOGY

The fungus survives either in the form of mycelium or sclerotia in the old infected roots or in the soil for several years. Primary infection takes place from diffused mycelium present in root debris or soil or by contact of new plant roots with dead roots. The fine roots are attacked first. Pathogen also produces toxins that help in killing the bark.

The disease is favored high soil moisture and temperature (15–25°C) in acid soil at pH 6.1 to 6.5. In ill drained soil, the disease develops quickly, and mycelium spread from infected to healthy roots through soil particles (Gupta, 1996).

2.1.4.5   MANAGEMENT

1. Sterilization of infected pits with 3% formaldehyde three weeks before plantation or soil solarization.
2. Removal of rotten roots and application of a disinfectant Chaubatia paste (Redlead 800 g + copper carbonate 800 g in 1-liter linseed oil) or Brassicol paint (10 g in 1-liter linseed oil) on the cut end and a healthy portion of root during November–December.
3. Drenching of tree basins thrice with 0.1% Carbendazim/Benomyle/ Thiophanate methyl fungicide at 15 days interval during the rainy season (Gupta and Thakur, 1992).
4. Spraying the whole plant and root system during dormancy with 20 ppm aureofungin has proven very effective.

2.1.5   *PHYTOPHTHORA ROOTROT/COLLAR ROT/CROWN ROT*

2.1.5.1   INTRODUCTION

This disease is referred to as *Phytophthora* Root Rot/Collar Rot/Crown Rot depending upon the plant part infected. *Phytophthora* Root Rot/Collar Rot/ Crown Rot is an occasional, but serious problem in many apple orchards. This disease can also infect most other tree fruits, including nut trees. It is especially severe on apple trees that are grown on Malling Merton 106 (MM.106) rootstocks. The earliest record of apple bark rot attributed to *Phytophthora omnivore* (Syn. *P. Cactorum*) was reported by Osterwalder in 1912. However, the trunk infection of Apple by *P. Cactorum* was first described by Bains in 1939 from England when several epidemics in Indiana orchards. In India, the fruit rot of Apple due to *P. Cactorum* was observed by Bose and Mehta in 1951, whereas, collar rot phase was described from Himachal Pradesh for the first time.

2.1.5.2 SYMPTOMS

Trees are often affected as they come into bearing, about 3 to 5 years after planting. Collar rot appeared mostly near the graft union, but may also originate on the lower trunk or at pruning wound. Cankers at the base of the main trunk can be recognized by the dark, sunken appearance of the bark. Inner bark tissues will appear reddish-brown instead of white. There is generally a sharp contrast between the healthy white tissues and the infected tissues in the rootstock portion of the tree (Agrios, 1997).

Typical symptoms of collar rot are the production of rough patches on the stem around the collar or crown region of the tree at soil level, which develop into cankers. Symptoms of collar rot are soften confused with white root rot because both the disease occur in soil. In the case of white root rot, primary symptoms started from finer roots and spread upward up to collar region while in the collar rot, primary symptoms are at the collar region, which spreads downwards to roots (Agarwal, 1961).

2.1.5.3 CAUSAL ORGANISM

Collar rot is caused by the soilborne pathogens *Phytophthora cactorum*, *P. cambivora*, and *P. cryptogea*. Other species, *P. syringae*, *P. megasperma*, and *P. drechsleri* have also been associated with this disease (Agrios, 1997).

2.1.5.4 DISEASE CYCLE AND EPIDEMIOLOGY

The fungus is a soil inhabitant and overwinters in the form of mycelia, oospores, or chlamydospores.

A soil temperature of 12 to 20°C with Ph of 5 to 6 is found to be the best for the survival of fungus.

2.1.5.5 DISEASE MANAGEMENT

1. Provide good soil drainage.
2. Keep graft union at least 30 cm above soil level at the time of planting.

Diseases of Apples and Their Management   29

3. The disease can be best controlled by superficial healing of the local-ized lesion with the help of blow-lamp or scarifying cankered lesion to the healthy tissue followed by the application of Chaubattia past or copper oxychloride paint (Gupta and Sharma, 1999).
4. Removal the affected bark and apply Chaubatia paste (Redlead 800 g + copper carbonate 800 g in 1-liter linseed oil) or Brassicol paint (10 g in 1-liter linseed oil) on healthy portion.
5. Drenching the soil by Dithane M-45@0.3% or Blitox/Fytolan/Blue copper @ 0.5–1.0% in 30 cm radius around the tree trunk (Agarwal, 1970).

2.1.6 ARMILLARIA ROOT ROT

2.1.6.1 INTRODUCTION

Armillaria root rot, also known as shoestring root rot, an oak root fungus disease, Mushroom root rot can affect several fruit crops, but it is most common in apple and peach trees. Its host range also includes numerous species of deciduous and evergreen trees, shrubs, and woody vines (Agrios, 1997).

2.1.6.2 SYMPTOMS

Foliage may turn yellow, then brown, and dry rapidly. Dark brown to black rhizomorphs, or shoestrings, appear at the soil line around the trunk of the tree. A creamy-white layer of fungus is often present between the bark and the wood. Honey-colored mushrooms may form in groups around the drip line of the tree or next to the trunk during moist periods.

2.1.6.3 CAUSAL ORGANISM

Armillaria tabescens is the causal organism of *Armillaria* root rot. It is referred to as the honey mushroom. Lacking an annulus on the stipe of the mushroom (basidiocarp), it is easily distinguished from other *Armillaria* species. Blackish, hardened mycelial extrusions are produced on the bark of infected roots (Agrios, 1997).

2.1.6.4 DISEASE CYCLE AND EPIDEMIOLOGY

Armillaria mellea can survive for several decades as mycelia and rhizo-morphs in stumps. Disease spread can occur via windblown basidiospores produced by the mushroom phase. However, it is believed that *Armillaria* more commonly spreads via mycelia when roots come into contact with each other and via rhizomorph growth in the soil.

2.1.6.5 MANAGEMENT

1. Avoid planting in sites where *Armillaria* is known to have been present. If a woodlot has recently been cleared for an orchard, allow a year or two for the tree roots left behind to completely decompose before planting new fruit trees.
2. Maintain tree health by following good cultural practices. Trees stressed by other biotic and/or abiotic factors may be more vulner-able to attack (Agrios, 1997).

2.1.7 SOUTHERN STEM BLIGHT

2.1.7.1 INTRODUCTION

Southern stem blight (or southern blight) can sometimes be a serious problem on young apple trees. This disease has an extremely wide host range, which includes herbaceous and woody ornamentals, vegetables, and weeds, as well as tree fruits. Southern blight is the most severe 1–3-year-old trees, generally only attacking trees in their first few years in the orchard.

2.1.7.2 SYMPTOMS

A coarse, white mycelial mat is often found at the base of an infected tree, progressing upward. Small, white sclerotia develop within the mycelium, later turning tan to brown. Leaves of an infected tree may exhibit a reddish or grayish purple discoloration, later drying and turning brown as the fungus girdles the crown, and the tree dies. Wilting and dieback develop as a result of the decay of lower trunk or crown tissues. Tree death usually occurs rapidly. On exam-ining the root system, mustard seed-sized sclerotia can be seen (Agrios, 1997).

Diseases of Apples and Their Management 31

2.1.7.3 CAUSAL ORGANISM

Southern blight is caused by the soilborne fungus, *Sclerotium rolfsii*. The fungus produces white mycelia and reddish-brown to dark brown or tan, hard, round sclerotia. No asexual spores are produced.

2.1.7.4 DISEASE CYCLE AND EPIDEMIOLOGY

This fungus survives as sclerotia in the soil, as mycelia in decomposing plant material, and as mycelia in previously infected plants (including weeds). Sclerotia enable the fungus to survive long periods of adverse conditions.

Warm summer temperatures (25–35°C), high soil moisture, good soil aeration, and plentiful organic debris promote a high incidence of disease (Agrios, 1997).

2.1.7.5 DISEASE MANAGEMENT

1. Avoid planting into previously infested sites. Do not plant apples where the disease has been severe on previous crops, such as clover, tomato, or soybean. Keep in mind that the fungus can also exist as sclerotia in old pasture soils from previously infested weeds.
2. Keep the soil around the tree base free of dead organic matter. Dead orchard weeds, including those killed by herbicides, may serve as a food base for the fungus.
3. Rootstocks vary in their susceptibility to southern blight. The most resistant apple rootstock currently used is M.9.
4. Drench the nursery with Thiram (3 g/L water) or Brassicol (4 g/L water) or aureofungin (4 g/L water) solution.

2.1.8 SOOTY BLOTCH AND FLY SPECK

2.1.8.1 INTRODUCTION

Sooty blotch and flyspeck are two different diseases. Both diseases commonly occur together on the same fruit superficial blemishes or discoloration. The presence of disease reduces the grade and market value of the fruit.

2.1.8.2 SYMPTOMS

Sooty blotch appears as sooty or cloudy blotches on the surface of the fruits. The blotches are olive green with an indefinite outline. The "smudge" appearance results from the presence of hundreds of minute, dark pycnidia that are interconnected by a mass of loose, interwoven hyphae. The sooty blotch fungus is generally restricted to the outer surface of the cuticle. In rare cases, the hyphae penetrate between the epidermal cell walls and the cuticle (Agarwal, 1967).

Flyspeck colonies on fruit surfaces are well-defined groupings of shiny, black, superficial pseudothecia. Colony size varies from 1–3 cm and from round to irregular. Conidiophores and conidia are produced within the colonies of pseudothecia during warm, moist weather (Agarwal, 1967).

2.1.8.3 CAUSAL ORGANISM

The fungus *Gloeodespomigena (Schw)* causes of sooty blotch, and *Schizothyriumpomi* (Mont and Fr.) (formerly *Microthyriellarubi*; anamorph *Zygophialajamaicensis*) is the causal organism of flyspeck (Saha, 2002).

2.1.8.4 DISEASE CYCLE AND EPIDEMIOLOGY

G. pomigena and S. pomi both fungi overwinter as pseudothecia on infected apple twigs and woody reservoir hosts. Mycelial growth that forms the sooty blotches can occur in the absence of free water at RH more than 90%. The optimum condition for conidial production for flyspeck pathogen is 16–21°C and RH above 96%. Symptoms development of both diseases is relatively slow, typically requiring 20–25 days in the orchard, but in 8–12 days under optimum conditions.

2.1.8.5 MANAGEMENT

1. Spray Zineb @0.2% or Captan @0.2% or Captafol @0.1% and Carbendazim @0.05 have been found highly effective in controlling of these diseases (Agarwal, 1967; Gupta and Sharma, 1981).
2. Post-harvest dips in commercial disinfectants were used to remove signs of both pathogen buffered sodium hypochlorite @ 200,500,

Diseases of Apples and Their Management   33

or 800 ppm solution or a mixture of hydrogen peroxide and peroxy-acetic acid at 60 ppm/80 ppm/120 ppm/160 ppm/360 ppm /480 ppm, then cleaning, brushing, and grading (Agarwal, 1967).

2.2 BACTERIAL DISEASES

2.2.1 *FIRE BLIGHT*

2.2.1.1 *INTRODUCTION*

Fire blight is one of the most devastating bacterial diseases affecting apple, pear, and other rosaceous plants. Disease causal organism *Erwinia amylovora* is a native pathogen of wild, rosaceous hosts. It was the first bacterium proven to be a pathogen of plants. Today, fire blight is an important disease of apples and pears in many parts of the world. This disease varies in severity from year to year, depending on temperature and precipitation. Additionally, fire-blighted wood can provide a suitable site for other diseases, such as black rot and white rot. The disease was first reported by William Denning in 1780 in the Hudson Valley of New York and the bacterial nature of the disease by Burrill in 1882. In India, it was reported in the year 1943. Since then, it has found to occur in some parts of Jammu and Kashmir, Himachal Pradesh, and Uttar Pradesh. The disease often appears in the epiphytotic form.

2.2.1.2 *SYMPTOMS*

Plant parts affected by fire blight appear as if scorched by fire. Infected blossoms may exhibit ooze and change color from red to brown to black as the disease progresses. Infected leaves turn brown to black and desiccate but remain attached to the branches. Vegetative shoots often wilt and take on the shape of a shepherd's crook, the pith of infected stems exhibiting a dark brown discoloration. The outer bark of infected branches and limbs is often sunken and darker than normal, whereas the inner tissues are water-soaked with reddish streaks while the pathogen is active, later turning brown. Fruit infected during the early season remain attached to the cluster base, yet remain small and appear shriveled and dark, whereas fruit infected as the disease progresses from the branches appear less shriveled and dark. Fruit infected following injury often develop red, brown, or black lesions and may exude an ooze that first appears clear or milky and later turns red to brown (Ogawa and English, 1991).

2.2.1.3   CAUSAL ORGANISM

Erwinia amylovora is the causal organism of fire blight. The rod-shaped bacterium is gram-negative and facultatively anaerobic. Isolation and tentative identification can be made using several selective or differential media; rigorous identification requires additional biochemical and molecular testing.

2.2.1.4   DISEASE CYCLE AND EPIDEMIOLOGY

E. amylovora overwinters in small twig cankers and dead wood to provide an initial source of inoculum early in the next season. Transferred by rain or insects, the bacterium penetrates host tissue at wounds or natural openings. Inoculum produced as ooze from fresh infections can serve as a secondary source of disease for later-season vegetative shoots, blossoms, and fruit. Lesion extension slows in late summer to autumn in response to less favorable conditions.

The severity of fire blight varies from season to season, dependent upon the interaction of a susceptible plant, a virulent pathogen, and favorable weather conditions. Plant susceptibility varies with plant age and horticultural practices employed; strains of *E. amylovora* vary in virulence toward plant genotypes. Temperature above 24°C and heavy rain favor infection and the rapid spread of disease. Weather conditions, particularly temperature and moisture, affect vector activity (primarily bees) and pathogen multiplication.

2.2.1.5   MANAGEMENT

1. During winter, all the blighted twigs, branches, and cankers should be cut out 10 cm below the last point of infection, and it burnt.
2. Prune the affected parts during the dormant season and apply 0.1% mercuric chloride solution on cut ends.
3. The tools should be disinfected after each in a mercuric chloride solution.

Diseases of Apples and Their Management 35

4. Resistant variety viz., red delicious, northwest greening, cox s orange pippins, Stayman, Winesap, Britemac, Carroll, Primegold, Pricilla, Quinte, Splendour, and Viking, should be planted in new areas.
5. Susceptible cultivars and rootstocks should be avoided.
6. Spraying the entire plant with Streptomycin 550 to 100 ppm in 1% glycerine, three times during pre-blossom and blossom period (Simon and Mullin, 2000).

2.2.2 CROWN GALL

2.2.2.1 INTRODUCTION

Crown gall affects woody and herbaceous plants from over 90 families, including apples grown for fruit production and ornamental use. Crown gall is variable in severity but gradually lowers tree vigor and may lead to tree death.

2.2.2.2 SYMPTOMS

Galls, varying in size, form on the crown, roots, trunk, or limbs. The texture of a gall can range from soft and spongy to hard, depending on the amount of vascular tissue it contains. Careful diagnosis of smaller galls is important, as they may be confused with excessive callus growth around wound sites or with nematode or insect-induced galls (Burr, 1995).

2.2.2.3 CAUSAL ORGANISM

The bacterium *Agrobacterium tumefaciens* is the causal organism of crown gall. It is a rod-shaped, gram-negative, aerobic, motile bacterium having one to six flagella. A large extrachromosomal piece of DNA, commonly referred to as a tumor-inducing (Ti) plasmid, is carried by *A. tumefaciens*. (Thomson, 1992).

2.2.2.4 DISEASE CYCLE AND EPIDEMIOLOGY

Wounds are necessary for the infection process and initiation of the disease cycle. *A. tumefaciens* enters through a wound, attaches to a susceptible plant cell, and inserts transfer DNA (T-DNA) from the Ti plasmid into the plant cell chromosome. Expression of the T-DNA results in the overproduction of plant hormones, which stimulates plant cells to divide, enlarge, and form a gall. The pathogen may move from galls to surrounding roots and soil, then disseminate to new plants or planting sites by rain, irrigation water, wind, insects, tools, and plant parts used for propagation.

2.2.2.5 MANAGEMENT

Good cultural and sanitation practices are key deterrents to crown gall. These include choosing a rootstock with low susceptibility, budding rather than grafting, developing management practices that minimize wounding, removing young infected trees as well as older galled trees, and dipping shears in rubbing alcohol for 10–15 seconds between cuts. Planting sites where galled plants were grown should be left fallow for several years (Simon and Mullin, 2000).

2.3 VIRAL DISEASE

2.3.1 APPLE MOSAIC VIRUS

2.3.1.1 INTRODUCTION

Apple mosaic virus is one of the oldest known and most widespread apple viruses. Infected apple plant produces low fruit yields. The disease can result in the loss of up to 40% of the crop, depending on the cultivar.

2.3.1.2 SYMPTOMS

Leaves with apple mosaic turn pale yellow or white along the veins and develop large yellow spots that later turn brown. Leaves fall prematurely. Small yellow spots develop on the leaves of less sensitive varieties or varieties infected with mild strains of the virus. Tree growth and yield are reduced. Besides varietal differences, mosaic symptoms vary with the strain of the virus and are more pronounced in a cool spring (Hadadi et al., 2011).

2.3.1.3 CAUSAL ORGANISM

Apple mosaic disease is caused by *Apple mosaic virus*. It is often found in mixed infections with several other viruses. There is no indication of field spread other than potentially through root grafting. ApMV, besides many *Malus* spp. and pear occurs naturally in more than 30 mostly woody hosts including hazelnut, *Prunus* spp., *Rubus* spp., *Rosa* spp., *Betula* spp., *Chaenomeles* spp., and *Aesculus* spp. (Posnette and Cropley, 1956).

2.3.1.4 TRANSMISSION

Apple mosaic is transmitted by budding, grafting, and root grafts of healthy trees with infected trees.

2.3.1.5 DISEASE CYCLE

Apple mosaic virus is localized in leaf and shoots tissue, and does not spread by either insects or pollen.

2.3.1.6 MANAGEMENT

1. Plant virus-free varieties and use virus-free grafting scion.
2. The use of certified virus-tested (and found to be free of all known viruses) planting material is the preferred strategy for protection from this disease.
3. Thermotherapy (24 to 32 days at 38°C) and/or apical meristem culture have been used to eliminate various viruses.
4. Diseased trees do not need to be removed, but should not be used as a source for scion material.
5. All varieties of apple are susceptible to apple mosaic, but some, such as 'Jonathan,' 'Golden Delicious,' and 'Granny Smith,' exhibit more conspicuous symptoms, but Winesap and McIntosh are only mildly affected.
6. This disease not very common to control and can easily be prevented (Simon and Mullin, 2000).

KEYWORDS

- **apple mosaic virus**
- **crown gall**
- **disease cycle**
- **fire blight**
- **relative humidity**

REFERENCES

Agarwal, R. K., & Sharma, V. C., (1966). White root rot disease of apple in Himachal Pradesh. *Indian Phytopathol., 19*, 82–86.

Agarwal, R. K., (1961). Problem of root rot of apple in Himachal Pradesh and prospectus of its control with antibiotics. *Himachal Hortc.,* 171–178.

Agarwal, R. K., (1967). Relative efficacy of fungicides for control of *Gloeodes pomigena* causing Sooty blotch disease of apple. *Proc. Acad. Sci., India, 37*, 171–178.

Agarwal, R. K., (1970). Relative importance of control method of *Phytopthoracollar* rot disease of apple. In: *Plant Disease Problems* (pp. 632–638). Publication, IARI, New Delhi.

Agrios, G. N., (1997). *Plant Pathology* (p. 635). Academic Press, New York, San Diego.

Anonymous, (2001). *Apple Scab: Integrated Pest Management for Home Gardner, Pest Notes* (pp. 1–3). University of California, Agriculture and Natural Resources.

Bains, R. C., (1935). Phytopthora trunk canker of apple. *Phytopathology, 15*, 5.

Bessey, C. M., (1877). *On Injurious Fungi: The Blight* (pp. 185–204). Iowa State Coll. Agric. Bienn. Rep. 1876–1877.

Bose, S. K., & Mehta, P. R., (1951). Record of disease occurrence. *Plant Proct. Bull.,* 346–348.

Bose, S. K., & Sindhan, G. S., (1976). *Work Done on the Disease of Temperate Fruits in U.P., and Their Control* (pp. 491–499). AICFI, Project Workshop, Ranikhet.

Burr, T. J., (1980). *Powdery Mildew of Apple*. Tree fruits IPM. Sheet No. 4 Published by New York State Agri. Exp. Stn. Geneva.

Burr, T. J., (1995). Crown gall. In: Ogawa, J. M., Zehr, E. I., Bird, G. W., Ritche, D. F., Uriu, K., & Uyemoto, J. K., (eds.), *Compendium of Pome and Stone Fruit Diseases* (pp. 52–53). American Phytopathological Society Press, Minnesota.

CDMS Chem. Search (2011). http://premier.cdms.net/webapls/formsloginRef.asp?/webapls (Accessed on 18 November 2019).

Gupta, G. K., & Agarwal, R. K., (1968). *Alternaria* blight of apple. *Pl. Prot. Bull. F.A.O., 16*, 32.

Gupta, G. K., & Thakur, V. S., (1992). *Integrated Management of Apple Scab Disease* (p. 23). ICAR and Dept. of Plant Patho. UHF, Nauni, Solan, H.P.

Gupta, G. K., (1979). Some observations of apple scab in Himachal Pradesh. *Indian Phytopathology, 32*, 172.

Gupta, G. K., (1996). Fungal disease of temperate fruit in India, In: Singh, S. J., (ed.), *Advances in Diseases of Fruit Crops in India* (pp. 331–344). Kalyani Publisher, Ludhiana.

Diseases of Apples and Their Management

Gupta, V. K., & Sharma, S. K., (1993). Fungal disease of temperate fruit. In: Chadha, K. L., & Pareek, O. P., (eds.), *Advances in Horticulture-Fruit Crops* (Vol. 3, pp. 1349–1372).

Gupta, V. K., & Sharma, S. K., (1999). *Diseases of Fruit Crops* (p. 3–27). Kalyani Publisher, Ludhiana.

Gupta, V. K., (1981). Development of powdery mildew of apple in relation to fungicides. *Indian J. Agric. Sci., 15*, 41–45.

Gupta, V. K., (1987). Apple scab and its management. *Indian Horticulture, 32*, 48–52.

Hadidi, A., Barba, M., Candresse, T., & Jelkmann, W., (2011). *Virus and Virus-Like Diseases of Pome and Stone Fruits*. St. Paul, MN: APS Press.

Jain, S. S., (1961). Root and collar rot diseases in Himachal Pradesh. *Himachal Hort., 2*, 19–23.

Jones, A. L., & Aldwinckle, H. S., (1990). *Compendium of Apple and Pear Diseases*. St. Paul, MN: American Phytopathological Society.

Joshi, N. C., Malik, A. G., Koul, M. L., & Anand, S. K., (1975). Some observations on epidemic of scab disease of apple in Jammu and Kashmir during (1973). *Indian Phytopathology, 28*, 288–289.

Nath, P., (1935). Studies on the disease of apple in Northern India. II A short Notes on apple scab due to *Fusicladium dendriticum* Puck. *J. of Indian Bot., 14*, 12–124.

Nattras, R. M., (1927). The white root rots of fruit trees caused by *Rosellinia necatrix* (Hart.) ZBerl. *Annu. Agric. Hirtic. Res. Sta. Long Ashton, 66*–72.

Odile, C., Wendy, M. S., Catherine, M., & Jacques, L., (2006). Apple scab: Improving understanding for better management. *Agriculture and Agri-Food Canada, Publication 10203E*, p. 22.

Ogawa, J. M., & English, H., (1991). *Diseases of Temperate Zone Tree Fruit and But Crops* (p. 461). Pub. 3345. Univ. Calif. Div. Agric. Natural Resources, Oakland.

Osterwalder, A., (1912). Abterben von veredlungen: Verursachtdurch *pytopthora omonivora* de Bary. *Landwirtschaftliches Jahrbuch der Schweiz, 26*, 321–322.

Pathak, V. N., (1986). *Diseases of Fruit Crops* (p. 309). Oxford & IBHCo. New Delhi.

Posnette, A. F., & Cropley, R., (1956). Apple mosaic viruses. Host reaction and strain interference. *Journal of Horticultural Science, 31*, 119–133.

Saha, L. R., (2002). *Hand Book of Plant Diseases* (pp. 260–270). Kalyani Publisher, Ludhiana.

Salmon, E. S., & Wormald, H., (1913). Report on economic mycology. *J. South East Agric. Coll., 22*, 453.

Schweinitz, L. D., (1822). *Gymnosporangium. Decand. Snopsis Fung. Carol. Sup. Schrift. Naturf. Ges., 1*, 74–75.

Simone, G. W., & Mullin, R. S., (2000). *1999–2000 Florida Plant Disease Management Guide: Fruit and Vegetables* (Vol. 3). Gainesville: University of Florida Institute of Food and Agricultural Sciences.

Thomson, S. V., (1992). Fire blight of apple and pear. In: Kumar, J., Chaube, H. S., Singh, U. S., & Mukhopadhyay, A. N., (eds.), *Plant Disease of International Importance: Diseases of Fruit Crops* (Vol. III, pp. 32–65). Prentice Hall, Englewood Cliffs, New Jersey.

William, W. T., Juliet, E. C., & David, A. R., (2004). *Powdery Mildew of Apple* (pp. 1–4). Cornell University, Cornell Cooperative Extension, The NYS Department of agriculture and Markets, the NYS Department of Environmental Conservation, and USDA-CSREES.

CHAPTER 3

Citrus Diseases and Their Management

V. JYOTHI[1] and M. E. SHILPA[2*]

[1]Scholar, Department of Plant Pathology, UAS, GKVK, Bangalore, Karnataka, India

[2]Scholar, Department of Agriculture Microbiology, UAS, GKVK, Bangalore, Karnataka, India

[*]Corresponding author. E-mail: kisshoreraj1333@gmail.com

3.1 INTRODUCTION

Citrus problems such as physiological disorders, pest, and disease damage and nutritional deficiencies are the main causes of low yield of citrus fruits, but disease having a negative impact on fruit quality as well as the marketing of the citrus fruits. Economic losses due to plant diseases may be severe, but fortunately, all the pathogens are not present in the severe form at a time.

3.2 BACTERIAL DISEASES

3.2.1 CITRUS GREENING

Citrus greening or huanglongbing is one of the most important and severe diseases of the citrus tree. It has reduced citrus fruit yields and quality in all types of citrus wherever it occurs in Asia, from China and the Philippines to the Arabian Peninsula, and to Africa. It is one of the diseases the rest of the citrus-producing countries are guarding against and bracing for its eventual spread to them.

3.2.1.1 SYMPTOMS

Citrus greening consists of smaller leaves, yellowing of the leaves of part or, usually, the entire canopy of the trees, reduced foliage, and severe dieback of twigs. Growing shoots that stand out from the normally green canopy. The most characteristic symptoms, however, are that infected trees produce fruit that fails to ripen, and instead remains green and imparts an unpleasant flavor to juice produced from such fruit. The cause of citrus greening is the fastidious phloem limited bacterium *Candidatus liberobacter asiaticus* in Asia and *C. liberobacter africanus* in Africa. The bacteria live in the host plant's phloem, where they impede the movement of nutrients. The African strain does not require as high temperature for optimum expression as the Asiatic strain. The pathogen is spread by vegetative propagation and also by two psyllid insect vectors. The Asian strain is spread primarily by *Diaphorinacitri*, whereas the primary vector for the African strain is *Triozaerytreae*, but both insect vectors can transmit either strain of the bacterium. The Asian Citrus Psyllid is small, measuring 2–4mm. These insects sit at a 45° angle. Nymphs are typically yellow and secrete a waxy substance when feeding on the plant.

3.2.1.2 HOSTS

All species and cultivars of citrus from all around the citrus growing areas are affected, such as orange, grapefruit, mandarin, tangelo, kumquat, lemon, lime, pomelo, trifoliate orange and tangelo, and native citrus.

3.2.1.3 SPREAD OF DISEASES

The long-distance spread can occur by the movement of infected citrus planting material, or by the movement of plant material infested with huanglongbing infected psyllids. Movement of other host plants such as orange jasmine (*Murraya* spp.) and curry leaf (*Bergerakoenigii*) also pose a risk of introducing huanglongbing infected Asiatic citrus psyllids. High-speed wind, tropical storms, and cyclones could also lead to the long-distance spread of Asiatic citrus psyllids from Indonesia and Papua New Guinea to northern Australia.

3.2.1.4 MANAGEMENT

Control of citrus greening depends on the exclusion of the pathogen from a citrus-producing area, the use of disease-free propagating material, removal

Citrus Diseases and Their Management

of infected trees as soon as they are detected, and attempts to control the insect vectors with insecticides or by biological control.

3.2.2 CITRUS CANKER

Citrus canker is one of the most feared of citrus diseases, affecting all types of important citrus crops. It causes necrotic lesions on fruit, leaves, and twigs. Losses are caused by reduced fruit quality and quantity and premature fruit drop. The disease is endemic in Japan and South-East Asia, from where it has spread to all other citrus-producing continents except Europe. In the United States, citrus canker was introduced into Florida in 1912, with infected nursery trees from Japan, and spread to all the Gulf States and beyond. It took 20 years, destruction by burning of more than a quarter-million bearing trees and more than 3 million nursery trees, many millions of dollars in expenses, and untold inconvenience and heartaches before citrus canker was eradicated from Florida. It took 20 more years (until 1949) to eliminate it entirely from the United States. Citrus canker, however, was again found in residential trees in the Miami area in October 1995, and the tree removal regulations were reinstated. Citrus canker has been eradicated from South Africa, Australia, and New Zealand. The latest outbreak and eradication of citrus canker in Australia occurred in 1991. In South America, citrus canker was found in Brazil in 1957. It subsequently spread to Uruguay, Paraguay, and Argentina and despite attempts to eradicate it. The disease has become permanently established there. Eradication efforts in Brazil, however, have kept the large citrus-producing areas of that country free of the disease.

3.2.2.1 SYMPTOMS

Lesions are produced on young leaves, twigs, and fruits. The lesions at first appear as small, slightly raised, round, light green spots. Later, they become grayish-white, rupture, and appear corky with brown, sunken centers. The margins of the lesions are often surrounded by a yellowish halo. The lesion size varies from 1 to 9 millimeters in diameter on leaves and up to 1 centimeter in diameter or length on fruits and twigs. Severe infections of leaves, twigs, and branches debilitate the tree, while severely infected fruit appears scabbed and deformed. Infected trees may suffer from low vigor and a reduction in fruit quality and quantity

3.2.2.2 SPREAD

Bacterial cells over the season in leaf, twig, and on fruit canker lesions. During warm, rainy weather, they ooze out of lesions and, if splashed by rain onto young tissues, bacteria enter them through stomata or wounds. Bacteria infect older tissues only through wounds. Several cycles of infection can occur on fruit; therefore, fruits often have lesions of many sizes. Free moisture and strong winds seem to greatly favor the spread of the bacteria. Citrus canker seems to be much more severe in areas in which the periods of high rainfall coincide with the period of high mean temperature, whereas it is not important in areas where high temperatures are accompanied by low rainfall.

3.2.2.3 MANAGEMENT

Burning and destroying all infected and adjacent trees to prevent the spread of the pathogen. In areas where the citrus canker bacterium is endemic, three to four sprays of copper fungicides are required for even partial control of the disease on susceptible trees. Control of citrus canker is obtained by using windbreaks, pruning diseased shoots in summer and autumn, forecasting of impending epidemics, and applying copper fungicide sprays. The use of disease-free planting materials, good sanitation, and hygiene practices may lead to less spread and establishment of the disease.

3.2.3 CITRUS (BACTERIAL) BLAST

3.2.3.1 OCCURRENCE

The disease is favored by cold and wind-driven rain.

3.2.3.2 SYMPTOMS

3.2.3.2.1 Blasted Leaves

The bacterium enters through injured tissues or wounds. Black lesions on leaf petioles usually start at the tip of leaf petiole, progresses into the leaf

Citrus Diseases and Their Management

axil. Wilt progress rapidly, curl, and dry on the tree. Leaf-blade breaks off, leaving leaf petiole stuck on the tree.

3.2.3.2.2 Twig Dieback

Twig lesions are covered with a reddish-brown scab. Small twigs and vigorous shoots are girdled and by the lesions, which later killed the twig and shoot tissues. In severe conditions, complete defoliation of the exposed side or entire trees occurs.

3.2.3.3 MANAGEMENT

Planting windbreaks and using bushy cultivars with relatively few thorns help prevent wind injury; pruning out dead or diseased twigs in spring after the rainy period reduces the spread of the disease; and scheduling fertilization and pruning during spring or early summer prevents excessive new fall growth, which is particularly susceptible to blast infection. Bordeaux sprays applied before the first rain may help prevent bacterial blast.

3.2.4 BLACK PIT

3.2.4.1 Symptoms

Lemons are most susceptible, infects fruit during cool rain or foggy periods, usually associated with winds. Pathogen needs a wound for infection. The disease does not spread from fruit to fruit in storage. Lesions stop expanding after several days and then darken the whole tissues. The disease is worse in warmer storage and with mature fruit.

3.2.4.2 MANAGEMENT

Prune out the diseased twigs in spring is the best possible cultural practices to reduce disease spread. Prevent excessive new fall growth by completing fertilization and pruning by late May. Chemical management is followed during Oct. Nov, 10-10-10 Bordeaux Full coverage, other fixed copper materials like Nordox, Kocide, etc.

3.2.5 CITRUS VARIEGATED CHLOROSIS

First reported in 1987 as affecting citrus trees in Brazil. In subsequent years, the disease spread rapidly and appeared to pose an immediate threat to the citrus industry in Brazil and possibly worldwide. A similar disease, known as "pecosita," seems to occur in Argentina. Citrus variegated chlorosis causes tree stunting, twig, and branch dieback, and reduced size and quality of fruit.

3.2.5.1 SYMPTOMS

Young leaves of affected trees appear mottled and chlorotic as though they were affected by zinc deficiency. In more mature leaves, the lower sides of the chlorotic areas produce small, light brown gummy lesions that later may become dark brown, somewhat raised, and necrotic. The entire foliage of trees becomes chlorotic to yellow. Fruit of affected trees is smaller, often no more than one-third the diameter of healthy fruit. The fruit rind is hard to the point that it damages juicing machines, and, therefore, processing plants reject batches that contain a significant number of affected fruit. Soon after a young citrus tree becomes infected with variegation chlorosis, tree growth slows down, the tree remains stunted, and twigs and branches die back, but the trees do not die. In some cases, trees may appear to recover. The pathogen of citrus variegation chlorosis is a strain of the xylem-limited fastidious bacterium *X. fastidiosa*. The bacterium grows in the xylem vessels of affected plants and reaches large numbers in them. The bacterium is spread by the vegetative propagation of infected budwood and, most likely, by xylem-feeding sharpshooter insects known to transmit other *X. fastidiosa* strains. The latter mode of transmission, although not yet proved, would explain the observed rapid spread of citrus variegated chlorosis within citrus orchards.

3.2.5.2 MANAGEMENT

The control of citrus variegated chlorosis remains difficult. The only effective means of control to date is through the use of pathogen-free budwood

Citrus Diseases and Their Management

in areas where the disease does not yet exist. Once introduced into an area, the disease seems to be spread rapidly to new trees by insect vectors, and its control becomes impossible.

3.3 FUNGAL DISEASES

3.3.1 *CITRUS BLACK SPOT*

The fungus causes a citrus black spot. The disease is severe mostly in the hot, humid, low-lying subtropical regions, and occurs sporadically in lower summer rainfall areas. The black spot of citrus, caused by the fungus *Guignardia citricarpa*, is important only as a fruit disease.

3.3.1.1 *SYMPTOMS*

Pre-harvest symptoms usually appear during the later stages of fruit development and severely affected fruit drop prematurely. A superficial lesion appears on the fruit rind. It typically begins as small orange or red spots with black margins and eventually becomes brown or black. Post-harvest lesions may also develop prior to and during shipment. All citrus cultivars viz. Sweet Valencia, lemons, mandarins, and grapefruits are mostly affected by cultivars.

3.3.1.2 *SPREAD*

Infections are caused by pycnidiospores and ascospores. The ascospores provide the primary source of inoculum on fruit. Pycnidiospores are washed from pycnidia onto lower-hanging susceptible fruit by rain. The release of ascospores, which are the main source of inoculum, is triggered by frequently occurring rainfall with temperatures in the range of 17 to 33°C.

3.3.1.3 *MANAGEMENT*

It is important to manage the inoculum source such as fallen fruit or twigs in a manner that reduces the chances of infecting other plants. Citrus Black

Spot can colonize and reproduce on dead twigs. Deadwood removal should be done. Although commercial control of black spot can only be achieved through the use of plant protection products, the efficacy thereof can be enhanced by mulching of the orchard floor with a suitable material such as grass cuttings. The use of copper and strobilurin fungicides are capable of penetrating the epidermis and the cuticle and killing the mycelium present.

3.3.2 CITRUS GUMMOSIS

Several factors, such as freeze damage, high water table, and salt accumulation, contribute to the disease. Gummosis is believed to be a condition of weak and injured trees and is reported to be infectious. This also known as brown rot gummosis is caused by one or more species of the fungus *Phytophthora*. This disease can affect the root system, the trunk below and aboveground, branches, leaves, blossoms, and fruit. It is especially troublesome during prolonged rainy periods. Trees with the bud union beneath or close to the soil and trees in poorly-drained locations are highly susceptible. Footrot becomes a more serious problem under unusual conditions such as those that occur following hurricanes.

3.3.2.1 SYMPTOMS

Infection of the lower areas of the trunk by *Phytophthora spp.* Results in dark, water-soaked areas in the active areas of infection. Often gum exudes profusely from active lesions. The dead bark frequently breaks away from the wood in vertical strips. Callus tissue begins to form on the margin of the surrounding healthy bark if the fungus becomes inactive because of unfavorable weather conditions. The disease may become active again when conditions become favorable. If the lesion encircles the trunk, girdling occurs and results in the death of the tree. Healing is slower if infection occurs below ground level. The fungus may attack young feeder roots, causing them to decay. Infection of lateral and fibrous roots can become widespread in wet soils. This infection results in poor health of the tree, a thin canopy, failure to make new growth, and poor fruit production. *Phytophthora* spp. Also may attack nursery stock and young orchard trees during rainy weather. Examination of the crowns of infected trees shows symptoms similar to those described for older trees.

Citrus Diseases and Their Management 49

3.3.2.2 SPREAD

Phytophthora fungi are present in almost all citrus orchards. Under moist conditions, the fungi produce large numbers of motile zoospores, which are splashed onto the tree trunks. The *Phytophthora* species causing gummosis develop rapidly under moist, cool conditions. Hot summer weather sloes disease spread and helps drying and healing of the lesions. Secondary infections often occur through lesions created by *Phytophthora*. These infections kill and discolor the wood deeper than gummosis itself.

3.3.2.3 MANAGEMENT

The only really effective method of control for *Phytophthora* is prevention. Once the fungus is introduced and becomes established in a grove or nursery, it is very difficult or impossible to eradicate. Preventative measures, therefore, have to be observed during all the stages of citrus production. All nursery practices have to consider sanitation as the main step for preventing *Phytophthora* infections, as well as other diseases. Firstly, the seeds used should be from certified sources and should be properly surface disinfected. The planting medium should be disinfected as well, including the germination mix and the bagging mix. At least the bagging mix should come from an area where no citrus has previously been planted. The nursery site should also be well-drained and should allow minimum access from outsiders. It is strongly recommended to have a disinfection pit at the entrance to disinfect shoes. Budding practices are as important as the management of the plants. It is recommended that budding be done no lower than 12 inches above ground level on the bag. One of the main factors to consider in planting a new site is the drainage characteristics of the soils. Poorly drained soils either because of low-lying topography or because of poor water conductivity due to high clay contents are not suitable for citrus. Farmers should select well-drained areas or improve drainage capabilities by putting drainage ditches in place. Planting citrus on cambered beds is a very effective way of preventing heavy losses from the disease. Cambering keeps the root system above the water table during periods of high rainfall. Rootstock varieties of citrus differ greatly in their degree of tolerance to *Phytophthora*. Some varieties are highly tolerant, while others are highly susceptible. Susceptibility may vary with age and be highly determined by soil and environmental conditions. There is also a major difference between foot rot susceptibility and root rot susceptibility. Some varieties may express good footrot tolerance but be root rot susceptible.

3.3.3 CITRUS SCAB

Citrus scab diseases occur in various parts of the world and can cause severe losses when they infect the fruit grown for fresh market, but in susceptible varieties, they can cause stunting of plants and can reduce the quantity and quality of fruit grown for processing. Citrus scab, or sour orange scab, caused by the fungus *Elsinoefawcettii* (anamorph *Sphacelomafawcettii*), is the most widespread and occurs wherever rainfall conditions are conducive to infection. Sweet orange scab, caused by *E. australis* (anamorph *S. australis*), occurs in South America, and Tyson's scab, caused by *S. fawcwttii* var. *Scabiosa*, occurs on lemons in Australia.

3.3.3.1 SYMPTOMS

Citrus scab diseases cause a distortion of young shoots by producing pustules consisting of a stroma of mycelium and dead host cells, plus hyperplastic host cells that have few or no chloroplasts. Scab stromata at first are pink to light brown but later become corky and turn yellowish, grayish-brown, or dark. Scab fungi produce small hyaline conidia in acervuli, and in some parts of the world (Brazil), they produce ascocarps with asci and ascospores. Scab fungi overwinter on the tree canopy. Their conidia can germinate and cause infection quickly, requiring only about 2.5 hours of wetness for initiating infection.

3.3.3.2 MANAGEMENT

1. *Choice of Cultivar:* Select a resistant species, hybrid, or cultivar. Plant in a sunny, drier location.
2. *Cropping System:* Intercrop citrus with other types of non-citrus plants or trees that are not prone to infection. Grow young plants under cover in nurseries and avoid overhead irrigation. Fungicides depending on location and disease severity, up to three fungicide applications per season may be required to control the disease, especially if leaves are heavily infected from the previous season. Start spraying before flowering, during the seasonal flush of leaves, with a second application at petal fall, and a third several weeks later during fruit formation.

Citrus Diseases and Their Management

3. *Irrigation:* Reducing or eliminating overhead irrigation of susceptible varieties during active growth. Period of the fruit inhibits infections and reduces the severity of the disease.
4. *Weed control:* Do not allow tall weeds to grow around citrus plants as they increase the relative humidity (RH) in the citrus tree canopy. High RH favors infection and disease development.
5. *Pruning:* Periodically thin trees to increase air circulation. Foliage will dry more rapidly after a rainfall, and fungicide sprays can penetrate the canopy more efficiently.

3.3.4 SOOTY MOLD

3.3.4.1 SYMPTOMS

Fungi that grow on the sweet excretions 'honeydew' of sap-sucking insects like aphids, mealybugs, or scales are commonly called sooty mold. Sooty mold is black and dry and looks just like soot. They occur on leaves, twigs, or fruits. They don't damage the plants directly but may block enough sunlight to interfere with photosynthesis. This, in turn, can stunt plant growth and eventually spoil the appearance of the fruit.

3.3.4.2 MANAGEMENT

To avoid sooty mold control, honeydew-producing insects as well as ants which tend to 'farm' aphids and mealybugs and actively transport them through the garden. An application of horticultural oil will loosen the mold deposits which can be washed from the plant with a jet of water.

3.3.5 CITRUS POWDERY MILDEW

Citrus powdery mildew is caused by the fungi *Oidiumcitri* and *O. tingitaninum.*

3.3.5.1 SYMPTOMS

White 'powdery' spores develop mostly on the upper leaf surface. Young leaves turn a pale whitish-grey-green, the ends of mildewed leaves can twist and curl upward. Young shoots can wither, and dieback and severe infection cause defoliation. White 'powdery' spores develop on young fruit. Infected fruit fall prematurely. Infection usually appears first on the new flush and immature growth. Plants can develop twig and branch dieback. All citrus cultivars can be affected, though some cultivars appear more susceptible than others. In India, the citrus varieties that are most susceptible are mandarins, sweet oranges, and tangerines.

3.3.5.2   SPREAD

This disease produces tiny, powdery spores that can survive on fallen leaves. It can be transported long distances by wind, on people (clothing, hands), equipment (e.g., pruning tools, mechanical harvesters or hedgers), or vehicles. Movement of infected citrus planting material poses a significant threat.

3.3.5.3   MANAGEMENT

Purchase healthy propagation material from reputable nurseries. Maintain good sanitation and hygiene practices.

3.4   VIRAL DISEASES

3.4.1   PSOROSIS

3.4.1.1   SYMPTOMS

A number of viruses may affect the growth and yields of citrus trees. However, symptoms are not always obvious. Infected trees, mostly orange and grapefruit, slowly decline; the main scaffold branches die, and trees become unproductive. The most distinguishing field symptom is scaling and flaking of the bark on the scion. Symptoms, including interveinal yellow flecking on young leaves, may appear in fall. During early stages, patches of bark on the trunk or scaffold branches show small pimples or bubbles, which later enlarge and break up into loose scales. Gumming often appears around the margins of a lesion. In advanced stages, deep layers of bark and the wood become impregnated with gum and die. Psorosis or bark scaling is caused by the citrus psorosis virus. Most virus and virus-like diseases can be avoided by the use of virus tested budwood.

3.4.1.2 MANAGEMENT

As with other graft-transmissible diseases, the use of disease-free budwood is the major method for preventing damage from psorosis. The Citrus Clonal Protection Program provides budwood free of major diseases to nurseries and growers. Where an old tree shows symptoms, scrape away the infected bark area to stimulate the formation of wound callus, which results in temporary recovery. Generally, a psorosis-infected tree will be less productive than healthy trees, and replacement is the best option.

3.4.2 TRISTEZA DISEASE COMPLEX

Tristeza occurs in almost all citrus-growing areas of the world. It affects practically all kinds of citrus plants but primarily orange, grapefruit, and lime. Severe strains of tristeza virus can cause severe losses of fruit quantity and quality and result in either a chronic or a quick decline and eventual death of infected trees. Tristeza symptoms consisting of a quick or chronic tree decline are particularly common and severe on trees propagated on sour orange rootstocks. Millions of citrus trees have been and continue to be killed in South Africa since 1910, in Argentina and Brazil since the 1930s, and in Colombia and Spain since the 1970s. Tristeza was first reported in Florida in the 1950s, but losses became serious after severe virus strains became widespread in the 1980s. Even more severe strains and more efficient insect vectors, however, have been moving north from South America through Central America and through the Caribbean islands, and they further threaten citrus production in the United States. In 1995, the brown citrus aphid *Toxoptera* citricida, considered to be the most efficient vector of severe (including stem pitting-causing) strains of citrus tristeza virus (CTV), was introduced into Florida. The following year, it spread to almost all citrus groves. This introduction poses an immediate threat to the 20 million citrus trees grafted on sour orange rootstock in Florida alone. It also threatens, however, potential catastrophic losses to Florida and the total citrus industry in the United States.

3.4.2.1 SYMPTOMS

CTV on the various citrus species varies primarily with the particular strain of the virus and with the rootstock on which the citrus scion is propagated. Most tristeza virus strains are mild and produce no noticeable symptoms on commercial citrus varieties; they are detected only by indexing on sensitive indicator hosts, such as Mexican lime, or by serological and nucleic acid tests. More severe strains cause a condition known as seedling yellows, consisting of severe chlorosis and dwarfing on seedlings of sour orange, lemon, and grapefruit, especially when they are kept under greenhouse conditions. In the field, young sweet orange, grapefruit, and other citrus trees growing on sour orange rootstock and inoculated with some of the severe strains of the tristeza virus develop a quick decline within a few weeks.

The leaves of trees developing quick decline turn yellow or brown and later wilt and fall off while the fruit continues to hang on the dead tree. Some severe strains, however, do not cause quick decline but instead either interfere with the growth of young trees, which remain severely stunted and fail to come into production, or cause trees to decline over several years (chronic decline), during which the trees grow poorly, become less productive, decline, and eventually die. Decline to induce tristeza virus strains infecting citrus trees on sour orange rootstocks to cause phloem necrosis at the graft union, which results in the accumulation of foodstuffs in an overgrowth of the scion above the union while few foodstuffs go through to the roots. As a result, the roots grow poorly or die, causing the decline of the aboveground parts of the tree. In addition to the mild and decline-causing strains of CTV, there are severe strains that cause stem pitting Infected trees exhibit deep longitudinal pits in the wood under the bark, in trunks, in branches, and even in twigs of infected grapefruit or sweet orange trees regardless of the rootstock grafting and by several species of aphids in the semi-persistent manner, i.e., the aphids require feeding for at least 30 to 60 minutes to acquire the virus and subsequently remain viruliferous for about 24 hours. The various aphid species vary greatly in their ability to transmit CTV. The most efficient aphid vector, *Toxopteracitricida*, known as the brown citrus aphid, colonizes, and affects only citrus but is 10 to 25 times more efficient as a CTV vector than any of the other aphids.

3.4.2.2 SPREAD

T. citricida can transmit CTV strains causing severe decline or stem pitting that the other aphid vectors do not transmit or transmit poorly. *T. citricida*

Citrus Diseases and Their Management 55

occurs in most citrus-growing areas but not yet in the Mediterranean countries. In the last 20 years, this aphid had been moving northward from South America through Central America and the Caribbean islands. By 1993 it had reached Cuba. In late 1995, *T. citricida* was found in Florida, and as expected, it spread throughout most of the citrus-growing areas within the next year.

3.4.2.3 MANAGEMENT

Citrus tristeza is difficult to control. Where the disease is absent, strict quarantine regulations should be enforced. Only tested budwood certified to be free of CTV should be used under all conditions, and any trees detected to carry severe strains of tristeza virus should be destroyed. If the disease already occurs in an area, considerable control can be obtained by avoiding grafting trees on sour orange and, instead, grafting on tristeza-tolerant rootstocks; using scion varieties tolerant to stem pitting also is recommended. In addition, trees can be cross protected from severe tristeza for fairly long on which they are grafted. Actually, these strains also cause stem pitting on the rootstocks themselves. Trees with stem pitting are stunted and set less fruit, the fruit is of smaller size and of poor quality, the twigs are brittle and break easily, and the trees decline but do not die for many years. The pathogen, CTV, consists of a thread-like particle 2,000 nanometers long by 12 nanometers in diameter. Each particle contains one positive-sense single-stranded RNA consisting of 20-kilo bases and a coat protein subunit with a molecular weight of 25,000. The tristeza virus RNA codes for 10 to 12 proteins, but the function of several of them is still uncertain. The largest protein (349 k) is a papain-like proteinase, methylesterase, and helicase. The 25 k is the coat protein periods by inoculating them with certain mild strains of the virus. Presently, considerable efforts are being made to genetically engineer citrus trees to express CTV genes, such as the coat protein gene, that might make the trees resistant to tristeza.

3.5 SPIROPLASMA DISEASES

3.5.1 *CITRUS STUBBORN DISEASE*

Citrus stubborn is present in hot and dry areas such as most Mediterranean countries, the southwestern United States, Brazil, Australia, and possibly South Africa. In some Mediterranean countries and in California, stubborn is regarded as the greatest threat to the production of sweet oranges and grapefruit. Because of the slow development of symptoms and the long survival of affected trees, the spread of stubborn is insidious and its detection difficult. However, yields are reduced drastically; the trees produce fewer fruits, and many of those are too small to be marketable.

3.5.1.1 SYMPTOMS

The stubborn disease affects leaves, fruits, and stems of all commercial varieties regardless of the rootstock. In general, affected trees show a bunchy, upright growth of twigs and branches, with short internodes and an excessive number of shoots. Some of the affected twigs die back. The trees show slight to severe stunting. The leaves are small, often mottled or chlorotic. Excessive winter defoliation is common. Affected trees bloom at all seasons, especially in the winter, but produce fewer fruits. Some of the fruits are very small and lopsided, frequently resembling acorns. Such fruits have abnormally thin rind from the fruit equator to the stylar end. The rind is often dense or cheesy. Some fruit show greening of the stylar end. Affected fruit tends to drop prematurely. Fruits are usually sour or bitter and have an unpleasant odor and flavor. Also, fruit from affected trees or parts of trees tends to have poorly developed and aborted seeds.

3.5.1.2 SPREAD

The pathogen is *Spiroplasmacitri*. It is found in the phloem. It can be cultured readily on artificial media. *Spiroplasmacitri* has also been found in or transmitted to plants of many dicotyledonous families and some monocots, including most crucifers and several stone fruits, such as peach and cherry. Some infected hosts, such as pea and bean, become wilted and die, whereas most others remain symptomless. The citrus stubborn disease is transmitted with moderate frequency by budding and grafting. It is spread naturally in citrus orchards by several leafhoppers, such as *Circulifer tenellus*, *Scaphytopiusnitridus*, and *Neoaliturus haemoceps*.

3.5.1.3 MANAGEMENT

The control of citrus stubborn depends on the use of spiroplasma-free budwood and rootstocks, as well as early detection and removal of infected

Citrus Diseases and Their Management 57

trees. Young citrus trees responded experimentally to treatment with tetracy-cline antibiotics, but this is not practiced commercially.

3.6 VIROID DISEASES

3.6.1 *EXOCORTIS*

Exocortis is the shelling of susceptible rootstocks. Exocortis is spread by a virus-like particle that kills the bark; the bark dries, cracks, and may lift in thin strips. Some droplets of gum often appear under the loose bark. Exocortis occurs worldwide. It affects trifoliate oranges, citranges, Rangpur, and other mandarin and sweet limes, some lemons, and citrons. It is important commer-cially when infected budwood of orange, lemon, grapefruit, and other citrus trees is grafted on exocortis-sensitive rootstocks. Such trees show slight to great reductions in growth, and yields are reduced by as much as 40%.

3.6.1.1 *SYMPTOMS*

Infected susceptible plants develop narrow, vertical, thin strips of partially loosened outer bark that gives the bark a cracked and scaly appearance when they are about 4 to 8 years old. Infected exocortis susceptible plants may also show yellow blotches on young infected stems, and some citrons show leaf and stem epinasty along with cracking and darkening of leaf veins and petioles. In the plant cells, the viroid is associated with nuclei and internal membranes of host cells and results in aberrations of the plasma membranes. All infected plants, including resistant cultivars grafted on such trees, usually appear stunted to a smaller or greater extent and have lower yields.

3.6.1.2 *SPREAD*

Citrus exocortis viroid (CEVd) consists of 371 nucleotides. CEVd is transmitted readily from diseased to healthy trees by budding knives, pruning shears, or other cutting tools, by hand, and possibly by scratching and gnawing of animals; CEVd is also transmitted by dodder and by sap to herbaceous plants. On contaminated knife blades, CEVd retains its infectivity for at least eight days. The viroid is highly resistant to heat inactivation and to almost all common chemical sterilants except sodium hypochlorite solution and ribonuclease. CEVd has been identified in the past by graft indexing on sensitive clones of Etrog citron, which develops leaf epinasty and bark splitting within a few months. In the past 10 years, CEV identification has been made by electrophoresis of infectious sap and by using radioisotope-labeled DNA probes complementary to CEV.

3.6.1.3 MANAGEMENT

It is best to remove infected trees because pruning clippers and saws, unless thoroughly disinfected with bleach, can transmit exocortis to other trees. Some rootstocks may be less susceptible than others to exocortis.

3.7 POST-HARVEST DISEASES

3.7.1 SOUR ROT OF CITRUS

3.7.1.1 SYMPTOMS

Sour rot is caused by a fungal pathogen *Geotrichumcitri-aurantii* (Feraris) Butler. It has been reported in most areas where citrus is grown, and it occurs on all cultivars, but it is particularly troublesome on fruits that are stored for long durations. The fungus only infects fruit through injuries and in particular deep injuries that involve the albedo tissue. Sour rot develops more frequently on mature to over-mature fruit with high peel moisture. The initial symptoms are water-soaked lesions, light to dark yellow, and slightly raising with the cuticle being more easily removed from the epidermis than lesions caused by green or blue mold. Decayed fruit tissue has a sour odor that attracts fruit flies, and these can spread the fungus to other injured fruit during storage. The fungus is present in soil and can reach the fruit surface from wind-blown or splash-dispersed soil and by fruit-soil contact. Fruits on the lower portion of the citrus tree contain higher populations of the fungus and soil from the field or from diseased fruit can contaminate drenching equipment, soak tanks, pallet bins, washer brushes, belts, and conveyors. Packed infected fruits allow the disease to spread to sound fruit in the container. The disease develops rapidly at warm temperatures, with an optimum at 27°C.

Citrus Diseases and Their Management

3.7.1.2 MANAGEMENT

Carefully harvesting and handling of fruit to minimize the fruit injuries; preventing fruit from contacting soil; equipment, rooms, and fruit containers should be sanitized routinely; continuous treatment of drenchers or soak tankers with chlorine; removal of rotten fruit after dumping on packinghouse line to minimize equipment contamination, particularly washer brushes; using sodium ortho-phenylphenate (SOPP), which is applied in a foam, spray or drench during washing; immediately storage of packed fruit at 10°C or less delays disease development

3.7.2 GREEN MOLD OF CITRUS

3.7.2.1 SYMPTOMS

Green mold is caused by a fungal pathogen *Penicillium digitatum* (Pers.:Fr) Sacc. It is identified by the mass of olive-green spores produced on infected fruit, and their prolific production ensures that this fungus is found wherever fruit is present, including field, packinghouses, equipment, degreening, and storage rooms, transit containers and in the marketplace. Infection takes place only through wounds where nutrients are available to stimulate spore germination, and fruit decay begins at these infected injury sites. The early infection area appears as a soft watery spot. As the lesion progresses, white mycelia develop, and these produce the green spores. The white mycelium develops into a broad zone surrounding this sporulating area. Within a few days, the entire fruit can be covered with green spores. Spoilage of fruit, caused by the spread of spores from diseased fruit onto adjacent fruit, can occur within the shipping container, but green mold spores will only infect damaged fruit in packed cartons.

3.7.2.2 MANAGEMENT

Careful harvesting and handling of fruit to minimize fruit injuries; drench fruit with chlorinated thiabendazole (TBZ), fludioxonil (Graduate) or imazalil if fruit cannot be packed within 24 hours after harvesting; apply a stringent daily sanitization with effective disinfectants (such as chlorine and quaternary ammonium) to the wet line through the wax brushes, and fruit bins after dumping; wash fruit with sodium ortho-phenylphenate (SOPP, 2%) and applying TBZ and/or imazalil on packing line in aqueous or wax treatments; prompt removal of any infected fruit from the packinghouse, and no repacking of packed cartons with fungicide-treated sporulating fruit in the packinghouse facility; and use low temperature to retard pathogen infection and disease development by immediately pre-cooling or storing fruit after packing at 10°C or less.

3.7.3 BLUE MOLD OF CITRUS

3.7.3.1 SYMPTOMS

Blue mold is caused by a fungal pathogen *Penicillium italicum* Wehmer. *P. italicum* infects citrus fruit via injuries to cause blue mold. Blue mold is recognized by the mass of blue spores produced in decayed fruit. Initial lesions are similar to the lesions of green mold, but the spores are blue in color and are surrounded by a narrow band of white mycelium encompassed by the water-soaked rind. Blue mold develops less rapidly than green mold under ambient conditions so that the green mold is often observed in mixed infections. Blue mold is more common in fruit held in cold storage for summer, and it can spread in packed cartons more readily than green mold. It occurs in all citrus-producing regions of the world, although it is not as prevalent as green mold under Florida conditions.

3.7.3.2 MANAGEMENT

Careful harvesting and handling to minimize injuries; stringent daily sanitization with approved sanitizers for packing line, equipment, fruit bins. Degreening and storage rooms should be sanitized when sporulating fruit are observed, and water in drenchers and soak tanks should be continuously chlorinated; prompt removal of any infected fruit in the packinghouse, and no repacking of packed cartons with fungicide-treated sporulating fruit in the packinghouse facility; monitoring of spore population in the packinghouse for concentration and resistance to post-harvest fungicides; spaying thiophanate-methyl (Topsin-M) pre-harvest, washing fruit with SOPP and/ or applying TBZ and/or imazalil on packing line in aqueous or wax treatments; and immediately pre-cooling or storing fruit after packing at 10°C or below.

Citrus Diseases and Their Management

3.7.4 PHOMPSIS END ROT OF CITRUS

3.7.4.1 SYMPTOMS

The fungus grows on tree deadwood, where it produces spores that spread by water to immature fruit during rainfall or irrigation. Infection of the young fruit produces small pustules. The fungus also becomes established in dead tissue of the button, where it lays dormant until harvest. As the button deteriorates during storage, the fungus grows from the surface into the base of the fruit through natural openings that occur in the abscission zone. Decay progresses evenly through the rind and core until the entire fruit is completely rotted, with no spread to adjacent fruit. This type of stem-end rot is dark to light brown in color and more prevalent in late-season non-degreened or cold storage fruit of all types. Ethylene degreening has no effect on Phomopsis stem-end rot. It is a serious post-harvest decay in humid subtropical and tropical areas.

3.7.4.2 MANAGEMENT

Using the good cultural practice to produce trees with minimal amounts of deadwood; applying TBZ, fludioxonil, and/or imazalil on the packing-house in aqueous or wax treatment immediately pre-cooling or storing fruit after packing at a temperature of 10°C or less use of biocontrol agent like *Streptomyces griseoviridis.*

3.7.5 BROWN ROT

3.7.5.1 SYMPTOMS

Brown rot occurs in both pre-harvest and post-harvest stages. Phytophthora species persist in the soil and are spread through rain splashes to fruit hanging on the lower canopy of the trees, thereby infecting the fruit. Most infections develop on the tree within 3 to 4 feet of the soil surface, although they might be found in higher locations as a result of wind-driven rains. Initial infection shows as light discoloration on any area of the fruit surface. As the decay develops, the lesion becomes light brown, firm, and leathery. Under humid conditions, decayed areas spread rapidly, and white mycelia may form on infected areas. Fruits with brown rot have a characteristic rancid odor. Brown rot spreads in packed containers from infected to healthy fruit.

3.7.5.2 SPREAD

Infects from the soil to low-hanging fruit when splashed up during rain or irrigation. Needs wet fruit for about 3 hours at 55–75°F to infect, and no wound needed. Spreads rapidly from fruit to fruit in storage.

3.7.5.3 MANAGEMENT

Using cultural practices to minimize long periods of fruit wetness by pruning to move low hanging branches, by proper irrigation and soil drainage management, and use of mowing and herbicide treatment to reduce ground vegetation avoiding harvest of fruit from poorly drained groves during rainy periods. Apply sprays of copper or Aliette in August or early September to the tree canopy in blocks with a history of brown rot or at the occurrence of disease outbreaks; and. Immediately pre-cooling or storing fruit after packing at a temperature of 10°C or less.

KEYWORDS

- brown rot
- citrus exocortis viroid
- citrus tristeza virus
- sodium ortho-phenylphenate
- sporulating
- thiabendazole

REFERENCES

Agrios, G. N., (1992). *Plant Pathology*. Academic Press, London.

Anonymous (2004). *Package of Practices for Horticultural Crops*. Directorate of Extension Education. SKUAST-Jammu (J&K).

Citrus Diseases and Their Management

Arden, F. S., & Alan, A. M., (1986). *Vegetables Disease and Their Control*. A Wiley Inter Science Publication, New York.

Arjunan, G., Karthikeyan, G., Dinakaran, D., & Raguchander, T., (1997). *Diseases of Horticultural Crops*. Devi Publications, Tiruchirapalli, Tamil Nadu.

Embleton, T., Jones, W. Labanauskas, C., & Reuther, W., (1967). Leaf analysis as a diagnostic tool and guide to fertilization. In: Reather, W., Webber, H. J., & Batchelo, L. D., (eds.), *Citrus Industry*, (Vol. 3, pp. 183–210). The University of California, USA.

Floyd, B. F., (1977). Dieback or exanthema of citrus tree. Proc. Fla. Fr. Gr. Assoc. (Jan) as cited by Hume, H. H., 1900. *Fla. Agric Exp. Stn. Bull.*, *53*, 157.

Hume, H. H., (1900). Some citrus troubles. *Fla. Agric. Exp. Stn. Bull.*, *53*, 157–161.

Knorr, L. C., (1973). *Citrus Diseases and Disorders* (pp. 26–27). University Press of Florida. Gainesville.

Mehrotra, R. S., (1999). *Plant Pathology* (pp. 750–753). Tata Mc Graw-Hill Publishing Co. Ltd. New Delhi.

Rajput, C. B. S., & Haribabu, S. R., (1985). *Citriculture* (pp. 273–355). Kalyani Publishers, New Delhi.

Saha, L. R., (2002). *Hand Book of Plant Diseases* (pp. 290–292). Kalyani Publishers, Ludhiana.

Singh, R. S., (1983). *Plant Diseases* (pp. 552–553). Oxford & IBH Publishing Co. Pvt. Ltd. New Delhi.

Zekri, M., (1996). Several disorders in citrus trees. *The Citrus Industry, 77*(2), 18–20, 53.

CHAPTER 4

Diseases of Grapes and Their Management

B. D. DEVAMANI

Department of Plant Pathology, UAS, GKVK, Bangalore, Karnataka, India,
E-mail: kisshoreraj1333@gmail.com

4.1 INTRODUCTION

Grape is the most widely grown fruit crop in India. Maharashtra is the leading grape-producing state followed by Andhra Pradesh and Punjab. Biotic constraints are one of the major constraints in grape production. These constraints may cause enormous losses, which invades on leaves, twigs, inflorescence, roots, and fruits reasonably ample loss considering total fruit reduction and production of poor quality fruit produced, rendered decreased rate in price (Agrios, 1992).

4.1.1 POWDERY MILDEW: UNCINULA NECATOR (I.S: OIDIUM TUCKERI)

Losses in fruits yield due to the powdery mildew may be up to 40–60%. In addition to the loss of fruit yield, infected berries tend to be higher in acid content than healthy fruits and are unsuitable for winemaking.

4.1.1.1 SYMPTOMS

- The disease attacks the vines at any stage of their growth. All the aerial parts of the plant are attacked. Cluster and berry infections usually appear first.
- Infection in floral parts, causing shedding of flowers and poor fruit set.
- Early berry infection results in shedding of affected berries.

- Powdery growth is visible on older berries, and the infection results in the cracking of the skin of the berries. Often infected berries develop a net-like pattern of scar tissues.
- Powdery growth mostly on the upper surface of the leaves.
- Malformation and discoloration of affected leaves. Leaf lesions appear late and don't cause much damage.
- Discoloration of the stem to dark brown.

4.1.1.2 PATHOGEN

The mycelium is ectophytic and produces bilobate or multilobate appressoria. The conidiophores are simple, multiseptate, and erect, bearing a chain of 3–4 conidia. Under Indian conditions, the perfect stage of the fungus is not found. When the mating types are present, cleistothecia can form on all infected tissues during the later part of the growing season (Arjunan et al., 1997).

4.1.1.3 MODE OF SPREAD AND SURVIVAL

The pathogen survives as dormant mycelium and conidia present in the infected shoots and buds.

Spread of the pathogen through air-borne conidia (Gessler et al., 2011).

4.1.1.4 FAVORABLE CONDITIONS

Cool, dry weather, and a maximum temperature in the range of 27–310°C with R.H. up to 90% favor disease development (November and December). An increase of R.H. by 1% increased disease incidence by 2.4%, and an increase of temperature by 10°C decreases the disease by 4.4% (Becker et al., 1993).

4.1.1.5 MANAGEMENT

- Clean cultivation of vines or removal and destruction of all diseased parts.
- Dustings of vines with 300 mesh Sulfur (1st when new shoots are 2 weeks old, 2nd prior to blossoming, 3rd when the fruits are half-ripe).
- Prophylactic spray with B.M. 1% or Lime sulfur at the dormant stage delays the development of the disease by decreasing initial inoculum.

Diseases of Grapes and Their Management 67

- Spray wettable sulfur @0.3% or karathane or calixin @0.1%.
- Morestan @0.03% sprayed at 4 days interval, starting from the last week of December to the 1st week of March (Hed et al., 2006).
- Grow resistant varieties like Chholth Red, Chholth white, Skibba Red, Skibba White, etc.

4.1.2 DOWNY MILDEW

4.1.2.1 SYMPTOMS

- Symptoms appear on all aerial and tender parts of the vine. Symptoms are more pronounced on leaves, young shoots, and immature berries.
- Irregular, yellowish, translucent spots on the upper surface of the leaves.
- Correspondingly, on the lower surface, dirty white, powdery growth of fungus appears.
- Affected leaves become yellow and brown and get dried due to necrosis.
- Premature defoliation.
- Dwarfing of tender shoots.
- Infected leaves, shoots, and tendrils are covered by whitish growth of the fungus.
- White growth of fungus on berries, which subsequently becomes leathery and shrivels. Infected berries turn hard, bluish-green, and then brown.
- Later infection of berries results in soft rot symptoms. Normally, the fully grown or maturing berries do not contact fresh infection as stomata turn non-functional.
- No cracking of the skin of the berries.

4.1.2.2 PATHOGEN

Plasmopara viticola is a causal organism of this disease biotroph. The intercellular mycelium of the fungus is coenocytic, thin-walled, hyaline, and produce spherical or pear-shaped haustoria. Sporangiophores arise from hyphae under high humid conditions. The branching of the sporangiophores is at right angles to the main axis and at regular intervals. From the apex of each branch, 2–3 sterigmata arise and bear lemon-shaped, papillate sporangia. Sporangial germination may be through zoospores or by germ tube based on humidity and temperature.

4.1.2.3 SPREAD AND SURVIVAL

Sporangia or zoospores are spread by wind, rain, etc. The pathogen survives as oospore in the infected leaves, shoots, and berries and also as dormant mycelium in infected twigs.

4.1.2.4 FAVORABLE CONDITIONS

- **Optimum temperature:** 20–22°C.
- **Relative humidity (RH):** 80–100%.

4.1.2.5 MANAGEMENT

- Collect and burn fallen leaves and twigs.
- Sanitation of the orchard.
- Vine should be kept high above the ground to allow the circulation of air by proper spacing.
- Pruning (April–May and September and October) and burning of infected twigs.
- Grow resistant varieties like Amber Queen, Cardinal, Champa, Champion, Dodridge, and Red Sultana ¾ The disease can be effectively managed by giving 3–5 prophylactic sprays with 1% B.M or Fosetyl-Al (Aliette) 0.2% or metalaxyl + mancozeb 0.3 to 0.4% or azoxystrobin or dimethomorph.

4.1.3 ANTHRACNOSE/BIRDS EYE DISEASE

It is especially serious on new sprouts during the rainy season. Among various foliar diseases of grapevine in India, anthracnose has the longest spell spread over the period from June to October.

4.1.3.1 SYMPTOMS

- Visible on leaves, stem, tendrils, and berries.
- Young shoots and fruits are more susceptible than leaves.
- Circular, grayish-black spots or red spots with yellow halo appear.

Diseases of Grapes and Their Management 69

- Later, the center of the spot becomes grey, sunken, and fall off, resulting in a symptom called 'shot hole.'
- Black, sunken lesions appear on young shoots.
- Cankerous lesions on older shoots. Girdling and death of shoots occur.
- Infection on the stalk of bunches and berries results in the shedding of bunches and berries, respectively.
- Sunken spots with the ashy grey center and dark margin on fruits (Birds eye symptom). In warm and wet weather, pinkish spore mass develops in the center of the spots.
- Mummification and shedding of berries.

4.1.3.2 ETIOLOGY

Anthracnose disease of grape is caused by *Elsinoe ampelina* (Anamorph—*Gloeosporium ampelophagum*), produces hyaline, single-celled conidia Teleomorph—*Elsinoe ampelina,* and produces hyaline four-celled ascospores (Pscheidt et al., 1989).

4.1.3.3 MODE OF SPREAD AND SURVIVAL

Survives as dormant mycelium in the infected stem-cankers.

The secondary spread is by means of conidia formed in the leaf and other plant parts, which are easily disseminated by wind and splashed rain. A continuous drizzle of rain for 2–3 days encourages the disease. No infection can take place in the absence of rain. Wind associated with a warm atmosphere (temp.), and heavy rains favor the disease spread.

4.1.3.4 FAVORABLE CONDITIONS

Warm wet weather, low lying, and badly drained soils favor disease development. Anab-e-shahi is the susceptible variety against this disease.

4.1.3.5 MANAGEMENT

- Removal of infected twigs.
- Selection of cuttings from disease-free areas and dipping them in 3% FeSO4 solution for ½ an hour before planting.

- Spraying Bordeaux mixture 1% or COC@0.2% or carbendazim@0.1%.
- Grow resistant varieties like Bangalore blue, Golden Muscat, Golden queen, and Isabella.

4.1.4 ALTERNARIA LEAF SPOT

This disease is caused by *Alternaria vitis.*

4.1.4.1 SYMPTOMS

- Appearance of patches mostly along the margin of leaves ¾ Individual spots appear rarely in the middle of the leaves.
- In the initial stage, minute, yellow spots appear on the upper surface of the leaves.
- Later, spots enlarge and form brownish spots with concentric rings in them.
- In severe cases of attack, leaves dry completely and defoliation occurs.

4.1.4.2 MODE OF SPREAD AND SURVIVAL

Survives as mycelium or conidia in infected plant debris. Secondary spreads through airborne conidia.

4.1.4.3 FAVORABLE CONDITIONS

High humidity, high rainfall and dew, and a heavy dosage of nitrogenous fertilizers favor disease development.

4.1.4.4 MANAGEMENT

- Destruction of crop debris;
- Selection of disease-free seed;
- Seed treatment with mancozeb@2g/kg seed;
- Foliar spray of mancozeb@0.25%.

Diseases of Grapes and Their Management

4.1.5 *BACTERIAL LEAF SPOT: PSEUDOMONAS VITICOLA*

This disease was the first time reported from S.V. Ag. College, Tirupati (1969–1970). Anab-e shahi and Thomson seedless varieties are highly susceptible.

4.1.5.1 SYMPTOMS

Symptoms appear on foliage and twigs. Initially, yellowish, circular, translucent spots develop along the midrib and veins of leaves. Necrotic areas along midrib and veins are the characteristic symptom. Leaves turn yellow, shriveled, and shed pre-maturely (Arjunan et al., 1997).

4.1.5.2 SPREAD AND SURVIVAL

Infected twigs show primary inoculums. The bacterium survives in the soil on plant debris. Secondary inoculums through wind splashed rain.

4.1.5.3 MANAGEMENT

- Prune out and destroy infected plant debris.
- Spraying antibiotics like streptomycin 100 ppm along with copper fungicides.

4.1.6 *RUST: PHAKOPSORA VITIS*

Grape rust is common throughout South-East Asia and other parts of the world. Cultivar black prince is highly susceptible.

4.1.6.1 SYMPTOMS

- The presence of clustered, small, yellow to orange, powdery spores on the underside of mature grapevine leaves. Small, dark spots also appear on the upper surface.
- Under severe infection, the entire leaf surface is covered by sori, and premature defoliation occurs

72 *Diseases of Fruits and Vegetable Crops*

- The disease eventually leads to weakening of the vine due to poor shoot growth ¾ Reduction in quantity and quality of fruit.

4.1.6.2   DISEASE CYCLE

- **Primary inoculum:** Teliospores.
- **Secondary inoculum:** Uredospores.
- Grape rust affects *Vitis* spp., including commercial and some ornamental grape varieties. It has also been found on native *Ampelocissus* species.

4.1.6.3   MANAGEMENT

Spray zineb@0.2% or dust sulfur@25kg/ha.

KEYWORDS

- **alternaria leaf spot**
- **anthracnose**
- *downy mildew*
- **pathogen**
- *Pseudomonas viticola*
- **relative humidity**

REFERENCES

Agrios, G. N., (1992). *Plant Pathology*. Academic Press, London.

Arjunan, G., Karthikeyan, G., Dinakaran, D., & Raguchander, T., (1997). *Diseases of Horticultural Crops*. Devi Publications, Tiruchirapalli, Tamil Nadu.

Becker, C. M., & Pearson, R. C., (1993). *Epidemiology and Control of Black Rot*. Reports to the New York Wine and Grape Foundation. New York State Agricultural Experiment Station.

Coombe, B. G., & Dry, P. R., (1992). *Viticulture* (Vol. 2). Practices. Winetitles. Adelaide, Australia.

Diseases of Grapes and Their Management

Ferrin, D. M., & Ramsdell, D. C., (1977). Ascospore dispersal and infection of grapes by Guignardia bidwellii, the causal agent of grape black rot disease. *Phytopathology, 67*, 1501–1505.

Gessler, C., Pertot, I., & Perazzolli, M., (2011). *Plasmopara viticola*: A review of knowledge on downy mildew of grapevine and effective disease management. *Phytopathologia Mediterranea, 50*, 3–44.

Hed, B., & Travis, J. W., (2006). Evaluation of alternative and organic fungicides for control of Niagara grape diseases. *Plant Disease Management Reports, 1*, SMF008.

Hoppmann, D., & Wittich, K. P., (1997). Epidemiology-related modeling of the leaf-wetness duration as an alternative to measurements, taking *Plasmopara viticola* as an example. Zeit. Pflanzenkrank. Pflanzenschutz, *Journal Plant Dis. Prot., 104*, 533–544.

Pearson, R. C., & Goheen, A. C., (1994). *Compendium of Grape Diseases*. American Phytopathological Society Press, St. Paul, MN.

Pscheidt, J. W., & Pearson, R. C., (1989). Effect of grapevine training systems and pruning practices on occurrence of Phomopsis cane and leaf spot. *Plant Disease, 73*, 825–828.

CHAPTER 5

Diseases of Custard Apple and Their Management

G. L. SHARMA[1*] and N. LAKPALE[2]

[1]*Department of Plant Pathology and Fruit Science, Indira Gandhi Krishi Vishwavidyalaya, Raipur, Chhattisgarh, India*

[2]*College of Agriculture, Indira Gandhi Krishi Vishwavidyalaya, Raipur, Chhattisgarh, India*

Corresponding author. E-mail: glsigau@rediffmail.com

5.1 INTRODUCTION

Custard apple (*Annona squamosa*) is a genus of *Annona* and a native of the tropical Americas and West Indies. Some of the other trees widely grown are *Annona reticulata*, *Annona cherimola* (cherimoya), and *Annona scleroderma*. The fruit varies in shape and color, spherical through conical, with a thick rind composed of knobby segments. The color of custard apple fruit varies from pale green through blue-green. When ripe, the segments of the fruit separate quite easily when lightly pressed between hands, exposing the flesh. Custard apple-growing regions in India include Assam, Bihar, Madhya Pradesh, Maharashtra, Odisha, Rajasthan, and Uttar Pradesh, Andhra Pradesh, Telangana, and Tamil Nadu. Approximately 55,000 hectares are dedicated to custard apple cultivation. Along with Maharashtra, Gujarat is another large custard apple growing state.

Custard apples have fewer serious pests and diseases, so you can grow custard apples with no or limited use of chemicals. Annona fruits grow well throughout the plains of India at elevations not exceeding 4,000 ft. It prefers a tropical climate, but with cool winters. The fruit tolerates a variety of conditions, from saline soils to droughts. Farmers usually cultivate the fruits on hills in barren lands. Erratic rains will, however, impede fruit quality. The tree displays yellow trumpet-shaped flowers that emit a pleasant sweet smell, with only a small number of flowers setting fruit. The fruits are variable in shape with the outer being covered in rounded knobs, with the inside containing a custard-like flesh. Custard apples have fewer serious pests and diseases, so you can grow custard apples with no or limited use of chemicals.

5.2 FUNGAL DISEASES

5.2.1 CYLINDROCLADIUM LEAF SPOT

5.2.1.1 SYMPTOMS

On the upper surface of leaves, dark purple colored irregular-sized spots appeared. Similar kinds of spots also seen on the upper part of the infected fruits, which later spread on the lower part also. These spots enlarge and later dry out and crack.

5.2.1.2 CAUSAL ORGANISM: CYLINDROCLADIUM COLHOUNII

Conidiophores have a single filament. The filament is septate, hyaline, and terminates in a narrowly clavate vesicle. Conidia are cylindrical, hyaline, septate (1 or 3), rounded at both ends. The perithecia were yellow in color primarily, turn orange to red at maturity. The perithecia contain asci and ascospores in an asci present in four numbers.

5.2.1.3 FAVORABLE ENVIRONMENT

- Temperature: 15°C–25°C.
- Relative humidity (RH): 70% and above.
- Intermittent rains.

5.2.1.4 SURVIVAL

In diseased plant debris or other hosts in range, pathogen survives in the form of conidia and dormant mycelium.

5.2.1.5 MANAGEMENT

1. Infected crop debris and the wild host should be destroyed.
2. Increase air circulation by proper training and pruning.

3. Rain splash dispersal of pathogen should be minimized by pruning lower or ground touching branches (skirts to 50 cm above), mulching under trees, and grassing inter-tree spaces.
4. Collect dead twigs and mummified fruits in each season.
5. Fruit infection should be regularly monitored during the season so that spraying can start before fruits get infected severely.
6. If required, spray of Copper oxychloride or Mancozeb @ 2.5 g/l of water on leaves only, never on the fruits.

5.2.2 ALTERNARIA LEAF SPOT

5.2.2.1 SYMPTOMS

The typical symptoms first appear along the leaf margins as small yellowish spots, which gradually enlarge and turn into brownish patches with concentric rings. These patches coalesce to each other and cause drying and defoliation of infected leaves. On fruits and rachis, dark brown to purplish patches appears just below its attachment with the shoots.

5.2.2.2 CAUSAL ORGANISM: ALTERNARIA SPP.

Mycelium frequently septate, intercellular, branched, light brown becomes darker with age. Conidiophores short, dark-colored, simple, septate. Conidia are borne singly or in chains (acropetal succession), dark-colored, muriform (presence of 5–10 transverse septa and few longitudinal septa), and obclavate.

5.2.2.3 FAVORABLE ENVIRONMENT

- Temperature: 15°C–25°C.
- Relative humidity (RH): 70% and above.
- Intermittent rains.

5.2.2.4 SURVIVAL

In diseased plant debris or other hosts in range, pathogen survives in the form of conidia and dormant mycelium.

5.2.2.5 MANAGEMENT

1. Infected crop debris and the wild host should be destroyed.
2. Increase air circulation by proper training and pruning.
3. Rain splash dispersal of pathogen should be minimized by pruning lower or ground touching branches (skirts to 50 cm above), mulching under trees, and grassing inter-tree spaces.
4. Collect dead twigs and mummified fruits in each season.
5. Fruit infection should be regularly monitored during the season so that spraying can start before fruits get infected severely.
6. If required, spray of Copper oxychloride or Mancozeb @ 2.5 g/l of water on leaves only, never on the fruits.

5.2.3 BLACK CANKER

5.2.3.1 SYMPTOMS

On fruits, black spots of irregular shape and size ranging from small specks to large patches. These spots have an indistinct "feathered edge." Inside these spots, tissues are not damaged more than 10 mm deep.

5.2.3.2 CAUSAL ORGANISM: PHOMOPSIS ANNONACEARUM

Mycelium septate, branched, and well developed. Conidiophores hyaline, simple/branched, septate, bears conidia singly. There are two types of conidia:

1. **Alpha:** Hyaline, fusiform, straight, aseptate; and
2. **Beta (Stylospores):** Hyaline, aseptate, filiform, curved, or bent. Pycnidium globose to irregular, wide ostiole, brown to black.

5.2.3.3 FAVORABLE ENVIRONMENT

*   Temperature: 15°C–20°C.
*   Wet weather provides at least six hours of continuous wetness.
*   Prolong rainy periods exist in the spring.

Diseases of Custard Apple and Their Management

5.2.3.4 SURVIVAL

The pathogen survives in diseased plant debris or other hosts in range and serves as the primary source of inoculum.

5.2.3.5 MANAGEMENT

1. Infected crop debris and the wild host should be destroyed.
2. Increase air circulation by proper training and pruning.
3. Rain splash dispersal of pathogen should be minimized by pruning lower or ground touching branches (skirts to 50 cm above), mulching under trees, and grassing inter-tree spaces.
4. Collect dead twigs and mummified fruits in each season.
5. Fruit infection should be regularly monitored during the season so that spraying can start before fruits get infected severely.
6. If required, spray of Copper oxychloride or Mancozeb @ 2.5 g/l of water in the initial stage of infection.

5.2.4 ANTHRACNOSE

5.2.4.1 SYMPTOMS

On leaves and unripe fruits, dark brown to black necrotic spots of varying size, later on, increase in size, coalesce with each other, and entire affected area become black. Generally, blossom-end is the initial point of infection and then progresses to cover the whole fruit surface. Affected fruits shriveled and either hang on the tree or fall down.

5.2.4.2 CAUSAL ORGANISM: COLLETOTRICHUM GLOEOSPORIOIDES

Mycelium is septate, inter, and intra-cellular branched. Conidiophore is Hyaline to brown, septate, branched at the base, smooth, short, packed in acervuli. Conidia is falcate (sickle-shaped), posses large oil globule in the center, hyaline, unicellular, borne singly on the tip of short conidiophores, size – 8–26 × 5–7 μm. Acervuli cushion-shaped provided with short conidiophores and sterile setae, size: 115–467 × 22–95 μm. Setae are straight, un-branched, tapered towards the apex, brown to black, smooth, thick-walled and septate, size – 40–90 × 4–6 mm.

5.2.4.3   FAVORABLE ENVIRONMENT

- Temperature: min. 10 and max. 35°C.
- Relative humidity (RH): 90%.
- Wind and splashing raindrops.

5.2.4.4   SURVIVAL

Pathogen survives in the soil, and in infected fruits left over the tree serves as a primary source of inoculums.

5.2.4.5   MANAGEMENT

1. Destroy infected leaves and fruits regularly.
2. Cut and destroy dead twigs before flowering.
3. Maintain adequate nutrient concentration in plants, particularly nitrogen and calcium.
4. If required, spray of copper oxychloride or Mancozeb @ 2.5 g/l of water in the initial stage of leaf infection.

5.2.5   FRUIT ROT

5.2.5.1   SYMPTOMS

Initially, on the surface of the fruits, purple-black spots or blotches irregular in shape with a distinct edge appeared. These patches become hard and cracked. Internal discoloration and extensive corky rotting produced.

5.2.5.2   CAUSAL ORGANISM: BOTRYODIPLODIA THEOBROMAE

Mycelium septate, branched, and well developed. Conidia produced on short conidiophores which are bicelled, straight, and brown in color measuring

Diseases of Custard Apple and Their Management 81

25 × 15 μm. Pycnidia flask-shaped, slightly sub-cuticular having a small neck and ostiolate, erumpent, stromatic, setose, and black walls measuring 250–300 μm in size. The pycnidia may be found in groups or singly.

5.2.5.3 FAVORABLE ENVIRONMENT

- Temperature: 26°C–32°C.
- Relative humidity (RH): 80%.

5.2.5.4 SURVIVAL

Pathogen survives in the infected plant parts, which serve as the primary source of inoculum.

5.2.5.5 MANAGEMENT

1. Infected crop debris and the wild host should be destroyed.
2. Increase air circulation by proper training and pruning.
3. Rain splash dispersal of pathogen should be minimized by pruning lower or ground touching branches (skirts to 50 cm above), mulching under trees, and grassing inter-tree spaces.
4. Collect dead twigs and mummified fruits in each season and destroy them.
5. Fruit infection should be regularly monitored during the season so that spraying can start before fruits get infected severely.
6. If required, spray of Copper oxychloride or Mancozeb @ 2.5 g/l of water in the initial stage of infection.

5.2.6 PINK DISEASE

5.2.6.1 SYMPTOMS

On branches patches of light pink or white fungal growth formed. On infected area secretion of gummy substances seen and affected branch sometimes die. This symptom is very common in the high-density orchard, where wet conditions persist for a long time.

5.2.6.2 CAUSAL ORGANISM: CORTICIUM SALMONICOLOR

Mycelium hyaline, thin-walled and sparsely septate. Hypha 7–15 μm in diameter. Asexual spores are hyaline, thin-walled, angular or round, size 8–15 × 5–10 μm. These spores readily germinate in water.

5.2.6.3 FAVORABLE ENVIRONMENT

- Long period of hot and damp weather.
- Vigorous trees are more susceptible than neglected trees.

5.2.6.4 SURVIVAL

Pathogen survives in the form of mycelia aggregates, which are embedded deep in the plant tissues are the primary source of inoculums. During humid and hot conditions, the pathogen becomes active, and silvery-white feathery mycelium is seen, which leads to the formation of the necator stage. Asexual spores are easily spread by water/rain splashes, wind or insects.

5.2.6.5 MANAGEMENT

1. Prune and destroy infected branches and twigs.
2. Scrap the bark of the infected area and apply Bordeaux paste or Copper oxychloride paste.
3. Increase air circulation by proper training and pruning.
4. Rain splash dispersal of pathogen should be minimized by pruning lower or ground touching branches (skirts to 50 cm above).
5. Spray tridemorph or oxycarboxin (1 ml/L.) or copper oxychloride or Bordeaux mixture (4:4:50) on infected areas.

5.2.7 FRUIT SPOT

5.2.7.1 SYMPTOMS

Dark purple to grey spot appears on the fruit surface in the indentation. Spot size is up to 15 mm, and these spots coalesce to form larger irregular areas on the fruit surface.

Diseases of Custard Apple and Their Management 83

5.2.7.2 CAUSAL ORGANISM: PSEUDOCERCOSPORA SP.

Mycelium internal, hyphae 1–4 µm wide, or forming swollen hyphal cells, up to 12 µm dia., sometimes in monilioid sequences, pale olivaceous to olivaceous-brown, thin-walled, smooth. Stromata well developed, immersed, 30–150 µm dia., composed of swollen hyphal cells, 2–10 µm dia., and dark olivaceous-brown. Conidiophores in large, dense fascicles, forming sporodochial conidiomata, erect, straight, subcylindrical, conic to somewhat geniculate-sinuous, unbranched, 5–40 × 3–7 µm, 0–2-septate, sub hyaline, olivaceous to olivaceous-brown, thin-walled, smooth. Conidia solitary, obclavate-cylindrical, 15–80 × 2.5–5 µm, 1–7-septate, sub hyaline to pale olivaceous-brown, thin-walled, smooth, apex obtuse, base obconically truncate, hila unthickened and not darkened.

5.2.7.3 FAVORABLE ENVIRONMENT

Prolonged wet weather.

5.2.7.4 SURVIVAL

The fungus probably survives in dormant spots on infected fruits until the onset of conditions is conducive to sporulation. Dispersal of conidia is primarily by rain-splash or raindrops or wind or through the transport of infected plant material and/or fruits from infected areas.

5.2.7.5 MANAGEMENT

1. Infected crop debris and mummified fruits should be destroyed in each season.
2. Increase air circulation by proper training and pruning.
3. Rain splash dispersal of pathogen should be minimized by pruning lower or ground touching branches (skirts to 50 cm above) and mulching under trees.
4. Fruit infection should be regularly monitored during the season so that spraying can start before fruits get infected severely.
5. If required, spray of Copper oxychloride or Mancozeb @ 2.5 g/of water in the initial stage of infection.

5.2.8 PURPLE BLOTCH

5.2.8.1 SYMPTOMS

The disease is characterized by the appearance of small purple-colored spots on the fruit surface and under prolonged wet conditions cover the entire fruit surface very quickly. Infected fruits fall down prematurely. Due to infection, the entire internal flesh portion becomes discolored.

5.2.8.2 CAUSAL ORGANISM: PHYTOPHTHORA PALMIVORA

Mycelium hyaline, coenocytic, and branched. Sporangiophore hyaline, branched, indeterminate, sympodial branching, nodulate, terminates in a sporangium. Sporangia lemon or pear-shaped, papillate at the apex, hyaline, single-celled, multinucleate (7–30 nuclei), thin-walled, size 22–32 × 16–24 μm, bears bi-flagellated zoospores. Oospore spherical, thick, and smooth-walled.

5.2.8.3 FAVORABLE ENVIRONMENT

- Temperature: 12°C–36°C.
- Relative humidity (RH): More than 90%.

5.2.8.4 SURVIVAL

The pathogen survives in dormant spots on infected fruits until the onset of conditions is conducive to sporulation. Dispersal of sporangia and zoospores is primarily by rain-splash or raindrops or wind or through the transport of infected plant material and/or fruits from infected areas.

5.2.8.5 MANAGEMENT

1. Infected crop debris and mummified fruits should be destroyed in each season.
2. Increase air circulation by proper training and pruning.
3. Rain splash dispersal of pathogen should be minimized by pruning lower or ground touching branches (skirts to 50 cm above), and mulching under trees.

Diseases of Custard Apple and Their Management 85

4. Fruit infection should be regularly monitored during the season so that spraying can start before fruits get infected severely.
5. If required, spray of Copper oxychloride or Mancozeb @ 2.5 g/l of water in the initial stage of infection.

5.3 BACTERIAL DISEASE

5.3.1 BACTERIAL WILT

5.3.1.1 SYMPTOMS

The disease is characterized by leaves of young plants show pale green color and droop almost vertically. Infected trees rapidly decline, and severe defoliation noticed. In matured/old trees, slow decline without yellowing of leaves occur. In basal trunk and large roots of infected trees, water-conducting tissues/vessels show dark discoloration. The cross-section of the infected portion shows extensive discoloration of the outer growth ring.

5.3.1.2 CAUSAL ORGANISM: RALSTONIA (PSEUDOMONAS) SOLANACEARUM

The bacterium is Gram-negative, rod-shaped, motile by single polar flagellum, aerobic, 0.5–1.5 μm in size.

5.3.1.3 FAVORABLE ENVIRONMENT

- Hot and humid weather.
- Temperature: 28°C–30°C.
- Use of superphosphate increases disease, whereas nitrogen application decreases.

5.3.1.4 SURVIVAL

Pathogen survives in the infected plant parts and in soil. Spread by irrigation water, implements, and soil from infected filed to the healthy field.

5.3.1.5 MANAGEMENT

1. Cultural practices like the use of healthy rootstock, destruction of weeds and volunteer plants, avoidance of surface water for irrigation and soil from infected fields to disease-free field through implements, and workers keep the disease under control.
2. The nursery should be grown in disease-free soil and should be disinfected before sowing with soil solarization or treated with chemicals like formaldehyde.
3. Use resistant rootstock like Cherimoya.

KEYWORDS

- **bacterial wilt**
- **fruit rot**
- ***Phytophthora palmivora***
- **pink disease**
- **solarization**
- **sporulation**

REFERENCES

Anonymous (1998). *Custard Apple Information Kit* (p. 26). Published by Queensland government. Total.

Anonymous (2014). *The AESA Based IPM Package-Custard Apple* (p. 26). Published by NIPHM, Hyderabad.

Gupta, V. K., & Sharma, S. K., (2000). *Diseases of Fruit Crops*. Kalyani Publishers, New Delhi.

https://www.deedi.qld.gov.au (Accessed on 18 November 2019).

https://www.farmer.gov.in/./ipm/IPM%20package%20for%20Custard%20Apple.pdf (Accessed on 18 November 2019).

https://www.ncipm.org.in/NCIPMPDFs/.S/CustardApple.pdf (Accessed on 18 November 2019).

https://www.vikaspedia.in/.custard-apple/custard-apple-diseases-and-symptoms (Accessed on 18 November 2019).

Hutton, D. G., & Sanewski, G. M., (1989). *Cylindrocladium* leaf and fruit spot of custard apple in Queensland. *Australasian Plant Pathology*, *18*(1), 15–16.

Vishwakarma, S. N., (2003). *Phaloan Ke Rog*. Published by GBPU&T, Pantnagar.

CHAPTER 6

Diseases of Guava and Their Management

B. D. DEVAMANI

Department of Plant Pathology, UAS, GKVK, Bangalore, Karnataka, India, E-mail: kisshoreraj1333@gmail.com

6.1 INTRODUCTION

Guava (*Psidium gaujava* L.) is the fourth most widely grown fruit crop in India. Uttar Pradesh is the leading state in guava production, followed by Andhra Pradesh and Bihar. One of the major constraints in guava production is diseases. It invades on leaves, twigs, inflorescence, roots, and fruits reasonably ample loss considering total fruit reduction and production of poor quality fruit produced, rendered decreased rate in price. Guava (*Psidium gaujava* L.), an important member of family *Myrtaceae* L., is assumed to be originated from the Southern part of Mexico. Besides good health, it is rich in vitamins A, B, C, and minerals (Baradi, 1975). Guava plant growth is related to nutrient acquisition, and guava branch architecture influences foliage fauna (Hasna et al., 2000). Unfortunately, this crop suffers from a number of diseases at all stages of its development, i.e., right from the nursery stage to grow-up plant.

6.2 FUNGAL DISEASES

6.2.1 WILT: FUSAIUM OXYSPORUM F.SP. PSIDII, F. SOLANI, AND VERTICILLIUM ALBOARUM

6.2.1.1 ECONOMIC IMPORTANCE

It was first reported in 1935 from Allahabad. Jhoty et al. in 1984 reported that seven thousand acres of land in A.P under guava cultivation was reduced to half the land value by the presence of the disease.

6.2.1.2 SYMPTOMS

- The affected plants show yellow coloration with slight leaf curling at the terminal branches, becoming reddish at the later stage, and subsequently, premature shedding of leaves takes place.
- Twigs become bare and fail to bring forth new leaves or flowers and eventually dry up. Fruits of all the affected branches remain underdeveloped, become hard, black, and stony.
- The entire plant becomes defoliated and dies. A few plants also show partial wilting, which is a very common symptom of wilt in guava.
- The finer roots show black streaks, which become prominent on removing the bark. The roots also show rotting at the basal region, and the bark is easily detachable from the cortex.
- The cortical regions of the stem and root show distinct discoloration and damage. The light brown discoloration is noticed in vascular bundles. Bark splitting can be seen in wilted plants in later stages.
- The disease can be categorized into slow wilt and sudden wilt. In slow wilt, the plant takes several months or even a year to wilt after the appearance of initial symptoms, and in sudden wilt, the infected plant wilts in 15 days to one month.

6.2.1.3 FAVORABLE CONDITIONS

- pH 6.0 is optimum for disease development. Both pH 4.0 and 8.0 reduces the disease.
- Disease is more in clay loam and sandy loam compared to heavy soil.
- Higher disease incidence in the monsoon period.
- Disease appears from august and increases sharply during September-October.
- The presence of nematode, *Helicotylenchus dihystera* favors the disease.

6.2.1.4 MANAGEMENT

6.2.1.4.1 Cultural

- Proper sanitation of orchard.
- Wilted plants should be uprooted, burnt, and a trench of 1.0–1.5m should be dug around the tree trunk. Treat the pits with formalin and cover the pit for three days and then transplant the seedlings after two weeks.

Diseases of Guava and Their Management

- While transplanting seedlings avoid damage to the roots.
- Maintain proper tree vigor by timely and adequate manuring, inter-culture, and irrigation. ¾ Intercropping with turmeric or marigold.
- Soil solarization with transparent polythene sheet during summer months.
- Application of oil cakes like neem cake, mahua cake, kusum cake supplemented with urea. Apply 6 kg neem cake + 2 kg gypsum per plant. ¾ Judicious amendments of N and Zn.

6.2.1.4.2 Host Plant Resistance

- Resistant variety: Apple guava.
- Guava species, *Psidium cattleianum* var. *Lucidum* and *Syzigium cumini* (Jamun) are resistant to wilt.
- *Psidium cattleianum (P. molle), P. quianense,* Chinese guava *(P. friedshthalianum)* and Philippine guava are recommended as resistant rootstocks.

6.2.1.4.3 Biological

Aspergillus niger strain AN 17, Trichoderma viride, Trichoderma harzianum, and *Penicillium citrinum* can used as biocontrol agents (BCAs).

6.2.1.4.4 Chemical

- Stem injection with 0.1% water-soluble 8-Quinolinol sulfate.
- Drench with 0.2% Benomyl or Carbendazim, four times in a year, and spray twice with Measystox and Zinc sulfate.
- Disinfestation of soil with Metam-sodium at 252 ml/10m2 area to control nematodes.

6.2.2 DIEBACK AND ANTHRACNOSE FRUIT ROT: COLLETOTRICHUM GLEOSPORIDES

6.2.2.1 DISEASE SYMPTOMS

6.2.2.1.1 Die Back Phase

- The plant begins to die backward form the top of a branch.

90   *Diseases of Fruits and Vegetable Crops*

- Young shoots, leaves, and fruits are readily attached, while they are still tender. The greenish color of the growing tip is changed to dark brown and later to black necrotic areas extending backward.
- The fungus develops from the infected twigs and then petiole and young leaves, which may drop down or fall, leaving the dried twigs without leaves.

6.2.2.1.2   Fruit and Leaf Infection Phase

- Fruit and leaf infection are generally seen during rainy season crop. Pin-head spots are first seen on unripe fruits, which gradually enlarge.
- Spots are dark brown in color, sunken, circular, and have minute black stromata in the center of the lesion, which produce creamy spore masses in moist weather.
- Several spots coalesce to form bigger lesions.
- The infected area on unripe fruits become corky and hardy and often develops cracks in case of severe infection.
- Unopened buds and flowers are also affected, which causes their shedding.
- On leaves, the fungus causes necrotic lesions, usually ashy grey and bear fruiting bodies at the tip or on the margin.

6.2.2.2   PATHOGEN

Conidia are hyaline, aseptate; oval to elliptical conidiophore is cylindrical. Acervulli are dark brown to black

6.2.2.3   SURVIVAL AND SPREAD

The infection spreads by wind-borne spores develop on dead leaves, twigs, and mummified fruits in the orchard. Dense canopy is congenial for the germination of spores due to high moisture conditions. Movement of planting material through infected foliage and transportation of fruits from high disease-prone areas also spread the pathogens.

Diseases of Guava and Their Management 91

6.2.2.4 FAVORABLE CONDITIONS

Closer planting without canopy management, dew, or rains encourages spore production, and its dispersal around canopy and temperature between 10 to 35°C with best 24 to 28°C favors the disease progress.

6.2.2.5 MANAGEMENT

Spray Mancozeb 0.25%.

6.2.3 GUAVA RUST: PUCCINIA PSIDII

6.2.3.1 SYMPTOMS

The pathogen can affect foliage, young shoots, inflorescences, and fruit of guava. Typical symptoms associated with this disease include distortion, defoliation, reduced growth, and if severe, mortality. On fully expanded leaves, dark bordered, roughly circular brown lesions with yellow halos develop.

6.2.3.2 MANAGEMENT

Control of guava rust is based on the use of fungicides. Scouting fields for the onset of disease or during the times of the year when environmental conditions are favorable for pathogen infection are recommended so that proper and timely fungicide applications can control the diseases.

6.2.4 FRUIT CANKER: PESTALOTIA PSIDII PAT.

6.2.4.1 DISEASE SYMPTOMS

- Symptoms generally occur on green fruits and rarely on leaves.
- Initially minute, brown or rust-colored, unbroken, circular, necrotic areas appear on fruits, which is an advanced stage of infection; tears open the epidermis in a circinate manner.
- The margin of the lesion is elevated, and a depressed area is noticeable inside. The crater-like appearance is more noticeable on fruits than

on leaves. In older cankers, white mycelium consisting of numerous spores are noticeable.

- In severe cases, raised, cankerous spots develop in great numbers, and the fruits break open to expose seeds.
- Infected fruits remain underdeveloped, become hard, malformed, and mummified and drop. Sometimes, small rusty brown angular spots appear on the leaves.

6.2.4.2   SURVIVAL AND SPREAD

The pathogen is primarily a wound parasite and avoids injury to fruits.

6.2.4.3   FAVORABLE CONDITIONS

Germination of spores is maximum at 30°C and do not germinate below 15°C or above 40°C with RH above 96%.

6.2.5   ALGAL LEAF AND FRUIT SPOT (CEPHALEUROS VIRESCENS KUNTZE)

6.2.5.1   DISEASE SYMPTOMS

- Alga infects immature guava leaves during the early spring flush.
- Minute, shallow brown velvety lesions appear on leaves, especially on leaf tips, margins, or areas near the midvein, and as the disease progresses, the lesions enlarge to 2–3 mm in diameter.
- On leaves, the spots may vary from specks to big patches, which may be crowded or scattered.
- On immature fruits, the lesions are nearly black. As fruits enlarge, lesions get sunken and get cracked frequently on older blemishes as a result of enlargement of fruits; lesions are usually smaller than leaf spots. They are darkish green to brown or black to color.

6.2.5.2   SURVIVAL AND SPREAD

The pathogen survives on infected plant debris. The disease is airborne and spreads by air and rain splashes.

Diseases of Guava and Their Management

6.2.5.3 FAVORABLE CONDITIONS

Wet, humid conditions promote the spread of the disease; zoospores spread by splashing water.

6.2.6 STYLER END ROT (PHOMOPSIS PSIDII DE CAMARA)

6.2.6.1 DISEASE SYMPTOMS

- The visible disease symptom appears as discoloration in the region lying just below and adjoining the persistent calyx. Such an area gradually increases in size and turns dark brown.
- Later, the affected area becomes soft. Along with the discoloration of epicarp, the mesocarp tissue also shows discoloration, and the diseased area is marked by being pulpy and light brown in color in contrast to the bright white color of the healthy area of the mesocarp.

6.2.6.2 SURVIVAL AND SPREAD

Pathogen survives in infected fruits and plant debris. Pathogen spread through the wind.

6.2.6.3 FAVORABLE CONDITIONS

Temperature 10 to 35°C (25°C optimum) is favorable for disease development.

6.2.6.4 MANAGEMENT (FRUIT ROT/FRUIT CANKER, ALGAL LEAF AND FRUIT SPOT)

Prune and destroy the dead twigs and fruits. Plant spacing and fertilizer régimes should be managed to avoid unnecessarily dense plant canopy. Prune old and non-productive branch, which may serve as a potential source of infection for managing fruit rot disease good field sanitation (maintain field free of infected dry or semi-dry twigs and mummified fruits of the previous harvest which may serve as primary inoculum. Algal leaf spot can be reduced by maintaining tree vigor with cultural techniques such as proper fertilization and irrigation, proper pruning to enhance air circulation within the canopy and sunlight penetration, managing weeds and wider tree spacing. Managing insect, mite, and other foliar diseases increase tree vigor and lessen susceptibility to algal disease. Apply Zineb 75% WP @600–800 g in 300–400 l of water/acre or Mancozeb 75% WP 20 g in 10 l of water/tree.

KEYWORDS

- **fruit canker**
- **guava**
- **guava rust**
- **pathogen**
- **solarization**
- **styler end rot**

REFERENCES

Baradi, E. I., (1975). Tropical fruits guava. *Abst. Trop. Agri., 1*, 9–16.

Chand, G., Srivastava, J. N., Kumar, S., & Kumar, S., (2013). Efficacy of different fungicides against anthracnose of mango (*Mangifera indica* L.) in Eastern Bihar. *International Journal of Agricultural Science, 9*(1), 204–206.

Das, S., & Raj, S. K., (1995). Management of root rot of sugarbeet (*Beta vulgaris*) caused by Sclerotiumrolfsii in field through fungicides. *Indian J. Agric. Sci., 68*(7), 543–546.

Hasna, M. K., Meah, M. B., & Kader, M. A., (2000). Assessment of yield loss in Guava owing to fruit Anthracnose. *Pakistan Journal of Biological Science, 3*(8), 1234–1236.

Ray, S. K., Das, S., Hasan, M. A., & Jha, S., (2006). *Efficacy of Different Fungicides for the Management of Anthracnose of Mango on Old and Rejuvenated Orchard in West Bengal* (p. 99). 8th Int. Mango Symp. Sun City, South Africa.

Singh, R. N., & Singh, G. P., (1982). Field evaluation of systemic and nonsystemic fungicides for the control of powdery mildew of pea. *Pesticides, 16*, 19–20.

CHAPTER 7

Diseases of Jackfruit Crops and Their Management

R. C. SHAKYWAR

Department of Plant Protection, College of Horticulture and Forestry, Central Agricultural University, Pasighat, Arunachal Pradesh, India, E-mail: rcshakywar@gmail.com

7.1 INTRODUCTION

The jackfruit (*Artocarpus heterophyllus*) is also known as jack tree, or sometimes simply jack is a species of tree in the mulberry family. It has pointed projections outside and soft flesh inside, which is intensely sweet and delicious in taste (Fang et al., 2008). It is consumed as fruit as well as vegetable. Jackfruit is a less exploited crop that is rarely grown as a mono-crop (Bose, 1990). Usually, it is grown as a mixed crop. Jackfruit is rich in vitamins and minerals (Anonymous, 2006). Vitamin B_6 content in Jackfruit fulfills about 25% of our daily requirement (AEC, 2003). The fat content of Jackfruit is very low at just about 0.64 grams (Jagadeesh et al., 2006–2008). However, water is not a nutrient; it is pertinent to mention that jackfruit is made up of 80% of water (Anonymous, 2011). Jackfruit contains a high amount of carbohydrate and calorie that provides energy instantly (Singh et al., 2000). It is rich in antioxidants that protect from cancer, aging, and degenerative disease (Anonymous, 2007; Chanda et al., 2009). Due to rich antioxidants, it is good for eyesight and protects from conditions like cataract and macular degeneration. It is a good source of potassium that maintains fluid and electrolyte levels in balance. Also, it improves bone and skin health (Arung et al., 2006a, b). While millions of households in Vietnam, Bangladesh, Malaysia, and elsewhere across South and Southeast Asia are incorporating the jackfruit into their dishes, India remains reluctant (Prakash et al., 2009). When warm, humid, wet weather coincides with the flowering and fruiting season, rotting can cause total loss of fruits in jackfruit trees. It is more likely to occur in high-rainfall areas or during and after stormy periods (Hossain, 2006). The texture of unripe jackfruit is like mutton, and so is its taste when cooked in spices, which makes it popular amongst vegetarians and nonvegetarians alike (Hossain, 2002). The seeds of Jackfruit are also eaten in boiled form as a snack in some countries. It is a healthy fruit with almost no fats but high energy (Singh et al., 2001). The taste and preparations make it a delectable fruit that is not too hard to include in the diet (Jagtap et al., 2010). Furthermore, jackfruit cultivation is also influenced by different factors such as plant pathogens; bacteria, algae, and phanerogamic plant-parasitic are the major constraints. These pathogens are a serious threat to the future prospects of jackfruit cultivation globally. Although, enhancement of the yield and quality of plants through integrated disease management is one of the major tasks. These are followings diseases (algal leaf spot, dieback, fruit rot/soft rot, leaf spot, Phytophthora soft rot, pink, and rust and rot combination, etc.) given below in details:

7.1.1 ALGAL LEAF SPOT

7.1.1.1 SYMPTOMS

Algal leaf spot is characterized by grayish, green, brown, or orange cushion-like blotches on the leaf surface. Some hosts may also have diseased twigs and branches that are girdled and stunted with reddish-brown fruiting bodies (Rekha et al., 2009). The spots are generally 1/2 inch or less in diameter, although they may coalesce to form larger colonies. Leaf tissue may die beneath the spots, and the leaves may yellow and drop prematurely. Algal leaf spot is sometimes called green scarf because the spots may have a crusty, fuzzy, or flaky appearance (Zheng et al., 2009).

7.1.1.2 CAUSAL ORGANISM: CEPHALEUROS VIRESCENS

7.1.1.3 ETIOLOGY

The causal organism is a green parasitic alga whose usual hosts are plants with leathery leaves such as cotoneasters, magnolias, hollies, rhododendrons, and viburnums.

Diseases of Jackfruit Crops and Their Management 97

7.1.1.4 FAVORABLE CONDITIONS

Algal leaf spot is a foliar disease most commonly seen in warm, humid climates or in greenhouses.

7.1.1.5 DISEASE CYCLE

During wet weather, the algae produce spores that are spread by wind and splashing rain. The spores infect leaf tissue causing small, greenish circular spots that may age to light brown or reddish-brown. The spots may appear raised and velvety. The algae will overwinter or survive other unfavorable environmental conditions in leaf spots, including those on fallen leaves.

7.1.1.6 MANAGEMENT

1. Encourage healthy plants with good cultural techniques.
2. This disease is most damaging to plants that are already slow-growing or weakened.
3. Clean up fallen leaves and remove diseased leaves from the plant.
4. Good sanitation practices will help control this disease.
5. Promote dry leaves by improving air circulation and drainage.
6. If necessary, selectively prune overcrowded vegetation. Avoid spraying water on the leaves.
7. Use fungicidal sprays containing copper, such as Bordeaux mixture/ copper oxychloride @ 0.25%.

7.1.2 DIE BACK

7.1.2.1 SYMPTOMS

Discoloration and darkening of the bark from the tip downwards are the main symptoms of the disease. It advances, and young green twigs start withering from the base towards veins of leaf edges. Infected leaves turn brown with their margins rolling upwards (Uddin et al., 2009). Infected twigs or branches shrivel due to necrosis and fall down, and there may be exudation of gum from affected branches, which may then attacked by shoot borers causing internal discoloration of the twigs. On infected twigs, erumpent acervuli of *Colletotrichum gloeosporioides* can be seen (Gupta and Tandon, 2004).

7.1.2.2 CAUSAL ORGANISM: BOTRYODIPLODIA THEOBROMAE

7.1.2.3 ETIOLOGY

Pycnidia are asexual fruiting bodies up to 5mm in diameter. Conidia are asexual spores thin-walled at first, which become thick-walled, and septate mycelium is present.

7.1.2.4 MODE OF SPREAD AND SURVIVAL

The primary source of inoculums is Dormant mycelia and the secondary source of inoculums is soil and airborne conidia.

7.1.2.5 EPIDEMIOLOGY

A Temperature of 25–30°C low relative humidity (RH) (80–85%) and the presence of susceptible hosts favors the development of the disease.

7.1.2.6 MANAGEMENT

1. Hot water treatment at 52°C for 4–5 min.
2. Fruits should be sprayed with a mixture of bioinoculants.
3. Pruning of infected twigs followed by spraying with Carbendazim (0.1%) or Thiophanate methyl (0.2%) or Chlorothalonil (0.2%) has been recommended.
4. Controlling shoot borer and shoot hole borers with suitable insecticides is also important in reducing dieback disease.

7.1.3 FRUIT ROT/SOFT ROT

7.1.3.1 SYMPTOMS

Young fruits and inflorescence are badly attacked by the fungus, and only fruits reach maturity. Female inflorescence and mature fruits are generally not infected. It is a *soft rot* disease-causing dropping of a large no. of affected fruits. Initially, the fungus appears as a grayish growth with abundant mycelia that gradually become dense, forming a black growth. The fungus advances slowly until the whole fruit or entire inflorescence becomes rotten and drops (Haq, 2006, 2011).

7.1.3.2 CAUSAL ORGANISM

Three species of plant pathogenic fungi of the genus *Rhizopus artocarpus, Rhizopus oryzae,* and *Rhizopus stolonifera.*

7.1.3.3 ETIOLOGY

Rhizopus stolonifer was first established in 1818 by the (Ehthrenberg) Vuillemin. Even he was the first person to discover the genus Rhizopus. This is one of the important species in the genus Rhizopus. It is recognized as black bread mold also because it generally attacks the bread and produces the black colored sporangium and spore. In fact, it has the whole mycelium of black color. This fungus is worldwide in distribution. They are mostly saprophytes growing on various things like bread, jams, pickles, cheese, moist foodstuffs, leather goods, soft fruits, and vegetables (IPGRI, 2000).

7.1.3.4 EPIDEMIOLOGY

Warm, humid, rainy conditions favor the development of Rhizopus rot. Wind, rains, and insects dislodge and spread the tiny fungal spores. When deposited on moist fruit surfaces, the fungal spores germinate, and infective mycelia enter and grow into the tissues. The infection produces a layer of black spores on the fruit surface to start a secondary cycle of infection and disease. Although wounds can predispose the fruit to infection, yet the unwounded flowers and young fruits are also susceptible (Kader, 2009).

7.1.3.5 MODE OF SPREAD

Rhizopus can survive on decaying plant litter or in the soil to initiate new infections.

100   *Diseases of Fruits and Vegetable Crops*

7.1.3.6 MANAGEMENT

1. Prune the tree to encourage good ventilation and reduce RH in the canopy.
2. Remove and destroy diseased fruits from trees and the ground.
3. Moreover, clean the decaying organic debris within and around the tree.
4. Ensure that water does not stagnate around the root zone.
5. Control weeds around the young trees.
6. Intercrop jackfruit with trees that are not susceptible to Rhizopus infection.
7. Keep ripe fruits away from contact with the soil or decaying organic material.
8. Avoid bruising and wounding of the fruits.
9. Wash fruits after harvest in clean water and dry them thoroughly before packing or transporting.
10. Examine critically and remove the fruits showing disease symptoms.
11. Avoid storing fruits in the hot and poorly ventilated container after harvest.
12. Spraying trees with young fruits using Captan (0.2%) or Bordeaux mixture (1.0%) or Copper oxychloride (025%) at an interval of three weeks during the months of January, February, and March is effective in controlling the disease.
13. Before fruit setting or 10 days after fruit setting, spray with carbendazim, e.g., Bavistin, or mancozeb, e.g., Indofil M-45 @ 2 gm/L water, 2–3 times at 10-day intervals. If spraying after flowering, spray with Tebuconazole, e.g., Folicur @ 0.05% or metalaxyl + mancozeb (e.g., Ridomil MZ-75) or Indofil M-45 @ 2.5gm/L water, 3 times at 15 day intervals (Ukkuru and Pandey, 2011).
14. None of the variety was reported resistant against this disease.

7.1.4 LEAF SPOT

7.1.4.1 SYMPTOMS

The spot on the leaf is irregular in shape and size, light to dark brown surrounded by diffused chlorotic yellow hallow, marginal leaf tissues become black, necrotic, and gradually spread to the leaf center (Mahrouf, 2004). Occasionally diffused yellow halo may also be seen. Fungus produces Acervulus, the asexual fruiting body where the circular, black lesions that occur rapidly increase in size and girdle the stem culminating in the death of the vine. Later on, the fungi move to the spikes of individual plants (Luke et al., 2007).

Diseases of Jackfruit Crops and Their Management   101

7.1.4.2   CAUSAL ORGANISM: COLLETOTRICHUM CAPSICI (PETCH)

7.1.4.3   ETIOLOGY

The pathogen has septate mycelia with inter and intracellular haustoria. Sexual spores are the ascospores borne in ascus, and the asexual spores are Conidia borne in sporangia (Loizzo et al., 2010). They mainly survived as dormant mycelia and spread through the air and soil-borne conidia. The primary source of inoculums: Ascospores borne in ascus and the chlamydospores. The sexual fruiting body is perithecia, a secondary source of inoculums: air and soil-borne conidia spread through the air and soil-borne conidia (Butani, 1978).

7.1.4.4   EPIDEMIOLOGY

The fungi require a temperature of 30–32°C, RH 90–95%, and the susceptible host for the occurrence of the disease.

7.1.4.5   MANAGEMENT

1. Destruction of diseased debris and spraying the plants with Bordeaux mixture (0.5%).
2. At the initial stage, infected leaves, spikes should be collected and destroyed.
3. Early diagnosis of the disease and spray with Carbendazim (1%) or Companion (0.2%) has been found to be effective.
4. Low nitrogen and increased potassium content in the soil make the plants more resistant to this disease.
5. Application of bio-control agents such as *Trichoderma viride* @ 5–10g/kg of FYM in the soil.
6. Cuttings for planting should always be taken from disease-free and healthy vines.

7.1.5   PHYTOPHTHORA SOFT ROT

7.1.5.1   SYMPTOMS

Infection takes place through black and blue or injured skin in rough-skinned varieties and through wounds in smooth-skinned verities. Water socked

102   *Diseases of Fruits and Vegetable Crops*

lesions appear within 48–78 hours of inoculation (Valavi et al., 2011). These become enlarged to form light brown spots with sporulating hyphae near the edge. Infected fruits develop *soft rot*. The damage caused to the bark of crown roots or bark of the trunk is called Phytophthora gummosis, collar rot, or foot rot (Karthy et al., 2009).

7.1.5.2   CAUSAL ORGANISM: PHYTOPHTHORA PALMIVORA

7.1.5.3   ETIOLOGY

The fungus produces abundant sporangia on agar under continuous fluorescent light. However, light is not required for sporangia production on an infected plant. Sporangia are usually produced in clusters sympodially. Sporangia are papillate and ovoid, with the widest part close to the base. They are easily washed off, and each detached sporangium contains a short pedicel. The average size of the sporangia is 50 x 33 μm with a length of about 1.6 times longer than it is wide. Sporangia germinate directly in a nutrient medium by producing germ tubes that develop into mycelial masses. In water, however, zoospores are released from germinating sporangia. Zoospores aggregate and form distinct patterns at 16°C in water (Kotowaroo et al., 2006).

7.1.5.4   EPIDEMIOLOGY

Heavy sporangial production occurs during rainy season water was the best, and *P. palmivora* thrives best at 25–28°C. A soil pH of 5.4–7.5 favors disease development. Rain and wind are the two major factors in the epidemiology of Phytophthora fungus. Rain splash is needed for the liberation of sporangia of the pathogen from the surface of infected fruit into the atmosphere and for the projection of the soil inoculums into the air (Costa et al., 2000). The wind is required for dispersal of the inoculums once it reaches the air. Therefore, windblown rain is essential for the initiation of the primary infection and the development of epidemics in jackfruit orchards (Shyamalamma et al., 2008).

7.1.5.5   MANAGEMENT

1. Infected fruit on the trees and those that have fallen to the ground should be removed to reduce the inoculums for aerial infection of fruit and stems and infection of seedling roots in subsequent plantings.

Diseases of Jackfruit Crops and Their Management 103

2. Phytophthora soft rot can be controlled by fungicides.
3. Application of a preventive fungicide such as Mancozeb @ 3 g/l of water or basic copper sulfate to the fruit column is effective in protecting soft rot from infection by *P. palmivora*.
4. Spraying with Benomyl @ 1 g/l is completely managed by the disease.

7.1.6 PINK DISEASES

7.1.6.1 SYMPTOMS

It is widespread in tropical and subtropical areas. The disease appears as a pinkish powdery coating on the stem. Pink color represents profuse conidial production of fungus. Young woody branches of the affected trees lose their leaves and show dieback symptoms. Pink encrustation is seen on the lower shaded side (Krishrtaveni et al., 2001).

7.1.6.2 CAUSAL ORGANISM: BOTRYOBASIDIUM SALMONICOLAOR/ CORTICIUM SALMONICOLOR

7.1.6.3 MANAGEMENT

The affected branches should be pruned and cut ends should be pasted with Bordeaux paste or copper oxychloride paste.

KEYWORDS

- **epidemiology**
- **etiology**
- **fungal diseases**
- **leaf spot**
- **physiological disorder**
- ***Phytophthora* soft rot**

REFERENCES

AEC (Agro-Enterprise Center), (2003). *A Report on Meeting of Underutilized Fruits in Nepal.* UTFANET, Southampton University, UK.

Anonymous (2006). *Jackfruit Artocarpus heterophyllus.* Chichester, England, UK: Southampton Center for Underutilized Crops Printed at RPM Print and Design.

Anonymous (2007). *Cultivation Practices for Fruits* (pp. 77–79). Published in the local language (*Kannada*) by Director of Extension, Dharwad.

Anonymous (2011). Biennial research report of All India Coordinated Research Project on tropical fruits. Compiled and Edited by Sidhu, A. S., Patil, P., Reddy, P. V. R., Satisha, G. C., & Sakthivel (*Tech. DOC. No. 100*). Published by the Project Coordinator (Tropical Fruits), IIHR, Bangalore, India.

Arung, E. T., Shimizu, K., & Kondo, R., (2006a). Inhibitory effect of artocarpanone from *Artocarpus heterophyllus* on melanin biosynthesis. *Biological and Pharmaceutical Bulletin, 29,* 1966–1969.

Arung, E. T., Shimizu, K., & Kondo, R., (2006b). Inhibitory effect of isoprenoid substituted flavonoids isolated from *Artocarpus heterophyllus* on melanin biosynthesis. *Planta Medica, 72,* 847–850.

Bose, T. K., (1990). *Fruits of India, Tropical and Subtropical* (pp. 487–495). Department of Horticulture, Bidhan Chandra Krishi Viswavidyalaya, Kalyani.

Butani, D. K., (1978). Pests and diseases of jackfruit in India and their control (a review). *Fruits, 33*(5), 351–357.

Chanda, I., Chanda, S. R., & Dutta, S. K., (2009). Anti-inflammatory activity of a protease extracted from the fruit stem latex of the plant *Artocarpus heterophyllus* Lam. *Research Journal of Pharmacology and Pharmacodynamics, 1,* 70–72.

Costa, W. A., J. M., & De Rozana, M. F., (2000). Effects of shade and water stress on growth and related physiological parameters of the seedlings of five forest tree species. *Journal of the National Science Foundation of Sri Lanka., Colombo, 28*(1), 43–62.

Fang, S. C., Hsu, C. L., & Yen, G. C., (2008). Anti-inflammatory effects of phenolic compounds isolated from the fruits of *Artocarpus heterophyllus. Journal of Agriculture and Food Chemistry, 56,* 4463–4468.

Gupta, A. K., & Tandon, N., (2004). *Review on Indian Medicinal Plants.* New Delhi, India: Indian Council of Medical Research.

Gupta, K., & Tandon, N., (1996). In: *Review on Indian Medicinal Plants* (pp. 182–200). New Delhi: Indian Council of Medical Research.

Halder, N. K., Farid, A. T. M., & Siddiky, M. A., (2008). Effect of boron for correcting the deformed shape and size of jackfruit. *J. Agric Rural Dev., 6*(1&2), 37–42.

Haq, N., (2006). *Jackfruit, Artocarpus heterophyllus, Southampton Center for Underutilized Crops* (p. 192). University of Southampton, Southampton, UK.

Haq, N., (2011). Thottappilly, G., Peter, K. V., & Valavi, S. G., (eds.), *Genetic Resources and Crop Improvement in the Jackfruit.* Publisher Stadium Press India Pvt. Ltd.

Harlan, J. R., (1971). Agricultural Origins: Centers and Noncenters. *Science, 174,* 468–474.

Hossain, A., & Haq, N., (2006). *Jackfruit (Artocarpus heterophyllus), Field Manual for Extension Workers and Farmers.* SCUC, Southampton University, UK.

Hossain, M. K., & Nath, T. K., (2002). *Artocarpus heterophyllus* Lam. In: Vozzo, J. A., (ed.), *Tropical Tree Seed Manual: Agriculture Handbook* (p. 721). Washington, DC: U. S. Department of Agriculture Forest Service.

Diseases of Jackfruit Crops and Their Management

IPGRI, (2000). *Descriptors for Jackfruit (Artocarpus Heterophyllus)*. International Plant Genetic Resources Institute (IPGRI), Rome.

Jagadeesh, S. L., Reddy, B. S., Basavaraj, N. Swamy, G. S. K., Gorbal, K., & Hegde, L., (2008). Inter tree variability in chips purpose jackfruit selections of Western Ghats of Karnataka, India. *Indian J. Genetics and Plant Breeding, 68*, 183–188.

Jagadeesh, S. L., Reddy, B. S., Basavaraj, N., Swamy, G. S. K., Gorbal, K., & Hegde, L., (2007). Inter tree variability for fruit quality in jackfruit selections of Western Ghats of India. *Sci. Hort., 112*, 382–387.

Jagadeesh, S. L., Reddy, B. S., Swamy, G. S. K., Gorbal, K., Hegde, L., & Raghaven, G. S. V., (2006). Chemical composition of jackfruit (*Artocarpus heterophyllus* Lam.) Selections of Western Ghats of India. *Food Chemistry, 102*, 361–365.

Jagtap, U. B., Panaskar, S. N., & Bapat, V. A., (2010). Evaluation of antioxidant capacity and phenol content in jackfruit (*Artocarpus heterophyllus* Lam.) fruit pulp. *Plant Foods for Human Nutrition, 65*, 99–104.

Kader, A. A., (2009). *Jackfruit: Recommendations for Maintaining Postharvest Quality.* Department of Plant Sciences, University of California. http://postharvest.ucdavis.edu/Produce/ProduceFacts/Fruit/jackfruit.shtml (Accessed on 18 November 2019).

Karthy, E. S., Ranjitha, P., & Mohankumar, A., (2009). Antimicrobial potential of plant seed extracts against multidrug-resistant Methicillin-Resistant *Staphylococcus aureus* (MDR-MRSA). *International Journal of Biology, 1*, 34–40.

Kotowaroo, M. I., Mahomoodally, M. F., Gurib-Fakim, A., & Subratty, A. H., (2006). Screening of traditional antidiabetic medicinal plants of Mauritius for possible alpha-amylase inhibitory effects *in vitro. Phytotherapy Research, 20*, 228–231.

Krishrtaveni, A., Manimegalai, G., & Saravanakumar, R., (2001). Storage stability of jackfruit (*Artocarpus heterophyllus*) RTS beverage. *J. Food Science and Technology Mysore, 38*(6), 601–602.

Loizzo, M. R., Tundis, R., Chandrika, U. G., Abeysekera, A. M., Menichini, F., & Frega, N. G., (2010). Antioxidant and antibacterial activities on food borne pathogens of *Artocarpus heterophyllus* Lam. (Moraceae) leaves extracts. *J. Food Science, 75*, 291–295.

Luke, S., Shyamalamma, S., & Narayanaswamy, P., (2007). Morphological and molecular analysis of genetic diversity in Jackfruit. *The Journal of Horticultural Science and Biotechnology, 85*(5), 764–768.

Mahrouf, A. R. M., (2004). *Socio-Economic Dimensions in Conservation and Use of Jackfruit Species Biodiversity in Sri Lanka Report of ADB-TFT Project.*

Patil, K. S., Jadhav, A. G., & Joshi, V. S., (2005). Wound healing activity of leaves of *Artocarpus heterophyllus. Indian Journal of Pharmaceutical Sciences, 67*, 629–632.

Prakash, O., Kumar, R., Mishra, A., & Gupta, R., (2009). *Artocarpus heterophyllus* (Jackfruit): An overview. *Pharmacognosy Review, 3*, 353–358.

Rekha, R., Warrier, B., Gurudev, S. R., Anandalakshmi, V., Sivakumar, S., Geetha Kumar, A. M., & Hegde, M. T., (2009). Standardization of storage conditions to prolong viability of seeds of *Artocarpus heterophyllus* Lam a tropical fruit tree. *ARPN Journal of Agricultural and Biological Science, 4*(2), 6–9.

SCUC, (2006). *Jackfruit Artocarpus Heterophyllus, Field Manual for Extension Workers and Farmers* (p. 27). SCUC, Southampton, U.K.

Shyamalamma, S., Chandra, S. B. C., Hegde, M., & Naryanswamy, P., (2008). Evaluation of genetic diversity in jackfruit (*Artocarpus heterophyllus* Lam.) based on amplified fragment length polymorphism markers. *Genetics and Molecular Research, 7*(3), 645–656.

Siddique, M. A., & Azad, A. K., (2010). *Prioritization of Research for Horticultural Crops* (p. 56). Final Report of a Study, BARC, New Airport Road, Dhaka.

Singh, I. S., Singh, A. K., & Pathak, R. K., (2001). *Jackfruit* (p. 15). Department of Horticulture. N.D. Univ. of Agric., & Tech, Narendra Nagar (Kumargarj), Faizabad.

Singh, R., Singh, G., & Chauhan, G. S., (2000). Nutritional evaluation of soy fortified biscuits. *J. Fd. Sci. Technol.*, *37*(2), 162–164.

Uddin, M. T., Rukanuzzaman, M., Khan, M. M. R., & Islam, M. A., (2009). Jackfruit (*Artocarpus heterophyllus*) leaf powder: An effective adsorbent for removal of methylene blue from aqueous solutions. *Indian Journal of Chemical Technology*, *16*, 142–149.

Ukkuru, M., & Pandey, S., (2011). Nutritional significance, recent developments in India on processing and value addition of jackfruit. In: Thottappilly, G., Peter, K. V., & Valavi, S. G., (eds.), *Jackfruit*. Publisher- Studium Press India Pvt. Ltd.

Valavi, S. G., Jacob, B., Amin, M. N., Magdalita, P. M., & Thottappilly, G., (2011). In: Thottappilly, G., Peter, K. V., & Valavi, S. G., (eds.), *Harvesting, Grading, Marketing and Economics of Jackfruit Production in the Jackfruit*. Publisher- Studium Press India Pvt. Ltd.

Zheng, Z. P., Chen, S., Wang, S., Wang, X. C., Cheng, K. W., & Wu, J. J., (2009). Chemical components and tyrosinase inhibitors from the twigs of *Artocarpus heterophyllus*. *Journal of Agricultural and Food Chemistry*, *57*, 6649–6655.

CHAPTER 8

Diseases of Litchi and Their Management

VINOD KUMAR

Senior Scientist, National Research Center on Litchi, Mushahari,
Muzaffarpur, Bihar, India, E-mail: vinod3kiari@yahoo.co.in

8.1 INTRODUCTION

Litchi (*Litchi chinensis* Sonn.) [Family Sapindaceae] is a tropical and subtropical fruit tree native to the Guangdong and Fujian provinces of China and now cultivated in many parts of the world. Litchi is extensively grown in China, India, Thailand, Vietnam, and the rest of tropical Southeast Asia, the Indian Subcontinent (Papademetriou and Dent, 2002), and more recently in South Africa, Brazil, the Caribbean, Queensland, California, and Florida (Crane et al., 2008). It is cultivated in about 84,000-hectare area in India, with production around 5.85 lakh metric tonnes and productivity about 7.0 metric tonnes/ha (NHB, 2015). The fresh fruit has a delicate, whitish pulp with a floral smell and a fragrant, sweet flavor. Since this perfume-like flavor is lost in the process of canning, the fruit is usually eaten fresh.

Diseases are one of the constraints to the production of litchi fruits. They indirectly reduce yield by debilitating the tree, and directly reduce the yield or quality of fruit before and after they are harvested. The occurrence and severity of the different diseases vary according to locations, age of trees, cultural practices, and orchard ecosystem. Most of the diseases are caused by fungal pathogens. No bacterial or viral infections have been reported so far. There are some reports on the occurrence of algal leaf spot and plant-parasitic nematodes. For management diseases, integrated crop management (ICM) module is more relevant that take care of disease as well as nutrient management aspect together. There is a strong interaction of pathogens with the nutrient acquisition, uptake, and translocation of nutrients, when crop performance of infected versus non-infected healthy trees is compared. Monitoring is an important tool to keep track of the pathogens and their potential damage. This provides knowledge about the current disease and crop situation and is helpful in selecting the best possible combinations of management methods. In the following sections, the diagnostic symptoms, causal organism, their mode of spread, and survival and management practices are discussed.

8.1.1 FUNGAL DISEASE

8.1.1.1 LEAF, PANICLE AND FRUIT BLIGHT (ALTERNARIA ALTERNATA)

Leaf blight is a prominent disease in nursery plants (Kumar and Anal, 2014). The symptoms start from the tip of the leaf as light brown to dark brown necrosis that advances towards both the margins of the leaf, leading to complete necrosis of the affected leaves that dries up subsequently (Figures 8.1A and 8.1B). Initial symptoms resemble to potassium deficiency, and hence often, it is confused with deficiency symptoms. In orchard trees, it causes leaf blight, similar to the nursery stage. Its association was also found with blighting of inflorescence/panicle of litchi for the first time in 2014 in the Bihar state of India (Kumar, 2015a), leading to no fruit sets in the affected portion of the tree (Figure 8.2). At a later stage, the same pathogen also causes fruit blight, often sporadic incidence is seen on trees (Figure 8.2), and postharvest fruit decay/rots. The pathogen survives on infected leaves under the tree canopy. Survival of the pathogen is through conidia on fallen leaves and also on senescing leaves under the tree canopy.

FIGURE 8.1A & B Variation in symptoms of leaf blight disease in nursery plants.

Diseases of Litchi and Their Management 109

A. alternata is an opportunistic pathogen on numerous hosts and has been isolated from almost all habitats (Farr et al., 1989; Murthy et al., 2003). Due to large spore and dark pigment (melanin), *A. alternata* spores are resistant to UV radiation and can float in the air enabling long-distance dispersal. Weather significantly influences incidence, abundance, and biodiversity of spores of *A. alternata* (Magyar et al., 2009). Litchi being cultivated over a large area in Bihar and adjoining states of India, the pathogen has the potential to spread very fast, causing heavy damage. Climate change, coupled with a change in the cropping system and pathogen population, maybe the cause of the increasing incidence of this disease. Therefore, the epidemiology of the disease needs to be investigated.

FIGURE 8.2 Symptoms of panicle (Left) blight leading to no fruit set, and fruit blight (Right).

For the management of the disease, orchard sanitation should be followed by removing and destroying infected leaves wherever possible. For leaf blight, prophylactic spray of copper oxychloride (0.25%) or mancozeb (0.15%) may be done. If disease severity increases, spray of difenoconazole (0.05%) or chlorothalonil (0.15%) or thiophanate methyl (0.1%) or azoxystrobin (0.023%) may be done. For panicle and fruit blight phase, one spray just after flowering and before fruit set, and another spray 20 days before harvest of fruits should be done. Induction of antifungal compounds using defense elicitors could be effective against Alternaria disease.

8.1.1.2 LEAF SPOTS (BOTRYODIPLODIA THEOBROMAE, COLLETOTRICHUM GLOEOSPORIOIDES, PESTALOTIA PAUCISETA, MICRODIPLODIA ITCHI)

Four different fungal pathogens cause leaf spots, but the incidence and severity are quite low in India. The leaves generally show mixed infection of these fungal pathogens. Except for *P. Pauciseta*, spots due to these pathogens appear in July and advance during the following 3–4 months. Older leaves show a higher percentage of infection than the younger leaves. The leaf spots caused by *Botryodiplodia* and *Colletotrichum* start either from the tip or the margin of the lamina. These spots are dark, chocolate in color in case of former while it is brown irregular in outline in case of latter. In the case of *Pestalotia*, spots (0.5–2.0 cm) are light-colored and appear on both sides of the leaves. Leaf spots caused by *Microdiplodia* are yellowish-brown to brick red mostly around the margin, which gradually becomes light brown and shows black dotted pycnidia in the central light brown zone of the developing spots (Pathak and Desai, 1971). Later the spots coalesce. All the pathogens survive on the leaves throughout the year and develop vigorously after the onset of the monsoon.

The occurrence of leaf spots are sporadic and currently seldom causes economic damage; however, wherever required, a foliar spray of mancozeb (0.25%), or copper oxychloride (0.25%) or thiophanate methyl (0.15%) may be done if the severity of the disease increases. In the climate change scenario, there is a probability of an increase in leaf spots with denser canopy combined with increased humidity.

8.1.1.3 TWIG BLIGHT (COLLETOTRICHUM GLOEOSPORIOIDES PENZ., AND GLOEOSPORIUM SP.)

'Twig blight' is another economically important emerging disease of litchi in India (Kumar et al., 2011, 2014a, b). The symptoms appear as the death of leaves on new shoots and a foliar blight and tip dieback, which is difficult to separate. The leaf blight appears as tan spots on the leaves. The afflicted leaves look as if they are scorched from the sun (Figure 8.3). The twig blight, along with an infestation of foliage feeding insect pest complex, not only severely hampers the young trees' growth but also reduces the potential fruit-bearing flushes in grown-up orchards.

For the management of this disease, infected leaves and twigs should be removed, particularly on young trees. Provide need-based spray either with copper oxychloride (0.25%) or carbendazim (0.1%) or chlorothalonil (0.15%) or difenoconazole (0.05%) or azoxystrobin (0.023%).

FIGURE 8.3 Symptoms of twig blight on a litchi tree.

8.1.1.4 WILT (FUSARIUM SOLANI)

Young trees of litchi, often below five-year age, wilt in less than a week's time (Kumar et al., 2011). The first symptoms appear as yellowing of foliage, drooping leaves followed by gradual wilting and drying, leading to complete death of the plant within 4–5 days (Figure 8.4). Some brown spots appear on the root crown and lateral roots' phloem, spreading later to the xylem. The pathogen survives as chlamydospores in soils and is spread by movement of sick soil from diseased plants to healthy plant rhizospheres.

Fusarium spp. is an invasive pathogen that poses a risk to the cultivation of many crops, including perennial fruit trees maximum and minimum ambient temperature, and the soil temperature has been found positively and significantly correlated with wilt incidence caused by *Fusarium* spp. In some crop pathosystems, but in litchi, such information is lacking. The result of simulated climate change conditions have shown that global climate changes could influence disease incidence caused by soil-borne pathogens, probably through plant-mediated effects (Chitarra et al., 2015).

FIGURE 8.4 Symptoms of wilt on young litchi tree.

Application of castor cake or neem cake as manures along with biocontrol agents (BCAs) like *Trichoderma harzianum, T. Viride, Pseudomonas fluorescens*, etc. Has been proved effective in managing the disease. In the absence of biocontrol agents, drenching rhizosphere soil with hexaconazole or carbendazim (0.1%) may be done. Growers are also advised not to plant litchi trees on waterlogged soils or in the low-lying field, often receiving flood water.

8.1.1.5 ANTHRACNOSE (COLLETOTRICHUM GLOEOSPORIOIDES)

Anthracnose has been a major disease of litchi fruits in India; however, in recent years, its incidence has decreased in India. It is more of a problem after harvest. In Florida, anthracnose caused by *C. gloeosporioides* is an extremely important factor in lowering the grade and quality of fruits and reducing yield (McMillan, 1994). The disease can affect leaves, twigs, and flowers of litchi as well. Infected fruit becomes unmarketable. The fungus may not always cause immediate disease, which sometimes only becomes apparent after harvest. The disease first appears as brown pinhead lesions usually on the top of semi-mature fruit. Circular dark-brown to black sunken lesions later become easily visible on mature fruits (Figure 8.5). More spots appear on the top and sides of the fruit and by the time of harvest, cover up to half of the fruit surface. Usually, superficial skin blemishes do not affect production and fruit quality, but marketability is affected. Outbreaks are common after warm wet weather.

FIGURE 8.5 Symptoms of anthracnose on litchi fruits.

The fruit is susceptible to infection from blossoming time until it is half-grown. Most of the decay on mature fruits develop from latent infection when the fruit is small. These incipient infections, which may be pinpoint in size on the growing fruit, develop into large coalescing brown spots as the fruit matures. During shipping or refrigerated storage, the infected fruits will develop a whitish mycelia mat over the skin.

Infected leaves and stems are the principal sources of inoculums. Conidia are spread by water splash, with free water required for infection. Infection occurs by direct penetration of the fruit skin. Immature fruits are infected in the orchard, with the pathogen then remaining quiescent until they mature. High temperatures after harvest favor the disease.

Maintenance of orchard hygiene, through pruning of deadwood and removal of infected leaves, twigs, and fruit, helps to reduce the incidence of the disease, while preharvest applications of fungicides such copper oxychloride (0.25%) or carbendazim (0.1%) or chlorothalonil (0.15%) or difenoconazole (0.05%) or azoxystrobin (0.023%) offer good control. Pre-harvest spray of fungicides helps in extending postharvest life, but while doing so, care must be taken for residual toxicity of chemical applied. The development and spread of the disease will be influenced by climate dynamics (high temperature and high relative humidity (RH)).

8.1.1.6 DOWNY BLIGHT (PERONOPHYTHORA EROXI)

This is an important disease of litchi in China, Vietnam, and Taiwan (Ann et al., 1984, 2012; Vien et al., 2001; Wang et al., 2009) but does not occur in India. *P. eroxi* is one of the most serious diseases of fruit in Guangdong Province (Chi, 1984). The pathogen attacks fruits, panicles, and new shoots, causing panicle rot, withering, and watery brown spots on fruits, which later sporulate, producing downy white sporangiophores (Figure 8.6). The pathogen also causes significant postharvest losses. Liu-Jin et al. (2006) reported that infection of *P. eroxi* induced the activities of polyphenol oxidase (PPO), eroxidise, and phenylalanine ammonia-lyase (PAL) in litchi pericarp, which accelerated the enzymatic browning of litchi pericarp. It can over-winter via mycelium and oospores in the remnants of leaves. The pathogen is polycyclic in nature, resulting in the dispersal of inoculums over an extended time period and wide area during the season. Crop losses due to this disease become extensive under the conditions of hot, humid, rainy weather that often occurs in Southern China during the period when the fruit develops (Figure 8.7).

Cleaning up the orchard by removing shaded, infected, and dead branches after harvest reduce disease. Chemical control is the primary method for controlling litchi downy blight from March to July each year. A spray of copper oxychloride during winter and copper sulfate in spring also helps to reduce the inoculums level. Spray with metalaxyl 8%+ mancozeb 64% (0.25%) or Al-fosetyl (0.16%) during flowering and fruit development if the disease severity is high.

Diseases of Litchi and Their Management

FIGURE 8.6 Symptoms of downy blight.

FIGURE 8.7 Postharvest fruit rots caused by various fungi (Left to right- *Aspergillus niger, Alternaria alternata,* and *Colletotrichum gloeosporioides).*

8.1.1.7 FRUIT ROTS (ALTERNARIA ALTERNATA, COLLETOTRICHUM GLOEOSPORIOIDES, ASPERGILLUS NIGER, ASPERGILLUS FLAVUS)

A wide range of fungi viz. *Alternaria, Colletotrichum, Botryodiplodia, Aspergillus, Fusarium, and Penicillium* sp. are reported to cause postharvest fruit rots if fruits are not handled properly. Liu et al. (2006) reported that in Guangdong and Hainan (China), *C. gloeosporioides* was the main pathogen causing postharvest decay which came mainly from fruits with the latent infection prior to harvest. A mean pathological to a tune of 23.2% during 2012 and 17.9% during 2013 in the supply chain of litchi in India has been reported (Kumar et al., 2016), highest being at the retail level. Initially, the disease symptoms are perceptible on the injured portion of the fruits. With the advance of the disease, the decayed areas get depressed. The rot gradually penetrates deep into the pulp. Ultimately rind of infected fruits cracks off, exposing the pulp, which subsequently is covered with thick cottony mycelium (Figure 8.8). Such affected fruits emit an odor of fermentation.

There is a likelihood of an increase in *postharvest* fruit decay in response to high field temperatures, developing microcracks, and sunburn, which provide entry to postharvest pathogens; however, currently, we do not have experimental evidence about how climate change may affect postharvest fruit decay. Temperature, CO_2, and ozone, directly and indirectly, affect the production and quality of fruit. Prolonged exposure to CO_2 concentrations could alter sugar level, protein, and mineral contents, leading to a loss of nutritional and sensory quality (Moretti et al., 2010). Increased levels of ozone in the atmosphere can lead to detrimental effects on postharvest quality of fruit.

For pathogens that initially infect fruit in the field, maintenance of orchard hygiene and the application of preharvest fungicides, such as those used for anthracnose, offer some control. For pathogens that primarily infect after harvest through wounds in the fruit skin, careful handling to avoid injury is important. One spray of carbendazim 15–20 days before harvest is recommended. Prompt precooling and maintenance of the optimum temperature and RH during storage and transport prevents fruit decay. Sulfur fumigation may be done if allowed by importing countries. For this, fruits are placed in a closed chamber where 50–100 g sulfur per m^3 of air is burnt for 20–30 minutes. For transportation, use Corrugated Fiber Board (CFB) boxes of 2 kg capacity properly unitized for stacking.

FIGURE 8.8   Rust pustules on a portion of the leaf.

8.2   ALGAL DISEASE

8.2.1   *ALGAL LEAF SPOTS (CEPHALEUROS VIRESCENS)*

Algal spot occurs in poorly managed orchards and affects a wide range of important fruits, including litchi (Mishra et al., 1974; Menzel et al., 1988; McMillan, 1994). Red, circular to semicircular spots (3.0–6.5 mm) appear mostly on the leaves and sometimes also on tender stems. Orange-yellow to pink velvety coating develops on the spots on the formation of sporangia (Figure 8.8). In the absence of sporangia, however, these spots remain green-brown. In older leaves, the lesions turn light brown to brick-red. The disease appears in the rainy season and continues till early winter. Pathogen survives on infected leaves. Zoospores are dispersed by rain and wind, and infect leaves through the stomata, with chains of algal cells developing in the leaf tissue.

So far in India, red rust affects more to longan (*Dimocarpus longan* Lour.) and mango (*Mangifera indica* L.) than litchi. With the change in rainfall and management practices that leads to the denser canopy, microclimate may change, and the disease may become important in some orchards. Pruning improves tree ventilation and reduces disease. For the management of this disease, foliar spray of copper oxychloride (0.3%) may be done in July and October if severity increases.

8.3 NEMATODE DISEASE

Several plants parasitic nematode species have been reported from India, but none of them are causing economic damage to litchi trees. A study conducted by Srivastava et al. (2000) during 1993–1996 in Doon Valley, Uttar Pradesh, India revealed the presence of *Trichodorus pakistanensis, Hemicriconemoides mangiferae, Helicotylenchus dihystera*, and *Xiphenema inequale* as the major species. *T. pakistanensis* was most dominant than the other species. Nath et al. (2009) reported nine plant-parasitic nematode species associated with soil and root of the litchi plants in North Tripura district, India, out of which *Hemicriconemoides litchi, Rotylenchulus reniformis,* and *Meloidogyne incognita* were found most abundant, frequent, and important nematode species. Analyses of rhizospheric soil samples of litchi rhizosphere at NRCL, Muzaffarpur revealed the presence of three species of plant-parasitic nematodes *viz.,* ring nematode (*Hemicriconemoides mangiferae),* lance nematode (*Hoplolaimus indicus)* and spiral nematode (*Helicotylenchus dihystera)* (Kumar, 2015b). Out of these, ring nematode was the most prominent nematode, but its population density too was far less than the pathogenic level of six nematodes per gram of soil. Nematodes at this population density cannot cause economic damage to litchi plants and hence do not require any nematicide application.

8.4 PHYSIOLOGICAL DISORDER

8.4.1 FRUIT CRACKING

This is the most important disorder occurring in almost all the important litchi growing countries of the world, causing losses as high as 5–70% (Menzel and Waite, 2005; Singh et al., 2012). Losses up to 14% (mean 3.8–5.9%) during 2012 and 2013 due to 'fruit cracking' at harvest of fruits in the Bihar state of India was reported by Kumar et al. (2016). All cracked fruits lose their value for the fresh market, and they are used for processing only (particularly for

Diseases of Litchi and Their Management

fruit juice) if they are not affected by fungus. Cracked fruits are susceptible to storage disease, have shorter storage as well as shelf-life. This disorder is associated with hot dry weather, drought, and low calcium concentrations. It has been observed that early ripening cultivars under poor management practices are more susceptible to this disorder. The other nutrient element found to be associated with fruit cracking is boron (deficient soils). A high concentration of abscisic acid and low gibberellins has been found in the fruit pericarp, seed, and aril of cracked fruits. Insects, hail, and the sun can damage the skin during cell expansion and induce cracking towards harvest. In general, it has been observed that fruit skin cracking often occurs when there is a sudden change in soil moisture, or there is drought soon after fruit set. Cell division is reduced, and the fruit skin becomes inelastic, and often splits when the aril grows rapidly before harvest. This can occur after irrigation or heavy rain, or just after an increase in RH.

It has been reported that the application of calcium @2 ml/l (liquid formulations) and gibberellins @20 ppm reduced the activity of cellulose and thereby reduced cracking (Sinha et al., 1999; Peng et al., 2001). Chandel and Sharma (1992) reported that the application of 2, 4-D, and NAA @20 ppm reduced fruit cracking. Boron sprays in the form of borax or boric acid @ 2g/L at the initial stage of aril development in conjunction with sufficient soil moisture in the root zone prevented fruit cracking significantly (Kumar and Kumar, 2008). Constant moisture and appropriate humidity are needed at the time of fruit maturity. Irrigation at 30–40% depletion of available soil moisture is quite helpful in reducing cracking of fruits. Mitra et al. (2014) reported that irrigating the orchard at 20% pan-coefficient and placement of dessert cooler or by irrigating the orchard through overhead sprinkler completely control fruit cracking. The soil moisture conservation in the root zone for a prolonged period can be achieved by providing mulching beneath the canopy (Sharma, 2006; Kumar and Kumar; 2008; Sandhu and Gill, 2013). Early cultivars or particularly those cultivars which have relatively thin skin, few tubercles per unit area, and rounded to flat in shape are less prone to cracking. Planting windbreaks (a row of tall-growing trees, such as seedling mango and *jamun*) around the orchard provides protection from desiccating hot winds that reduce the malady.

8.4.2 SUNBURN

Light brown blotches appear on the portion of the fruit skin facing direct sunrays. In severe cases, more than half of the surface area becomes discolored, blotchy light brown (Figure 8.10). The blotches become intense in a few days, and the blotchy area dries up, blocking aril growth and finally destroying the market value of the fruit. The symptoms appear more on the southwest side than the north-east side as on the latter side, fruits remain almost in the shade except during the early hours of the day (Figure 8.9).

FIGURE 8.9 Symptoms of cracking of fruits of litchi.

The occurrence of sunburn on fruits is a serious problem in litchi producing countries like India, South Africa, Australia, and Thailand (Sanyal et al., 1990; Menzel et al., 2002). The damage caused due to sunburn occurs up to 19.2% in different varieties. Sunburn affected fruits up to 44.5% (mean 14.9%) during 2012, and up to 27% (mean 10.4%) during 2013 at harvest of fruits in the Bihar state of India was reported by Kumar et al. (2016). Besides environmental factors, cultivars, hormonal, nutritional, and soil moisture factors have also been found associated with this disorder (Sharma and Ray, 1987; Sanyal et al., 1990; Menzel et al., 2002; Menzel and Waite, 2005). The fruits in orchards have scanty irrigation during the fruit development stage favor sunburn (Menzel and Waite, 2005).

FIGURE 8.10 Symptoms of sunburn on litchi fruits.

Irrigation at regular intervals during the fruit growth and ripening stage reduces sunburn. Planting windbreak around the orchard provides protection from desiccating hot winds. Irrigation through sprinkler system during hot hours increases humidity, cools the orchard atmosphere, thus decreases the incidence of sunburn. In light and sandy soil, only light irrigation with increased frequency (4–5 days interval) is found beneficial. The trees provided with sufficient quantities of organic manures such as compost, FYM, cakes, green manure, vermicompost, and irrigation applied at regular interval during fruit development and ripening stage prevents fruits from sunburn. Raising windbreak around the orchard has also been found to reduce sunburn problems. Planting of maize or sugarcane around the litchi orchard and daily irrigation in such a border crop creates a congenial microclimate, which reduces sunburn.

8.5 PERSPECTIVE

Climate change will impact the litchi fruit industry and regions through a change in disease dynamics besides increased incidence of physiological disorders such as cracking and greater potential for downgrading product quality because of the increased incidence of sunburn. Assessment of disease dynamics, strengthening surveillance of diseases, and development of ICM strategies are the need of the hour. Specific forecasting tools may be developed, especially for Alternaria disease and studies on the management of postharvest spoilage. Clonal selection, hybridization, and biotechnological tools may not only help in quality fruit production but also may provide multiple stress tolerance in litchi. Applying irrigation through the under-canopy and over-head sprinkling of water creates a more favorable humid micro-climate within the orchard boundary that restricts both sunburning as well as skin-cracking to a minimum of 3 to 4% level thus improving quality production. Windbreaks around orchard will be useful. More emphasis should be given on the development and adoption of organic production practices.

KEYWORDS

- **algae**
- **corrugated fiberboard**
- **fruit cracking**
- **integrated crop management**
- **litchi**
- **nematode**

REFERENCES

Ann, P. J., & Ko, W. H., (1984). Blossom blight of litchi in Taiwan caused by *Peronphythora litchii. Plant Dis., 68*, 826.

Ann, P. J., Tsai, J. N., & Yang, H. R., (2012). First report of leaf and stem downy blight of longan seedlings caused by *Peronophythora litchii* in Taiwan. *Plant Dis., 96*(8), 1224. doi: 10.1094/PDIS-01-12-0009-PDN.

Chandel, S. K., & Sharma, N. K., (1992). Extent of fruit cracking in litchi and its control measure. *South Indian Hortic., 40*(2), 74–76.

Chi, P. K., Pang, S. P., & Liu, R., (1984). On downy blight of *Litchi chinensis* Sonn. I. The pathogen and its infection process. *Acta Phytopathol. Sin., 14*(2), 113–119.

Chitarra, W., Siciliano, I., Ferrocino, I., Gullino, M. L., & Garibaldi, A., (2015). Effect of elevated atmospheric CO_2 and temperature on the disease severity of rocket plants caused by fusarium wilt under phytotron conditions. *PLoS One, 10*, e0140769. doi: 10.1371/journal.pone.0140769.

Crane, J. H., Balerdi, C. F., & Maguire, I., (2008). *Lychee Growing in the Florida Home Landscape*. Fact Sheet HS-6, University of Florida. https://edis.ifas.ufl.edu/mg051 (Accessed on 18 November 2019).

Farr, D. F., Bills, G. F., Chamuris, G. P., & Rossman, A. Y., (1989). *Fungi on Plant and Plant Products in the United States*. APS Press, St. Paul. MN, USA.

Kumar, K. K., & Kumar, R., (2008). *Managing Physiological Disorder in Litchi*. Technical Bulletin -5, ICAR-National Research Centre on Litchi, Muzaffarpur, Bihar, India.

Kumar, R., & Nath, V., (2013). Climate resilient adaptation strategies for litchi production. In: Singh, H. P., Rao, N. K. S., & Shivashankar, K. S., (eds.), *Climate-Resilient Horticulture: Adaptation and Mitigation Strategies* (pp. 81–88). Springer India, New Delhi. doi: 10.1007/ 978-81-322-0974-4_8.

Kumar, V., & Anal, A. K. D., (2014). Leaf blight of litchi (*Litchi chinensis* Sonn.) caused by *Alternaria alternata* (Fr.) Keissler: An important disease of nursery plants. In: *Proceedings of the National Symposium on Plant Pathology in Genomic Era and the 66th Annual Meeting of Indian Phytopathological Society* (p. 78). Raipur, India.

Kumar, V., (2015a). Studies on alternaria disease of litchi. In: *Annual Report 2014–2015* (pp. 21). ICAR-National Research Centre on Litchi, Muzaffarpur, Bihar, India.

Kumar, V., (2015b). Studies on plant parasitic nematodes associated with litchi. In: *Annual Report 2014–15* (p. 25). ICAR-National Research Centre on Litchi, Muzaffarpur, Bihar, India.

Kumar, V., Anal, A. K. D., & Nath, V., (2014a). Prevalence of some threatening pests and disease of litchi (*Litchi chinensis* Sonn.) in Bihar state of India. *J. Appl. Hort., 16*, 235–240.

Kumar, V., Kumar, A., & Nath, V., (2011). Emerging pests and diseases of litchi (*Litchi chinensis* Sonn.). *Pest Manag. Hort. Ecosyst., 17*, 11–13.

Kumar, V., Kumar, A., Nath, V., & Kumar, R., (2014b). New threats of insect pests and disease in litchi (*Litchi chinensis* Sonn.) in India. *Acta Hortic., 1029*, 417–424. doi: 10.17660/ ActaHortic.2014.1029.53.

Kumar, V., Purbey, S. K., & Anal, A. K. D., (2016). Losses in litchi at various stages of supply chain and changes in fruit quality parameters after harvest. *Crop Prot., 79*, 97–104. doi: 10.1016/j.cropro.2015.10.014.

Liu, A., Chen, W., & Li, X., (2006). Developments of anthracnose on harvested litchi fruits and the effects of the disease on storage of the fruits. *Acta Phytophylacica Sinica, 33*(4), 351–356.

Liu-Jin, Liu, A., & Chen, W., (2006). Physiological changes of litchi fruit infected by *Peronophythora litchi. J. Fruit Sci., 23*(6), 834–837.

Magyar, D., Frenguelli, G., Bricchi, E., Tedeschini, E., Csontos, P., Li, D. W., & Bobvos, J., (2009). The biodiversity of air spora in an Italian vineyard. *Aerobiologia, 25*, 99–109.

McMillan, R. T. Jr., (1994). Diseases of *Litchi chinensis* in south Florida. *Proc. Fla. State Hortic. Soc., 107*, 360–362.

Menzel, C. M., & Waite, G. K., (2005). *Litchi and Longan: Botany, Production and Uses*. CABI Publishing, Wallingford, Oxfordshire, UK.

Menzel, C. M., Bagshaw, J., Campbell, T., Greer, N., Noiler, J., Olesan, T., & Waite, G. K., (2002). *Lychee Information Kit*. Queensland Department of Primary Industries. Nambour, Australia.

Menzel, C. M., Watson, B. J., & Simpson, D. R., (1988). The lychee in Australia. *Queensl Agric. J., 114*, 19–26.

Mishra, B., Prakash, O., & Misra, A. P., (1974). A serious leaf spot disease of litchi caused by *Cephaleuros virescens. Indian J. Mycol. Plant Pathol., 3*, 219–220.

Mitra, S. K., Dutta, R. S. K., & Mandal, D., (2014). Control of fruit cracking and sunburning in litchi by irrigation and moisture conservation. *Acta Hortic., 1024*, 177–181. doi: 10.17660/ActaHortic.2014.1024.20.

Moretti, C. L., Mattos, L. M., Calbo, A. G., & Sargent, S. A., (2010). Climate changes and potential impacts on postharvest quality of fruit and vegetable crops: A review. *Food Res. Int., 43*, 1824–1832. doi: 10.1016/j.foodres.2009.10.013.

Murthy, K. K., Shenoi, M. M., & Sreenivas, S. S., (2003). Perpetuation and host range of *Alternaria alternata* causing brown spot disease of tobacco. *Indian Phytopath., 56*, 138–141.

Nath, R. C., Sinha, B. C., Mukherjee, B., & Dasgupta, M. K., (2009). Community analysis and diversity of plant parasitic nematodes associated with litchi plantations in North Tripura district. *Pest Manag. Hort. Ecosyst., 15*(1), 51–59.

NHB, (2015). *Second Advance Estimate of Area and Production of Horticulture Crops (2013–14)*. National Horticulture Board, Ministry of Agriculture, Government of India, Gurgaon, India. http://nhb.gov.in/area-pro/2nd_Advance_Estimates_2013–14.xls (Accessed on 18 November 2019).

Papademetriou, M. K., & Dent, F. J., (2002). *Lychee Production in the Asia-Pacific Region.* Food and Agricultural Organization of the United Nations, Office for Asia and the Pacific, Bangkok, Thailand. ftp://ftp.fao.org/DOCREP/FAO/005/AC684e/ac684e00.pdf (Accessed on 18 November 2019).

Peng, J., Xi, J. B., Tang, X. D., Wang, Y. G., Si, X. M., & Chen, J. S., (2001). Effect of Ca(NO$_3$) and GA spray on leaves on the fruit cracking of 'Nuomici' litchi. *Acta Horticulturae Sinica, 28*, 348–350.

Sandhu, S., & Gill, B. S., (2013). *Physiological Disorders of Fruit Crops* (p. 189). New India Publishing Agency, New Delhi.

Sanyal, D., Ahsan, A., Ghosh, B., & Mitra, S. K., (1990). Studies on sun burning and skin cracking in some varieties of litchi. *Indian Agriculturist, 34*(1), 19–23.

Sharma, R. R., (2006). Physiological disorders in tropical and subtropical fruits: causes and control. *(In) Fruit Production-Problems and Solutions* (pp. 301–325). International Book Distributing Co., Lucknow, UP, India.

Sharma, S. B., & Ray, P. K., (1987). Fruit cracking in litchi- A review. *Haryana J. Hort. Sci., 16*, 11–15.

Singh, H. S., Nath, V., Singh, A., & Pandey, S. D., (2012). *Litchi: Preventive Practices and Curative Measures.* SPSS Publications, Delhi.

Sinha, A. K., Singh, C., & Jain, B. P., (1999). Effect of plant growth substances and micronutrients on fruit set, fruit drop, fruit retention and cracking of litchi cv. *Purbi. Indian J. Hortic., 56*, 309–311.

Srivastava, N., Rawat, V. S., & Ahmad, M., (2000). Distribution of plant parasitic and soil nematodes associated with litchi fruit trees in Doon valley (U. P.) India. *Indian J. Nematol., 30*(1), 100–101.

Vien, N. V., Benyon, F. H. L., Trung, H. M., Summerell, B. A., Van, N. K., & Burgess, L. W., (2001). First record of *Peronophythora litchii* on litchi fruit in Vietnam. *Australas. Plant Path., 30*(3), 287–288.

Wang, H. C., Sun, H. Y., Stammler, G., Ma, J. X., & Zhou, M. G., (2009). Baseline and differential sensitivity of *Peronophythora litchii* (lychee downy blight) to three carboxylic acid amide fungicides. *Plant Pathol., 58*(3), 571–576. doi: 10.1111/j.1365-3059.2008.01990.x.

CHAPTER 9

Major Diseases of Mangoes and Their Management

SUPRIYA GUPTA,* PANKAJ RAUTELA, C. S. AZAD, and K. P. SINGH

Department of Plant Pathology, College of Agriculture, G.B. Pant University of Agriculture and Technology, Pantnagar, Uttarakhand, India

*Corresponding author. E-mail: gupta.supriya15@gmail.com

9.1 INTRODUCTION

Mango (*Mangifera indica* L.) is an important fruit that suffers from several diseases at all stages of its development, i.e., from nursery to the consumption of fruits. All the plant parts, namely, trunk, branch, twig, leaf, petiole, flower, and fruit, are attacked by different pathogens. Some of the diseases are very severe on plants and have become a limiting factor in profitable mango production.

Many types of agents cause diseases in mango, and of these fungi cause the largest number of diseases, while bacteria, algae, angiospermic parasites, and nutritional deficiencies are the other causal agents of mango maladies. Major diseases of mango and their control measures are discussed below.

9.2 FUNGAL DISEASES

9.2.1 POWDERY MILDEW (OIDIUM MANGIFERAE BERTHET)

Powdery mildew is one of the most serious diseases of mango affecting almost all the varieties, first reported from Brazil in 1914 by Berthet, later on, it has been recorded from several countries of the world. In India, it has been reported from Uttar Pradesh, Punjab, Maharashtra, Gujarat, Madhya Pradesh, Jammu and Kashmir, Haryana, Andhra Pradesh, Karnataka, and Tamil Nadu. Prakash and Srivastava (1987) reported 30–90% losses in

Lucknow. The disease is reported to cause crop loss as high as 70–80% depending upon the weather conditions on an individual plant basis.

9.2.1.1 SYMPTOMS

The characteristic symptom of the disease is the white superficial powdery fungal growth comprising a large number of conidia borne on conidio-phores on leaves, stalks of panicles, flowers, and young fruits, which is blown away by even a slight disturbance caused by winds. The sepals are relatively more susceptible than the petals. Young fruits are covered entirely by the mildew epidermis of the infected fruit cracks, and corky tissues are formed. Purplish brown blotchy areas appear on the skin of older fruits. The affected flowers and fruits do not grow in size and drop pre-maturely before attaining pea-size, reducing the crop load considerably or might even prevent the fruit set.

9.2.1.2 CONDITIONS FOR DISEASE DEVELOPMENT

Rains or mists accompanied by cooler nights during flowering are congenial for the disease spread. Warm temperature with heavy morning dew and cloudy weather favor disease development (Kulkarni, 1924). The minimum, optimum, and maximum temperature for conidia germination are stated to be 9, 22, and 30.5°C, respectively (Uppal et al., 1941). The disease appearance and incidence are greatly influenced by the sunshine (hours) per day, and it is inversely proportional to the disease having a correlation coefficient of $r = 0.5943$ (Gupta, 1985). The other factors influencing the disease were relative humidity (RH), low, and intermediate humidity (20–65%) are excellent for maximum germination of spores Palti et al. (1974). Gupta (1989) also advocated that the atmospheric temperature is an important factor for the appearance and development of the disease. But overall disease development is favored by high humidity (Pinkasetal, 1973).

9.2.1.3 MANAGEMENT

In India, the disease has been controlled by spraying of wettable sulfur, dinocap, carbendazim, benomyl, tridemorph, tridemephon, bitertanol, oxythioquinone, thiophanate methyl, vigil, etc. (Datar, 1981, 1986; Gupta and Dang, 1981;

Major Diseases of Mangoes and Their Management 127

Joshi and Chauhan, 1985). The disease can be managed by pruning of diseased leaves and malformed panicles and following three sprays of fungicides at 15 days interval recommended for effective control of the disease:

- Wettable sulfur 0.2% (2 g Sulfex / lit. water) at the panicle size of 7.50–10.00 cm.
- Dinocap 0.1% (1 ml / g Karathane / lit. water) after 15–20 days of the first spray.
- Tridemorph 0.1% (1 ml Calixin / lit. water) after15–20days of second spray.

Wettable sulfur (0.2%) can be used in all three sprays, and the number of sprays may be reduced as per the appearance time of disease.

9.2.2 ANTHRACNOSE (COLLETOTRICHUM STATE OF GLOMERELLA CINGULATA STON, SPAULL, AND SCHRENK)

The anthracnose disease is of widespread occurrence and is the most important disease wherever mango is grown. Anthracnose on mango was first reported from Puerto Rico in 1903 and subsequently from various countries of the world (Prakash et al., 1996). In India, the disease was reported by various workers (Prakash and Misra, 1988). It is widely distributed in Bihar, Punjab, Maharashtra, Gujarat, Kerala, Uttar Pradesh, Rajasthan, and Karnataka. Losses due to anthracnose have been estimated to be 2–39% (Prakash et al., 1996).

9.2.2.1 SYMPTOMS

Flower blight, fruit rot, and leaf spots are among the symptoms of this disease (Arauz, 2000). Infections on the panicles (flower clusters) start as small black or dark-brown spots. These can enlarge, coalesce, and kill the flowers, greatly reducing yield. On leaves, anthracnose infections start as small, angular, brown to black spots. Later on, the lesions get blighted and rupture. Infected leaves often show a 'shot hole.' Young leaves are more prone to attack than the older ones. Leaves and young panicles infected with gall midge (insect) stimulate the activity of the fungus resulting in a heavy incidence of the disease (Prakash and Srivastava, 1987). The leaves droop down, slowly drying up and ultimately fall-off, leaving a black scar on the twig. The disease produces elongated black necrotic areas on the twigs. Gummosis is usually the aftereffect of the disease (Prakash et al., 1996). Black spots develop on panicles as well as on fruits. The severe infection destroys the entire inflorescence resulting in no setting of fruits. Young infected fruits develop black spots, shrivel, and drop off. On older fruits, black spots are produced. Initially, the spots are round, but after coalescence, they form large irregular botches or even cover the entire fruit. The spots have large extensive rotting.

9.2.2.2 CONDITIONS FOR DISEASE DEVELOPMENT

The severity and prevalence of the disease are influenced by excessive rains, heavy dews, or high humidity during critical infection period (Doidge, 1932). The temperature range for disease development 10–30°C, and the R.H. 95–97% are highly congenial. The optimum temperature for infection of anthracnose is around 25°C. Continuous wet weather during flowering causes serious blossom blight. RH above 95% for 12 hrs is essential for infection and development of *C. gloeosporioides* on mango fruit. The infection progresses faster in wounded tissues, and in ripe fruits (Prakash et al., 1996).

9.2.2.3 MANAGEMENT

Since the fungus has long saprophytic survival ability on dead twigs, the diseased twigs should be pruned and burnt along with fallen leaves for reducing the inoculum potential.

- Trees may be sprayed twice with Bavistin (0.1%) at 15 days interval during flowering to control blossom infection. (Prakash and Misra, 1988)
- Its foliar infection can be managed by two sprays of Copper oxychloride (0.3%), while latent infection of the pathogen on fruits could be reduced by pre-harvest sprays of Thiophanate methyl or Carbendazim (0.1%).
- Post-harvest infection of this pathogen can be managed by a post-harvest dip of fruits either with hot water alone (45 ± 20°C) or hot water in combination with fungicides, Thiophanate methyl or Carbendazim (0.05%).

Major Diseases of Mangoes and Their Management 129

9.2.3 DIE BACK (LASODIPLODIA THEOBROMAE (PAT.) GRIFFON AND MOUBLE (BOTRYODIPLODIA THEOBROMAE PAT.).

Dieback is one of the serious diseases of mango. The disease on the tree may be noticed at any time of the year, but it is most conspicuous during Oct.-Nov November (Prakash and Singh, 1976a). It is a destructive disease of mango and is known to occur in India and other mango growing countries. About 30–40% mango plantation has been found affected by the disease creating a serious problem in the Moradabad region of Uttar Pradesh, India (Prakash and Singh, 1976; Prakash and Srivastava, 1987; Ploetz and Prakash, 1997). The disease is prevalent not only in Uttar Pradesh, Rajasthan, but also in other mango growing states of India *viz.* Delhi, Tamil Nadu, Punjab, Haryana, Orissa, Gujarat, and Maharashtra.

9.2.3.1 SYMPTOMS

The disease is characterized by drying back of twigs and branches from the top downwards, particularly followed by complete defoliation, which gives the tree an appearance of scorching by fire. The onset of dieback becomes evident by discoloration and darkening of the bark. The dark area advances, and young green twigs start withering first at the base and then extending outwards along the veins of leaf edges. The affected leaf turns brown, and its margins roll upwards. At this stage, the twig or branch dies, shrivels, and falls off, which is the characteristic symptom of the advanced stage of the disease. This may be accompanied by the exudation of gum. In old branches, brown streaking of vascular tissue is seen on splitting it longitudinally. The areas of cambium and phloem show brown discoloration, and yellow gum-like substance is found in some of the cells. The disease is known to be enhanced by a beetle, *Xyleborusaffinis* (Batista, 1947).

9.2.3.2 CONDITIONS FAVORABLE FOR DISEASE DEVELOPMENT

RH above 80% and temperature of 25 to 31.5°C, and rains enhance disease development. Trees damaged by gummosis, insects, drought, and lack of nutrition favor disease development. High summer temperature predisposes the mango plants to the attack of the pathogen through reducing the vitality of the plant (Das Gupta and Zachariah, 1945). Disease development is

130   *Diseases of Fruits and Vegetable Crops*

favored by rains, RH (approx. 80%), and maximum and minimum temperatures of 31.5 and 25.9°C.

9.2.3.3 MANAGEMENT

- Prune the diseased twigs and spray with copper oxychloride (0.3%) or Bordeaux mixture (5:5:50) on infected trees (Prakash and Raoof, 1985).
- Pruning should be done in such a way that the twigs are removed 2–3 inches below the affected portion. In small plants, pruning of twigs is followed by pasting of copper oxychloride.
- Sprays of carbendazim (0.1%) or methyl thiophenate (0.1%) or chlorothalonil (0.2%) at fortnightly interval during the rainy season is important.

9.2.4 MANGO MALFORMATION (FUSARIUM MONILIFORME VAR. SUBGLUTINANS SHELDON)

Mango malformation was first recognized in 1891 by Maries (Watt, 1891) from Darbhanga, Bihar. Malformation is widely prevalent in northern India, particularly in the states of Punjab, Delhi, and western U.P., where more than 50% of the trees suffer from this malady.

9.2.4.1 SYMPTOMS

There are two types of symptoms, namely, floral malformation and witches broom or bunchy top or vegetative malformation with a proliferation of infected tissue. The flowering panicles turn into a compact mass of flowers. This compact mass is very hard and not soft like a normal panicle. An individual flower is greatly enlarged and has a large disc. The percentage of bisexual flowers in malformed panicles is very low. In the bunchy top, the disease appears on the young plants in the nursery beds when they are 4–5 months old, and compact leaves are formed in a bunch at the apex of shoot or in the leaf axil. A similar bunch consisting of small rudiments crowded together on short shootlets is seen in vegetative malformation in which the growth of the shoot let is arrested. The affected seedlings develop excessive vegetative branches that are of limited growth, swollen, and have very short internodes.

Major Diseases of Mangoes and Their Management 131

The vegetative malformation is more pronounced in young seedling and seedling trees. The malformed heads dry up in black masses and persist on the trees for a long time. The malformed inflorescences contain more of endogenous cytokinins than healthy ones (Nicholson and van Staden, 1988).

9.2.4.2 FACTORS RESPONSIBLE FOR INCIDENCE

The severity varies considerably from year to year. A tree, once affected, cannot escape the disease in subsequent years (Mallik, 1963). The disease is serious in the north-west region where temperatures lie from 10–15°C during December-January (winter) before flowering. The disease is mild in the areas where temperatures lie from 15–20°C, sporadic from 20–25°C and nil beyond 25°C. Puttarudiah and Channabasavana (1961), and Singh et al. (1961) reported that the occurrence of malformation differed according to the age of the plants. They observed more diseases in young plants than in old ones. About 91% incidence in 4–8 years old plants and 9.6% in older plants was reported (Singh et al., 1961). Age of the flowering shoot also influences the incidence of MF, as reported by Varma (1983).

9.2.4.3 MANAGEMENT

Considerable incidence reduction has been reported by spraying 100–200 ppm NAA during October (Majumdar, 1977).

- Deblossoming of early emerged/infested panicles.
- Use of disease-free planting material and a prophylactic spray of insecticides and fungicides can keep the orchards healthy.

In areas with less than 5–10% infection, pruning of diseased plants should be made compulsory; pruning of diseased parts along the basal 15–20 cm apparently healthy portions. This is followed by the spraying of Bavistin (0.1%) or Captaf (0.2%).

9.2.5 SCAB (ELSINOE MANGIFERAE BITANCOURT AND JENKINS)

Scab is found in all mango producing countries of the world and found prominently on young mango seedlings. The disease is first observed in

Cuba and Florida in the 1940s. The disease can cause significant damage to young seedlings in nurseries, but it is not a problem in commercial orchards (Prakash and Srivastava, 1987; Ploetz and Prakash, 1997).

9.2.5.1 SYMPTOMS

The scab fungus attacks leaves, twigs, panicles, blossoms, and blotches on the bark of stems. Young succulent leaves are most susceptible to the disease. Lesions differ in size and color depending upon the age of the plant. The symptom produced by the pathogen is almost similar to anthracnose, but lesions produced on leaves are smaller than anthracnose, slightly angular, elongate, 2–4 mm in diameter, brown but during the rainy season, down the surface is covered by delicate velvety growth. The disease may cause crinkling, distortion, and premature shedding of leaves under severe conditions. Sometimes irregular shot holes are also observed on leaves. The blotches on the stem bark are grayish and irregular in shape. On young fruits, the infection is grey to grayish brown with dark irregular margins., Spots also enlarge, and the center may become covered with cracked and fissured, corky tissues as the fruit attains in size. Conidia of the fungus are produced on the fruit until it reaches maturity.

9.2.5.2 CONDITIONS FAVORABLE FOR DISEASE DEVELOPMENT

No specific information is available on the epidemiology of mango scab but as assumed on the basis of disease development by other species of Elsinoe, that high humidity and free moisture are needed for the formation of spores and initiation of infection by *E. mangiferae* (Ploetz and Prakash, 1997).

9.2.5.3 MANAGEMENT

- Spray application of fungicides for the control of anthracnose has generally been considered effective for scab control.
- Frequent spray of copper fungicides (0.3%) to protect new flushes of growth are effective for scab control in nurseries (Prakash and Misra, 1988).

Major Diseases of Mangoes and Their Management

9.2.6 SOOTY MOULD OR BLACK MILDEW (MELIOLA MANGIFERAE)

Black mildew is also known as sooty mold or sooty blotch (Plotze and Prakash, 1997). It is very common wherever honeydew or sugary substances secreting insects viz. mango hopper, scales, coccids, and mealybugs are found. The fungus in the true sense is non-pathogenic, because it does not enter the host tissue and absorb nutrients. It draws the substances not from the host directly but from the sweet substances known as 'honeydew.'

9.2.6.1 SYMPTOMS

The disease in the field is recognized by the presence of a black velvety thin membranous covering, i.e., sooty mold on the leaf surface. In severe cases, the trees turn completely black due to the presence of mold over the entire surface of twigs and leaves. The disease is common in the orchards where mealybug, scale insect, and hopper are not controlled efficiently. The severity of infection depends on the honeydew secretion by these insects. During flowering time, if the fungus infects the blossoms, fruit setting is affected, and sometimes even small fruits are fall. Mature fruits having black patches are also detracting considerably from the appearance and marketability. The fungus is essentially saprophytic and is non-pathogenic because it does not derive nutrients from the host tissues and grow restricted to plant surfaces.

9.2.6.2 CONDITIONS FAVORABLE FOR DISEASE DEVELOPMENT

The disease is exaggerated in dense orchards where penetration of light intensity is low. Trees exposed to sunlight have less incidence, while trees in the center of the orchard, especially those growing dense, have 95% incidence. Honeydew substance secreted by the insects is found to be a condition favorable for the development of sooty mold. The incidence of insects on the shoot is directly associated with disease severity. High humidity, however, proved to be congenial for the growth of the fungus (Singh and Singh, 1972; Misra and Prakash, 1993).

9.2.6.3 MANAGEMENT

Destruction of insects secreting honeydew substances by suitable insecticides. Spraying of Elosal (900 g/450 lt.) at 10–15 days intervals proved to be quite effective (Singh and Singh, 1972). Pruning of affected branches and their destruction prevents the spread of the disease. Spraying of 2% starch is found effective. It could also be controlled by spray of Nottasul + Metacin + gumacasea (0.2% + 0.1% + 0.3%).

9.2.7 PHOMA BLIGHT (PHOMA GLOMERATA (CORDS) WOLL. HOCHAPF)

Phoma blight is a widespread disease of mango and has been reported for the first time by Prakash and Singh (1977) at Central Mango Research Station, Lucknow (India). Later, the disease has been found to be prevalent in many other mango growing areas of India (Prakash and Srivastava, 1987; Prakash, 1996; Ploetz and Prakash, 1997).

9.2.7.1 SYMPTOMS

The symptoms of the disease are noticeable in old leaves only. Initially, the lesions are minute, angular, irregular, yellow to light brown, scattered overleaf lamina. As the lesions enlarge, their color changes from brown to cinnamon, and they become almost irregular. Fully developed spots are characterized by dark margins and dull grey necrotic centers. In case of severe infection, such spots coalesce forming patches measuring 3.5–13 cm in size, resulting in defoliation and complete withering of infected leaves.

9.2.7.2 MANAGEMENT

The disease could be kept under control by a spray of copper oxychloride (0.3%) just after the appearance of the disease and subsequent sprays at 20-day intervals. Benomyl (0.2%) followed by Copper oxychloride (0.3%) has been found effective against the disease (Prakash, 1978, 1979).

9.2.8 BLACK BANDED (RHINOCLADIUMCORTICOLUM MASSEE (PERFECT STATE PEZIOTRICHUMCORTICOLUM (MASSEE) SUBRAMANIAN)

The disease has been reported from Goa, West Bengal, Karnataka, Maharashtra, Bihar, Orissa, Andhra Pradesh, Kerala, Andaman and Nicobar Islands and Tamil Nadu in mild to severe form (Ploetz and Prakash, 1997).

Major Diseases of Mangoes and Their Management 135

9.2.8.1 SYMPTOMS

The disease is noticed on the midribs and veins of leaves, twigs, and branches of mango as black velvety fungal growth. Since the disease is seen into black color bands, hence named as black banded. The infected portions of the bark usually show the mycelial growth and clusters of conidiophores, which confined to the upper layer of bark (Prakash and Srivastava, 1987).

9.2.8.2 MANAGEMENT

Application of Bordeaux paste and spraying of Bordeaux mixture (1%) and Copper oxychloride (0.3%) is recommended for the control of the disease followed by gunny rubbing.

9.2.9 LEAF BLIGHT (MACROPHOMINA MANGIFERAE HINGORANI AND SHARMA)

It was first reported from Brazil to be caused by *Macrophomina* sp. by Lacca (1922). In India, it was reported by Patel et al. (1948). Later, it has been reported by various workers from Delhi, U.P., and Gujarat. The disease has been subsequently found to be present throughout the year in Delhi and other parts of India (Prakash and Srivastava, 1987).

9.2.9.1 SYMPTOMS

The disease usually appears as yellowish pinhead like spots on leaves and twigs of the affected plants. Soon after, spots enlarge and become light brown to dark brown in color with slightly raised and brown purplish margins and later ash-colored due to the appearance of pycnidia. Spots are round to start with but later become oval or irregular in size depending upon environmental conditions. Due to the complete drying of leaf, the infection travels downwards towards the petiole. On the stem, the lesions are elliptic, which later girdles the stem at the point of infection. On fruits, water-soaked circular lesion is produced, which enlarge rapidly and cause rotting.

9.2.9.2 MANAGEMENT

Removal and destruction of infected parts are helpful in reducing the disease inoculum. Spraying of Burgandy mixture, lime sulfur, and dithane have been recommended by Hingorani et al. (1960).

9.1.10 PINK DISEASE (ERYTHRICIUM SALMONICOLOR (BERK. AND BROOME)

The disease is widespread and destructive in many tropical and subtropical regions of the world where heavy rainfall occurs. The disease is also known as 'thread blight,' 'rubellosis,' and 'cobweb' (Prakash and Srivastava, 1987; Prakash, 1996; Ploetz and Prakash, 1997). It causes a reduction in leaf canopy, which creates a large area open through which sunlight can penetrate, scalding inner branching and causing the bark to crack.

9.2.10.1 SYMPTOMS

The disease appears as white, feltymycelial thread of *Erythricium salmoni* color on the twigs and branch crotches (Lim and Khoo, 1985). By then, get established in the internal tissues and interferes with the transport of nutrients by invading bark. Severely infected bark gets shredded, and the wood exposed. Leaves turn yellow and dry, shoots and branches of the affected plants wilt and dry. Roots are not infected. Under favorable conditions, the mycelial threads coalesce to form a rough, pink crust on the bark surface. This stage usually coincides with the penetration of the bark and wood by the fungus. The pink color on the tissues represents profuse conidial production by the fungus and hence, the name 'pink disease.' In advance cases, the fungus may produce a pustular or nectar stage. These pustules are orange-red and arranged systematically in rows along the stem.

9.2.10.2 MANAGEMENT

The disease can be kept under control by cutting and removing the affected branches, and thus eliminating the entire infected end. Such ends should be protected with Bordeaux paste. The disease can also be controlled by lime sulfur and oil-based coppers. Besides this, wide tree spacing and proper free air circulation in the canopies, sunlight penetration helps in reducing the disease.

Major Diseases of Mangoes and Their Management 137

9.2.11 GREY BLIGHT PESTALOTIOPSIS MANGIFERAE (HENN.)

The Pathogen does not kill the plant entirely, but photosynthesis activity is undoubtedly reduced. It is a weak parasite capable of infecting injured tissues, and healthy fruits in contact with the diseased ones.

9.2.11.1 SYMPTOMS

The disease is characterized by the presence of brown spots on the lamina of mango leaves. These spots may develop from the margins of the tip. Spots are light brown and minute, which gradually increase in size and become dark brown. Some of the spots enlarge and form large lesions with the grayish-white or light olive grey center with tan-colored margins. These spots may coalesce to form larger grey patches on the leaf lamina. At this stage, black spots of acervuli may become visible to the naked eye in the central region and more on the upper surface of the leaf and never extend beyond the midrib. If the infection starts from the tip, it advances regularly on either side of the midrib. In the final stage, the infected portion gets detached from the leaf. In case of severe infection, leaves are defoliated.

9.2.11.2 MANAGEMENT

The following spray may be helpful to reduce the incidence of grey blight:

- First spray of wettable sulfur (0.2%), + Zineb (0.2%), after completion of heavy showers;
- Second spray of wettable sulfur (0.2%) before flowering;
- Third spray of Carbendazim (0.3%) at pea stage; and
- Final spray of Zineb (0.2%) before maturation of stone.

9.3 BACTERIAL DISEASES

9.2.1 BLACK SPOT (XANTHOMONAS COMPESTRIS PV MANGIFERAE INDICAE)

It is one of the serious diseases and is reported in many countries. In India, it is prevalent in Andhra Pradesh, Maharashtra, Karnataka, Kerala, Tamil

Nadu, U.P., Bihar, Delhi, Haryana, and Madhya Pradesh and probably in several other mango growing areas

9.3.1.1 SYMPTOMS

It affects leaves, petioles, fruits, twigs, and tender stems. Numerous small angular water-soaked lesions appear in groups towards the tip of the leaf blade. These are first light yellowish but later with age turn dark brown to black angular, cankerous, and raised and get surrounded by a distinct halo. Several lesions are often found to coalesce, forming large necrotic spots. In severe infections, the leaves turn yellow and drop off. On young fruits, water-soaked lesions develop, which also turns dark brown to black. Infected fruits may show skin cracking and badly affected ones dry prematurely. They often burst open, releasing highly contagious gummy ooze containing bacterial cells.

9.3.1.2 CONDITIONS FAVORABLE FOR DISEASE DEVELOPMENT

In India, the disease remains dormant during November to March due to low temperature (11.8–22°C), and the leaf infection is considerably reduced by the fall of the infected leaves. Kent mangoes showed the close relationship of the rainy season with the incidence of bacterial leaf spot. One day's rainfall did not affect the disease appearance, whereas four days of rainfall led to the 36.9% disease incidence. Twig canker persists and initiates fruit infection. The disease spread is rapid during the rains and becomes severe in July-August.

9.3.1.3 MANAGEMENT

Shekhawat and Patel (1975) recommended orchard sanitation by way of removal of infected materials and seedling treatment as preventive measures. Five application of Bordeaux mixture (4:4:50) with spreader was recommended by Wager (1937), Sprays with fungicides such as copper-containing materials and agrimycin (Viljoen and Kotze, 1972) have also been advocated. Agrimycin-100 proved best in managing disease development. Streptomycin sulfate, followed by aureofungin, has been recommended by Prakash and Raooff (1985) to control bacterial disease of mango. Monthly

Major Diseases of Mangoes and Their Management 139

sprays of Bavistin (1000 ppm) or copper oxychloride (3000 ppm) were also found effective.

9.3.2 MANGO BACTERIAL CANKER DISEASE (XANTHOMONAS CAMPESTRIS PV. MANGIFERAE INDICAE (PATEL, MONIZ, AND KULKARNI) ROBBS, RIBIERO AND KIMURA)

Bacterial cancer is an important bacterial disease and causes losses that are as high as 100% in certain cultivars. In India, it was first observed from Poona and Dharwar as leaf spot disease (Patel et al., 1948a and b) Canker incidence was noticed first time in polyembryonic cultivars of mango at the Experimental Research Station, Rehmankhera at Central Institute for Subtropical Horticulture, Lucknow (formerly known as Central Mango Research Station) during the year 1978 and thereafter, it created an alarming situation in the years 1980, 1982, 1983, and 1988. It is also reported from Karnataka (Bangalore), Maharashtra (Ratnagiri, Nagpur, Dapoli, Raigarh), West Bengal (Malda), Bihar (Dholi, Sabour), Goa (Panaji), Delhi (IARI and Badarpur), Gujarat (Navsari, Junagarh, Gandevi), Andhra Pradesh (Sangareddy), Tamil Nadu (Rameshwaram), Kerala (Thiruvananthapuram), Andaman, and Nicobar Islands (Port Blair), Punjab (Gangian), Haryana (Faridabad), Rajasthan (Jaipur), Orissa (Bhubaneshwar) and Madhya Pradesh (Jabalpur). Thus, the disease is spreading fast with low to high magnitude, and its occurrence is gradually extending in the new areas. Its widespread and severity posed much losses of mango fruits.

9.3.2.1 SYMPTOMS

The disease affects all the aboveground parts of the plant, i.e., leaves, petioles, twigs, branches, and fruits. Lesions on leaves are angular to irregular, dark brown to black, cankerous on the lower side, but occasionally on both sides and surrounded by a chlorotic halo. On young leaves, the halos were larger and distinct, while on older leaves, it was narrow and could be observed only against light. Under severe infections, the leaves turn yellow and dropped off. Cankers on petioles are raised and dark brown to black in color, while on twigs and branches are raised with longitudinal fissures. Exposing the vascular tissues mostly filled with a gummy substance that oozes outward. The infection was deep-seated, black discoloration of underlying tissues with cracked bark. Lesions on fruits are raised and dark brown to black, which gradually develop into cankers. Under the favorable condition, lesions increase in size and sometimes cover complete fruit. Such lesions often burst extruding gummy substances containing bacterial cells of the pathogen. Fruits may drop off, if the infection comes at the stem end.

9.3.2.2 CONDITIONS FOR DISEASE DEVELOPMENT

The development of the pathogen in the field is favored by high RH (above 90%) and temperatures between 25–30°C (Kishun and Sohi, 1983). Pathogen has been found to be more active under field conditions from July to September than from November to March. Though the temperatures from April onwards remain favorable (2830°C), fresh infections do not occur until it rains. Maximum and minimum temperature between 30–40 and 17.326.0°C, RH 68–100%, evening RH 25–68%, and high wind velocity during the month (April-May) have been found favorable for the disease build-up (Prakash et al., 1994). In and around Lucknow, the development of symptoms on fruits was observed when fruits reached near to maturity whereas, Viljoen and Kotze (1972) had observed bacterial canker symptoms during the entire life of the fruit.

9.3.2.3 MANAGEMENT

Regular inspection of orchards, sanitation, and seedling certification are recommended as preventive measures against the disease. The selection of stones from healthy fruits for rootstock is advisable (Prakash et al., 1994). Two sprays of streptocycline (200–300 ppm) at 20 days intervals (Bose and Singh, 1980), Streptomycin sulphate (250 ppm) followed by Aureofungin (Prakash and Raoof, 1985) and 3 sprays of Streptocycline (200 ppm) at 10 days intervals (Misra and Prakash, 1992) reduced the fruit infection. Streptocycline (300 ppm) and Copper oxychloride (0.3%) were found more effective in controlling bacterial canker (Prakash et al., 1994). Kishun and Sohi (1984) reported that the disease can be reduced by monthly spray-ings of Bavistin (1000 ppm). Stem injection of Bavistin (1000 ppm) in 3 to 5-year-old mango seedlings has also been found effective (Kishun, 1985, 1988a, b). B. subtilis and *B. amyloliquifaciers* have been very effective and maybe further exploited for biocontrol of the pathogen (Pruvost and Luisett, 1991).

Major Diseases of Mangoes and Their Management 141

9.4 ALGAE DISEASES

9.4.1 RED RUST (CEPHALEUROS VIRESCENS KUNZE)

Red rust disease, caused by an alga, has been observed in mango growing areas and is a common algal disease on the mango in the tarai and in the other humid regions of India. Its occurrence has been reported from Bihar, Karnataka, and Uttar Pradesh, West Bengal, Maharashtra, Gujarat, Punjab, Haryana Orissa, Goa, and other states (Prakash, 1996)

9.4.1.1 SYMPTOMS

The disease is readily recognized by the presence of the rusty red fructification of the alga on the surface of the leaves, veins, petiole, and young twigs. Initially, the spots are greenish-grey in color and velvety in texture and finally turn reddish-brown. Spots are circular to irregular in shape, erumpent, measuring 2 mm in diameter when coalesce form larger and irregular spots, after shedding of spores, the algal matrix remains attached to the leaf surface, leaving a creamy white mark at the original rust spot. The disease is more common in closely planted orchards.

9.4.1.2 CONDITIONS FOR DISEASE DEVELOPMENT

Rainwater has been found to be a source for the spread of the alga infection, and it has been found on the increase during the rainy season (Thrimurthy et al., 1981). Growth and spread of *C. virescens* have been studied by Prakash and Misra (1988). Maximum temperature above 30°C, minimum being around 25°C with high RH, and frequent moderate rains with high wind velocity are conducive for the growth and spread.

9.4.1.3 MANAGEMENT

- Pruning the canopy, mowing beneath trees, and using wider row spacing, which increases air circulation and sunlight penetration, helps reduce conditions that favor the pathogen. Pruning and manuring of host trees are also beneficial.

142   *Diseases of Fruits and Vegetable Crops*

- Prakash and Singh (1979) recommended the Bordeaux mixture (5:5:50) followed by Copper oxychloride for the control of the algal disease.
- Sprays with fungicides viz. Difolatan, Bordeaux mixture, and Copper fungicides algicides such as Fentin acetate have also been reported effective in managing the algal infection.

KEYWORDS

- **black banded**
- **black spot**
- **leaf blight**
- **phoma blight**
- **pink disease**
- **sooty mold**

REFERENCES

Arauz, L. F., (2000). Mango anthracnose: Economic impact and current options for integrated management. *Plant Dis., 84*, 600–611.

Batista, A. C., (1947). A serious disease of mango. *Thesis* (Vol. 19, pp. 212–215). Pernambuco College of Agric. Pernambuco.

Berthet, I. A., (1914). *Molestia de Mangueira. Boletin de Agric. (Sao Palo) XV, 8–10*, 818–819.

Das Gupta, S. N., & Zachariah, A. T., (1945). Studies on the diseases of *Mangifera indica* Linn. part V. on the dieback disease of the mango tree. *Journal of Indian Botanical Society, 24*, 101–110.

Datar, V. V., (1981). Chemical management of powdery mildew of mango. *Third Int. Symp. on Pl. Patho. (Abs.)* (pp. 150, 151). New Delhi.

Datar, V. V., (1986). Management of powdery mildew of mango with fungicides. *India Phytopathology, 39*, 271–272.

Doidge, E. M., (1932). Black spot of mangoes. *Farming South Africa, 7*, 89–91.

Gupta, J. H., (1985). Perpetuation and epidemiology of powdery mildew of mango caused by *Oidiummangiferae* Berth. *Second Inter. Symp. Mango (abstr.)* (p. 62). Bangalore.

Gupta, J. H., (1989). Perpetuation and epidemiology of powdery mildew of mango. *Acta Horticulture, 231*, 528–533.

Gupta, P. L., & Dang, J. K., (1981). Occurrence and control of powdery mildew of mango in Haryana. *Indian Phytopathology, (1980, publ. 1981), 33*, 631–632.

Hingorani, M. K., Sharma, O. P., & Sohi, H. S., (1960). Studies on blilght disease of mango caused by *Macrophominamangiferae. Indian Phytopath., 13*, 137–143.

Joshi, H. U., & Chauhan, H. L., (1985). Effective control of powdery mildew of mango. *(Abs.) Second Int. Symp. on Mango* (p. 63). Bangalore, India.

Kishun, R., & Sohi, H. S., (1984). Control of bacterial canker of mango by chemicals. *Pesticides, 18*, 32–33.

Kishun, R., (1985). Stem injection of chemicals for control of bacterial canker of mango. *Proc. 2nd Int. Symp. Mango* (p. 60). Bangalore.

Kishun, R., (1988a). Role of mango stones in survival of *Xanthomonas campestris* pv. *Mangiferae indicae*. In: *Advances in Research on Plant Pathogenic Bacteria* (pp. 33–35). Today and Tomorrow's Printers and Publishers, New Delhi.

Kishun, R., (1988b). Stem injection of chemicals for control of bacterial canker of mango. *Acta Horticulture, 231*, 518–552.

Kulkarni, G. S., (1924). *Report of the Work Done in Plant Pathology Section During the Year 1922 to 1923* (pp. 167–171). In Ann. Rept. Agri Bombay Presidency for year 1922–1923.

Lacca, A., (1992). *Bot. Agric. Sao Paulo xxiii, 82* (p. 17). (Cited from the report on the export of mango to Europe in 1932 and 1933, Bull. Dep. ld. Rec. Agric. Bombay, 170 of 1932, 1935).

Lim, T. K., & Khoo, K. C., (1985). *Diseases and Disorders of Mango in Malaysia*. Tropical Press, Kuala Lumpur, Malaysia.

Majumdar, P. J., (1977). *The Statesman*. New Delhi.

Mallik, P. C., (1963). Mango malformation symptoms, causes and cure. *Punjab Hort. J., 3*, 292–299.

Misra, A. K., & Prakash, O., (1992). Bacterial canker of mango incidence and control. *Indian Phytopathology, 45*, 172–175.

Misra, A. K., & Prakash, O., (1993). Host range and efficacy of different chemicals for the control of sooty mold of mango. *National Academy of Sci., 63*(B) II, 233–235.

Nicholson, R. D., & Van Staden, J., (1988). Cytokinins and mango flower malformation. I. Tentative identification of the complement in healthy and malformation inflorescences. *J. Plant Physiol., 132*, 720–724.

Palti, J., Pinkas, Y., & Chorin, M., (1974). Powdery mildew of mango. *Plant Disease Reporter, 58*, 45–49.

Patel, M. K., Kulkarni, Y. S., & Moniz, L., (1948a). *Pseudomonas Mangiferae indicae* pathogenic on mango. *Indian Phytopathology, 1*, 147–152.

Patel, M. K., Moniz, L., & Kulkarni, Y. S., (1948b). A new bacterial disease of *Mangifera indica*. *Current Science, 17*, 180–190.

Pinkas, Y., Arenstein, A., & Riebenfeld, A., (1973). Powdery mildew of mango. *Hassadeh, 53*, 685–687.

Ploetz, C. R. L., & Prakash, O., (1997). Foliar, floral and soil borne diseases. In: Litz, R. E., (eds.), *The Mango* (pp. 281–325). CAB, International, Wallingford, UK.

Prakash, O., &. Raoof, M. A., (1985). Bacterial canker in mango (abst.). *2nd Inter. Symp. Mango* (p. 59). Bangalore.

Prakash, O., & Misra, A. K., (1988). Growth of red rust *C. virescens* Kunz of mango during the unusual drought year 1987 in Kakori and Malihabad, mango belt of India. *XXI Congress Brasileiro de Fitopatologia Bras., 13*, 121.

Prakash, O., & Singh, U. N., (1976b). *New Disease of Mango* (pp. 300–302). Proc. Fruit Res. Workshop, Hyderabad.

Prakash, O., & Singh, U. N., (1977). Phoma blight, a new disease of mango (*Mangifera indica* L.). *Plant Dis. Reptr., 61*, 419–421.

Prakash, O., & Singh, U. N., (1979). Fungicidal control of red rust of mango. *Indian Journal of Mycology and Plant Pathology, 9*, 175–176.

Prakash, O., & Singh, U. N., (1982). Evaluation of various fungicides for the control of powdery mildew of mango caused by *Oidiummangiferae*. *Pesticides, XVI*, 171–178.

Prakash, O., & Srivastava, K. C., (1987). *Mango Diseases and Their Management-A World Review* (p. 165). Today and Tomorrow Pub., New Delhi.

Prakash, O., (1978). *A New Mango Disease and Its Crue* (pp. 30–31). Symposium on plant disease problems held at Jaipur.

Prakash, O., (1979). Chemical control of phoma blight in mango. *Indian J. Mycology and Pl. Patho.*, *9*(2), 184–185.

Prakash, O., (1996). Principal diseases of mango causes and control. In: *Advances in Diseases of Crops in India* (pp. 191–256). Kalyani Publisher, Ludhiana.

Prakash, O., Misra, A. K., & Ram, K., (1996). Some threatening diseases of mango and their management. In: *Management of Threatening Plant Diseases of National Importance* (pp. 179–205). Malhotra Publishing House, New Delhi.

Prakash, O., Misra, A. K., & Raoof, M. A., (1994). Studies on mango bacterial canker disease. *Bio. Memoirs, 20*, 95–107.

Pruvost, O., & Luisetti, J., (1991). Effect of time of inoculation with *Xanthomonas campestris* pv. *Mangiferae indicae* on mango fruits susceptibility, epiphytic survival of *X. c.* pv. *Mangiferae indicae* on mango fruits in relation to disease development. *Journal of Phytopathology, 133*, 139–151.

Puttarudiah, M., & Channabasavana, G. P., (1961). Mango bunchy top and the eriophyid mites. *Curr. Sci., 30*, 114–115.

Sekhawat, G. S., & Patel, P. N., (1975). Studies on bacterial canker of mango. *Z. Pflakrankh, Pflacchrtz, 82*, 129–138.

Singh, L. B., Singh, S. M., & Nirvan, R. S., (1961). Studies on mango malformation-I, review, symptoms, extent, intensity and cause. *Hort. Advances, 5*, 197–217.

Singh, S. P., & Singh, R. K., (1972). Studies on sooty mold of mango (*Mangifera indica* L.). Bihar. *Proc. 3rd Int. Symp. Subtropical and Tropical Horticulture* (p. 121). Bangalore.

Unusual Drought Year, (1987). In Kakori and Malihabad, mango belt of India. *XXI Congress Brasileiro de Fitopatologia Bras., 13*, 121.

Uppal, B. M., Patel, M. K., & Kamat, M. N., (1941). Powdery mildew of mango. *J. Univ. Bombay, 9*, 12–16.

Varma, A., (1983). Mango malformation. *Exotic Plant Quarantine Pests and Procedures for Introduction of Plant Materials*, pp. 173–188.

Viljoen, N. M., & Kotze, J. M., (1972). Bacterial black spot of mango. *Phytoparasitica, 4*(3), 93–94.

Wager, V. A., (1937). *Mango Diseases in South Africa* (pp. 12–14). Fmg. S. Africa.

Watt, G., (1891). The mango tree. In: *A Dictionary of the Economic Products of India* (Vol. 5, p. 149). Govt. Printing Press, Calcutta, India.

PART II
Vegetables

CHAPTER 10

Diseases of Potato Crops and Their Management

SHAILBALA

Junior Research Officer (Plant Pathology), Sugarcane Research Center, Bazpur Road, Kashipur, G.B. Pant University of Agriculture and Technology, Uttarakhand, India, E-mail: shailbalasharma10@gmail.com

10.1 INTRODUCTION

The use of fungal and bacterial antagonists, along with other disease management strategies, may emerge out as an alternative approach to chemical control in the near future. The most effective and sustainable control of the disease is obtained when different disease management strategies are integrated together. Some of the control measures include the use of pesticides while others depend on host resistance, cultural practices, etc. integrated in a manner that the crop can be grown effectively with the minimum damage to the crop and also the environment. So information about potato diseases along with management practices is furnished below, which will help to formulate an integrated disease management approach for minimizing the build-up and losses caused by potato diseases.

10.1.1 BACTERIAL DISEASES

In India, about 80% of the potato crop is cultivated under subtropical and 20% under temperate to sub-temperate climate. The cultivation of this crop is often affected due to the attack of various diseases caused by bacteria. The major bacterial pathogens of potato are confined to the genera *Erwinia, Corynebacterium* (*Clavibacter*), *Pseudomonas*, and *Streptomyces*. Certain species within these genera are responsible for soft rot, ring rot, wilt, and common scab disease, respectively (Chatterjee and Vidaver, 1986).

Erwinia species are of the greatest agricultural importance in potato disease of temperate regions. In tropical regions, the major bacterial pathogens of potato are probably *Pseudomonas solanacerum* rather than *Erwinia*, although *Erwinia* is still regarded as a significant potato pathogen even in such climate. There are some bacterial diseases that directly affect the quantity as well as the quality of potato tubers and responsible for the reduction in potato yield.

10.1.1.1 BLACK LEG AND SOFT ROT (ERWINIA CARATOVORAF. SP ATROSEPTICA (VAN HALL) DYE AND ERWINIA CARATOVORAF. SP CARATOVORA (JONES) BERGEY ET AL.)

Blackleg and soft rot are common diseases during storage and transition. Blackleg disease of potato is not common in India. Three soft rot Erwinias, *E. caratovora*f. Sp *atroseptica*, *Erwinia caratovora*f. sp *caratovora* and *E. chrysanthemi* are associated with potato causing tuber soft rot and blackleg (Perambelon, 2002). It occurs only rarely in the Shimla hills of Himachal Pradesh, Kumaon hills of Uttarakhand, Nilgiri hills, and Bihar plains. Stem bases of diseased plants typically show an inky-black to light-brown decay that originates from the seed piece and can extend up the stem from less than an inch to more than two feet. Leaves of infected plants tend to roll upward at the margins, become yellow, wilt, and often die. Aerial stem rot (also called bacterial stem rot or aerial blackleg) is initiated by soft-rot bacteria from sources external to the seed piece. Stem infection can occur through wounds or through natural openings such as leaf scars, which enlarge into a soft, mushy rot that causes entire stems to wilt and die.

Potato tubers with soft rot have tissues that are very soft as well as watery, and diseased tissue is cream to tan-colored, often has a black border separating diseased from healthy areas. Under low humidity, the initial soft rot lesions may become dry and sunken. Under high humidity, the lesions may enlarge and spread to a larger area. Soft-rot decay is generally odorless in early stages, but later a foul odor and slimy decay usually develop as secondary decay, bacteria invade infected tissues. Brown liquid ooze comes out from the rotted tubers. The tuber skin remains intact. Pathogen survives in the rotted tubers. Such rotted tubers when used as seed serve as one source of primary inoculum. Contaminated irrigation water, aerosols of rains, farm implements, soil micro-fauna, etc., helps in the secondary spread.

10.1.1.1.1 Management

Use only certified and disease-free seed tubers. If possible, use whole seed tubers that do not have to be cut. Avoid injury to tubers and sort out injured tubers. Clean all equipment used for cutting seed tubers thoroughly. Water management is essential with proper drainage facility. During crop growth, monitor irrigation as well as nitrogen fertility to minimize vine growth that will promote leaf wetness within crop canopy. Balanced use of fertilizers helps to reduce the favorable conditions for disease development.

Harvest tubers before soil temperature rise above 28°C, and vine must be completely dead, which will ensure skin maturity. Hold newly harvested potatoes at 12.5–15.5°C with 90–95% relative humidity (RH) for the 1–2 weeks to promote wound healing. The disease can be minimized if tubers treated with 3% Boric acid for 30 min and dry under shade. Store the produce either in well-ventilated cool stores and cold stores (Reddy, 2010).

10.1.1.2 BACTERIAL WILT AND BROWN ROT (RALSTONIA SOLANACEARUM)

It is the most destructive bacterial disease of potato. It is the first bacterial disease recorded in India from the Pune district of Maharashtra by Cappel in 1892. Bacterial wilt is caused by *Pseudomonas solanacearum* Smith and was first described as *Bacillus solanacearum* by E.F. Smith in 1896, which was later changed to *Pseudomonas solanacearum* (Smith, 1914). The name was further changed to *Burkholderia solanacearum* (Yabuuchi et al., 1992) and is currently known as *Ralstonia solanacearum* (Yabuuchi et al., 1995). The disease affects both above and underground parts of the crop. The characteristic symptoms of the disease are sudden wilting and finally collapse of the entire plant. In advance stage, if the base of the stem of the affected plant is cut transversely and squeezed, the bacterial mass is seen to ooze out as a dull white slimy mass on the cut surface.

The brown rot refers to the browning of the xylem in the vascular bundles. The browning is often visible from the surface of the infected stem as dark patches or streaks. The name ring disease is derived from the fact that a brown ring in the tuber due to discoloration of the vascular bundles. The skin of the infected tubers is often discolored. Bacteria causes vascular rot and pitted lesions in the tubers. In severely affected tubers, the eye buds are blackened. If the infected stems or tubers are cut across and squeezed, grayish-white bacterial ooze comes out of the vascular ring. The lesions on tubers are produced due to infection through lenticels. Initially, water-soaked spots develop, which enlarge forming pitted lesions. Infected soil and infected tubers are two distinct sources of inoculum which is responsible for the spread and carryover of disease.

10.1.1.2.1 Management

Bacteria have a wide host range. The pathogen infects more than 200 plant species, including solanaceous crops, grasses, shrubs, etc. Complete removal of pathogen from the field is difficult, but through some practices, we can manage the disease. So proper crop rotation with potato, finger millet, wheat, garlic, onion, cabbage, etc. must follow to reduce the soil-borne inoculum. After harvest, the field should be plowed to expose soil to the summer heat of May and June in the plains. Infected plant residue should always be removed and burnt. Avoid replanting of potato tubers in the infected soil. Rain or irrigation water should not be allowed to flow from infested fields to healthy plots.

Do not cut the seed tubers. Cutting spreads the pathogen even to healthy tubers. Restrict post-emergence tillage practices to a minimum level. Practice full earthing up immediately after planting. Application of stable bleaching powder @ 12 kg/ha has been found to reduce bacterial wilt by 80% when applied in furrows at the time of planting (Shekhawat et al., 1988b). Seed tubers should be taken from the disease-free field and treated with Strepto-cycline @ 0.02% for a minute. Certain bacteria like *Pseudomonas fluorescens*, *Bacillus polymyxa*, *Bacillus* spp. and Actinomycetes have been found to delay the development of *R. solanacearum* and reduce the incidence of bacterial wilt (Sivamani et al., 1987; Reddy, 2010).

10.1.2.3 SCAB OF POTATO (STREPTOMYCES SCABIES (THAXTER) WAKSMAN AND HENRICI)

In India prior to 1960, it was restricted to hilly regions only. Scab is a disease of potato tubers that results in lower tuber quality due to scab like surface lesions. This disease does not because yield losses, but it changes the tuber shape, thereby reducing the market value and increasing peeling losses. Pathogen mainly attacks the potato tubers causing two types of symptoms, i.e., shallow scab and deep pitted scab symptoms.

Diseases of Potato Crops and Their Management

Shallow scab symptoms are characterized as superficial roughened areas, slightly raised, or sometimes sunken below the level of healthy skin. The lesion is corky tissue that arises from abnormal proliferation of tuber periderm when the tuber is exposed to pathogen. Individual scab lesions are circular but may coalesce into large scabby areas. In deep pitted scab, lesion is 2–4 mm or more in-depth and darker than shallow scab. Insects may be involved in creating deep-pitted lesions. The term "common scab" generally refers to the response of the disease to soil pH. Pathogen is seed-borne and soil-borne in nature. It infects the young tubers in the field through lenticels. Infection is more in dry soil. In soils with a pH above 5.5, *Streptomyces scabies* is usually responsible for common scab and is capable of causing all the types of scab lesions.

10.1.1.3.1 *Management*

Always use disease-free seed and certified seed. Field must be irrigating regularly from tuber initiation stage until it reaches the size up to one cm in diameter. Crop rotation with radish, beets, and carrots help to minimize the inoculum and create an adverse condition for pathogen spread and disease development. Avoid soil application of animal waste, which favors scab development. Adopt a green manure to keep the disease check. Indian mustard as green manure and crop rotation effectively reducing common scab disease (Larkin and Griffin, 2007). Maintain soil pH levels between 5.0 and 5.2 by using gypsum. Plow the field in the month of April and expose the soil in high summer days during May and June. When applied before planting, some soil amendments such as sulfur and triple superphosphate help to reduce soil pH, which makes the soil less favorable to disease development. Local farm advisor can provide information on amounts that are appropriate for your soil conditions. Give tuber treatment with Boric acid (3% for 30 min) before or after cold storage and dry in the shade before storage or planting.

10.1.1.4 *RING ROT OF POTATO (CLAVIBACTER MICHIGANENSIS SUBSP. SEPEDONICUS (SPEIDKERMANN AND KOTTHOFF) DAVIS ET AL., SYN: CORYNEBACTERIUM SEPEDONICUM (SPIEDKERMANN AND KOTTHOFF) SKAPTASON AND BURKHOLDER)*

Ring rot is one of the most feared diseases of the potato industry, especially for seed producers. The bacterium overwinters in tubers in storage or in the ground and can multiply rapidly once it becomes established. Initial aboveground symptoms are wilting/yellowing and rolling of lower leaves. The yellowing appears at the leaf margin. Interveinal mottling may occur, followed by necrosis. Symptoms progress up the plant until the entire stem becomes wilt and dies. Affected plants have bunchy or rosette terminal leaves.

The disease gets its name from its characteristic tuber symptoms. The vascular ring breaks down in infected tubers changing in color from cream to brown as the decay progresses. The discolored ring is usually most evident when the tuber is cut crosswise at the stolen end. When infected tubers are squeezed, creamy, odorless bacterial exudates comes out in cheese-like ribbons. External swellings and cracks may also be evident in infected tubers.

The bacterium enters the tuber through wounds. One of the most common times for infection to occur is when tubers are being cut for seed. The pathogen spreads through infected plants via the vascular system (xylem). Potato is the only host that is naturally infected, although the bacterium can colonize sugar beetroots without causing the disease. While potato tubers most often become infected at or before planting, symptoms do not develop until after mid-season.

10.1.1.4.1 Management

Always use certified disease-free seed, although the use of certified seed tubers will not guarantee total freedom from ring rot bacteria. Seed lots known to be contaminated with ring rot bacteria should never be planted. Sanitation practices must be followed. Dispose of all infected tubers away from potato production areas. Thoroughly disinfect cutting knives, crates, and machinery used for handling, planting, harvesting, and grading and use new bags for clean seed. After cleaning, sanitize all storehouses and equipment with disinfectant. Do not plant the potato in a field with volunteer plants and have a history of the disease for two cropping seasons.

10.1.2 FUNGAL DISEASES

Fungal diseases both in the field and during storage can be limiting factors in sustainable and profitable potato production wherever they are grown. Potatoes can be stored for long periods for table and processing markets but are plagued by storage disease problems. There are many diseases caused by fungi that are important and require a variety of management

Diseases of Potato Crops and Their Management 153

practices to reduce them to tolerable economic levels. Such diseases include late blight (*Phytophthora infestans*), early blight (*Alternaria solani*), black scurf (*Rhizoctonia solani*), silver scurf (*Helminthosporium solani*), pink rot (*Phytophthora erythroseptica*), dry rot (*Fusarium* spp.), charcoal rot (*Macrophomin aphaseolina*), etc. These diseases have both a field as well as storage component and disease management inputs may be necessary throughout the season for disease control. In view of large number of fungal diseases, their wide distribution and the capacity of diseases to cause extensive crop damage, it is apparent that losses in potato production would be much higher in the absence of disease management practices.

10.1.2.1 LATE BLIGHT (PHYTOPHTHORA INFESTANS (MONT.) DE BARY)

Late blight is one of the serious diseases of potato wherever it is grown and caused the worst ever famine "The Irish famine" in 1845. Pathogen attacks on foliage, stem, and tubers. Infection appears as pale green, irregular spots on the tips and margins of the leaves which in moist weather enlarge rapidly with central tissue turning necrotic and dark brown or black. Often, the spots have a purplish tinge. On the lower side of the leaves, a white mildew (cottony growth) ring forms around the dead areas. In dry weather, the water-soaked areas dry up and turn brown. On stems and petioles light brown lesions develop, elongate, and enlarge often encircling the stem/petiole. The whole vines may be killed as well as blackened and the disease spread rapidly killing the entire crop in a few days when the favorable conditions exist.

Tubers, in soil are readily infected by rain splashes from blighted foliage. Initially, the tubers show a shallow, reddish-brown dry rot that spreads irregularly from the surface through the flesh. Under humid and cloudy weather, the lesions on all plant parts show white cottony growth of fungus. Night temperature below dew point for at least four hours, cloudiness on the next day, minimum temperature 10°C and minimum rainfall during 24 hrs of at least 0.1 mm, are the principal factors governing the occurrence of late blight (Van Everdingen, 1926). The seed tubers and refuge piles in the temperate regions and cold-stored tubers in subtropical regions serve as the primary source of inoculum. The pathogen mainly perpetuates from one season to another season through seed tubers (Hector, 1926). The amount of tuber infection which goes to stores through seed tuber range 0.01–3%, which is more than enough to initiate the late blight epidemic in the next crop season (Bhattacharya et al., 1990).

10.1.2.1.1 Management

The use of disease-free and certified seed is essential to lower down the amount of initial inoculum. Some resistant varieties, i.e., Kufri Badshah, K. Naveen, K. Sinduri, K. Kuber, K. Jeevan, K. Jyoti, etc. should be grown. All the blighted tubers must be removed and buried in the soil. Seed lots must be select carefully to avoid bringing in late blight on the seed. It is especially important to contact seed growers prior to winter to be assured that the seed lots of interest are available and will be sorted and graded with some specifications. Plants should be adequately hilled to avoid tuber infection. Cull piles/potatoes must be destroyed through composting, burial, chopping, freezing, or feeding to livestock prior to new crop emergence. Monitoring of cull disposal sites is essential, and emerging sprouts should treat with fungicides.

Remove infected tubers before storage and scouting of fields, especially low lying areas, field borders, weedy patches, and any place where lack of air movement or shading allows leaves to remain wet for prolonged periods, should be continued. Apply protectant fungicides before late blight appears. Make the first fungicide application prior to plants touching within the row. Continue fungicide applications at intervals determined by the potential for late blight. Apply protectant fungicides Mancozeb @ 2.5 g/l, Curzate @ 2 g/l and Ridomil MZ-72 @ 2 g/l with onset of disease at 10 days interval. Application of 1 or 2 sprays of systemic fungicide mixture (Metalaxyl, Ridomiletc) in areas where blight comes every year in epiphytotic form (Reddy, 2010). The management approach should be focused both on preventive or prophylactic as well as reducing infection rate which may be achieved by the use of different component including host resistance, cultural adjustments, and need-based use of fungicides (Shekhawat, 2000).

10.1.2.2   EARLY BLIGHT (ALTERNARIA SOLANI (ELL. AND MARTIN) JONES AND GROUT)

Early blight mainly infects leaves and tubers. Infection on the leaves appears in the form of brown spots, which may be angular, oval, or circular. The spots may or may not have concentric rings. The concentric rings are more prominent in large blotchy spots and give them a target board effect. The disease can be distinguished from late blight by the absence of white cottony fungal growth on the lower side of leaves. When a spot appears on vein, a part of it gets necroses. The fungus produces several toxins, including alternaric acid in the host which leads to yellowing of the leaves.

Diseases of Potato Crops and Their Management 155

Lesions on tubers are circular to irregular in shape, slightly sunken and often surrounded by a raised purple to dark brown border. Plants susceptible to biotic stresses are prone to early blight than healthy plants. The conidia and mycelium in the soil or in debris of infected plants serve as the primary source of inoculum. Generally, the disease is favored by moderate temperature (17–25°C) and high humidity (75%). Intermittent dry and wet weather is more conducive for early blight. The incidence and severity of disease is generally high in the crop receiving imbalance amount of fertilizer specially nitrogen (Figures 10.1a–k).

10.1.2.2.1 Management

Crop rotation is the best way to avoid soil-borne inoculum. Field sanitation practices should be followed. The use of moderately resistant varieties, i.e., Kufri Naveen, K. Sinduri, K. Jeevan, etc. is effective. Dead haulms should be raked together and burn immediately after crop harvest. The crop must be given balanced doses of fertilizers, especially nitrogen. The crop, if sprayed with 1% urea at 45 days and given one more subsequent spray after 8–10 days, may easily escape the severe outslaught of early blight. Cultivation of solanaceous crop being collateral hosts, nearby potato field must be avoided. Fungicidal sprays are quite effective in controlling the disease. Spray the potato crop with fungicide Mancozeb @ 2.5 g/l at 7–10 days interval or Copper fungicide at least 3–4 weeks after planting to lower down the disease incidence (Reddy, 2010).

10.1.2.3 BLACK SCURF (RHIZOCTONIA SOLANI KUHN (TELOMORPH: THANETOPHORUS CUCUMERIS (FRANK) DONK)

This fungal disease is prevalent both in the plains and hills and responsible for yield losses up to 25%. Actually, infection develops on all parts of the plant, including foliage (Prasad and Agrawal, 1983). The fungus attacks young sprout through epidermis and produces dark brown lesions, thereby killing the sprout before emergence, which results in patchy germination (Dutt, 1979). Elongated reddish-brown lesions develop on the stem at or below the soil surface. The lesions enlarge and may girdle the stem. The affected plants lack vigor. When the girdling is complete, the foliage curls and turns pinkish to purplish. Aerial tubers are commonly seen.

156 *Diseases of Fruits and Vegetable Crops*

FIGURE 10.1 (a–k) Early blight.

Diseases of Potato Crops and Their Management 157

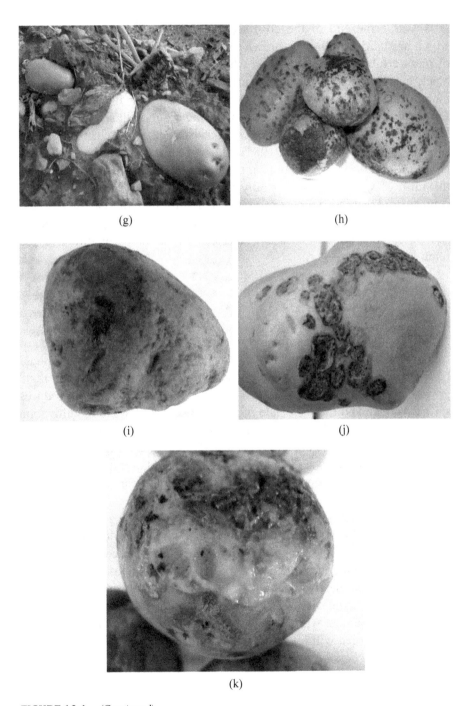

FIGURE 10.1 *(Continued)*

Towards the end of the crop season, the fungus produces sclerotia, which are hard, small, dark brown to black resting bodies on the surface of mature tubers. These sclerotia when get deposited continuously form black encrustation on the tuber surface. Others symptoms on the tuber include skin cracks, dry-core symptoms where crater like depression is formed on the lenticels (CPRI, 1981), pitting along with shape deformity, hard dry rot with browning of internal tissue (Thirumalachar, 1953) and seed piece decay (Chaudhary, 1983). The perfect stage of pathogen *Thanatephorous cucumeris*, develops on the green stem near the soil surface in the form of whitish grey mat only under specific weather condition. In subtropical areas, seed tubers serve as the main source of disease, whereas in temperate regions, the fungus survives in the soil through the year act as a potential source of disease. The affected tubers look unattractive and fetch low prices.

10.1.2.3.1 Management

A combination of tuber disinfection and improved cultural practices successfully check the incidence and severity of the disease. Seed potatoes should be free from sclerotia. Delay/late planting until conditions favor the rapid growth of potatoes favors disease suppression. Shallow planting reduces sprout damage and the incidence of stem cankers. Hot weather cultivation has been found effective in reducing the inoculum levels in the fields. The increase of the organic matter of soil through green manuring, oil cakes, and sawdust will decrease the inoculation in the soil and harvest crops promptly to avoid the development of sclerotia. Three to four years of crop rotation must be followed.

Growing maize or dhaicha (*Sesbania aegyptiaca*) for green manure as rotation crops also checks the inoculum and disease build-up. Rapeseed mustard and canola as green manure and crop rotation, effectively reducing *Rhizoctonia solani* (Larkin and Griffin, 2007). Continuous seed treatment for 3–4 years completely check the disease. Treated seed tubers are usually free from black scruff disease. Seed treatment with Boric acid @ 3% for 30 min or seed treatment with bio-agents *Trichoderma harzianum* or *T. viride* @ 4–6 g/kg of seed will lower down the initial inoculum. Field application of *Trichoderma harzianum* and *T. viride* @ 100 g/m^2 will also be effective to manage the disease.

Diseases of Potato Crops and Their Management 159

10.1.2.4 POTATO WART (SYNCHYTRIUM ENDOBIOTICUM (SCHILBERZKY) PERCIVAL)

The disease is caused by a fungus, which is an obligate parasite. All underground parts except root exhibit wart symptoms. The abnormal growth activity starts on buds, the underground stem, stolon, and tubers, which lead to the development of warts. The outgrowth is spherical, green, or greenish-white in color if exposed above, and it gives cream color when underground. In advanced stages, the warts are darker in color or sometimes black.

The wart consists of distorted, proliferated branched structures grown together into a mass of hyperplastic tissue. These masses vary from less than 1 g to more than 50 g in fresh weight (Hampson, 1996). The resting spores are released upon decomposition of the warty tissues. The resting structure is thick-walled and may remain viable in the soil for almost 30 to 40 years. Wart disease is much worse during wet seasons than in dry soil conditions (Singh and Shekhawat, 2000). The main survival structures of this pathogen are resting spores present in host debris or soil and remain viable in the soil for almost three to four decades. The wart pathogen is readily seed and soil-borne. The warted tubers used as seeds are the main source of disease spread.

10.1.2.4.1 Management

Domestic quarantine must be followed. Transfer of infected seed from one area to other areas within the country should be restricted. Infected tubers are the chief source of disease spread, so the use of immune potato varieties, i.e., Kufri Jyoti, K. Muthu, K. Kanchan, K. Sheetman, K. Sherpa, K. KhasiGaroetc will lower down the infection. Prolong crop rotation (more than 5 years) must be practiced. Infected seed, crop debris, etc. should be destroyed. Soil should be sterilized with Formaline 5% in the nursery, but no effective soil treatments are available and cannot be applied on a large field scale. Crop sanitation practices reduce disease incidence. Warted lumps and potato peels should not be thrown in the field or in the manure pit but destroyed by burning. Rouge out plants of susceptible varieties by burning and destroying warty tubers. Diseased potatoes should not be used as seed. These tubers may be consumed or fed to the cattle only after boiling.

10.1.2.5 CHARCOAL ROT (MACROPHOMIN APHASEOLINA (TASSI) GOLD)

The fungus is further sub-divided into two main sub-species, namely *M. phaseolina* subsp. *typica* and *M. phaseolina* subsp *sesamica* on the basis of growth characters, pattern of sclerotial formation, and distribution in culture media (Thirumalachar, 1953). The disease causes three types of symptoms, i.e., stem blight, charcoal rot, and dry tuber rot (Paharia, 1960). The fungus may grow from infected seed tuber up the stem to the soil surface and kill the plant. A soft, dark-colored shallow rot develops on lower stem area. Secondary organisms frequently follow primary infection by charcoal rot pathogen. Initial symptoms appear in the form of black spots (2–8 mm in diameter) (Singh et al., 2002). Black spots appear around the lenticels and eyes, which enlarge into patches extending deep into the tuber flesh. The pathogen infects through lenticels, eyes, stolen, and wounds made by larvae of the tuber moth to cause black sunken lesions and later blackening of internal tissues. The skin of the tubers remains unaffected, but the tuber flesh gets blackened to a depth of 2 to 5 mm.

In dry rot, black sunken areas develop on the tuber, which turns into cavities underneath the skin. The cavities are filled with fungus mycelium and sclerotia. In the advanced stage, secondary pathogens attack the tuber, which ultimately is reduced to a foul-smelling wet rotting mass (Pushkarnath, 1976). Both tubers and soil may serve as the primary source of disease. The fungus utilizes both organic and inorganic sources of nitrogen. The best carbon and nitrogen sources are starch and asparagines (Paharia and Sahai, 1968). The organism has a wide host range covering both solanaceous and non-solanaceous crops (Thirumalachar, 1953; CPRI, 1958–60). The disease becomes economically important only when soil moisture is high, and the soil temperature is above 82°F (28°C).

10.1.2.5.1 Management

Maize, green gram, sunhemp, and cowpea as green manure crops are suitable for the management of charcoal rot (Bhattacharya et al., 1977; Bhattacharya and Malhotra, 1979). When potato crop is planted year after year in the same field, the survivability of the pathogen and their build-up gradually increase over the year. So crop rotation with sunhemp and maize for several years are the best for controlling the disease. Certified seed tubers should be used. No potato cultivars are resistant, but *Solanum chacoense* clones and the hybrids

Diseases of Potato Crops and Their Management 161

derived from them have been found to possess a high degree of resistance to charcoal rot (Paharia et al., 1962). The early season cultivars may escape damage in an infested field.

Adjustment in the harvesting dates effectively manages the disease by escaping the most favorable soil temperature for pathogen and disease development. Early harvesting of the crop should be preferred. Harvesting the crop early before the soil temperature reaches 28°C can check the disease. Avoid injuring of tubers during harvest. Clean cultivation is essential. Repeated field plowing at 10–12 days interval in May–June expose the sub-soil and plant debris to the high temperature, which are deleterious to the pathogen (Shikka et al., 1971). Seed treatment with bio-agents *Trichoderma harzianum* and *T. viride* @ 4–6 g/kg of seed will manage the disease up to some extent.

10.1.2.6 FUSARIUM WILT AND DRY ROT (FUSARIUM OXYSPORUM SCHLECHTEND EX FN., F. SOLANI (MART) SACC. F.SP. EUMATII (CARPENTA) SNYDER AND HANS, F. MUCONEVA (FR. EX FR.) SACC., F. AVENACCUM (FR. EX FR.) SACC., F. ACUMINATUM ELLIS AND EVERH., ETC.)

The characteristic symptoms include wilt, stem rot, and damping-off of seedlings. On tuber, pathogen produces spots, necrosis, dry rot and seed piece decay. Both stems and tubers at stolen end show vascular browning. Stem rot leads to yellowing, rotting, and rosetting of leaves. Tubers and stolen may also develop brown lesions (Singh, 1986). It may cause damping-off of seedlings if planted early in the season when the temperature is high. *Fusarium acuminatum* causes rotting of the stems that may extend up to the growing tip (Rai and Singh, 1981). Infected tubers and soil are primary source of inoculum. The dry rot is an important disease of storage.

The skin of the dry rot infected tubers first becomes brown then turns darker and develops wrinkles. These wrinkles are often arranged in irregular concentric circles (Khurana, 2000). In the later stage of infection, a hole may be observed in the center of the concentric ring with whitish or pinkish growth of fungal mycelium. On cutting these affected tubers, whitish or brownish tissues are seen with one or more cavities. Eventually, the infected tubers loose water and become dry, hard, and shriveled. Dry rot development is affected by storage condition, tuber age, size, wounds, etc. Seed tubers are contaminated with chlamydospores, which survive in the soil as resting spores. Infected tubers and soil are the primary sources of inoculum. However, infected tubers are more important because in soil, the fungus remains viable only for 9–12 months (Mann and Nagpurkar, 1922).

10.1.2.6.1 Management

The crop should be harvested at full maturity; otherwise, immature tubers are more prone to injuries. The use of certified seeds is essential. Use only clean and healthy seed tubers for planting and tuber washing followed by drying under shade substantially reduces the infection. Sanitation practices, i.e., removal of crop debris, weed control, etc., reduce pathogen from field. If possible, the use of cut tuber during planting should be avoided. Otherwise, seed pieces may be treated with fungicide Mancozeb (1kg in 450 liters water) for 10 minutes and dried for 24–48 hrs before planting. Tuber damage and injury must be avoided during harvest and storage. Tubers should be stored in cold stores in plains.

Crop rotation and soil management must be followed to arrest the buildup of the pathogen population. Keep tuber in chamber at 20°C with high humidity. Soil amendment by using crop straw and oilcakes helps to reduce the disease. Singh et al. (1988) used four crop straws, i.e., gram, soybean, pigeon pea, and cluster bean to reduce the disease incidence. Adjustment in planting and harvesting dates is essential to manage the disease. In wilt symptoms of this disease, fungus attacks the plants when they are 40–60 days old. Therefore, the disease is more severe in early planted crop (Singh, 1986). By delaying the planting time, disease could be reduced by 36% (Singh et al., 1990c). Seed should be treated with bio-agents *Trichoderma harzianum* and *T. viride* @ 4–6 g/kg of seed.

10.1.2.7 POWDERY SCAB (SPONGOSPORA SUBTERRANEA (WALLR.) LAGERHEIM)

Powdery scab is an important disease of potatoes in many of the world's cool/temperate regions. The pathogen attacks all underground parts of the plants. It appears as pimple-like spots which are circular, smooth, and light brown present on the surface of young tubers and ultimately rupture, exposing a cavity containing a brown powdery mass of spore balls. The cavity is surrounded by the remanents of the ruptured periderm. The spots may be isolated, crowed, or even coalesce (Dutt and Pushkarnath, 1960). Occasionally warty protuberances are produced in some cultivars (Bhattacharya and Raj, 1978).

Sometimes, the affected tubers show the destruction of the flesh due to renewed activity of the pathogen below the primary pustules resulting in hollowed out patches. These types of lesions are called cankers. Sometimes

Diseases of Potato Crops and Their Management 163

small galls occur on the roots of affected plants. The fungus overwinters through spores in soil and on infected seed tubers. The pathogen thrives, and powdery scab becomes severe, particularly when cool, damp, soil condition prevails. Soil water has a major effect on the development of disease (Harrison et al., 1997). The spores germinate during crop season and produce zoospore in soil. Fungus takes 16 days to enter the tissue and 18 days to produce plasmodium (Bhattacharyya and Raj, 1981).

10.1.2.7.1 Management

Control of powdery scab includes all measures that can be taken to avoid excessive water around the tubers. These are related to soil tillage, seedbed preparation, planting, and ridging, soil compaction, irrigation, and drainage. Contaminated seed serves as a carrier. Resistant varieties should use in the field where scab is a major problem. A crop rotation scheme for three to four years using apparently disease-free seeds and disease-free soil will keep disease level low. Crop rotation with radish, beets, and carrot is very effective for the management of disease. Indian mustard as green manure and crop rotation effectively reducing powdery scab disease (Larkin and Griffin, 2007). Maintain soil pH levels between 5.0 and 5.2 by using acid-producing fertilizers such as ammonium sulfate etc. Avoid moisture stress during the 2 to 6 weeks following tuberization. Seed treatment with Boric acid (3%) for 30 min is effective for control of disease.

10.1.2.8 COLLETOTRICHUM ROOT ROT AND TUBER ROT (COLLETOTRICHUM COCCODES (WALLR.) HUGES)

The fungus attacks the root system, tubers, underground portion of stem and stolen. Disease symptoms include root and collar distortions, extensive foliar wilting and tuber rot. Vascular discoloration is observed up to 5 cm up the stem. Affected plants often bear aerial tubers. Freshly harvested tubers exhibit white mycelial growth on their surface and small circular or oval spot with irregular margins. In severe cases, tubers may rot completely. The disease is known to be serious during dry and warm seasons. Fungus also attacks other solanaceous crops like tomato, eggplant, chilies, etc. Infected tubers and soil harbor the pathogen till the next crop season. The sclerotia have been reported as a source of inoculum and can survive in the soil for up to 8 years.

10.1.2.8.1 Management

Crop rotation with nonhost crops and elimination of wild host will reduce level of inoculum and survival of pathogen in the soil. Planting resistant varieties are recommended, but known resistant varieties are very limited. Soil fumigation must be followed. Avoid planting in poorly drained soil, if possible. Use tillage practices to bury plant refuse and encourage decomposition. Tubers should harvest as soon as possible after vine kill; otherwise, high temperature and condensation on the tuber surface promote disease. Use good crop production practices, such as timely irrigation and adequate fertilization to reduce the crop stress.

10.1.2.9 PINK ROT (PHYTOPHTHORA ERYTHROSEPTICA PETHYB.)

Plants and tubers both are affected by the fungus. This minor disease is mostly observed in poor drained parts of the plants. Disease is characterized by stunting and wilting appear on potato vine growing late in growing season. Wilting starts from the base of the stem and progress upwards causing leaf yellowing, drying, and defoliation. Vascular discoloration and blackening of underground stem may also observe. Root may turn brown to black. The plants appear chlorotic with hard and dry margins of the leaves and blackening of the stem at and below the soil level. The pith is completely destroyed by the fungus.

Symptoms on tubers are more obvious and characteristic of the disease. Tuber decay begins at or near the stem or stolen end of the tubers. Infected tissue become rubbery but not discolored in early stage of infection. When infected tubers are cut open, the rotted portion is delimited by the dark line visible through tuber skin. The tubers are spongy and show blackening of tissue under the lenticels. On cutting, the tuber flesh turns pink changing to black after half an hour's (Phadtare, 1978; Rai, 1979). The spread of pink rot may continue in storage. The pathogen survives for long periods in the soil and becomes inactive when soil is saturated with water. Soil-borne oospores are considered as the primary source of inoculum.

10.1.2.9.1 Management

One of the most successful and important approaches to manage the disease is to reduce the amount of inoculum in soil by removing crop debris, volunteer,

Diseases of Potato Crops and Their Management 165

and cull potatoes from the field. Crop rotation with the non-host plant can reduce the amount of inoculum in the soil. This soil and seed-borne disease is enhanced by excessively wet soil conditions and is controlled by improving drainage. The disease can be eliminated by fumigating infested soil, complemented with Metalaxyl 5G at planting time and the use of healthy potato seed tubers. Pink rot is most frequently seen in mature plants approaching harvest. It is much worse when saturated soil is accompanied by warm temperatures. Avoid prolonged saturation of soils during irrigation, provide good drainage, and avoid harvesting wet tubers. Maintain good airflow, avoid the accumulation of moisture on tubers, and maintain a low temperature during storage because the fungus is inactive below 40°F (4.4°C).

10.1.2.10   SILVER SCURF (HELMINTHOSPORIUM SOLANI)

Silver scurf is a relatively new problem related to potato growing areas. It is considered a minor disease. Symptoms consist of tan to silvery gray, circular to irregular lesions formed on the tuber periderm. Lesions generally have a definite margin; vary from pinhead size to patches that cover most of the tuber surface. As the disease progresses, individual lesions coalesce. Under high humidity, the fungus sporulates on the infected tissue and form blackish mold. This fungus causes a surface blemish resulting in the tubers looking "dirty." This fungus likely attacks all potato cultivars but causes the most economic damage to those that are fresh market.

While the silver scurf pathogen infects potatoes in the field, the greatest damage occurs in storage, particularly with increasing time in storage and is especially visible on smooth skin cultivars. Damage often does not show up until the tubers have been stored for a month or more. With time in storage, cracking of the epidermis results which increases water loss and shrinkage of tubers. Both soil and infected tubers serve as the source of primary inoculum. The pathogen is not known to attack any plant species other than potato (Srivastava, 1965; Dutt, 1979).

10.1.2.10.1   Management

Silver scurf problems start with infected seed. Use seed that is free or relatively free of silver scurf. Always test seed to be grown for the fresh market lots for silver scurf infection before seed purchase. *H. solani* does not survive long periods of time in soil. Therefore, practicing crop rotation by not planting potatoes for at least two years or more in the same field will greatly reduce the chance of daughter tuber infection originating from fungal spores surviving in the soil. Tubers under vines that die early are more likely to become infected than tubers under green plants because the periderm on tubers matures after plants die. Therefore, use cultural and chemical practices to keep vines healthy until frost or vine kill. Seed cutting and handling equipment should be cleaned and disinfected between lots to kill spores left behind from the previous lot when increasing seed-tuber generations.

10.1.3 VIRAL DISEASES

Viral diseases of potato considered the main cause of potato degeneration. Almost 30 viruses are now reported on potato in India but only few are responsible for significant losses (Khurana and Singh, 1986). A lower incidence of viruses up 10% does not much reduce the potato tuber yield (Khurana and Singh, 1988) but a higher incidence coupled with early or severe infection causes serious depression in the tuber yield (Khurana and Singh, 1986).

Commercial propagation of potato is normally done vegetatively using 'seed' tubers. Therefore, 'Degeneration' of seed stocks due to viruses is common and results in the continuity of several viral pathogens. The losses in potato yield due to one or more virus(s) infecting potatoes vary from low to very high. Nearly all viral diseases, although seldom lethal, reduce plant vigor and yield potential. Frequently two or more viruses may be present within the plant at the same time. Viral diseases seldom cause decay in storage. There are some common viral diseases of potato which cause substantial yield loss.

10.1.3.1 POTATO LEAF ROLL

It is the most serious virus disease of potato especially in tropical areas. Losses may reach up to 90%. Disease is caused by potato leaf roll virus. The rolling of leaves is characterized by curling of leaflet margins inward thus forming a trough in which the midrib is at bottom. In secondary infection, this rolling of leaves starts in lower leaves and progresses upward throughout the plant. The rolled leaflets are stiff and rigid. They are thick and leathery sometimes with pink margin.

In some cultivars, a reddish or purple discoloration develops on the margins and underside of the leaves. Internal necrosis is a symptom of tubers. Leaf rolling is a result of disturbances in the phloem translocation

Diseases of Potato Crops and Their Management 167

system where it causes necrosis and abnormal formation of a carbohydrate which blocks starch transport from the leaves to the tubers. In nature transmission of virus occurs through infected tubers and aphid vectors. The virus is readily tuber borne and also efficiently transmitted in a persistent manner. *Myzus persicae* is the main aphid vectors.

10.1.3.1.1 Management

Only virus-free potato seeds should be used. Seed tubers should only be multiplied in low aphid or aphid free areas coupled with practices for ensuring sanitation, rouging of infected plants, if any, and dehaulming the crop as soon as the vector aphid population exceeds the critical limit of 20 aphids per 100 compound leaves. To avoid tuber infestation from infected foliage, the foliage must be destructed mechanically or chemically before harvest. Potato tubers and weeds are sources of infection which can also harbor viruliferous aphids. Building up of incoming aphid vectors be controlled by application of suitable systemic insecticides. Therefore, infected potato plants and host weeds should be eliminated within and around the field. Elimination of infection sources is only effective when carried out in the entire neighborhood.

10.1.3.2   POTATO MOSAIC

Disease is categorized according to the symptoms into mild/latent mosaic, vein banding severe mosaic, rugose mosaic, and crinkle.

10.1.3.2.1 Mild or Latent Mosaic

It is caused by potato virus X (PVX). Major symptoms is interveinal mosaic, sometimes little dwarfing in potato is also noticed. Some strains of PVX produce no visible symptoms of latent mosaic, although yields may be reduced 15% or more when compared to virus-free plants. PVX is the most widespread of all the potato viruses. It is also referred to as latent mosaic, potato latent virus, and potato mottle virus. Several strains of the virus exist. PVX infects several solanaceous crops, including potato, tomato, and tobacco. Systemic infection occurs in these crops. The virus survives between seasons in infected tubers. Plants produced from these tubers are also infected. It is mechanically transmitted by plant-to-plant contact (leaves, shoots, and roots), machinery, cutting tools, and animals. Chewing insects such as grasshoppers have also been suspected as a means of spreading the disease. There must be wounding and an exchange of plant sap for infection to occur.

10.1.3.2.2   Vein Banding Severe Mosaic

Vein banding severe mosaic is caused by potato virus Y (PVY). Disease is characterized as mild to severe mottle and streak or leaf drop streak with necrosis along the veins of underside of leaflets. When necrotic lesions form on the petioles, the leaves may dry up but cling to the vine or drop off leaving plants with a bushy top growth and lower stem devoid of foliage. *Myzus persicae* is the most efficient vectors. PVY is a common pathogen of potato worldwide and can reduce yields from 10 to 80%. The virus has a broad host range that includes many solanaceous cultivars and weed species. It also affects members of the Chenopodiaceae and Leguminosae. However, the most important reservoirs of the virus are potato tubers and volunteer potatoes. PVY is vectored in a non-persistent manner by at least 30 species of aphids. Plants grown from infected tubers may show the same symptoms as well as stunting, but they are less severe than those on plants infected with early current-season inoculum.

10.1.3.2.3   Rugose Mosaic

Rugose mosaic disease causes very severe damage to the individual plants. Disease is caused by PVX and PVY. PVY is the most severe of the mosaic viruses. Symptoms include mottling or yellowing of leaflets, leaf crinkling and sometimes leaf drop. Veins on the underside of leaves often show necrotic areas as black streaks. Infected plants may be stunted. Leaf mottling may be masked at low (below 50°F or 10°C) or high (above 70°F or 21°C) temperatures but at high temperatures the disease can be identified by the crinkling and rugosity of the foliage. A severe crinkling of the leaves occurs when potato viruses Y and X occur in the same plant. Some strains of PVY cause tuber necrosis symptoms. The foliage is not only mottled but is also severely wrinkled, puckered, and markedly reduces in size. The leaflet margins are rolled downward and entire plant is severely dwarf. *Myzus persicae* is capable of efficiently transmit potato virus.

Diseases of Potato Crops and Their Management

10.1.3.2.4 Crinkle of Potato

Disease is caused by potato crinkle virus i.e., combination of PVX and potato virus A (Solanum virus 3). It is characterized by yellowish patches on the foliage which are bigger and more prominent. The color becomes more pronounced and is accompanied by rust brown spot beginning near the tip of the leaves. The foliage is brittle and easily injured. Pathogen is transmitted by sap inoculation and by insect vector *Myzus persicae.*

10.1.3.2.5 Management

Field sanitation should be strictly followed right from planting till harvesting. Removal and destruction of infected plants and eradication of susceptible weed hosts help to prevent the spread of contagious virus. The use of certified and resistant varieties, i.e., KufriMegha (resistant to PVX and PVY) for seed purpose, is essential. Detop (remove foliage) the plant in the third or fourth week of December when aphid population start building up and then leave the potato tubers in soil to mature. Dehaulming the seed crop at the right time for multiplication of quality nucleus seed is also recommended (Mandahar et al., 1990).

Disinfect all the field equipments by dipping in or washing them either with 3% Trisodium phosphate or 1% Sodium hypochlorite solution. In the nursery, aphids can be controlled by application of insecticide Carbofuran @ 1 kg a.i./ha in the nursery bed at the time of sowing followed by 2–3 foliar sprays of Phosphamidon @ 0.05% at an interval of 10 days. Spray of systemic insecticide, i.e., Monocrotophos 36 SL @ 0.4 g/l regularly checks insect vectors of crinkle disease of potato.

10.1.3.3 POTATO STEM NECROSIS

Stem necrosis disease of potato was first reported in India in 1989 (Khurana et al., 1989). Disease is caused by groundnut bud necrosis virus or tospovirus like tomato spotted wilt virus (TSWV) (Khurana and Garg, 1998). Infected plants died before the end of growing season. The wide range of disease symptoms includes necrotic and chlorotic spots on leaves and stem, veinal necrosis, leaf droop, hanging, blackening of stem, and concentric ring spots in affected stems. Initially, a few light brown necrotic elliptical spots appear on young stem. The spots are observed after 18–20 days of plant emergence.

The lesion increase in size and extend in linear direction and then engirdle the entire stem. The disease appears as dark brown necrotic areas at the nodes and may cause girdling of the petiole leading to leaf drooping and necrosis. Abundant lesion appears on the stem and coalesce covering greater area within a short period of 7–10 days.

TSWV in nature is almost exclusively transmitted from plant to plant by several species of thrips (Thysonoptera: Thripidae) belong to family thripidae. Potato stem necrosis is a serious problem in early planted potato crops in central India. Disease incidence up to 90% was recorded in some parts of Madhya Pradesh and Rajasthan (Khurana et al., 1999). Potato yield losses due to disease vary greatly from place to place and year to year and may change from 15–30% (Khurana et al., 1989).

10.1.3.3.1 Management

Strict sanitation in the field helps to prevent the spread of the virus. Planting of essentially disease-free seed stocks from approved or reliable sources will lower the virus spread. Early planting of potato crops should be avoided. Two sprays of insecticide Monocrotophos @ 0.10% at 15 days interval from the initiation of the disease infection resulted in lower disease incidence and produced significantly higher yield. Up to 90% of disease incidence was recorded in early planted potato crop (Khurana et al., 1997). Delay in planting dates will lower down the both disease incidence and disease index (Patel et al., 2010). Use of resistance cultivars is helpful for reducing the disease. Injury of seed tubers must be avoided which will control the spread of the virus. Treat the seed tubers with Boric acid @ 3% for 30 min before planting and spray the crop with insecticide Monocrotophos @ 0.1% at 21 and 35 days after planting will control the insect vectors and lower down the disease incidence.

10.1.3.4 SPINDLE TUBER OF POTATO

Spindle tuber of potato is caused by Viroid. The potato spindle tuber viroid was the first viroid to be identified. Symptoms are often not visible during the first season but become progressively more severe in the following generations. Infected plants are stunted and have an upright or erect appearance. Tuber symptoms are more obvious but do take several generations to appear. Affected tubers are small and deformed becoming cylindrical and elongate ('spindle'). They are often pointed and can show growth cracking on larger

Diseases of Potato Crops and Their Management 171

tubers. Eyes will often become more prominent and sprouting is slower than with healthy tubers. As the name implies, infected tubers may be spindle or oblong-shaped or tend to be more rounded instead of the normal shape for a given variety. Prominent eyebrows are another important characteristic.

10.1.3.4.1 *Management*

Disease management can be divided into two parts, i.e., prevention of infection and viroid eradication. Prevention of infection includes all the measures to prevent the introduction of viroid into potato crops. Only disease-free tubers should be grown in fields. Certification schemes, including testing may be required to provide a further guarantee that the planting material is free from viroid. This viroid is mechanically transmitted, so prevent viroid introduction via human activity. Precaution during the processing of seed tubers at the time of planting (use of cutting knives) is also important. Viroid eradication includes cleaning of equipments, glasshouse, etc. Rouge out infected plants before they can serve as a source of inoculum. Crop rotation with non-host crop help to eliminate infected volunteer plants.

10.1.4 *MYCOPLASMAL DISEASES*

The diseases caused by mycoplasma can severely affect yield, tuber sprouting, and quality of potato. These mycoplasma resembles with bacteria are of indefinite shape and lack a cell wall. Very high disease incidence is sometimes recorded and immediate yield loss in potato crop in some ecological situations can be extreme. The disease tends not persist in stocks, only a few percentage of tubers giving rise to disease plants. Late infection do not affect yield. The disease is possibly more significant on its other hosts, i.e., tomatoes, etc. They are transmitted by leafhoppers and overwinters on weeds, grasses, and cereals but transmission is also possible via a parasitic plant, i.e., dodder. There are three common mycoplasmal diseases, i.e., purple top roll, marginal flavescence and witch broom noticed in potato grown areas.

10.1.4.1 *PURPLE TOP ROLL*

This disease is characterized by rolling and purple or pink coloration of the basal part of leaflets of the top leaves along with stunting, chlorosis, profuse auxiliary shoots with aerial tubers and swelling of nodes. There is no wilting of infected plants and no necrosis of the tubers takes place. Infected tubers seem to be the important source of perennation and transmission.

10.1.4.2 MARGINAL FLAVESCENCE

Symptoms of marginal flavescence are chlorosis on the margins of upper leaves. The chlorosis intensifies, leaf blade become thick, rough, and puckered. Growth is stunted because of short internodes with small leaves having narrow leaflets partly overlapping each other. Infected plants produce few small tubers close to stem, i.e., on short stolen. Tubers carry the pathogen and serve as source of primary inoculum.

10.1.4.3 WITCH BROOM

Symptoms of witch broom disease appear as extreme stunting of plant and numerous filamentous stems with simple leaves. Very small tubers are formed by affected plant. The pathogen is transmitted by leafhopper and perennates through tubers.

10.1.4.4 MANAGEMENT

The disease causes little problem in well managed crops. The disease can be effectively controlled by planting only disease free, healthy, and certified seeds. Potato tubers are main source of inoculum so use of certified as well as true potato seed (TPS) will be helpful in disease reduction. Hot water treatment of tubers at 50°C temperatures for 10–15 min has also been found effective in eradicating the pathogen.

10.1.5 NEMATODES

Damage to potato from nematodes and nematode mediated disease can have a substantial economic impact on potato production. Several nematode species can reduce yield and quality of commercial potatoes. Only few species inhibits in particular area where the majority of potato is grown. Two major nematodes of economic importance for potatoes in developing

Diseases of Potato Crops and Their Management

173

countries are cyst nematodes particularly in the highland tropics and root knot nematodes mainly in the lowland tropics. Nematode control is also made difficult by worldwide range of pathogenic variability and the fact that nematodes attacks many other plants and live several years in the soil. In many potatoes growing areas, as fallow periods have been eliminated and rotation have intensified and this problem has increased. So plant host resistance is potentially one of the most effective and economic means for controlling nematodes.

10.1.5.1 POTATO CYST NEMATODES (GLOBODERA PALLIDA AND GLOBODERA ROSTOCHIENENSIS)

Potato cyst nematode is one of the major problems throughout the world causing an average loss of about 9% (Khurana, 2000). Potato cyst nematode is popularly called as golden nematode in India. It has been confined to the Nilgiri and Kodaikanal (Tamil Nadu) due to internal quarantine. The plant appears as suffering from poor nutrition. At hotter part of the day, plants show temporary wilting. Typical symptoms are stunting of growth with unhealthy foliage.

Premature yellowing and the development of poor root system reduce the size and number of tuber. Main source of survival of nematode is the cyst. After harvest, the cyst can remain in soil. The larval hatch is initiated and larva move actively in soil and invade fresh roots form the giant cells to draw a food. Female increase in size and become spherical shape are seen sticking to the roots and later turn into cyst.

10.1.5.1.1 Management

Once the potato cyst nematode gets established in the soil of any locality, it cannot be completely eradicated and has to be managed by adopting several plant protection strategies. Field sanitation practices should be followed which helps to reduce the soil and tuber inoculum. Seed should not be taken from infested field. A crop rotation for 3–4 years keeps the nematode problem on low level. Potato, tomato, and brinjal should not be grown in infested field for 3–4 years. Growing non-host crops like radish, garlic, beetroot, green manure crop, etc., brings down the cyst population by more than 50%.

174   *Diseases of Fruits and Vegetable Crops*

Some practices include fallowing, removal of volunteer plants, etc., eliminate pathogen from field. Growing resistant varieties i.e., Kufri Swarna, K. Thenmalai, etc. lower down the cyst and propagule population to economically manageable level. Movement of water and soil from infested field should be avoided. Application of insecticide Carbofuran 3 G @ 2 kg ai/ha at the time of planting further brings down nematode population (Reddy, 2008).

10.1.5.2 ROOT KNOT NEMATODE (MELOIDOGYNAE SPP.)

This is not a very common problem on potato in India except in certain pockets during certain years. Several species, i.e., *Meloidogyne arenaria* (Neal) Chitwood; *M. hapla* Chitwood, *M. incognita* (Kofoid and White, Chitwood, *M. javanica* (Treub) Chitwood etc. are involved in causing root knots. Key morphological characteristics are described in detail by Eisenback and Triantaphyllou (1991) and Eisenback et al. (1981).

In general, aboveground symptoms include stunted, yellowed, chlorotic, and/or dead plants. Infected plants are likely to wilt earlier under temperature or moisture stress. Small galls or knots are formed on potato roots, but they often go under noticed. Reduction in size and number of tubers reduced the yield and warty pimple like outgrowths formed on tubers result in qualitative reduction. Root-knot nematode larvae invade roots or tubers, establish feeding sites, and develop into the adult stage. Adult female is swollen, sedentary, and lay eggs in a gelatinous matrix on or just below the root surface. These eggs hatch, and larvae invade other roots and tubers. Root-knot nematodes feeding reduce the vigor of the plant and causes blemishes on tubers.

10.1.5.2.1 Management

Some practices like use of certified planting material, cleaning soil from equipment before moving between fields, keeping irrigation water in a holding pond so that any nematodes present can settle out and pump water from near the surface of the pond, preventing /reducing animal movement from infested to uninfested field, avoid seed tubers from infested area, etc., will help to prevent the spread of nematode. Crop rotation can be very useful in reducing nematode population.

Field that are left fallow but kept weed free usually have an 80–90% per year reduction in root-knot population. Infested tubers left in the field after

Diseases of Potato Crops and Their Management 175

harvest can be a source of inoculum. Destroy potato plants that subsequently emerge from these tubers to restrict nematode reproduction. Late planting of autumn crop and early planting of spring crop in north western plains reduces nematode damage while in hills early planting of summer crop in 4th week of March is ideal. For heavy infestation, apply insecticide Carbofuran @ 3 kg ai/ha while for moderate infestation 2 kg ai/ha is sufficient. It should be applied in two doses first half at planting time and second half at earthening up time. Hot water treatment of potatoes tubers at 46–47.5°C for 120 minutes gave effective control of root-knot nematode infection (Reddy, 2008).

10.1.6   NON-PARASITIC DISEASE

A disease is an interaction between a host (potato), pathogen (fungi, bacterium, virus, nematode, mycoplasma), and environment that impairs productivity or usefulness of the crop. Adverse environmental effects are sufficient to initiate the disease in the absence of an infectious entity. Some plant diseases in which no foreign organism or parasite are associated with the cause of disease are categorized as non-parasitic diseases. Nonparasitic plant disorder is often called as a functional disorder or physiological disturbance. Blackheart of potato is one of the most important non-parasitic disease and responsible for potato degeneration under the storage condition.

10.1.6.1   BLACK HEART OF POTATO

The black heart of potato is important storage, transit, and market disease of potatoes as a result of poor oxygen supply and usually occurs in tubers stored in poorly ventilated rooms with closely packed conditions. It can occur in the field also. High soil temperatures and waterlogged soils contribute to blackheart development in the field. Dark grey to purplish or inky black discoloration occurs in the central tissues of the tuber. The discoloration occurs in an irregular pattern, usually with distinct line between healthy and affected tissue.

In advanced stages, the affected tissues may dry out and separate thus forming cavities. They are surrounded by disordered tissue and are referred to as cat's eye (Hiller and Thornton, 1993). Tubers will not recover once disease has occurred. Potato stored in poorly ventilated room results discoloration and disintegration of cells due to adverse enzymatic action continues after the supply of oxygen has diminished. Some potato varieties tolerate conditions of low oxygen better than others but all are susceptible to injury and subsequent development of black heart (Hooker, 1981).

10.1.6.1.1 Management

To avoid this disease, the tubers should not be stored at temperature above 33°C. Storage of tubers should be made in well-ventilated rooms. The bag of tubers should not be piled very high upon each other. Field should be properly irrigated. Avoid planting potatoes in fields that are poorly drained.

10.1.6.2 HOLLOW HEART

It is characterized by an internal split or cavity resulting from rapid growth induced by an abundance of moisture and food supply. This problem may occur in the field as well as during transit and storage. This is characterized by development of cavities of various sizes in the center of tubers. It is caused by unfavorable oxygen relation. It may occur in the field when soil temperature rises above 32.2C during growth and maturity of tuber (Bhat, 2016).

10.1.6.2.1 Management

Tubers should not be stored or transported at temperature above 32.2°C. Avoid excessive nitrogen fertilizer in the field.

10.1.6.3 GREENING

This disorder occurs when the tubers are exposed to light, either in the field or in storage resulting in the formation of chlorophyll in the peridermal layers of tubers exposed to light leads to greening which markedly reduces the products acceptability (Bhat, 2016).

10.1.6.3.1 Management

The plants should be properly hilled to prevent exposure of tubers to light. Avoid exposure of tubers to direct sunlight.

Diseases of Potato Crops and Their Management

Although a wealth of information has been generated on potato diseases and their eco-friendly management in India, even after lot more needs to be done for its effective management. In today's agricultural economy, it is really critical than ever that potato growers produce healthy and high yielding crops. In a vegetative propagated crop like potato, diseases are of significant importance and it is hard to get rid of once crop is infected by the biotic and abiotic stresses. They pose problems in areas where recommended package of practices are not followed. Indiscriminate use of chemical to manage these biotic and abiotic stresses attaches several concern regarding environment degradation, emergence of new pest, food safety, and increased cost of production. Clean and healthy seeds from clean field or pathogen free soil, prevention of entry and infection by pathogen in a standing crop and precaution during harvesting and storage of the produce, etc., are the basic requirement for management of potato diseases.

An ideal schedule for controlling diseases is to integrate measures covering all the disease management requirements. Now new concept is emerging, i.e., multiple disease management. There is need to work in this direction too. Research on disease management revolves around host resistance also. Different breeding program for potato cultivars with field resistance through biotechnology will be last longer. Plant pathologists have now learnt to use the modern tools of genetic engineering and tissue culture in the management of plant disease. Over-dependence on fungicides cannot be substituted for long. Despite of all efforts in developing resistant varieties, use of chemical would remain the necessary evil. However, their efficacy can be enhanced by developing need-based application system.

KEYWORDS

- greening
- hollow heart
- non-parasitic disease
- potato virus X
- potato virus Y
- true potato seed

REFERENCES

Bhat, K. L., (2016). *Physiological Disorders of Vegetable Crops* (p. 258). Daya publishing house, New Delhi.

Bhattacharya, S. K., & Malhotra, V. P., (1979). Control of charcoal rot of potato through crop rotations. *J. Indian Potato Assoc.*, *6*, 199–204.

Bhattacharya, S. K., & Raj, S., (1978). Studies on powdery scab of potatoes. I. Factors affecting disease development. *J. Indian Potato Assoc.*, *5*, 1–6.

Bhattacharya, S. K., & Raj, S., (1981). Studies on powdery scab of potatoes. II. Spore germination and histo-pathological studies. *J. Indian Potato Assoc.*, *8*, 118–123.

Bhattacharya, S. K., Bahal, V. K., & Bist, B. S., (1977). Effect of crop rotation on potato black scruf incidence. *J. Indian Potato Assoc.*, *4*, 1–4.

Bhattacharya, S. K., Singh, B. P., Sharma, V. C., Bombawale, O. M., Arora, R. K., & Singh, P. H., (1990). Mode of survival and source of primary inoculum of late blight of potato. *Intern. J. Tropical Plant Dis.*, *8*, 78–88.

Chatterjee, A. K., & Vidaver, A. K., (1986). Genetics of pathogenicity factors: Application to phytopathogenic bacteria. *Adv. Plant Pathol.*, *4*, 1–218.

Chaudhary, R. G., (1983). Cause of potato seed decay in low hills of Arunachal Pradesh. *J. Indian Potato Assoc.*, *10*, 111–115.

CPRI, (1958–1960). *Annual Scientific Report, Central Potato Research Institute* (pp. 94–101). Shimla.

CPRI, (1981). *Annual Scientific Report, Central Potato Research Institute* (pp. 65–69, 74). Shimla.

Dutt, B. L., & Pushkarnath, (1960). Resistance of potato varieties to powdery scab. *Indian Potato J.*, *2*, 78–82.

Dutt, B. L., (1979). *Bacterial and Fungal Diseases of Potato* (pp. 1–17). ICAR, New Delhi.

Eisenback, J. D., & Triantaphyllou, H. H., (1991). In: Nickle, W. R., & Marcel, D., (eds.), *Root Knot Nematodes: Meloidogyne Species and Race* (pp. 191–274). In manual of agricultural nematology. Inc. New York.

Eisenback, J. D., Hirschmann, H., Sasser, J. N., & Triantaphyllou, A. C., (1981). *A Guide to the Four Most Common Species of Root Knot Nematodes (Meloidogyne Species) with a Pictorial Key*. A Coop. Publ. Depts. Plant Pathology and Genetics and U. S. Agency for International development, Raleigh, NC.

Hampson, M. C., (1996). A qualitative assessment of wind dispersal of resting spores of *Synchytriumendobioticum*, the causal agent of wart of potato. *Plant Dis.*, *80*, 779–782.

Harrison, J. G., Searle, R. J., & Williams, N. A., (1997). Powdery scab of potato: A review. *Plant Pathol.*, *46*, 1–25.

Hector, G. P., (1926). Appendix II. Annual report of the economic botanist to the govt. of Bengal for the year 1924–1925. *Ann/Dept. of Agric. Bengal, 1924–1925*, 5–9.

Hiller, L. K., & Thornton, R. E., (1993). Management of physiological disorders. In: Rowe, R. C., (ed.), *Potato Health Management* (pp. 87–94). APS Press. St Paul, MN.

Hooker, W. J., (1981). Compendium of potato diseases. *American Phytopathological Soc* (p. 125). St. Paul, MN.

Indian Horticulture Database, (2014). *Crop Wise Area and Production Estimates for Horticultural Crops*. Ministry of Agriculture, Government of India, Gurgaon: 4–5.

Khurana, S. M. P., & Garg, I. D., (1998). Present status of controlling mechanically and non-persistently aphid transmitted potato viruses. In: Hadidi, A., et al., (eds.), *Plant Virus Disease Control* (pp. 593–609).

Khurana, S. M. P., & Singh, M. N., (1986). Viral and mycoplasmal diseases of potato. *Rev. Trop. Plant Pathol., 33*, 123–184.

Khurana, S. M. P., & Singh, M. N., (1988). Yield loss potential of potato virus X and potato virus Y in Indian potatoes. *J. Indian Potato Assoc., 15*, 27–29.

Khurana, S. M. P., (2000). *Diseases and Pests of Potato- A Manual* (pp. 19–21, 45–47). CPRI, Shimla.

Khurana, S. M. P., Garg, I. D., Singh, B. P., & Gadewar, A. V., (1999). Important diseases of potato in India and their management. In: Upadhayay, R. K., Mukherjee, K. G., & Dubey O. P., (eds.), *IPM Systems in Agriculture: Cash Crops* (Vol. 6, pp. 263–298). Aditya book Pvt. Ltd., New Delhi.

Khurana, S. M. P., Pandey, S. K., Singh, R. B., & Bhale, U. M., (1997). Spread and control of the potato stem necrosis. *Indian J. Virol., 13*, 23–28.

Khurana, S. M. P., Phadtare, S. G., Garg, I. D., Singh, M. N., & Bharadwaj, V. P., (1989). Potato stem necrosis epidemic due to tomato spotted wilt virus in India. *Proc. IV Intl. Plant Virus Epidemiology Workshop* (Vol. 10, p. 30). Montpellier, France.

Larkin, R. P., & Griffin, T. S., (2007). Control of soil borne potato diseases using Brassica green manure. *Crop Protec., 26*(7), 1067–1077.

Mandahar, C. L., Khurana, S. M. P., & Garg, I. D., (1990). Disease management. In: Mandahar, C. L., (ed.), '*Plant Viruses Pathology*' (Vol. II, pp. 273–293). CRC Press, Boca Raton, F1. (USA).

Mann, H. H., & Nagpurkar, S. D., (1922). Further investigations of the *Fusarium* blight of potatoes in western India. *Agric. J. India,* 567–576.

Paharia, K. D., & Sahai, D., (1968). Nutritional requirements of potato-isolate of *Macrophomin aphaseoli* from potato. *Indian J. Microbiol., 10*, 107–110.

Paharia, K. D., (1960). Charcoal rot of potato in India. *Indian Potato J., 2*, 1–11.

Paharia, K. D., Pushkarnath, & Deshmukh, M. J., (1962). Resistance of potato varieties to charcoal rot. *J. Indian Potato Assoc., 4*, 84–87.

Patel, D. B., Patel, N. A., & Modi, V. M., (2010). Influence of different dates of potato planting on stem necrosis disease. *International J. Plant Protec., 3*, 404–405.

Perambelon, M. C. M., (2002). Potato diseases caused by soft rot *Erwinia*: An overview of pathogenesis. *Plant Pathol., 51*(1), 1–12.

Phadtare, S. G., (1978). Pink rot of potato- a new report from India. *J. Indian Potato Assoc., 5*, 174–175.

Prasad, B., & Agrawal, H. O., (1983). New foliar diseases due to *Rhizoctonia* and *Chaetomium* spp. *J. Indian Potato Assoc., 10*, 116–120.

Pushkarnath, (1976). *Potato in Sub-Tropics* (p. 289). Orient Longman, New Delhi.

Rai, R. P., & Singh, B. P., (1981). A new disease of potato incited by *Fusarium acuminatum* Ell. and Ev. *Current Sci., 50*, 1037–1038.

Rai, R. P., (1979). Pink rot of potato in Shimla hills. *J. Indian Potato Assoc., 6*, 36–40.

Reddy, P. P., (2008). *Diseases of Horticultural Crops: Nematode Problem and Their Management* (p. 379). Scientific Publishers, Jodhpur, India.

Reddy, P. P., (2010). Bacterial and viral diseases and their management. In: *Plant Protection in Horticulture* (Vol. 3, p. 288). Scientific Publishers, Jodhpur, India.

Reddy, P. P., (2010). *Plant Protection in Horticulture: Fungal Diseases and Their Management* (Vol. 2, p. 359). Scientific Publishers, Jodhpur, India.

Secor, G. A., & Gudmestad, N. C., (1999). Managing fungal diseases of potato. *Canadian J. of Plant Pathology, 21*, 213–221.

Shekhawat, G. S., (2000). Management of potato diseases through host resistance. *J. Mycol. Pl. Pathol., 30*, 143–150.

Shekhawat, G. S., Kishore, V., Sunaina, V., Bahal, V. K., Gadewar, A. V., Verma, R. K., & Chakrabarti, S. K., (1988b). *Latency and Management of Pseudomonas Solanacearum* (pp. 117–121). Annual Scientific Report, Central Potato Research Institute, Shimla.

Sikka, L. C., Srivastava, S. N. S., Singh, A. K., & Bharadwaj, V. P., (1971). Integrated approach to control *Rhizoctoniasolani* on potato. *Indian Phytopath, 24*, 54–57.

Singh, B. P., (1986). Studies on *Fusarium* wilt and dry rot of potatoes (*S. tuberosum*). *PhD Thesis* (p. 145). AMU, Aligarh.

Singh, B. P., Arora, R. K., & Khurana, S. M. P., (2002). *Soil and Tuber Borne Diseases of Potato* (p. 74). Tech Bull. No 41(revised), C. P. R. I. Shimla.

Singh, B. P., Bhattacharya, S. K., Saxena, S. K., & Nagaich, B. B., (1990c). Managing *Fusarium* wilt of potato by adjusting date of planting. *J. Indian Potato Assoc., 17*, 75–77.

Singh, B. P., Nagaich, B. B., & Saxena, S. K., (1988). Studies on the effect of organic amendments on *Fusarium* wilt of potato. *J. Indian Potato Assoc., 15*, 60–67.

Singh, P. H., & Shekhawat, G. S., (2000). Wart disease of potato in Darjeeling hills. CPRI, Shimla. *Tech Bull. No., 19*(Revised), p. 73.

Sivamani, E., Anuratha, C. S., & Ganamnickam, S. S., (1987). Toxicity of *Pseudomonas fluorescens* towards bacterial plant pathogens of banana (*Pseudomonas solanacearum*) and rice (*Xanthomonas campestris*pv. *oryzae*). *Current Sci., 56*, 547–548.

Smith, E. F., (1914). Bacteria in relation to plant diseases. *Carnegie Inst. Wash., 3*, 309.

Srivastava, S. N. S., (1965). The occurrence of silver scurf of potato in India. *Sci. and Cult., 31*, 537.

Thirumalachar, M. J., (1953). Pycnidial stage of charcoal rot inciting fungus with a discussion on its nomenclature. *Phytopathology, 43*, 608–610.

Van Everdingen, E., (1926). Het. verbandtusschen de weergesteldhieden de aarolppelziekte, *Phytopthorainfestans* (the relation between weather conditions and potato blight, *Phytophthora infestans*) Tijdschr. *Plantenziekten., 32*, 129–140.

Yabuuchi, E., Kosako, Y., Oyaizu, H., Yano, I., Hotta, H., Hashimoto, Y., Ezaki, T., & Arakawa, M., (1992). Proposal of *Burkholderia* gen. nov. and transfer of seven species of the genus *Pseudomonas* homology group II to the new genus, with the type species *Burkholderiacepacia* (Paaleroni and Holmes, 1981 comb. Nov. *Microbiol. Immunol., 36*, 1251–1275.

Yabuuchi, E., Kosako, Y., Yano, I., Hotta, H., & Nishiuchi, E., (1995). Transfer of two *Burkholderia* and an *Alcoligenes* species to *Ralstonia* gen. nov. Proposal of *Ralstoniapickettii* (Ralston, Palleroni and Doudoroff, 1973) comb. Nov. and *Ralstoniaeutrapha* (Davis, 1969) comb. *Nov. Microbiol. Immunol., 39*, 897–907.

CHAPTER 11

Diseases of Tomato Crops and Their Management

PANKAJ RAUTELA, SUPRIYA GUPTA, C. S. AZAD, and R. P. SINGH

Department of Plant Pathology, College of Agriculture G.B. Pant University of Agriculture and Technology, Pantnagar, Uttarakhand, India

*Corresponding author. E-mail: azadbau81@gmail.com

11.1 INTRODUCTION

India is the second-largest producer of fruits and vegetables in the world. Area and production of vegetables in India amount to 9,396 thousand hectares and 162,897 thousand metric tonnes, respectively, with the productivity of 17.3 mt/ha. Out of which tomato is grown over 882thousand hectare area with a production of 18,735.9 thousand metric tonnes and productivity of 21.2 mt/ha.

Production in India is limited by losses caused by insect pests and diseases to the extent that the per capita consumption of vegetables is only 25–33% of the minimum daily requirement. The major obstacle in the successful harvesting of tomato is disease incidences. In the absence of adequate scientific know-how, the quality production of tomato fruits may not be up to the desired level. Diseases can be classified into two groups. The first are those caused by infectious microorganisms that include fungi, bacteria, viruses, and nematodes. The second group includes those caused by non-infectious physical or chemical factors, such as adverse environmental factors, nutritional or physiological disorders, and herbicide injury. The major diseases of tomato and their management are described in the following sections.

11.1.2 FUNGAL AND FUNGAL-LIKE DISEASES

11.1.1.1 LATE BLIGHT OF TOMATO

Late blight is an extremely destructive disease of tomato. It can infect and destroy the tomato plants at any stage of development and is capable of rapid spread resulting in the complete destruction of these crops.

11.1.1.1.1 Symptoms

The disease appears on tomato leaves as pale green, water-soaked spots, often beginning at leaf tips or edges and spread downward and inward depending upon the weather. The circular or irregular purplish leaf lesions are often surrounded by a pale green border that merges with healthy tissues. Lesions enlarge rapidly and turn dark brown to purplish-black (Figure 11.1). During periods of high humidity and leaf wetness, a cottony, whitish or grayish mold growth is usually visible on lower leaf surfaces at the edges of lesions which disappear with dry weather. When moist weather persists, infected areas on stem appear brown to black, and entire foliage may be killed in a short time (Figure 11.2).

FIGURE 11.1

Late blight can also develop on tomato fruit, resulting in large, firm, brown, leathery-appearing lesions, often concentrated on the sides or upper

Diseases of Tomato Crops and Their Management

fruit surfaces (Figure 11.3). If conditions remain moist, abundant white mold growth will develop on the lesions, and secondary soft-rot bacteria may follow, resulting in a slimy, wet rot of the entire fruit.

FIGURE 11.2

FIGURE 11.3

11.1.1.1.2 Cause and Conditions for Development

Late blight is caused by the fungus *Phytophthora infestans*. The fungus is basically a cool climate pathogen but has a tremendous capacity to adapt to the environment, thus becoming widespread in almost all environments suitable for tomato production. The spores of the pathogen are carried by the wind, splashed rain, and animals from diseased plants in one field to healthy plants in neighboring fields.

Disease development is favored by cool, moist weather. Night temperature below dew point for at least 4 h, minimum temperature of 10°C, accompanied by rain and fog or heavy dew in the next 24 hrs are ideal for forecasting of late blight occurrence in severe form. Lesions may appear on leaves within 3–5 days of infection during the growing season may partially decay before harvest.

11.1.1.1.3 Management

1. Use disease-free seed or certified seed for planting.
2. Apply sanitation practices.
3. Apply fungicides as needed throughout the growing season. Several forecast techniques have been developed that predict when a spray will be necessary based on environmental conditions. If RH >85% for more than 85 hrs, temperature 7.2–26.6°C for more than 115 hrs, the disease will appear within seven days. Apply protective spray of mancozeb @ 2.5 kg/ha or copper oxychloride @ 3 kg/ha at 7 day interval.
4. After disease appearance spray cymoxanil 8%+ mancozeb 64% (curzate) @ 2–2.5 kg/ha at 7–10 days interval. Never repeat the same chemical. It should be used alternatively.

Use resistant varieties. Pant Bahar and PusaRubi are resistant varieties of tomato.

11.1.1.2 EARLY BLIGHT OF TOMATO

Early blight is a very common disease of tomato. The disease can occur over a wide range of climatic conditions and can be very destructive if left uncontrolled, often resulting in complete defoliation of plants. The disease is equally serious on the hills as well as in the plains.

11.1.1.2.1 Symptoms

The first symptoms usually appear on older leaves and consist of small, scattered circular or irregular, pale brown to black, dead spots. As the spots enlarge, concentric rings may form, which gives the lesion a characteristic "target-board" appearance. Often there is a narrow, yellow halo around each spot (Figure 11.4). These spots may grow together, causing infected leaves to turn yellow and die. These spots coalesce together to form larger pustules. In severe form, the leaf becomes dry and defoliated.

FIGURE 11.4

On tomato, stem infections can occur at any age resulting in small, dark, slightly sunken areas that enlarge to form circular or elongated spots with lighter-colored centers. Concentric marks, similar to those on leaves, often develop on stem lesions (Figure 11.5). If infested seeds are used, seedlings may damp off soon after emergence. When large lesions develop at the ground line on seedlings, the plants may become girdled known as "collar rot," and plants may die. Some plants may survive with reduced root systems and usually, produce few or no fruits.

Blossom drop and spotting of fruit, along with loss of young fruit, may occur when early blight attacks tomatoes at the flowering stage. On fruits, early blight causes dark, leathery sunken spots, usually at the point of stem attachment (Figure 11.6). Affected areas often show concentric markings like those on leaves and covered with velvety black masses of spores (Figure 11.7). Fruits can also be infected in the green or ripe stage through growth cracks and other wounds. Infected fruits often drop before they reach maturity.

FIGURE 11.5

FIGURE 11.6

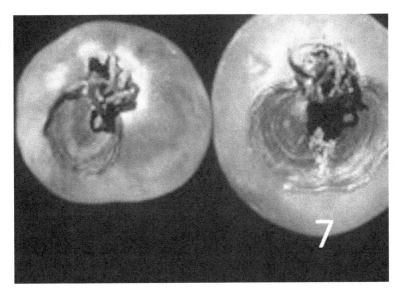

FIGURE 11.7

11.1.1.2.2 Cause and Condition for Development

Early blight is caused by the fungus, *Alternaria solani*, which survives in dry infected leaf or stem tissues on or in the soil for a year or more. It can also be carried on tomato seed. Spores form on infected plant debris at the soil surface or on active lesions in a wide range of temperatures, especially under alternating wet and dry conditions. They are easily carried by air currents, windblown soil, splashing rain, irrigation water, and insects. Infection of a susceptible leaf or stem tissues occurs in warm, humid weather with heavy dews or rain followed by warm and dry weather. Early blight can develop quite rapidly in mid to late season and is more severe when plants are stressed by poor nutrition, drought, or other pests.

11.1.1.2.3 Management

1. Practice 3 years of crop rotation with non-susceptible crops to allow infested plant debris to decompose in the soil.
2. Remove and destroy crop residue at the end of the season.
3. Cultivation of solanaceous crops in nearby fields must be avoided.
4. Use disease-free certified seed.

5. Maintain fertility at optimal levels-nitrogen and phosphorus deficiency can increase susceptibility to early blight. Spray urea @ 1% at 45 days and another after 10 days.
6. Irrigate early in the day to promote rapid drying of foliage.
7. Seed treatment with thiram @ 2.5 g/kg seed before sowing.
8. Spray mancozeb @ 0.25%, copper oxychloride @ 0.3% or chlorothalonil @ 0.2% starting from appearance of disease, 5–6 spray at 10–15 day intervals should be given.
9. Use resistant varieties. DARL-30, ArkaRakshak, and ArkaSamrat.

11.1.1.3 FUNGAL WILT

Wilt pathogens infect the vascular system of the plant. The pathogen or the toxins produced by the pathogen can be moved throughout the entire plant along with the plant sap. A pathogen invading the roots may cause disease symptoms at the top of the plant.

11.1.1.3.1 Symptoms

All vascular wilts, regardless of the pathogen, have some symptoms in common. The leaves or other parts of the infected plants lose turgidity, turn lighter green to greenish-yellow, droop, and wilt, turn yellow then brown, and die. In cross-sections of infected stems, discolored brown areas appear as a complete or interrupted ring of discolored vascular tissues.

11.1.1.3.2 Fusarium Wilt

Fusarium wilts can affect many vegetables that include tomato, potato, pepper, eggplant, crucifers (radish, cauliflower, and cabbage), cucurbits (cucumber, squash, and melon), beans, and peas.

In tomato, symptoms begin as slight vein clearing on outer leaflets and drooping of leaf petioles. Later the lower leaves wilt, turn yellow and die and the entire plant may be killed, often before the maturity of the plant. In many cases, single shoot wilt before the rest of the plant shows symptoms, or one side of the plant is affected first (Figure 11.8a, b). If the main stem is cut, dark, chocolate-brown streaks may be seen running lengthwise through the stem (Figure 11.9). This discoloration extends upward for some distance and is especially evident at the point where the petiole joins the stem.

Diseases of Tomato Crops and Their Management

FIGURE 11.8

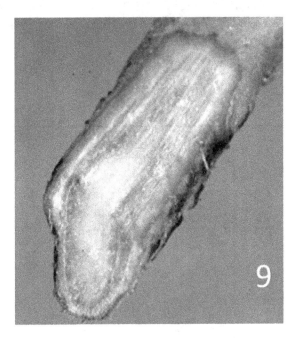

FIGURE 11.9

11.1.1.3.3 Verticillium Wilt

In the case of Verticillium wilt, older leaves are usually the first to develop symptoms, which include yellowing, wilting, and eventually dying and dropping from the plant (Figure 11.10). First, the bottom leaves become pale, then tips and edges die, and finally leaves die and drop off. V-shaped lesions at leaf tips are typical of Verticillium wilt of tomato (Figure 11.11).

FIGURE 11.10

Figure 11.11.

Infected plants usually survive the season but are somewhat stunted, and both yields and fruits may be small depending on the severity of the attack.

A light tan discoloration in the stem, usually confined to lower plant parts may be found (Figure 11.12). The discoloration is typically lighter in color than with Fusarium wilt. Sometimes symptoms on only one side of the plant are seen.

FIGURE 11.12

11.1.1.3.4 Cause and Conditions for Development

Fusarium wilt in tomato is caused by *F. oxysporum f. sp. lycopersici*. Verticillium wilt is caused by the fungi *Verticillium albo-atrum* and *V. dahliae*. They are soil-borne and can persist for many years. *V. albo-atrum* is a cool weather organism, although the disease is retarded by the higher temperatures that favor fusarium wilt. Visible symptoms may appear to be more severe when high temperatures exist, due to restricted water movement in the plant brought about by damage done to the water-conducting vessels earlier in the growing season.

11.1.1.3.5 Management

1. Practice long crop rotation (4 to 6 years) with cereals and grasses.
2. Apply field sanitation. Remove infected plant material after harvest.

192 *Diseases of Fruits and Vegetable Crops*

3. Maintain a high level of plant vigor with appropriate fertilization.
4. Use biological seed treatment and soil application of bio-control agents.
5. Seed treatment with carbendazim @ 1g/kg seed.
6. Soil drenching with 0.05% carbendazim suspension.
7. Use resistance varieties. Roza, Columbia, Roma, HS-110, Sel-28, and Ace varieties of tomato are resistant against fusarium wilt.

11.1.1.4   DAMPING-OFF OF VEGETABLES

Damping-off, a disease of germinating seeds and seedlings, generally refers to sudden plant death in the seedling stage. It is fatal to young seedlings and becomes colonized resulting in plant losses and delayed planting.

11.1.1.4.1   Symptoms

Damping-off fungi may attack the seed prior to germination, or they may attack after the seed has germinated, but before the seedling has emerged above the soil line. In pre-emergence damping-off, the infected seed becomes soft and mushy, turning a brown to black color, and it eventually disintegrates.

Post-emergence damping-off affects seedlings that have already emerged from the soil. These seedlings may develop a dark stem rot near the soil surface, which will cause them to be discolored and begins to shrink. Supportive strength of the invaded portion is lost, and the seedling topples over. In older seedlings, the new developing rootlets are infected, resulting in root rot and infected plants show symptoms of wilting and poor growth (Figure 11.13). The plant lives for a while; it is stunted and pale, and will eventually die. White fungal growth may be seen on the surface of the seedlings. Damping-off usually occurs in small patches at various places in the seedbed or field (Figure 11.14). The disease spots often increase from day to day until the seedlings harden. Seedlings are extremely susceptible to about two weeks after emergence. As the stem hardens and increases in size, the injury no longer occurs. Some seedlings are not killed at once, but the roots are severely damaged, and the stem is girdled at the ground level. Such plants remain stunted and often do not survive to transplant.

Diseases of Tomato Crops and Their Management 193

FIGURE 11.13

FIGURE 11.14

11.1.1.4.2 Causes and Conditions for Development

Pre-emergence damping-off, which consists of decay of the germinating seed or death of the seedling, is caused by *Pythium* spp. and *Phytophthora* spp.

while *Rhizoctonia* spp. is mostly responsible for post-emergence damping-off, which occurs after the seedlings have emerged from the soil but while small and tender. These fungi are soil-borne, and exudates derived from host plants stimulate the growth of these fungi. Excess soil moisture accelerates the disease development. Infection is favored by heavy soils, poor aeration, low pH, heavy seedlings resulting in dense planting, low light, and the presence of weeds. The fungi survive for long periods in soil and in host residue in the form of oospores and sporangia. The spread of damping-off fungi depends primarily on the mechanical transfer of mycelia, sclerotia, or resting spores in infested soil particles or infected plant tissue.

11.1.1.4.3 Management

1. Deep plowing and seedbed management should be used.
2. Solarization of soil in nursery beds in warm weather for 4–6 weeks before the sowing of seeds is beneficial.
3. Thinning of seedlings in seedbeds to permit good air circulation.
4. Adopt sanitary practices in the nursery and follow the rotation of nursery bed sites every year.
5. Light and frequent irrigation and stagnation of water should be avoided.
6. Treat the seed with captan or thiram @ 2.5g/Kg or carbendazim @ 1 g/Kg of seed, before sowing or treat the seed with bioagents pant bioagent-1 or pant bioagent-2 @ 10g/Kg at least 24 hrs before sowing the seeds.
7. Drench the nursery bed with mancozeb (0.25%) or carbendazim (0.05%) or suspension of bioagents (1.0%) when damping-off appears.

11.1.1.5 POWDERY MILDEW

Powdery mildew can cause a 10–90% loss in tomatoes. The extent of loss depends on the environmental conditions, date of disease onset, and effectiveness of disease control.

11.1.1.5.1 Symptoms

Powdery mildews are characterized by spots or patches of white to grayish powdery growth. White, powdery fungal growth develops on leaf surfaces,

Diseases of Tomato Crops and Their Management

petioles, and stems. It usually develops first on crown leaves, on shaded lower leaves, and on leaf undersurfaces. Yellow spots may form on upper leaf surfaces opposite powdery mildew colonies. Older plants are affected first. Infected leaves usually wither and die. The whole stem of the infected plant is blackened in a severe attack. Severely infected plants may reduce yield, shortened production times, and fruit that has little flavor.

11.1.1.5.2   Cause and Conditions for Development

Powdery mildew is caused by two different fungi, one is caused by *Leveilullataurica* (anamorph: *Oidiopsis taurica*) which grows endophytically and the other is caused by *Oidium lycopersici* which produce epiphytic conidiophore bearing mycelium that grows superficially on host surface powdery mildew fungi are host specific, obligate parasites and cannot survive in the absence of living host plants, except as cleistothecia. They survive the winter as dormant mycelium on perennial plants or as spores in thick-walled fruiting structures.

Powdery mildew develops quickly under favorable conditions because the length of time between infection and the appearance of symptoms is usually only 3–7 days. Powdery mildew fungi thrive under conditions of high relative humidity (RH), warm temperatures, low light, high fertility, and succulent plant growth. Free moisture on leaf surfaces inhibits infection by these pathogens. The minimum and maximum temperatures for the conidial formation and host penetration are 10°C and 32°C, respectively, and the optimum being 26–28°C. Spores and fungal growth are sensitive to extreme heat and direct sunlight. Mature foliage is most readily infected; very young leaves are nearly immune.

11.1.1.5.3   Management

1. Adopt field sanitation practices like cleaning and burning of diseased crop debris.
2. Use cultural practices that avoid excessive succulence, overcrowding, shading, overwatering, or excess fertilization, especially with nitrogen.
3. Spray dinocap @ 0.05% or wettable sulfur @ 0.2% or calixin @ 0.1%. Sulfur dust may be applied @ 20–25 Kg/ha but not when the temperature is above 30°C.
4. Plant resistant/ tolerant cultivars.

11.1.1.6 SEPTORIA LEAF SPOT OF VEGETABLES

Septoria leaf spot is one of the most common and destructive diseases of tomato. The disease causes rapid defoliation in warm and moist weather. It is of worldwide occurrence, its severity and extent of damage depend on crop and environmental conditions.

11.1.1.6.1 Symptoms

Septoria leaf spot is a disease of the foliage and stems. Although the symptoms may appear on the leaves and stems at any stage of plant development, they usually become evident after plants have begun to set fruit. Small circular spots are first observed as water-soaked areas on the undersurface of the lower leaves. As the spots enlarge, they develop dark brown margins and sunken, white grey centers. Yellow haloes often surround the spots. The number and size of the spots depend on the susceptibility of the host. In susceptible cultivars, the spots may be enlarged whereas, in less susceptible cultivars, there may be numerous spots of smaller size. The centers of the spots on the upper surface of the leaf show minute black fruiting bodies. Severely infected leaves die and drop off. The spots may also develop on floral calyx and on the stem but very rarely on the fruits. Severe leaf spotting and defoliation is common in severe infection. Favorable weather permits infection to move up the stem, causing a progressive loss of foliage from the bottom of the plant upward. Plants appear to wither from the bottom up. The loss of foliage causes a decrease in the size of the fruits and exposes fruit to sunscald. Spotting of the stem and blossoms may also occur.

11.1.1.6.2 Cause and Condition for Development

Septoria lycopersici is responsible for causing leaf spot in tomato. The fungus lives in the soil on infected debris of tomato and weeds. Spores formed on crop debris splash onto foliage and start the disease. Lower, weaker leaves catch an infection from the soil. Wind and rain spread spores produced in the dark bodies formed on leaf spots to adjacent uninfected leaves. Seeds and transplants may also carry the fungus.

Free water or dew on the leaf surface is essential for infection. The disease is favored by moderate temperatures and abundant rainfall. Two to three day's cloudy weather or drizzle provides favorable conditions. The

Diseases of Tomato Crops and Their Management 197

optimum temperature for growth of the fungus and disease development is 25°C with a minimum of 15 and a maximum of 27°C.

11.1.1.6.3 Management

1. Seed has been implicated as a source; therefore, the use of disease-free seed is necessary.
2. Rotate tomatoes with cereals, corn, or legumes for four years where the disease has been severe.
3. Apply proper sanitation practices like deep plowing to bury all plant refuse and control weeds. Fallen leaves should be removed from the field or burnt on the spot.
4. Treat the seeds in hot water for 25 minutes at 50°C or thiram @ 2.5 g/kg seed.

Spray mancozeb (0.25%) or chlorothalonil (0.2%) 3–4 sprays at 10–15 days interval.

11.1.2 BACTERIAL DISEASES

11.1.2.1 BACTERIAL WILT OF TOMATOES

Bacterial wilt/brown rot is one of the most destructive diseases of tomatoes, causing extensive damage to the crop.

11.1.2.1.1 Symptoms

Characteristic symptoms of bacterial wilt are rapid and complete wilting of normally grown-up plants. Lower leaves may drop before wilting. The disease starts with stunting, yellowing of the foliage, wilting, and finally, the entire plant collapse. Pathogen is mostly confined to the vascular region; in advance cases, it may invade the cortex and pith and cause yellow-brown discoloration of tissues. Vascular bundles of the stem seen to be brown, and this browning is often visible from the surface of the stems of infected plants as dark patches or streaks. Infected plant parts when cut and immerse in clear water, a white streak of bacterial ooze is seen coming out from cut ends. The development of adventitious roots from the stem is enhanced. In humid weather and high-temperature sudden drooping of the leaves without yellowing and rotting of the stem occurs in tomato. The roots appear healthy and are well developed; however, brown discoloration is present inside.

11.1.2.1.2 Cause and Conditions for Development

Bacterial brown rot or wilt in most of the vegetables is caused by *Ralstonia solanacearum* (formerly *Pseudomonas solanacearum* and, more recently, *Burkholderia solanacearum*). *R. Solanacearum* can survive through the soil, seed tubers, and cultivated hosts. Soil is considered the most important source of primary inoculum. The optimum temperature for growth of the bacterium is 35–37°C, with a maximum of 41 and a minimum of 10°C. Soil moisture, more than 50%, favors disease development. Organic manures promote the activity of the bacterium, while inorganic fertilizers decrease its activity.

11.1.2.1.3 Management

1. Follow crop rotations for three years with wheat, maize, oat, barley, sun hemp, finger millet, and vegetables like cabbage, onion, and garlic.
2. The field should be plowed to expose the soil to summer heat in the month of May to June.
3. Uproot the diseased plants as soon as the disease appears.
4. Treat the seedlings of tomato with 0.02% streptocycline for 30 minutes.
5. Drench the soil with streptocycline @ 0.01% or copper oxychloride @ 0.3% at early stage of plant showing symptoms.

Use resistant varieties like Shakti, ArkaRakshak, and ArkaSamrat of tomato.

11.1.2.2 BACTERIAL SPOT

It is a serious problem in tropical and subtropical regions of high humidity and rainfall.

11.1.2.2.1 Symptoms

On leaves, small, irregular areas with a grassy appearance develop first. These areas dry out and form slightly raised dry spots that are grayish brown,

Diseases of Tomato Crops and Their Management   199

particularly at the center. The bacteria often ooze from these spots and, when dry, form a glistening, cream-colored film on and around the lesions. Where infection is severe, the dried-out parts may combine and kill large areas of the leaf. Marginal and tip burns on leaves have often been noted. On flowers, water-soaked dark-brown-to-black areas develop, which later dry out and turn grey. Flower infection often causes blossoms and young fruit to wither and fall. Main stems are occasionally attacked. Spots are often elongated but retain the same grayish, scab-like characteristics as other affected parts. Small, water-soaked areas develop on green fruit. These dry out and form slightly raised and wrinkled, brown-to-grey, scab-like bodies, making the fruit unmarketable and susceptible to secondary rots.

11.1.2.2.2 Causes and Conditions for Disease Development

Xanthomonas campestris pv. *vesicatoria,* causes bacterial spot in tomato, the bacterium is strictly aerobic, gram-negative, rod-shaped and possess a single polar flagellum. It is seed-borne and overwinters on the surface of seeds, in infected debris, in reservoir hosts, and in soil. The disease is favored by warm (24–30°C) humid conditions

The bacterium is disseminated within the field by water droplets, clipping of transplants, and aerosols; it penetrates through stomata or enters through insect puncture or mechanical injury. The bacteria, carried in water droplets, can form new spots on the leaves, stems, or fruit where the droplet comes to rest. In wet weather, particularly if strong winds are blowing, the disease may spread rapidly through a crop from a few affected plants. Once the bacterium is established in the soil, it may persist for 2 or 3 years.

11.1.2.2.3 Management

1. Use disease-free seed or transplants seed that has been hot water treated.
2. At least 3-year crop rotation with nonhost crops should be followed.
3. Spray plants with streptomycin before transplanting. After transplanting, apply a mixture of mancozeb plus copper before the occurrence of disease. Protection is most needed during early flowering and fruit setting periods.

11.1.2.3 BACTERIAL SPECK

The disease is of worldwide occurrence and is a serious problem of tomato cultivation and can cause complete loss of production when the infection is in transplants while loss under mild infection can reach up to 12%.

11.1.2.3.1 Symptoms

Tiny, dark spots develop on affected leaves, extend through the tissue, and if numerous, cause yellowing of the surrounding leaf areas. Stem signs occur as brown or black lesions. Affected fruit has many small, black, slightly raised spots less than 2 mm in diameter. The spots do not extend deeply into the fruit.

11.1.2.3.2 Cause and Conditions for Disease Development

The disease is caused by *Pseudomonas syringae* pv. *tomato,* a gram-negative rod-shaped bacterium. Infected seed, seedlings, and contaminated soil are primary sources of this pathogen. The bacterium is seed-borne, and seedlings are very susceptible. The disease is favored by wet conditions and temperatures of 18–24°C. The disease spreads rapidly in seedbeds, and many seedlings can be infected by the time they are planted out. The bacterium can be disseminated by irrigation water, rain splashes, and contaminated tools and implements. The bacterium overwinters in seed and in plant debris for 30 weeks and in the soil for 30 days.

11.1.2.3.3 Management

1. Disease-free certified seed should be used.
2. Proper sanitation practices like deep plowing should be followed, to bury all plant refuse and control weeds. Fallen leaves should be removed from the field or burnt.
3. Rotation with nonhost crops should be done.
4. Spray of copper fungicides should be done as a preventive spray or on the appearance of symptoms.

Diseases of Tomato Crops and Their Management   201

11.1.2.4   BACTERIAL CANKER

Bacterial canker is a destructive disease of tomato present throughout the tomato growing regions of the world and cause serious losses in tomato crop.

11.1.2.4.1   Symptoms

Infected seedlings may be killed, stunted or malformed, or may show no signs until they are transplanted. In larger plants, initial signs are scorched or 'firing' markings on leaflets and wilting of lower leaves. Symptoms on leaves may be unilateral, the leaflets on one side of the rachis being wilted while those on the opposite side appear healthy As the infection advances through the vascular system, wilting progress until the whole plant collapses. Parts of the pith may collapse and become hollow. Internal browning of vascular tissue and pith cavities can be seen by snapping off a leaf at a node (the point where the leaf joins the stem). In wet conditions, brown, raised cankers may form on the stems and fruit. The fruit cankers have pale halos and are called 'bird's-eye' spots.

11.1.2.4.2   Cause and Conditions for Development

Bacterial canker is caused by *Clavibacter michiganensis* subsp. *michiganensis.* The bacteria can live from season to season in or on the seed, diseased plant debris in soil, infected seed or seedlings is the primary source of this pathogen. Within the field, dissemination is from open cankers mainly through raindrop splashes. Seeds are infected through the systemic invasion of fruits.

11.1.2.4.3   Management

1. 2–3-year crop rotation with nonhost crop should be followed
2. Soil solarization for 6weeks in summer eliminates the bacterial population in soil
3. Use healthy and disease-free seed and transplants
4. Seed should be disinfected by submersion in water at 52°C for 20 minutes and with copper compounds. Treatment of tomato plants

with beta-amino butyric acid (BABA) offers protection against bacterial canker

11.1.3 VIRUS DISEASES OF TOMATO

11.1.3.1 TOMATO YELLOW LEAF CURL VIRUS (TYLCV)

11.1.3.1.1 Symptoms

All begomoviruses cause signs that include bright-yellow-to-chlorotic mosaic on leaves, usually with some leaf distortion or leaf curling. Plants that are infected with either TYLCV or ToLCV at an early growth stage become severely stunted. Leaflets are reduced in size and misshapen. Emerging leaves are cupped downward. Leaves developing later are upright with yellowing between veins; their leaf margins roll upward. Plants infected when they are young lose vigor, flowers abort, and they stop producing marketable fruit. When infections occur in older plants, any fruit already presents ripen normally, but no further fruits are formed.

11.1.3.1.2 Cause and Conditions for Disease Development

Begomoviruses: tomato yellow leaf curl virus (TYLCV), tomato leaf curl virus (ToLCV). Old infected crops are a major source of new infections. Weeds such as the common thorn apple and nightshades can be infected by TYLCV. Whitefly *Bemisiatabaci* transmits the virus to healthy plants in a persistent manner. The spread of disease to new areas and the host is based on the introduction and movement of the vector.

11.1.3.1.3 Management

1. Use healthy, bold seeds and virus free seedlings and plant early in the season.
2. Uproot the infected plants as soon as they are noticed in the field.
3. Apply hot water treatment for 25 min at 50°C.
4. Use reflective mulches to deter whiteflies when plants are young.
5. Use resistant/tolerant varieties. Tomato varieties resistant to leaf curl are Hissar, Anmol, H-24, and H-36.

Diseases of Tomato Crops and Their Management 203

11.1.3.2 TOMATO SPOTTED WILT VIRUS (TSWV)

11.1.3.2.1 Symptoms

First symptoms appear as bronzing of upper leaf surface on young tomato plants, sometimes accompanied by small dark spots followed by down cupping a distortion of the leaflet. As the disease develops, the affected tissues blacken and shrivel until they shoot looks as though it has been scorched by flame, there is general stunting with growing stem tips generally dying back. Fruit set is poor in infected plants, young green fruit develop mottled spots of yellow to light green color with raised center, while on mature tomatoes characteristic orange to red rings appear resulting in irregular ripening of fruit.

11.1.3.2.2 Cause and Conditions for Disease Development

Tospoviruses: tomato spotted wilt virus (TSWV) is responsible for causing the disease. The virus is transmitted by several species of thrips in a persistent manner, the most important being *Thripstabaci* and *Frankliniella occidentalis*. The virus can only be acquired by larvae of insect feeding on infected hosts. TSWV onset and severity is directly connected to conditions that favor the early appearance, high activity and increase of vector populations.

11.1.3.2.3 Management

1. Use healthy, bold seeds and virus free seedlings and plant early in the season.
2. Uproot the infected plants as soon as they are noticed in the field.
3. Apply hot water treatment for 25 min at 50°C.
4. Use resistant/tolerant varieties.

11.1.3.3 TOBACCO MOSAIC VIRUS (TMV)

11.1.3.3.1 Symptoms

Symptoms appear on new growth 10–20 days after infection. As mottling, malformation, threading, curling, bleaching of the veins, and occasionally necrosis of the leaves that sometimes become fern-like or pointed. Mottling varies in severity, and severely mottled leaves may have a puckered appearance. Mottling and blemishing of fruits accompanied by reduced yield are common when infection occurs early in the season, leading to stunting and severe chlorosis of plants.

11.1.3.3.2 Causes and Conditions for Development

Strains of tomato mosaic virus (ToMV) and tobacco mosaic virus (TMV) are responsible for causing the disease. The viruses are mechanically transmitted, entering the plants through wounds usually caused by a handling of plants or by insect feeding. Contaminated tomato seeds, plant debris, and infected tobacco products that contaminate workers' hands serve as a source of the initial inoculum. Seed produced from infected plants usually carries the virus on the seed coat, but sometimes the virus is within the seed coat or endosperm. It is not present in the embryo. Seedlings produced from infected seed are infected at germination. In seedbeds, the disease can spread from a few infected seedlings during transplanting. The virus is very resistant and can remain infective in dead plant material for several years. Infective root debris may be found several meters deep where a diseased crop has been grown. Weeds may carry the virus over from year to year.

11.1.3.3.3 Management

1. Use healthy, bold seeds and virus free seedlings and plant early in the season.
2. Uproot the infected plants as soon as they are noticed in the field.
3. Apply hot water treatment for 25 min at 50°C.
4. Milk powder (20% wt/vol) is effective at inactivating to bamoviruses. Use it to wash hands and hand tools, and dip hands in milk every 5 minutes when handling plants.
5. For the management of mosaic, spray insecticide oxydemeton methyl or dimethoate @ 0.1% to control insect vector three to 4 times at a 10-day interval.
6. Use resistant/tolerant varieties.

11.1.4 NEMATODE DISEASES

11.1.4.1 ROOT-KNOT OF TOMATOES

Root-knot is a common disease of tomato impacting both the quality and quantity of marketable yields. In addition, root-knot nematodes interact with other plant pathogens, resulting in increased damage caused by other diseases. It has a very wide host range, i.e., potato, tomato, brinjal, chili, okra, carrot, colocasia, etc.

11.1.4.1.1 Symptoms

Root-knot is very distinctive because of the galls or swellings produced on roots and underground portions of stems, which often completely ruin the crops. Newly emerged leaves appear yellow, distorted, and crinkled along the margins, tend to droop, and sudden wilting is there. In most instances, the root-knot is characterized by smaller swellings, and more uniformly distributed infection on the lateral feeding roots. When the galls formed by the root-knot nematode are broken open, shiny white bodies about the size of a pinhead, the enlarged female nematodes, are usually found. Underground symptoms are most characteristic of the disease with root galls two to three times the diameter of healthy roots, which give the root system a knobby appearance.

11.1.4.1.2 Cause and Conditions for Development

Root-knot is caused by various *Meloidogyne spp.* such as *M. hapla, M. incognita,* and *M. javanica* Juveniles emerging from eggs in the soil penetrate between and through cells at the center of the root, usually near the growing tip. Feeding by the nematodes cause an increase in root cell numbers and size. Enlarged cells, called giant cells, serve as food sources for female root-knot nematodes. The nematode can survive in soil as a dormant egg stage for a few months and on susceptible plant hosts indefinitely. The disease is most severe in light sandy soils and at relatively warm soil temperatures. Optimum temperature for root-knot development is 25–30°C.

11.1.4.1.3 Management

1. Adopt crop rotation for at least 3 years with non-host crops, flooding of soil, deep plowing during summer.

206   *Diseases of Fruits and Vegetable Crops*

2. Destroy the crop residues, particularly roots after the harvest of the crop.
3. Use soil solarization in nursery beds to increase the soil temperature. In nursery apply carbofuran 3G @ 2g a.i./sq.m.
4. Use healthy seedlings for planting.
5. Intercropping of tomato with marigold is beneficial.
6. Combined application of neem-based formulation of *Pseudomonas chlamydosporia* and *P. fluorescens*, or *P. fluorescens* and *Trichoderma harzianum*, or *P. chlamydosporia* and *T. harzianum* @ 40 g/sq. m.
7. Apply organic amendments in the soil like neem cake, sawdust, @ 25q/ha with fertilizer at least three weeks before transplanting.
8. Soil application of nematicide carbofuran 3 G @ 1–2 kg a.i./ha, seven days before transplanting.
9. Plant resistant varieties wherever possible. Tomato varieties Ramya, Selection-120, Nematax, Karnataka Hybrid, NT-3, NT-12, Y-220, and NTDR-1 are resistant.

11.1.5 DISORDERS

11.1.5.1 BLOSSOM-END ROT OF TOMATOES

Blossom-end rot is a serious disorder of tomato; it can be very damaging, and severe losses may occur if preventive control measures are not undertaken.

11.1.5.1.1 Symptoms

Blossom-end rot begins with a small watery bruise on the blossom end (the end that is not attached to the stem) of the fruit. Symptoms may occur at any stage in the development of the fruit, but most commonly are first seen when the fruit is one-third to one-half full size. Symptoms appear only at the blossom end of the fruit. Initially, a small, water-soaked spot appears, which enlarges and darkens rapidly as the fruits develop.

On tomato and brinjal, blossom-end rot usually begins as a small water-soaked area at the blossom end of the fruit. This may appear while the fruit is green or during ripening. As the lesion develops, it enlarges, becomes sunken and turns black and leathery. In severe cases, it may completely cover the lower half of the fruit, becoming flat or concave. Secondary pathogens commonly invade the lesion, often resulting in the complete destruction of the infected fruit. Secondary molds often colonize the affected area, resulting in a dark brown or black appearance.

11.1.5.1.2 Cause and Condition for Development

Blossom-end rot is a physiologic disorder associated with a low concentration of calcium in the fruit. Calcium is required in relatively large concentrations for normal cell growth. When a rapidly growing fruit is deprived of necessary calcium, the tissues break down, leaving the characteristic dry, sunken lesion at the blossom end. Blossom-end rot is induced when the demand for calcium exceeds supply. This may result from low calcium levels or high amounts of competitive cations in the soil, drought stress, or excessive soil moisture fluctuations which reduce uptake and movement of calcium into the plant, or rapid, vegetative growth due to excessive nitrogen fertilization. It is a common problem in acidic soil. The disease is especially prevalent when rapidly growing, and succulent plants are exposed suddenly to a period of drought. When the roots fail to obtain sufficient water and calcium to be transported up to the rapidly developing fruits, the latter become rotted on their basal ends. Another common predisposing factor is cultivation too close to the plant; this practice destroys valuable roots, which take up water and minerals.

11.1.5.1.3 Management

1. Maintain the soil pH around 6.5. Liming of soil supply calcium and increase the ratio of calcium ions to other competitive ions in the soil.
2. Use nitrate form of nitrogenous fertilizer as a nitrogen source. Avoid over-fertilization as side dressings during early fruiting, especially with ammoniacal forms of nitrogen.
3. Avoid drought stress and wide fluctuations in soil moisture by using mulches, proper drainage, and/or irrigation.

Spray calcium chloride @ 0.2–0.5% at fruit setting.

11.1.5.2 CATFACE

11.1.5.2.1 Symptoms

Extreme malformation and scarring at the fruit's blossom end. Cavities that are lined with scar tissue lie between puckered or swollen areas. Locules (seed-containing compartments) of the fruit are sometimes exposed. Damage causes uneven ripening, and some signs make the fruit unmarketable.

11.1.5.2.2 Cause and Conditions for Development

Growth disturbances during flowering (pistillate formation) are thought to be the cause of the cat's face. Various factors before and during flowering (including periods of prolonged, unseasonal cool to cold temperatures; thrips feeding on young fruit; and excessive nitrogen fertilizer) may aggravate the problem and affect fruit development.

11.1.5.2.3 Prevention

1. Use tolerant cultivars or those that are not known to exhibit cat face.
2. Ensure optimal irrigation and nutritional and temperature management in greenhouses to reduce losses.
3. Prevent soils from becoming waterlogged.

KEYWORDS

- **beta-amino butyric acid**
- **tobacco mosaic virus**
- **tomato leaf curl virus**
- **tomato mosaic virus**
- **tomato spotted wilt virus**
- **tomato yellow leaf curl virus**

REFERENCES

Agrios, G. N., (2005). *Plant Pathology* (p. 902). Academic Press, San Diego, CA.

Anonymous (2015). *National Horticulture Board Database* (pp. 177–185). NHB Publication. New Delhi.

Fullelove, G., Wright, R., Meurant, N., Barnes, J., O'Brien, R., & Lovatt, J., (1998). *Tomato Information kit: Agrilink, Your Growing Guide to Better Farming*. Department of Primary Industries, Queensland Horticulture Institute: Brisbane.

Jones, J. B., Jones, J. P., Stall, R. E., & Zitter, T. A., (1991). *Compendium of Tomato Diseases* (p. 73). APS Press, St. Paul.

Naika, S., Jeude, J. V. L. D., Goffau, M. D., Hilmi, M., & Dam, B. V., (2005). Agrodok 17. *Cultivation of Tomato Production, Processing and Marketing* (p. 6). Digigrafi, Wageningen, Netherlands.

Panagopoulos, C. G., (2000). Diseases of vegetable crops. In: *Vegetable Disease* (pp. 15–189). Stamoulis, Athens.

Persley, D., Cooke, T., & House, S., (2010). *Diseases of Vegetable Crops in Australia*. CSIRO Publishing: Collingwood, Victoria.

Roberts, P. D., Murphy, J. F., & Goldberg, N. P., (1999). Fungal and bacterial diseases. In: Albajes, R., Gullino, M. L., Van Lenteren, J. C., & Elad, Y., (eds.), *Integrated Pest and Disease Management in Greenhouse Crops* (pp. 40–47). Kluwer Academic Publishers, Norwell, USA.

Sherf, A. F., & MacNab, A. A., (1986). *Vegetable Diseases and Their Control* (2nd edn., p. 728). John Wiley & Sons, NY.

Singh, R. S., (2005). *Plant Diseases* (p. 720). Oxford and IBH Publishing Co. Pvt Ltd., New Delhi.

Zitter, T. A., (1991). Tomato mosaic and Tobacco mosaic. In: Jones, J. B., Jones, J. P., Stall, R. E., & Zitter, T. A., (eds.), *Compendium of Tomato Diseases* (p. 39). APS Press. Available online: http://vegetablemdonline.ppath.cornell.edu/factsheets/Viruses_Tomato.htm (Accessed on 18 November 2019).

CHAPTER 12

Diseases of Elephant Foot Yams and Colocasia Crops and Their Management

SHAILBALA

Junior Research Officer, Plant Pathology, Sugarcane Research Center, G.B. Pant University Agriculture and Technology, Uttarakhand, India, E-mail: shailbalasharma10@gmail.com

12.1 INTRODUCTION: ELEPHANT FOOT YAM

Amorphophallus paeonifolius Blume, (syn: *Amorphophallus campanulatus,* elephant foot yam) is one of the important aroid tuber vegetable crops cultivated for its corms in the tropical and subtropical region of Asia (Ravi et al., 2009). In the year 2012, elephant foot yam occupied 36 thousand ha area, which is 8.41% of the total area under tuber crop with 364 thousand tonnes production, which is 3.36% of production under tuber crop from India. The productivity of this crop is 10 tonnes/ha (Reddy, 2015).

Elephant foot yam belongs to the Araceae family has a great scope as a cash crop. The plant derives its name from the fact that the plant from which the tuber is derived is quite huge and resembles the foot of an elephant to a great extent. It is popularly known as "oal" in Bengali, *suran* or *jimikand* in Hindi, *senai kizhangu* in Tamil, *suvarna gedde* in Kannada, *chena* in Malayalam, *oluo* in Oriya, *kanda gadda* in Telugu, etc. Elephant foot yam is a tuberous, stout, indigenous herb grown widely in India. In our country, it is grown mostly in West Bengal, Andhra Pradesh, Karnataka, Kerala, Maharashtra, Tamil Nadu, Uttar Pradesh, Punjab, Bihar, Assam, and Orissa.

Tubers of elephant foot yam contain 18% starch, 1–5% protein, and up to 2% fat. Leaves contain 2–3% protein, 3% carbohydrates, and 4–7% crude fiber. Tubers and leaves are quite acrid due to the high content of oxalates. Acridity is usually removed by boiling fairly for a long time. Elephant foot yam is a remunerative and profitable stem tuber crop. The crop is gaining popularity due to its shade tolerance, easiness in cultivation, high productivity, steady demand, and a reasonably good price, but this crop is attacked by bacteria, fungus, virus, nematode, etc. So it is important to know about common diseases of elephant foot yam and their management.

12.1.1 BACTERIAL DISEASES

Bacterial diseases constitute an important factor limiting the growth and cropping of cultivated plants. This crop is seriously threatened to extinction in southeastern Nigeria and other countries as a result of its high susceptibility to bacterial disease.

12.1.1.1 BACTERIAL LEAF BLIGHT (XANTHOMONAS AMORPHOPHALLI)

This disease is an important and widespread disease of elephant foot yam, which causes considerable yield loss. Symptomatic plants ranged from chlorotic and stunted to completely blighted as the disease progressed. Symptoms shown by infected plants are blight, defoliation, wilting, and dieback.

Three sprays of 100 ppm Agrimycin at 15 days interval after the first appearance of the disease significantly reduced the disease incidence and increased the yield. Seed tuber dip treatment with 0.05% Agrimycin for 12 hrs decreased the disease incidence (Reddy, 2010). Plant extracts of lemongrass (*Cymbopogon citrates),* black pepper seed (*Piper guineense*), orange peel (*Citrus sinensis*) effectively reduce the disease severity (Opara et al., 2013).

12.1.2 FUNGAL DISEASES

Lasiodiplodia (syn. Botryodiplodia) theobromae, Phytophthora colocasiae, Sclerotium rolfsii, Fusarium sp., *Rhizopus* sp., *Aspergillus* sp. and *Penicillium* spp., etc. causes 4–5% loss in elephant foot yam (Annual report, 2014–2015). Interaction between host, pathogen, and environment leads to diseased, infected, and disfigured tubers, ultimately affect the production, productivity, and market value of tubers. Some of the important fungal diseases affecting elephant foot yam production are collar rot and tuber rot.

Diseases of Elephant Foot Yams and Colocasia Crops

12.1.2.1 COLLAR ROT (RHIZOCTONIA SOLANI KUHN. AND SCLEROTIUM ROLFSII SACC)

The fungus causes damage by rotting the collar region of the stem, which ultimately topples down. The pathogen mainly attacks the collar region and produces a water-soaked lesion. The whole plant soon turns yellow. A thick white mycelial growth can be seen on the affected portion. Sclerotia (fruiting body) are also found on the mycelial growth. The pathogen is soil-borne in nature. The stem shrinks and collapses due to rotting. Observationally, collar rot grades into "basal stem rot," and with some pathogens is the first phase of "basal stem rot," often followed by "root rot." Collar rot is most often observed in seedlings grown in infected soil. Injury to the collar region during intercultural operations and infestation of roots by nematode predisposes the plant to collar rot disease. Finally, the disease is responsible for the heavy reduction in yield and the qualitative degradation of the crop. The waterlogged situation, poor drainage, and mechanical injury at the collar region favor disease incidence.

The disease can be managed by the use of disease-free planting material. Destruction of infected crop debris, summer deep plowing, crop rotation, proper drainage, etc. is essential to lower down the inoculum. Application of neem cake oil as well as bio-agent *Trichoderma harzianum* in the soil improves the soil condition and suppresses the pathogen growth. Soil drenching with fungicide Captan @ 2 g/l of water apply twice at 25–30 days interval. The antagonist mixture of *T. harzianum* and *Pseudomonas fluorescens* in a 1:1 ratio was found to be the best for the management of collar rot disease (Singh et al., 2006). Yadava (2004) also reported that the 3:1 mixture of *T. harzianum* and *P. fluorescens* gave the maximum reduction of collar rot.

12.1.2.2 TUBER ROT (PHYTOPHTHORA COLOCASIAE, SCLEROTIUM ROLFSII SACC, BOTRYODIPLODIA THEBROME PAT., FUSARIUM SPP., RHIZOPUS SPP.)

Tuber rot could be a serious problem in this crop if infected planting material use for raising a crop. Injury during harvest and transport further aggravates this problem. Infestation in roots and tubers by root-knot nematode act as a predisposing factor for infection by a large number of fungi. Elephant foot yam stored at high temperature with poor aeration may be helpful to these rotting fungi for infection (Reddy, 2015).

Sanitation by removal and burning of infected tubers, deep plowing, ridge planting, proper drainage, and controlled irrigation are helpful in managing the disease. Tubers need to be stored properly during the offseason for using them as seed material for next season. Seed treatment with fungicide Carbendazim @ 0.05% and store on the cemented floor will help to lower down the infection (Reddy, 2015).

12.1.3 VIRAL DISEASES

In India, elephant foot yam (*Amorphophallus paeonifolius*, formerly known as *Amorphophallus campanulatus*), produces an edible corm which is widely cultivated in a number of states from West Bengal to Tamil Nadu. Its importance to rural livelihoods is highlighted by the fact that it is included in the National Germplasm Evaluation Programme for tuber crop improvement. This crop is mainly propagated through vegetative propagules (corms), and so any virus infecting the crop has quarantine relevance as the virus is vertically transmitted. Thus correct identification of viruses transmitted through vegetative propagules has significance.

12.1.3.1 AMORPHOPHALLUS MOSAIC

The disease is caused by Amorphophallus mosaic virus. The chief characteristic symptoms are mosaic mottling on the leaves and distortion of leaf lamina. The pathogen causes more proliferation of lateral buds, separation of buds from the mother corms, and poor growth of roots. Corms produced by the mottled plants are much smaller than those without mottled leaves. Primary spread of disease is through infected planting material. Secondary spread of the disease is through insect vectors, *Myzus persicae* Sulz., *Aphis gossypii* Glover, *A. craccivora* Koch. and *Pentalonia nigronervosa* coq. Natural association of Konjac mosaic virus (KoMV) with mosaic disease of elephant foot yam in India was also reported and identified by enzyme-linked immunosorbent assay (Padmavathi et al., 2012).

The use of disease-free planting material from a healthy plant is important to avoid disease. Rouge out the diseased plant as soon as the first symptom is noticed. Spraying of systemic insecticide Monocrotophos @ 0.05% or any broad-spectrum insecticide should be undertaken to prevent secondarily spread of the virus.

Diseases of Elephant Foot Yams and Colocasia Crops

12.1.3.2 DASHEEN MOSAIC VIRUS (DSMV)

The mosaic, mottling, leaf puckering, rolling, leaf narrowing, distortion of leaf lamina, and stunting, etc. symptoms are noticed in the crop. Corm produced by mottled plants is much smaller than those showing no symptoms. The disease is primarily spread through infected planting material and secondarily spread through different species of aphid. The virus is transmitted in a non-persistent manner by aphid vectors *Aphis gossypii, Myzus persici, Pentalonia nigronervosa*, etc. Virus is also transmitted by mechanical inoculation and grafting.

Genus potyvirus (DsMV) has been reported from around the world, mainly in the tropics, where it infects members of the Araceae (Brunt et al., 1996). Hosts of DsMV include important root and tuber crops such as *Alocasia, Colocasia, and Xanthosoma* or ornamentals *Caladium, Dieffenbachia, and Philodendron*. On the Students' Instructional Farm of BCKV in Mohanpur, leaves of some elephant foot yams showed mosaic symptoms with parallel chlorotic streak (Pandit et al., 2001). Further study of diseased sampled confirms their identity as DsMV. This was the first report of Dasheen mosaic virus (DSMV) infecting elephant foot yam in India.

The use of disease-free planting material is the best way to avoid disease. Tubers free from the virus should be used. Spraying of systemic insecticide should be undertaken to prevent secondarily spread of the virus. Hot air treatment of tubers at 55°C for 10 minutes before planting followed by 2 sprays of Monocrotophos @ 0.05% or any broad-spectrum insecticide at 60 and 90 days after planting reduce disease incidence.

12.1.4 NEMATODE DISEASE

The destructive plant-parasitic nematodes are one of the major limiting factors in the production of elephant foot yam. *Meloidogyne incognita* and *Pratylenchus* spp are important nematode species damaging the crop.

12.1.4.1 ROOT-KNOT NEMATODE (MELOIDOGYNE INCOGNITA)

This crop is highly susceptible to root-knot nematode. The root produces typical root-knot symptoms. In corm, the gall appears as an irregular projection or outgrowth which harbor adult females as well as eggs. The area of infestation in the tuber becomes discolored. In high severity, the infested area dried seems dry rot. Infested corm was deformed and smaller in size. More diseases were noticed in cormels as compare to corm (Reddy, 2008).

Continuous cropping of particular crop results in a buildup of the nematode in the field. Hence crop rotation is essential. Farmers rotate Amorphophallus with resistant colocasia in some places. Taking up cassava and sweet potato in between susceptible crops are also found to be effective in reducing root-knot nematode. Cassava leaves and their dry powder have found to be nematicidal. Incorporation of powder to be an effective way of reducing nematode problems. In case of heavy infestation, apply insecticide Carbofuran @ 3.0 kg ai/ha in two split doses, first at the time of planting and second at the time of earthening up. Application of neem cake/farmyard manure/compost enriched with *Trichoderma harzianum/Paecilomyces lilacinus* gave effective control of nematodes.

12.1.4.2 LESION NEMATODE (PRATYLENCHUS SPP.)

The surface of the infested tubers becomes black in color with cracks. The blackening extended further deep inside the tubers reducing the edible portion. The infested tubers were smaller in size (Reddy, 2008).

Use of nematode free planting material is an effective means of controlling or reducing damage by the nematode. Hot water treatment at 46°C–52°C for 15–30 minutes gives good control of nematode. Use the varieties which show some resistance to the lesion nematode.

12.1.5 COLOCASIA (COLOCASIA ESCULENTA (L.) SCHOTT)

Colocasia belongs to the family Araceae. In the Hindi belt, it is known as ghuiyan or arvi. In the year 2012 in India, colocasia occupied approximately 52 thousand hac area producing 654 thousand tonnes with 13 t/ha productivity. It covers 12.14% area and 6.02% of production under tuber crops (Reddy, 2015). The underground parts consist of one or more large edible central corms and large number of cormels commonly known as tubers. Colocasia contains oxalic acid and its salt, which can decrease calcium absorption and promote kidney stone formation. Fortunately, boiling the corm can reduce these negative effects by at least 50%. An irritant which causes intense discomfort to the lips, mouth, and throat is commonly associated with eating colocasia of unknown variety. This acridity is caused partly by the microscopic needle-like calcium oxalate crystals (raphides) together with probably a protease. Diseases of colocasia significantly reduce the number of functional leaves and have led to a yield reduction of about 50% worldwide (Jackson, 1999). *C. esculenta* is affected by a number of infectious diseases caused by fungi, bacteria, nematodes, and viruses as well as non-infectious or abiotic factors.

12.1.5.1 BACTERIAL DISEASES

The cultivation of colocasia is often affected due to the attack of diseases caused by bacteria. There are some bacterial diseases, i.e., bacterial leaf spot and bacterial soft rot disease, which directly affect the quantity as well as the quality of tubers and responsible for yield loss.

12.1.5.1.1 Bacterial Leaf Spot (Xanthomonas campestris pv. Dieffenbachiae)

Disease begins as a small water-soaked spot on the lower surface of leaves. Spots enlarge, and a substantial area of brown, dead tissue surrounded by yellow halos develops on the upper leaf surface. Pronounced water soaking continues on the lower leaf surface, sometimes with a cream to light yellow bacterial exudate (sticky ooze) in the center of the water-soaked lesion. Apparently, infection occurs readily through hydathodes. The pathogen can invade vascular tissues of leaves, causing water-soaked streaks along major veins. Once inside the vascular tissue, bacteria may be able to reach the corms and cormels, thus contaminating the planting material. The main mode of pathogen movement is probably by splashing rain. Temperature ranges from 29–32°C favor disease development. The entry of pathogens is facilitated by wounding (Reddy, 2010).

This bacteria is easily transmitted whenever workers, tools, farm machinery, etc. come in contact with infected, sodden plant material so growers should stay out of fields when plants are wet. Careful handling of corms is also required.

12.1.5.1.2 Bacterial Soft Rot (Erwinia carotovora (Jones), E. Chrysanthemi)

Bacterial soft rot disease is characterized as strong smelling watery soft rot ranging in color from white to dark tan. Soft rot decay is generally odorless in early stages but later a foul odor. Wounds and bruises caused by feeding of insects and other animals and those inflicted at harvest are the most common infection courts for this disease. Abundant moisture is required for invasion of the bacteria (Ooka, 1994).

Only healthy tubers should be used for seed. Careful handling of corms to minimize injury at harvest, air drying of corms, and storage at low temperatures of only the sound corms. Period culling of rotten tubers from store reduces the chances of contamination of healthy tubers.

12.1.5.2 FUNGAL DISEASES

Fungal and oomyceteous plant pathogens of colocasia have been reported to cause losses in colocasia fields. Although, this crop is susceptible to at least 23 pathogens, only a few cause a serious reduction in growth and production (Ooka, 1990). According to Ooka (1994), among these diseases, fungal diseases of *C. esculenta* are the most significant. Diseases caused by fungi are the most prominent, aided by climatic conditions that favor the growth of the plant. Plant rot disease, reported to be caused by a species of the genus *Phytophthora*, now plagues crops and subsequently kills the entire plant in cases of severe fungal infection throughout the United States and West Africa (Hao, 2006).

12.1.5.2.1 Pythium Root Rot (Pythium aphanidermatum Fitzpatrick, P. Graminicola Subramaniam, P. Splendens Brown, P. Ultimum Trow, etc.)

Pythium root rot is probably the most widely distributed disease of the crop. This disease was probably spread with the introduction of crop. *Pythium aphanidermatum, P. graminicola, P. splendens* have been observed to causes losses of up to 80% in Hawaii (Trujillo, 1967). The primary symptoms appear as chlorosis of leaves, and the affected plants gradually wilt. The normally firm flesh of the corm is transformed into soft, mushy, often malodorous mass. In the wetland, the root system is destroyed except for a small fringe near the apex of the corm.

Diseased plants are easily removed from the soil by hand. The plant becomes stunted, with leaf stalk shortened and leaf blades curled and crinkled, yellowish, and spotted. Upon the demise of the main corm, the lateral cormels develop roots and remain clustered around the cavity left by the

Diseases of Elephant Foot Yams and Colocasia Crops 219

disintegration of the main corm. When the corm is cut open, there is usually a sharp line of demarcation between the healthy and diseased tissue. The fungus is soil-borne in nature. The first report of corm and root rot of *C. esculenta* caused by *Ovatisporangium vexans* and *Rhizoctonia solani* complex in Turkey reported by Dervis et al. (2014).

The severity of the disease may be reduced by soil drenching with fungicide Captan @ 0.2%. Seed should be selected carefully to avoid those showing any disease symptoms. The selected seed should be dipped into Captan suspension to provide them protection for a few days after planting (Trujillo, 1967). Summer deep plowing, soil solarization, and crop rotation practices must be followed to reduce the infection.

12.1.5.2.2 *Phytopthora Blight of Colocasia (Phytophthora Colocasie Racib)*

Phytophthora colocasiae, foliar oomyceteous diseases agent, is a major limiting factor in taro production worldwide (Brooks, 2008). Reduction in corm yield may be up to 50% due to this disease. Serious outbreaks of *Phytophthora* leaf blight in Samoa in 1993 and in the last few years in Cameroon, Ghana, and Nigeria continue to demonstrate the devastating impact of this disease on the livelihoods and food security of small farmers and rural communities dependent on the crop. The spread of disease to new geographical areas also poses a major threat to neighboring countries and colocasia growing regions still free from the disease (Singh et al., 2012). This disease is especially a serious problem in the humid tropics where the rainfall is greater than 2,500 mm per annum, and there is little seasonal variation.

The disease is characterized by the formation of large brown lesions on the leaves of the infected plants. Symptoms on leaves initially occur where water droplets accumulate form small brown spots surrounded by halos on the upper surface of leaves (Nelson et al., 2011). These spots expand very quickly and form large brown lesions (Hunter et al., 1998). Under the wet condition, the entire leaf can be destroyed within a few days of the initial appearance of symptoms (Misra et al., 2008). The underside of the leaves have spots that look water-soaked or grey, as they expand, blight forms, and the leaf is destroyed within a few days (Nelson et al., 2011). As the lesion expands, sporangia develop most actively at the margin of the lesion and progress to attack the healthy tissues (Singh et al., 2012).

One characteristic feature found on the leaves is the formation of bright orange droplets oozing out from above and below water-soaked leaf surfaces

(Singh et al., 2012). As a result, the droplet dry out during the day and become crusty. Another sign of pathogen infection are masses of sporangia that form a white powdery ring around the lesion. Symptoms on petioles include gray to a brownish-black lesion that can occur anywhere on the petiole. Petioles become soft and may break as the pathogen destroys the host (Singh et al., 2012). In North India, the disease first appears in the month of August and September.

The fungus is characterized by oospore and coenocytic hyphae (Mishra et al., 2008). Oospores have very thick walls that provide a durable survival structure. The fungus perennates through oospores in the diseased leaf debris in the soil and through the infected corms or on leaf debris left in the field after harvest. However, inoculum does not survive for very long on host tissues. Upon infection, oospores that overwinter on leaf tissue and petioles give rise to sporangiophores, which have lemon-shaped sporangia at their tips. Sporangia can infect colocasia leaves either directly by germ tubes or indirectly by producing zoospore depends on weather conditions (Mishra et al., 2008). Indirect infection occurs under unfavorable or very wet conditions by releasing zoospore from sporangia. Zoospore loses their flagella, become cysts, germinate, and feed on the host by germ tube and produce more sporangia to continue the disease cycle (Jackson, 1980).

The pathogen grows voraciously in areas with high humidity and very heavy rainfall, in addition, an optimal pH, i.e., 6.5 and temperature of 28°C. A cool, wet, and humid condition favor both the asexual and sexual reproduction of pathogen. The maximum sporangial production occurs at 21°C when the relative humidity (RH) is 100%. RH lower than 90% totally inhibits sporulation. The viability of the zoospore is lost rapidly at RH less than 90%. Trujillo (1965) found that the blight epidemic occurred when night and day temperatures ranged between 20–22°C and 25–28°C, respectively with 65% RH during day and 100% RH during night.

Agronomic methods that have given partial success include a careful choice of planting material, planting at high density, intercropping with other crops rather than growing it as a sole crop, and crop rotation. Field removal of infected leaves has also been useful, but it is extremely laborious. An integrated control approach combining cultural and chemical methods seems to be the best at present. The ultimate solution must lie in the breeding and release of resistant cultivars. So always grow resistant cultivars depend on the area. Since the disease is carried from season to season by seed corms, selection of disease-free seed should be given top priority.

In addition, sanitation of the field, including the destruction of badly affected plants and culled corms should also be practiced. Prophylactic three spray of fungicide Mancozeb @ 2 g/l with stickers like Sandovit @ 1ml/ l of solution at 10–15 days interval. Metalaxyl is also reported to inhibit mycelial growth of the fungus. Many bacteria, including actinomycetes and some fungi present as leaf surface micro-flora are antagonistic to the pathogen, and some have been shown to check the disease on leave (Narula and Mehrotra, 1987).

12.1.5.2.3 Storage Rot (Aspergillus Niger Van Tiegham, Botryodiplodia Thebrome Pat., Fusarium solani (Mart.) Sacc., Rhizopus Stolonifer Sacc, Sclerotium rolfsii Sacc.)

Fungi are the main microbial pathogens that cause the storage rot of colocasia corms (Nwachukwu and Osuji, 2008; Maduewesi and Onyike, 1981). Fungi causing storage rot have been reported in the USA, India, Egypt, the Pacific Islands, and Nigeria (Onyike and Maduewesi, 1985). The rot due to *Aspergillus niger and Botryodiplodia theobromae* was extensive resulting in complete maceration of colocasia tissue. These fungal organisms have also been reported to be the major cause of storage rots of other root and tuber crops which includes yam, cassava, and sweet potatoes (Yusuf and Okusanya, 2008; Okigbo et al., 2010; Okigbo and Nwakamma, 2005; Amusa and Baiyewu, 1999; Ogaraku and Usman, 2008; Banito et al., 2010). Due to mechanical injury and natural wounds on the surface of the corms, there is an attack by many pathogens or saprophytes, which result in rotting of an edible portion of plant during storage. The organisms causing 8–10% post-harvest loss in colocasia were identified as *B. theobromae, P. colocasiae, S. rolfsii, Fusarium* sp., *Rhizopus* sp., *Aspergillus* sp. and *Penicillium* sp (Annual report, 2014–2015).

It is essential to always avoid the mechanical injuries. Dipping of roots in the solution of fungicide Blitox-50 @ 2 g/l of water also helps to avoid saprophytic growth. Potassium sorbate (0.1 mg/ml) protected colocasia from fungal rot.

12.1.5.2.4 Brown Leaf Spot (Cladosporium Colocasie Sawada)

Leaf spot disease caused by *Cladosporium colocasiae Sawada* has been reported in Ghana (Awuah, 1995). The pathogen attacks both wetland and upland areas of colocasia and occurs mainly on the older leaves. On the upper surface the spot appears as a diffuse light yellow to copper color area. On the lower leaf surface the spots are dark brown due to superficial hyphae, sporophores, and conidia of the fungus. The lesions are generally 5–10 mm in diameter. The spots are diffuse in appearance (Reddy, 2010).

Spray the crop with fungicide Mancozeb @ 2 g/l with stickers like Sandovit @ 1 ml/l of solution. The spraying should be started before the appearance of disease and should be repeated thrice at 10–15 days interval. Metaxyl is also inhibited mycelial growth of the fungus.

12.1.5.2.5 Sclerotium Rot (Sclerotium rolfsii Sacc.) Sexual Stage: Corticium rolfsii (Curzi)

The pathogen causes drying of petioles and a corm rot. The rotted tissue is brown and soft but not watery and tends to be stringy (Reddy, 2010). Characteristic white growth and numerous dark brown sclerotia that resemble cabbage seed are observed. The disease is more severe in over mature colocasia. Sclerotia abundantly produced on infected corms persist in the soil, causing serious outbreaks of the disease in warm, wet weather following a significant dry spell. Affected plants are usually stunted, and the corms are rotted at the base where abundant sclerotia of the pathogen develop. The pathogen may survive saprophytically on plant debris or as sclerotia in the soil. When sufficient moisture is present, sclerotia germinate and infect young or old root, dead leaf petioles, and over mature corm. This disease is usually serious during the warm wet period.

Prompt harvesting of colocasia and crop rotation helps to avoid the disease. Field sanitation practices, i.e., removal and burning of infected planting material are helpful to reduce the infection. Soil drenching with quintozone or dicloran has been recommended for control (Reddy, 2010).

12.1.5.3 VIRAL DISEASE

Colocasia is a vegetative propagated crop, so this crop is more prone toward virus attack. Alomae and bobone virus disease and DSMV disease are an important and widespread disease of colocasia causes substantial yield losses.

12.1.5.3.1 Alomae and Bobone Virus Disease Complex

Alomae-bobone complex is very serious and can lead to the death of the plants. Alomae is a systemic necrosis that leads to the death of the plant. The first symptom is a feathery mosaic, and even young leaves fail to open properly and have a crinkled appearance. The leaves become abnormally thickened, and the vein prominent. Later on, the leaves do not open and begin to die. The necrosis spreads from tips to the leaves and down the petiole, and finally, the whole plant dies. Symptoms of bobonae are similar to those of alomae but comparatively milder. The affected plant tends to be more stunted, and the leaves are twisted and curled. The foliage does not become necrotic but remain green and eventually recover.

The alomae virus disease is caused by a complex of two or more viruses acting together. The two viruses that are definitely involved are the taro large bacilliform virus, which is transmitted by the plant hopper (*Tarophagus proserpina*), and the taro small bacilliform virus, which is transmitted by the mealybug (*Planococcus citri*) (Rodoni, 1995). Neither virus is transmissible by mechanical contact, nor does their host range seems limited to aroids only. The full-blown alomae disease occurs when these two viruses (and possibly others) are present. Severe cases of alomae can result in total crop loss, while bobone can cause up to 25% yield loss.

The disease is controlled by pulling out diseased plants in the field and by careful selection to ensure disease-free planting material. So it is always advisable to rouge out infected plants as they provide a reservoir of virus particles. Ultimately, control will have to rely on breeding and disseminating resistant cultivars.

12.1.5.3.2 Dasheen Mosaic Virus (DSMV) Disease

DSMV disease occurs worldwide. It is characterized by chlorotic and feathery mosaic patterns on the leaf, distortion of leaves and stunted plant growth. The disease is not lethal, but the yield is depressed. In some cases, more striking symptoms will appear. Feather like the pattern of light yellow tissue, may develop along veins. Severely infected plants may be noticeably slower than surrounding plants in the field. In severely affected plants, petioles are short, and leaves are dwarfed and fail to unfurl properly. This disease is caused by a stylet-borne, flexuous, rod-shaped virus that is spread by aphids. The virus is transmitted non-persistently by the aphids, *Myzus persicae*, *Aphis gossypii*, and *Pentalonia nigronervosa*. The virus is also sap transmissible. The virus is carried in infected planting stock.

Control is only through the use of virus-free planting material, field sanitation, and quarantine measures. The virus can be eliminated from planting stock through the use of meristem tip tissue culture. Insecticidal control of aphids is of little value in DSMV control. Aphids can transmit the virus before they are killed by the insecticide (Reddy, 2010).

12.1.5.4 NEMATODE

Nirula (1959) was the first to report root-knot nematode on colocasia from India. Nematode *Meloidogyne incognita, M. javanica* are noticed on colocasia crop and cause considerable loss in tuber yield.

12.1.5.4.1 The Root-Knot Nematode (Meloidogyne Incognita, M. Javanica)

Nematode infested plant first shows yellowing, then turn brown and finally dies back. Galls on root and swelling, as well as malformation on the corm are characteristic of attack by this nematode. The severe attack will stunt the plant with no corm formation or deformed and galled corms of little market value. Rotting of corms associated with nematode occurs during storage. Nematode can be carried over from one crop to next in wide range of other host plant and weeds. Nematode can easily spread through infested corm and cormels used for propagation (Reddy, 2010).

Nematode population severely damages the colocasia grown in dry land. So crop grows in very wet or flooded condition can suppress the nematode population. Fumigation is desirable for control of root-knot nematode in heavily infested soil. Always use nematode infested free planting material or resistant varieties in the field. Treatment of corm with water at 50°C for 40 minutes kills the nematode in the corm (Byars, 1917).

12.1.5.5 PHYSIOLOGICAL DISORDER

The plant requires all the essential nutrients in balanced proportions, and deviation from this may result in physiological disorders. This may be due to deficiency or toxicity of nutrients.

12.1.5.5.1 Metsubre

It is a nutritional disorder of colocasia and is supposed to be due to calcium deficiency. The effected corms have smooth or concave top, slightly brownish in color and are of varying size (Bhat, 2016).

Application of calcium will control this problem. Balanced use of fertilizer will also helpful.

Diseases of Elephant Foot Yams and Colocasia Crops 225

12.1.5.5.2 Loliloli

Starch present in normal corms, is deficient or absent in those with loliloi, a term used in Hawaii to describe a physiological disorder of colocasia. While the normal corm is firm, crisp, and resilient to touch, loliloli corm is soft, spongy, and water exudes when affected parts are squeezed. Loliloli corm is the result of withdrawal of starch from corm (Ooka, 1994).

Use of nitrogenous fertilizer after the corm has formed and the natural growth decadence of the plant started should be avoided to reduce the chances of loliloli colocasia occurring.

Elephant foot yam and colocasia are considered as under noticed tuber crops, so there is a need to do a lot of work. Definitely, there is an urgent need to intensify the breeding program towards the development of high yielding disease-resistant cultivars. If possible, the focus should be given on the development of transgenic plants of these crops. Integrated disease management packages, along with refinement, must be popularized. Safe control measure needs to be developed to ensure a safe environment, soil fertility, good planting material, economic viable, so that use of bio-control agents, i.e., *Trichoderma harzianum,* and *Pseudomonas flourescens* can be encouraged. There is a need to develop good storage techniques too. In this way, we can easily manage the diseases of these tuber crops.

KEYWORDS

- **colocasia**
- **dasheen mosaic virus**
- **elephant foot yam**
- **konjac mosaic virus**
- **physiological disorder**

REFERENCES

Amusa, N. A., & Baiyewu, R. A., (1999). Storage and market diseases of yam tubers in southwest Nigeria. *Ogun. J. Agric. Res. (Nig).*, *11*, 211–225.

Annual Report (2014–2015). *ICAR-Central Tuber Crops Research Institute* (p. 161). Sreekariyan Thiruvananthapuram, Kerala.

Awuah, R. T., (1995). Leaf spot of Taro (*Colocasia esculenta* (L.) Schott) in Ghana and suppression of symptom development with Thiophanate methyl. *African Crop Science Journal*, *3*(4), 519–523.

Banito, A., Kpemoua, K. E., Bissang, B., & Wydra, K., (2010). Assessment of cassava root and stem rots in eco-zones of Togo and evaluation of the pathogen virulence. *Pak. Journal Bot.*, *42*(3), 2059–2068.

Bhat, K. L., (2016). *Physiological Disorders of Vegetable Crops* (p. 258). Daya Publishing house, New Delhi.

Brooks, F. E., (2008). Detached-leaf bioassay for evaluating taro resistance to *Phytophthora colocasiae*. *Plant Disease, 92*, 126–131.

Brunt, A. A., Crabtree, K., Dallwitz, M. J., Gibbs, A. J., & Watson, L., (1996). *Viruses of Plants* (p. 1484). CAB International, Wallingford, UK.

Byars, L. P., (1917). A nematode disease of dasheen and its control by hot water treatment. *Phytopathology*, *7*, 66.

Dervis, S., Soylu, S., & Serce, C., (2014). Corm and root rot of *Colocasia esculenta* caused by *Ovatisporangium vexans* and *Rhizoctonia solani*. *Romanian Biotechnological Letters*, *19*(6), 9868–9874.

Hao, S., (2006). "Rain, pests and disease shrink taro production to record low." *Honolulu Advertiser*, p. C1.

Hunter, D., Pouono, K., & Semisi, S., (1998). The impact of taro leaf blight in the pacific islands with special reference to Samoa. *Journal of South Pacific Agriculture*, *5*, 44–56.

Jackson, G. V. H., (1980). *Diseases and Pests of Taro* (p. 51). South Pacific Commission, Noumea, New Caledonia.

Jackson, G. V. H., (1999). Taro leaf blight. Published by the plant protection service of the secretariat of the pacific community. *Pest Advisory Leaflet.*, *3*, 2.

Maduewesi, J. N. C., & Onyike, R. C. I., (1981). Fungal rotting of cocoyams in storage in Nigeria. In: Terry, E. R., Oduro, K. A., & Cavenesseds, F., (eds.), *Tropical Root Crops: Research Strategies for the 1980's* (pp. 235–238).

Misra, R. S., Sharma, K., & Mishra, A. K., (2008). *Phytophthora* leaf blight of taro (*Colocasia esculenta*): A review. *The Asian and Australian Journal of Plant Science and Biotechnology*, *2*(2), 55–63.

Narula, K. L., & Mehrotra, R. S., (1987). Bio-control potential of phytophthora leaf blight of colocasia by phyllosphere microflora. *Indian Phytopath.*, *40*, 384–389.

Nelson, S., Brooks, F., & Teves, G., (2011). *Taro Leaf Blight in Hawaii* (p. 14). Plant disease bulletin No. PD-71. College of Tropical Agricultural and Human Resources. University of Hawaii, Manoa, HI, USA.

Nirula, N. K., (1959). Root knot nematode on colocasia. Bangalore. *Curr. Sci., 28*, 125–126.

Nwachukwu, E. O., & Osuji, J. O., (2008). Evaluation of plant extract for antifungal activity against *Sclerotum rolfsii* causing cocoyam cormel rot in storage. *Research J. of Agriculture and Biological Sciences*, *4*(6), 781–787.

Ogaraku, A. O., & Usman, H. O., (2008). Storage rot of some yams (*Dioscorea* spp) in Keffi and Environs. *Nasarawa State, Nigeria. PAT., 4*(2), 22–27.

Okigbo, R. N., & Nwakamma, P. T., (2005). Bio-degradation of white yam (*Dioscorea rotundata* Poir) and water yam (*Dioscorea alata* L.) slices dried under different conditions KMITL. *Sci. & Tech. Journal*, *5*(3), 577–586.

Okigbo, R. N., Agbata, C. A., & Echezona, C. E., (2010). Effect of leaf extract of *Azadirachta india* and *Chromolaena odorata* on post harvest spoilage fungi of yams in storage. *Current Res J. of Soil Sci.*, *2*(1), 9–12.

Onyike, R. C. I., & Maduewesi, J. N. C., (1985). Variability in pathogenicity of isolate and sensitivity to Benlate, Diathane M. 45 and PCNB of cocoyam rot fungi. *Nig. J. of Plant Prot.*, *9*, 74–81.

Ooka, J. J., (1990). Taro diseases. In: *Proceedings of Taking Taro into the 1990s: A Taro Conference* (pp. 51–59). Komohana Agricultural Complex, Hilo, Hawaii. Honolulu, Hawaii: University of Hawaii. Research Extension Series, Hawaii Institute of Tropical Agriculture and Human Resources No. 114.

Ooka, J. J., (1994). Taro diseases: A guide for field identification. *Pest Management Guidelines* (p. 13). HITAHAR Research Extension Series no. 148, University of Hawaii.

Opara, E., Theresa, C. N., & Isaiah, C., (2013). Potency of some plant extracts and pesticides on bacterial leaf blight disease of cocoyam (*Colocasia esculenta*) in Umudike, South eastern Nigeria. *Greener J. of Agricultural Sciences*, *3*(5), 312–319.

Padmavathi, M., Srinivas, K. P., Hema, M., & Sreenivasulu, P., (2012). First report of Konjac mosaic virus in elephant foot yam (*Amorphophallus paeoniifolius*) from India. *Australasian Plant Disease Notes*, *8*(1), 10.

Pandit, M. K., Nath, P. S., Mukhopadhyay, S., Devonshire, B. J., & Jones, P., (2001). *New Disease Reports*, *3*, 12.

Ravi, V., Ravindran, C. S., & Suja, G., (2009). Growth and productivity of elephant foot yam (*Amorphophallus paeonifolius*) (Dennst. Nicolson): An overview. *J. Root Crops.*, *35*, 131–142.

Reddy, P. P., (2008). *Diseases of Horticultural Crops: Nematode Problem and Their Management* (p. 379). Scientific Publishers, Jodhpur, India.

Reddy, P. P., (2010). Bacterial and viral diseases and their management. In: *Plant Protection in Horticulture* (Vol. 3, p. 288). Scientific Publishers, Jodhpur, India.

Reddy, P. P., (2010). *Plant Protection in Horticulture: Fungal Diseases and Their Management* (Vol. 2, p. 359). Scientific Publishers, Jodhpur, India.

Reddy, P. P., (2015). *Plant Protection in Tropical Root and Tuber Crops* (p. 335). Springer New Delhi Heidegerg, New York Dordrecht, London Springer, India.

Rodoni, B., (1995). *Alomae Disease of Taro* (Vol. 15, p. 12). Australian Centre for Int. Agric. Res, Canberra, *Research Notes.*

Singh, D., Jackson, G., Hunter, D., Fullerton, R., Lebot, V., Taylor, M., Losefa, T., Okpul, T., & Tyson, J., (2012). Taro leaf blight—A Threat to food security. *Agriculture*, *2*(3), 182–203.

Singh, R., Singh, P. P., & Singh, V., (2006). Integrated management of collar rot of *Amorphophallus paeoniifolius* Blume caused by *Sclerotium rolfsii* Saccardo. *Vegetable Science*, *33*(1), 45–49.

Trujillo, E. E., (1965). Effect of humidity and temperature on Phytophthora blight of taro. *Phytopathology*, *55*, 183–188.

Trujillo, E. E., (1967). Diseases of genus colocasia in the Pacific area and their control. In: *Proceedings of the International Symposium on Tropical Root Crops* (Vol. 2, pp. 13–19). University of West Indies, St. Angustine, Trinidad.

Yadava, R. S., (2004). *Studies on Epidemiology and Management of Leaf Blight and Collar Rot of Elephant Foot Yam (Amorphophallus paeonifolius L.)* (p. 86). MSc (Ag) Thesis, N.D. Univ. Agric. & Tech. Faizabad.

Yusuf, C., & Okusanya, B. A. O., (2008). Fungi Associated with the storage rot yam (*Dioscorea rotundata* Poir) in Yola, Adamawa state. *J. of Sustainable Dev. In Agric. and Environ.*, *3*(2), 99–103.

CHAPTER 13

Diseases of Garden Peas (*Pisum sativum* L.) and Their Management

RAMESH NATH GUPTA

Department of Plant Pathology, Bihar Agricultural College, BAU, Sabour, Bhagalpur, Bihar, India, E-mail: rameshnathgupta@gmail.com

13.1 INTRODUCTION

Garden pea (*Pisum sativum* L.) is an important pulse crop belonging to family *Fabaceae* and sub-family *Papilionaceae*. It is an important Rabi, herbaceous, frost-hardy annual crop. It has a global importance but mainly grown, particularly in Asia, Europe, and North America. In India, its cultivation mainly confined to northern and central parts.

13.2 ORIGIN AND DISTRIBUTION

The origin and progenitor of *Pisum sativum* L. is not clearly well known. The Mediterranean region, eastern, and Central Asia and Ethiopia have been indicated as the center of origin. Recently the Food and Agriculture Organization (FAO) designated Ethiopia and western Asia as the center of diversity, with a secondary center in southern Asia and the Mediterranean region. Peas were popular with the ancient Greeks and Romans and the word 'pea' is derived from the Latin word 'pisum.' The first cultivation of pea appears to have been in western Asia from where it spread to Europe, China, and India. Presently it is found in all temperate and tropical countries.

13.3 CLIMATIC REQUIREMENTS

Garden pea may be grown in varied climatic and weather conditions. It requires cold and dry climate and the longer cold spell helps enhancing yield. Seeds can germinate even at high temperature but the process is slow. The optimum temperature requires for germinations 20–22°C.

13.4 SOIL REQUIREMENTS

Garden peas can be grown on all types of soils but it prefers well-drained sandy loam soil. The soils should rich an organic matter as it enhances better growth by supplying nutrient sat as lower rate. It grows best at pH of 6.5 and does not thrive in highly acidic or alkaline soils or saline soils. The ideal soil is clay loam. If soil conditions are good, its cultivation becomes very easy and successful. Garden pea is grown mainly as a Rabi crop which normally sown in October and November and harvested in the month of February and March. Being a leguminous crop has capacity of fixing atmospheric nitrogen in the soil. In spite of high yielding varieties and improved agronomic practices the productivity of pea is low. Diseases are one of the most limiting factors responsible for low yield per hectare. Among the diseases, powdery mildew (*Erysiphe polygoni*), Downy mildew (*Peronospora pisi*), wilt (*Fusarium oxysporum* f.sp. *pisi*), Rust (*Uromycesfabae*) and *Ascochyta* blight (*Ascochyta* spp.) are the most important diseases of pea in India (Sharma, 1998).

13.5 POWDERY MILDEW

Powdery mildew is one of the most important worldwide distributed airs borne disease of garden pea. Its prevalence has been reported from Australia, Canada, China, India, Japan, Malaysia, Russia, Spain, South Africa, Sudan, Tanzania, Turkey, USA, and many other countries (Singh and Singh, 1988). The disease is widespread and often economically important in semi-arid regions of the world. The disease was first reported in India from Dehradun by Butler in 1918. The disease is usually more destructive during late season crop, which attains a serious threat and causing huge losses in both quality and quantity of the produce. The disease can reduce the number of pods per plant, number of seeds per pod, plant height, biomass, and ultimately 25–50% yield loss. Munjal et al. (1963) observed that in severely infected crop, there was reduction of 21–31% pod number, 25–47% in pod weight. Griton and Ebert (1975) reported that 50% reduction in yield was due to powdery mildew infection. However, the reduction in yield is mainly due to reduced photo-synthetic activity in infected plants by the attack of this pathogen. Epidemic development of the disease is very often, fast, and it progresses as compound

Diseases of Garden Peas (Pisum sativum L.) 231

interest. In dry and warm condition the disease becomes more destructive, while downy mildew flourishes in humid weather condition.

13.5.1 SYMPTOMS

First symptom appears on the upper surface of the lower leaves as a very small, slightly discolored spots. These soon give rise to white powdery areas which continue to enlarge as white patches. These white patches combine together to form larger whitish floury areas. As the disease advances, the upper leaves of the plants also show similar symptoms. In severe conditions, both leaves surface adversely infected by the pathogen. Due to the presence of conidia and conidiophores on leaves looks as dusted with flour and the whole crop in the field appears as white from a distance. At the end of conidial formation or in a later stage, the leaf surface shows yellow to brownish patches, and finally, foliage dies. In severe conditions, entire tendrils and petioles are covered with white powdery mass. Pods are also infected in all stages and show white floury patches consisting of white powdery mass. Pods become small in size, shriveled, pod number, and pod weight are also reduced multiple infections may cover the entire aboveground plant. Severely infected plants are unthrifty and have poor yield and quality. Small, oval, black fruiting structures may form in mature lesions.

13.5.2 CAUSAL ORGANISM

The causal agent of the disease is *Erysiphe pisi*, although other species such as *Erysiphe trifolii* and *Erysiphe baeumleri* have also been reported causing this disease. According to Paul and Kapoor (1983), the various species described under the name *Erysiphe polygoni* on different hosts in India were found to comprise eight species viz., *E. polygoni, E. pisi, E. berberidis, E. betae, E. heraclei, E. martii, E. rananculi* and *E. salviae*. The pathogen is ectoparasitic in nature and withdraws their nutrition from host surface. The fungus overwinters on infected plant debris and in alternate hosts. Air current helps in spreading the fungus locally and over long distance.

13.5.3 DISEASE CYCLE AND EPIDEMIOLOGY

Powdery mildew is an air-borne disease of worldwide distribution. Cleistothecia develop on dead plant debris and serve as source of primary inoculums for the next season. Ascospores formed in these fruiting bodies are released by decay of the fruiting bodies and blown by wind to lower leaves where cause infection and produce powdery mass of spores. Secondary spread occurs through windblown conidia. Rainfall suppresses the disease by washing off the spores and causing them to burst instead of germinates. Free moisture on plants will also restrict germination of the spores and does not promote the epidemic. Epidemics of powdery mildews frequently start from foci of infection. Spatial distribution of fungal plant pathogens is determined by components of the disease cycle such as pathogen survival, source of primary inoculums, mode, and amount of inoculums dispersal. Host plant genotypes with quantitative resistance may have less disease because attacking pathogens have reduced infection efficiency, longer latent period or reduced propagule production. One or more of these components can reduce the disease progress (temporal increase) and may spread (spatial increase) in the field. Powdery mildew develops quickly in warm and dry condition for 4–5 days, particularly at flowering and podding stage. If infection occurs earlier than four weeks from maturity, yield losses due to powdery mildew arise from the infection covering stems, leaves, and pods, which will lead to shriveled seeds (Yarwood et al., 1954).

13.5.4   MANAGEMENT

13.5.4.1   CULTURAL

Cultural practices should be applied to avoid the favorable condition for infection and spread of the powdery mildew. Collection and destruction of plant debris and avoidance of close planting are helpful in reducing disease incidence. Pratibha and Amin (1991) reported that sowing of pea from late September to early October, late November or early December show more powdery mildew and give reduced yields. The adjustment of date of sowing may be important in avoiding or reducing powdery mildew infection. The disease does not develop fast under sprinkler irrigation system (Hagedorn, 1984) and it may help to reduce powdery mildew because spores are washed off the plant.

13.5.4.2   HOST RESISTANCE

Genetic resistance is acknowledged as the most effective, economical, and environmental friendly method of disease control. However, only three genes

Diseases of Garden Peas (Pisum sativum L.) 233

(*er1*, *er2* and *Er3*) have been described so far in pea germplasm and only *er1* has been widely used in breeding programs. The use of polygenic resistance or combining several major genes could enhance the durability of the resistance. Many workers have investigated varietal resistance of pea to powdery mildew, but very few varieties have been reported to be resistant (Singh et al., 1988). Resistance in powdery mildew is controlled by a single gene pair 'er' (Azmat et al., 2010). Fondevilla et al. (2006) reported that in pea, two single recessive genes, er1 and er2 have been identified for resistance to powdery mildew caused by *E. pisi*. Gupta and Mate (2009) reported that conidia of *E. polygoni* varied in size according to susceptibility of the cultivars.

13.5.4.3 BIOLOGICAL

Introduction of resistance in pea with nonpathogenic pea powdery mildews (*Oidium* sp., *Phyllactinia corylea*) against a subsequent infection with *Erysiphe pisi* resulted in reduced conidial germination, appresorium formation and secondary branch development up to 12 days after inoculation (Singh et al., 2003). Application of aqueous extract of vermicompost also inhibits spore germination and development of powdery mildews on pea (*Pisum sativum* L) in the field at very low incidence (0.1–0.5%). *Trichoderma harzianum* (0.5%) was found effective and economical for controlling the disease and giving better seed yield (Surwase et al., 2009). Plant extracts and oils are better than chemical because they are environment-friendly and safe from residual effect. It adversely affects the germination of conidia or conidial density (Singh et al., 1984). Turmeric extract significantly reduced the disease severity and increased the number of pods, number of grains, and grain yield (Shabeer and Irfan, 2006). The botanical NSKE (5%) was found highly effective and economical for controlling the disease and giving higher seed yield (Surwase et al., 2009).

Maurya et al. (2004) reported that neem bark methanol extract (NBM) and motha (*Cyperus rotundus*) rhizome ethyl acetate extract (CRE) significantly reduced disease intensity. Ginger, tulsi, mahua, and cashew nut extracts were also effective at concentration 2000 ppm *in vivo*.

13.5.4.4 CHEMICAL

Chemical control is feasible with a choice of protective and systemic fungicides. Among the large number of fungicides tested, high level of control of the disease was observed with the use of tridemorph, dinocarp, and wettable sulfur as foliar spray but sulfur is the most economical fungicide. Foliar sprays of EBI fuhgicides like propiconazole, hexaconazole, tebuconazole, diclobutrzol, triadimenol, enarimol, etc. have been found effective at lower doses than sulfur and posses good antisporulant activity (Rana et al., 1991; Gupta and Sharma, 2004). Tilt (propiconazole) was the most effective treatment, which not only increased seed yield but also reduced disease severity significantly as compared to unsprayed check followed by kerathane (0.1%) and carbendazim (0.1%) during the crop seasons for two years (Prasad and Dewivedi, 2007). The fungicides hexaconazole (0.05%) and Propiconazole (0.05%) were found very effective and economical for controlling the disease and giving higher seed yield (Surwase et al., 2009). Sulfur has a significant chemical cost advantage over systemic fungicides and is acceptable to biodynamic/organic growers.

13.6 DOWNY MILDEW OF PEA

Downy mildew of pea was first discovered by Berkeley in England in 1846 (Chupp and Sherf, 1960), and since then, it has been found in all parts of the world. Dixon (1981) has also reported that downy mildew of peas is widely distributed all over the world. Olofsson (1966) and Biddle et al. (1988) reported yield losses of 30% in Sweden and 45% in the UK, respectively. In India, the disease is prevalent throughout the Indo-i plain and causes considerable yield loss.

13.6.1 SYMPTOMS

Downy mildew causes a different kind of symptoms on pea plants. Three different infection types with different symptoms can be recognized during a crop cycle. The symptom is first visible on the developing third and fourth leaves, and it first appears on the lower leaves then spread to upper leaves. Scattered yellow to brown patches of indeterminate shape appear on the leaves surface. Systemic infection in seedlings causes stunted growth with conidia sporulation, which often covers a major part of the plant surface, and this is caused by oospores in the soil which infect germinating seeds. These infections can seriously reduce the plant population. A lower frequency of infection (50%) was obtained when placed 3 cm above the seed level, and infection was even more reduced (1%) when the oospores were placed 3 cm

Diseases of Garden Peas (Pisum sativum L.)   235

below the seeds. Later in the season, top systemic symptoms can develop with stunting and sporulation occurring over the entire surface of the top of plants. Taylor et al. (1990) showed that systemic infection could also originate from leaf infection. Top systemic infection is the result of direct infection of the top meristem. This type of infection is found more frequently in varieties with reduced stipule size, which may be determined by the gene st (Taylor et al., 1990). In these varieties, the top meristem is not protected by the stipules, which wrap around the apex in varieties with normal-sized stipules. Following infection, the mycelium develops in the intercellular spaces penetrating the stem, the leaf stalks, and even the pods through the veins (Kosevskii and Kirik, 1979).

Local foliar and tendril lesions with conidia sporulation on the abaxial foliar surface is a typical symptom. Local infections on tendrils may develop from conidia present on the plant surface. Pod infection causes yellow lesions on the pod surface and epithelial proliferations on the endocarp. Pod infection develops from conidia deposited on young pods rather than by mycelial growth through the peduncle and pedicel (Mence and Pegg, 1971). Oospores are formed within the yellow lesions. Pod infection often causes distorted pods, seed abortion, and brown discolored small peas with a bitter taste. Pod infections directly affect pea quality and are, therefore, a serious expression of the disease. The seeds beneath the lesions are aborted and undersized. Later in the season, oospore develops in the senescent tissues and can survive in the soil for up to 15 years (Van der Gaag and Frinking, 1997).

13.6.2   *CAUSAL ORGANISM*

Downy mildew of pea is caused by *Peronospora pisi* Syd. and belongs to kingdom- chromista, phylum- Oomycota, class- Oomycetes, order- Peronosporales and family- peronosporaceae. Mycelium of the fungus is hyaline, aseptate, intercellular, branched with finger-shaped haustoria. Sporangiophores first appear as simple elongating hyphae from stomata on abaxial leaflet surface then branching from a single axis to produce multiple terminal sporangia. Gametangia developed extensively on inner surface of pods from smooth, bulbous hyphae, adhering to the host epidermis. Species of *Peronospora* produce conidia that lack modification in the apical region, the operculum which do not contain zoospores and germinate by germ tubes (Shaw, 1981). *P. viciae* also produces oospores, which have a typical reticulate pattern of the exosporium. These species are also capable of regular and predictable production of oospores in large numbers. Both heterothallic and homothallic isolates of these two species have been found. Sexual reproduction is probably important for the adaptation of the fungus to various host genotypes by recombination of virulence genes. The vegetative stage is probably diploid like in other species of *Peronospora* (Tommerup, 1981). The fungus is an obligate parasite which can only grow on living plant tissue. Forma specialis pisi can only infect *Pisum* species and not species of the genus *Vicia* within the tribe *Vicieae* (Campbell, 1935).

13.6.3   DISEASE CYCLE AND EPIDEMIOLOGY

The disease is both seed and soil-borne in nature. In the seed, the fungus may be present as a contaminant, or the seed infection may extend beneath the seed coat. Oospores on germination, the infected seeds might act as the source of perpetuation; however, the main sources of primary inoculums are the oospores on in the diseased crop debris where these can survive up to two years. Oospores on germination produce a germ tube that infects the seedlings systemically. Oospores in the soil also act as a source of primary inoculums early in the season. The oospores can survive for a long time in the soil. Oospores survive for 10–15 years in the soil (Olofsson, 1966). Secondary spread of the disease occurs through sporangia disseminated by the moist wind. Under favorable temperatures for sporangial germination, four hours of leaf wetness is required for Infection. High humidity also helps in the dissemination of sporangia. Cool nights coupled with foggy weather or the presence of dew favor the disease development. Optimum temperature for infection to occur is 16°C.

13.6.4   MANAGEMENT

13.6.4.1   CULTURAL

Field sanitation, destruction of infected crop debris and 3–4 years crop rotation with nonhost crops are important in reducing the primary inoculums. Proper drainage and wider spacing create micro conditions that are unfavorable for disease development. Since the pathogen is also seed-borne in nature, it is always recommended to use disease-free seed. Seed treatment in hot water at 50°C for 25 minutes has also been found effective to eradicate the seed-borne inoculums. Watering early in the morning so that leaf

Diseases of Garden Peas (Pisum sativum L.) 237

surfaces dry out rapidly and avoid it in the evening, which can lead to high humidity or leaf wetness and persists throughout the night.

13.6.4.2 HOST RESISTANCE

During recent years improved varieties have been reported from India, and it is imperative that they are protected against the major diseases of pea. Pea varieties are known to differ in their reaction to downy mildew disease. Bains et al., (1993) reported the pea varieties PWR-3 and Bonneville have resistance under Punjab condition. Variation in resistance between pea cultivars has been reported by Stegmark (1988). Some cultivars are completely resistant to some isolates but are fully susceptible to others. Race-specific resistance of pea was found in several cultivars, but there is no genotype with complete resistance to all known pathogen races (Matthews and Dow, 1983). The cultivar 'Dark Skin Perfection' is more resistant to downy mildew than some other cultivars used for the production of peas for canning and freezing (Stegmark, 1988).

13.6.4.3 CHEMICAL

The pathogen is known to overwinter in the form of oospores in the infected plant tissues and in the seed. Seed borne inoculums should be eradicated through seed dressing with systemic acylalanine fungicides like metalaxyl, cymoxanil, and fludioxanil are very effective against systemic seedling infections (Brokenshire, 1980). However, later in the season, the pod infection can still be severe. There is no real effective fungicide treatment against pod infection in peas. In the long run, the current acylalanine fungicides may become ineffective due to the development of tolerance by the pathogen. The secondary spread of the disease should be checked by spraying fungicides like carbendazim + chlorothaonil, Carbendazim, + mancozeb. To avoid buildup of fungicide resistant strains, it is always better to rotate the fungicide and apply in mixture with non systemic.

13.7 FUSARIUM WILT

Wilt of pea caused by *Fusarium oxysporum* f. sp. *pisi.* (Linford) Snyder and Hansen is one of the most important diseases in pea growing areas. The *Fusarium* wilt of pea was first reported in 1918 by Bisby in Minnesota

(Chupps and Sherf, 1960). In India, the occurrence of Pea wilt report was made available by Patel et al. (1949) from Bombay. Peshne (1966) made a comparative study on morphology, physiology, and pathogenicity of *Fusarium oxysporum* f. sp. *pisi*. Wilt of pea causes serious losses worldwide and is one of the major yield reducers. Sharma et al. (1998) found that soil born disease like root rot and wilt are the limiting factors in producing early crop of pea. Maheshawari et al. (1983) made a survey of wilt and root rot complex of pea in the various pea growing areas in Northern India and reported 13.9 to 95% yield loss. Losses up to 60% were reported in the crop during the years of epiphytotics in Himachal Pradesh (Kumar, 1983).

13.7.1 SYMPTOMS

Plants are susceptible to the disease at any stage of crop growth. Symptoms of pea wilt have been described by different workers from time to time. Linford (1928) observed symptoms as a distortion and wilting of leaflets by sudden collapse of the plants and extensive cortical decay of the roots. The first and the foremost characteristics symptoms is a downward incurving of the margins of younger leaves and stipules, accompanied by a slight yellowing of the leaves and a superficial grayness suggesting an excessive development of waxy bloom. The lower internodes increase in diameter and the entire stem becomes rigid (Schroeder and Walker, 1942). Sukapure et al. (1957) reported symptoms of pea wilt as rolling of leaves, upper parts of the plants may be pale and the growth of terminal bud is checked. Stem and upper leaves may become more rigid than normal and the roots crisp, while the lower leaves turn pale and commence to wither. Sometimes, the entire plant becomes yellow and the lower leaves wither progressively upwards; however, after the collapse of a few basal leaves, the upper part of the plant wilts abruptly and may become dry while still green in color. After such wilting the stem shrivels downwards from the tip to basal internodes, which remains firm and turgid till the end discoloration of vascular system and partial wilting is characteristic symptom of *Fusarium* wilt. If pods are formed, they contain only a few shrunken, immature seeds and dry earlier.

13.7.2 CAUSAL ORGANISM

Wilt is an important soil born disease, caused by *Fusarium oxysporum* f. sp. *pisi* (Linford) snyder and Hansen is one of the most important diseases

Diseases of Garden Peas (Pisum sativum L.)

of pea. The fungus belongs to kingdom fungi, Phylum Ascomycota, Order Tuberculariales and Family Tuberculariaceae. The fungus hyphe is septate, delicated, white to peach colored, usually with a purple tinge. Microconidia are borne on simple phialades arising laterally on hyphae or from short, less branched conidiophores. These are oval ellipsoid to cylindrical or curved and measures 5–12 x 2.2–3.5 micrometer. Macroconidia are borne on elaborately branched conidiophores or on the surface of sporodochia. These are thin-walled, 3–5 septate, fusoid-subulates pointed at both ends and measure 27–46 x 3–4.5 micrometer. Chlamydospores are both terminal and intercalary.

13.7.3 DISEASE CYCLE AND EPIDEMIOLOGY

Fusarium oxysporumf. sp. *Pisi* is mainly soil-borne pathogen that is commonly found in soil. It can survive in soil as chlamydospores without pea crop for more than 10 years. The pathogen is not a seed-borne in nature, but it can be carried on the seed coat. Anwar et al. (1994) isolated the pathogen from both ingeminated seed and abnormal seedlings. The fungus causes infection on the fibrous roots or epicotyls region and grows inter and intercellular in the cortex and ultimately concentrates in the xylem vessels. The mycelium may grow systemically through the vascular system and reach to the seed causing infections. The infected seeds germinate and resulting in production of abnormal seedlings. After the death of the plant, the fungus continues to grow and sporulate on the stem cortex, resulting in production of soil-borne inoculums. The spores penetrate the pea plant through the root hairs and fibrous roots. The pathogen mainly enters through cut surface and the exposed stele is necessary for infection. It grows upward through the stem often well into the upper branches in the xylem. This process interferes with the passage of water from the roots to the stems, leaves, and pods resulting in yellowing, dwarfing, and wilting of plants.

The pathogen is soil-borne and also survives on the seed. The spores can be carried from one field to another on farm equipment, on crop debris and in wind or water-borne soil. The disease is more prevalent in alkaline soil. Favorable condition for plant growth reduces fungus growth and does not permit the disease to progress rapidly. The reduction in shoot length could be used to supplement the visual severity rating for *fusarium* wilt in field pea (Neumann and Xu, 2003). The pathogen establishes in some areas quickly than others and is serious under moist conditions. A soil temperature of 23 to 27°C is most favorable for disease development.

13.7.4 MANAGEMENT

13.7.4.1 CULTURAL

The use of disease-free seed, practicing field sanitation, and long crop rotations are important in reducing the primary inoculums of the pathogen. Early planting of pea crop has also been suggested to lower the disease levels. Soil amendment and repeated cropping with different non-host crops such as wheat, oats, maize, and sorghum reduced the wilt population. Minimum disease incidence and maximum grain yield were observed during the 1st and 15th November when the soil temperature are low for the development of wilt and it was very severe when the planting were done in first week of September (Sharma and Sharma, 2003). Use of balanced fertilizers reduces the wilt complex in pea (Sagar and Sugha, 1998). Minimum disease incidence and apparent infection rate were observed when seeds were sown at 8 cm distance spacing as compared to other spacing (Verma and Dohroo, 2005a).

13.7.4.2 HOST RESISTANCE

The only economic control in wilt infected soil is to grow resistance or tolerant varieties of pea against wilt. Virin and Walker (1939, 1940) evolved a system of numerical disease evaluation based on the symptoms of disease by calculating the disease indices. In this method, the index represents the average number of days from sowing to the particular stage of the disease concerned. Ramphal and Choudhary (1978) observed soil inoculation with wilt pathogen was the most effective method of producing pea wilt disease. They screened pea cultivars and reported Kalanagni as immune and Alaslea as a resistant variety. Datar (1983) tested 36 cultivars against *Fusarium oxysporum* f. sp. *pisi* in field condition, out of which five cultivars namely BR-12, Khapera Khada, 15/1, 4–3 and 479–13 were found resistant and rest of them were found susceptible. Kumar and Kohli (2001) reported that sixteen cultivars of pea (*Pisum sativum*) were screened against *Fusarium oxysporum* f. sp. *Pisi*, in which Arkel was most susceptible while accession DPR-3 was found most resistant. Inheritance of wilt resistance revealed that single dominant gene is governing the resistance. "Sanjunichi" and "Tsurunashiakahana" among the pea cultivars were found resistant cultivars (Mashita and Fukaya, 2006).

Diseases of Garden Peas (Pisum sativum L.) 241

13.7.4.3 CHEMICAL CONTROL

Seed dressing with ceresan, dexon, captan, carbendazim, and benomyl has been found effective for the control of pea wilt (Gupta et al., 1989). Seed treatment of pea with Bavistin and Benlate gave complete control of *Fusarium oxysporum* f. sp. *pisi.* (Utikar et al., 1978). Shukla et al. (1979) tested the efficacy of some seed dressing fungicides for the control of pea wilt and reported that Bavistin have the best result in improving germination, reducing mortality and giving significant higher yield. Maheshwari et al. (1981) reported that seed dressing with Benomyl and Dithane M-45 and soil treatment with Phorate + Captan reduced plant mortality against *Fusarium oxysporum* f. sp. *pisi* and increased yield. The pathogen was also controlled by soaking pea seeds in a 1:1 combine's suspension of Captafol (0.1%) and Captan (0.2%) (Gangopadhyay and Kapoor, 1979). Sinha and Upadhyay (1990) tested 11 chemical compounds and found that pathogen growth was completely inhibited by Emisan-6 and Sulfex (80% S) at all concentrates tested. Wang et al. (1995) reported that seed treatment with fungicide like organomercurials, thiram or carbendazim in known to reduce seed borne inoculums of pathogen and has been recommended. Pandey and Upadhyay (1999) reported Bavistin was highly effective fungicide while *T. viride* and *T. harzianum* best among all bio-control agents. Integration of all bio-agents with Bavistin was not beneficial but bioagent + Thiram was highly effective. Verma and Dohroo (2002) conducted field experiments and reported that Bavistin treatment resulted in the highest mean seed germination, lowest pre emergence rot and highest yield with the lowest wilt incidence. Maheswari et al. (2008) reported that Carbendazim proved most effective fungitoxicant for checking the mycelial growth of *Fusarium oxysporum* F. sp. *lentis* (5.6 mm) followed by Captan (9.9 mm), Hexaconazole (12.5 mm) and Diniconazole (16.44 mm). Phosphoorganic insecticides are powerful cutinase inhibitors and inhibited cutinase released by *Fusarium oxysporum* F. sp. *Pisi* structures for infection (Koller et al., 1982). Carbendazim and thiophanate-methyl applied as seed treatment were highly effective in increasing fresh and dry weight, root, and shoot length, nitrogen content in pea plants and also improved seed germination and plant survival (Neweigy et al., 1985).

13.7.4.4 BOTANICALS

Sharma et al. (2003) reported the antifungal activity of different plant extracts against *Fusarium oxysporum* f. sp. *pisi in vitro* condition the leaf extracts of *Datura stramonium* and *Azadirachta indica* had superior anti-fungal activity. Verma and Dohroo (2003) studied fungitoxic effect of different plant extracts against *Fusarium oxysporum* f. sp. *pisi in vitro* and found the leaf extract of garlic showed 100% inhibition followed by *ocimum* extract. Devi and Paul (2003, 2005) the fungitoxic activity of extract of 10 plant species was evaluated against the pea wilt caused by *Fusarium oxysporum* f. sp. *pisi*. The *Ranuculus muricatus* extract completely inhibited the growth of the wilt pathogen followed by Lantana, *Ocimum*, and *Datura*. Sahni and Saxena (2009) reported the antifungal activity of ethanolic extracts of medicinal plants were evaluated against *Fusarium oxysporum* f. sp. *pisi* by "Modified disc technique" and various plant extracts resulted inhibition on the growth of mycelium however bark of *Euphorbia nerifolia* exhibited absolute toxicity against the test fungus.

13.7.4.5 ESSENTIAL OIL

Sharma et al. (2003) reported the antifungal activity of different plant oils were evaluated against *Fusarium oxysporum* f. sp. *pisi in vitro* condition. The oils of *Datura stramonium* and *Azadirachta indica* have considerable antifungal activity against the wilt pathogen. Abo-El Seoud et al. (2005) reported essential oils of fennel, peppermint, caraway, eucalyptus, geranium, and lemon were tested for their antimicrobial activities against *Fusarium oxysporum* and essential oils of fennel, peppermint were selected as an active ingredient for the formulation of biocides due to their efficiency in controlling the tested *Fusarium oxysporum*. Sitara et al. (2008) reported essential oils extracted from the seed of neem, mustard, black cumin and asafetida were evaluated for their antifungal activity against *Fusarium oxysporum, F. moniliforme, F. nivale*, and *F. semitectum*.

13.7.4.6 BIOLOGICAL CONTROL

Pre-inoculation application of *Gliocladium roseum* provided good control of wilt of pea (Saksirirat et al., 1994). Velikanov et al. (1994) reported that *Trichoderma aureoviride, Trichoderma harzianum, T. viride* and *Gliocladiumvirens* were found to be hyperparasitic on both *Fusarium oxysporum* and *Fusarium solani*. Verma and Dohroo (2003) studied the efficacy of the fungal antagonists *Trichoderma harzianum* and *Trichodermaviride* against Fusarium wilt of pea caused by *Fusarium oxysporum* f. sp. *pisi in vitro*. *Trichoderma*

Diseases of Garden Peas (Pisum sativum L.) 243

harzianum and *Trichoderma viride* showed the maximum growth inhibition of the wilt pathogen. Seed treatment with antagonists like *Trichoderma viride* and *Trichoderma harzianum* was found effective in the management of pea wilt both under glasshouse and field conditions (Verma and Dohroo, 2005c).

13.8 RUST

It is an important disease of garden pea particularly in wet areas of pea cultivation throughout the world. The rust fungus was first identified by Jordi in 1904 (Buchheim, 1922). In India, pea rust pathogen *U. viciae fabae* by Sydow and Butler on *Vicia faba* from Pusa, Bihar in 1906. The first report of its occurrence on pea in India was by Butler (1918). Two species of *Uromyces* have been reported to cause rust of pea. One of them *Uromyces pisi* (Persoon) de Bary, has been reported from several European countries (Deutelmoser, 1926; Jorstad, 1948). It is a heteroecious species having its aecial stage on *Euphorbia cyparissias* and rarely occurs in India, other species *U. viciae-fabae* (Pers.) has been found to cause pea rust in India (Prasada and Verma, 1948). The disease may cause epidemic proportions under favorable weather conditions resulting in considerable yield losses.

13.8.1 SYMPTOMS

The rust pustules appear on all green above-ground parts of the plant. The minute raised yellow rust pustules to appear on above-ground parts of the plants and most preferably on underside of the leaves and less abundant on the pods and stems. All the four stages develop on green part of the host, including the pods. The first symptoms appeared with the development of aecia. The yellow spots have aecia appear first on the undersurface of the leaves, stems, and petioles persist for longer time The formation of the aecial stage is preceded by a slight yellowing, which gradually turns brown. The uredopustules are powdery light brown in appearance. Late in season dark brown to black teleutopustules develop on the leaves but most commonly on stem and petioles.

13.8.2 CAUSAL ORGANISM

The disease is caused by two species of *Uromyces* viz, U. pisi (Pers.) de Bary and *U. viciae-fabae* (Schroet). Later one is worldwide distributed pathogen of pea and also reported from faba bean (*Vicia faba* L.), lentil (*Lens culinaris* Medic.) and sweet pea (*Lathyrus sativus* L.) (Shroff and Chand, 2010). The fungus is an autoecious with aeciospores, urediospores, and teliospores on the surface of host plant and completes its life cycle on the same host. In India, urediospores are converted to teliospores under field condition during the month of March due to the higher temperature. Urediospores are short-lived while teliospores can survive in plant debris from one season to another (Hebblethwaite, 1983). Germination of teliospores takes place between 17 to 22°C temperatures and at the start of next season producing basidiospores which initiated new infection cycle (Joshi and Tripathi, 2012). The disease is favored by high humidity, cloudy, and rainy weather condition. Disease development in field condition is favored between 20°C to 22°C (Kushwaha et al., 2006). On peas the fungus starts its life cycle with the formation of pycnia, aecia, uredia, and finally telia. The yellow spots having aecia in round or elongated clusters are the earliest visible symptoms of the disease on the leaves. Pycnia occurs in small groups associated with the aecia. The aecia are cupulate and 0.3 to 0.4 mm in diameter. The peridium is short and whitish. The aeciospores are round to angular or elliptical with the hyaline wall 1 micron thick and verrucose. The aeciospores measures 14 to 22 microns in diameter. The uredia develop on both the sides of the leaves and on other parts of the plant. They represent a powdery light brown appearance. The urediospores are round to ovate, light-brown, echinulate with 3–4 equatorial germ pores and measure 20–30 (22–28) × 16–26 (19.22) microns. The telia occur in the same sorus as the uredia are developed from the same mycelium. The teliospores are dark-brown or black, subglobose, ovate or elliptical with rounded or flattened apex which is considerably thickened and appears papillate and measure 25–38 (40) × 18–27 micron. The mycelium in intercellular, branched, septate having yellowish or orange-red oil drops in the cytoplasm knob-like haustoria are formed in the host cell. The teliospores represent the sexual stage of the fungus and on germination give rise to a promycelium from each cell. The diploid nucleus passes into the promycelium, undergoes meiosis, and four haploid nuclei are formed. Each nucleated cell of the promycelium then produces short sterigmata at the tip of which swells to form a globular basidiospore in which the single nucleus form the cell of the promycelium moves. The basidium in the uredinales thus consists of two stages: the probasidium or hypobasidium (teliospore and epibasidium (promycelium). Polymorphism is the *Uromyces* spp. exhibited by development of following types of spore in the sequence listed:

Diseases of Garden Peas (Pisum sativum L.) 245

1. **Stage 0:** Pycnia or spermogonia bearing pycniospores or spermatia and they are produced on a haploid thallus resulting from infection by basidiospores. This is the only monocaryotic stage of the rust mycelium on the host. They contain a palisade of sporogenous cells which cut off at their tips single-celled, uninucleate pycniospores in sweet nectar. The spore laden nectar is exuded from the pycnia and carried by insects. The spermatia affect fertilization by fusion with receptive or flexuous hyphae present in the mouth of pycnia of the apposite mating type. The pycnia are highly variable in shape, being globose, conical, hemispherical, lens shape or undefined shape without proper delimiting boundary.

2. **Stage I:** Aecia bearing aeciospores and are formed as a result of dicaryotization of the monocaryotic mycelium. They are associated with pycnia and also act as repeating spores or uredia. The aecia contain a palisade of binucleate sporogenous cells at their base. These cells invariably bear unicellular, binucleate, hyaline aeciospores in chains. They germinate to produce a dicaryotic mycelium initiating the dicaryon which bears the uredia and telia.

3. **Stage II:** Uredia or uredinia bearing urediospores or uredinio-spores. The uredia are sori produced by binucleate thallus and bearing one-celled urediospores singly on pedicel. The urediospores are binucleate with hyaline or colored walls and are perforated by conspicuous pores (germ pores). The wall is echinulated with pointed conical spines.

4. **Stage III:** Telia bearing teliospores or teleutospores and which are vary enormously. The spores may be single or multicellular, smooth, stalked or sessile and embedded in the host or free. The young telio-spore is binucleate but at maturity it has a single diploid nucleus representing the diplophase in the life cycle.

5. **Stage IV:** Basidium or promycelium bearing basidiospores. It represents the transition of diplophase to haplophase by the germination of teliospore, meiosis of diploid nucleus and formation of haploid basidiospores.

13.8.3 DISEASE CYCLE AND EPIDEMIOLOGY

The rust pathogen is mainly soil-borne in nature as teliospores survive in crop debris. In India, the rust appears to survive on weed hosts belonging to *Lathyrus, Vicia*, etc., and the spores are windblown to the main crop (Singh, 1987). Wild hosts may serve as primary or secondary source of infection. Aeciospores in *U. viciae-fabae* were found to be repeating spores and play an important role in outbreak of pea rust in north India. Inoculation of pea plants either by aeciospores or urediospores resulted in the production of aeciospores (Kushwaha et al., 2006).

Very little information is available on the effect of environmental factors on development of pea rust. Prasada and Verma (1948) reported that at relatively low temperature 17–22°C results in formation of secondary aecia while at 25° the infection causes development of uredia. No infection by aeciospores occurs at 30°C. These spores remain viable for 8, 6, 4, 3 and 2 weeks at temperatures of 3–8°C, 10–12°C, 17–18°C and 30°C, respectively. No viability is retained after 6 weeks showing that the aeciospores do not survive during the off season for the crop. Singh (1998) reported that optimum temperature for germination of uredospores is 16–25°C and no germination occurs at 28–29°C. In the plain of North India, where warm-season sets towards the end of March. These spores remain viable for 16–17 weeks, when stored at 3–8°C and only for 2 weeks at 36–37°C. Thus these spores do not survive in the hot summer interning two successive crop seasons. The teliospores of the rust fungus have been found to have no dormancy and can germinate at 12–22°C soon after their formation. Batra and Stavely (1994) working on *Uromyces appendiculatus* (Pers.) reported that the urediospore germinate at 15°C to 24°C while the teliospores in crop debris germinate at 10°C to 15°C under favorable conditions the spore complete the infection cycle within next 5–10 days. Recently Shroff and Chand (2010) reported that infection process of *Uromyces viciae fabae* started after deposition of aeciospores on the surface of pea leaves at a temperature 25–30°C and relative humidity (RH) of 99–100%. Kushwaha et al. (2006) observed that aeciospores in *Uromyces fabae* were found to be repeating spores and play an important role in pea rust outbreaks in the North Eastern Plain Zone (NEPZ) of India. Among the different growth stages of pea, the pod formation stage was highly susceptible and production the maximum number (744) aecidia/leaf at 20–25°C. Urediospore production mainly coincided with the senescence of the pea plants. Atmospheric temperatures around 20°C maximum and 5°C minimum with high RH (60–70% mean weekly) and light shower or drizzle favors for *U. viciae-fabae* development and spread. Maximum Temperature 25°C and minimum 7–8°C and less or more rains disfavor rust spread (Mittal, 1997).

Diseases of Garden Peas (Pisum sativum L.) 247

13.8.4 MANAGEMENT

13.8.4.1 CULTURAL PRACTICES

Cultural practices mostly affect the environmental conditions favorable for host plant and natural enemy of pathogen and make unfavorable to pest and pathogens. The cultural practices may be very important tool in avoiding or reducing the inoculums/disease without any unwanted side effects like pollution. Generally, the disease appears late in crop season, and hence, the losses can be reduced by early planting. Delayed sowing after 5 October had increased the severity of rust severity and decreased grain yield. Cultivar 'khaparkheda' gave the highest seed yield (1.54 t/ha) and had the lowest severity (Sangar and Singh, 1994). Singh et al. (1996) reported that three pea cultivars were sown on 4 dates between 5 October and 4 December in a field trail and the severity of rust was increased as sowing delayed. Field sanitation to destroy diseased crop debris and long crop rotations avoiding broad beans, *Vicia, lathyrus* should be followed for minimizing losses from the disease (Singh, 1987). Efforts should be also made to locate and destroy the weed hosts.

13.8.4.2 HOST RESISTANCE

The evolvement of rust-resistant varieties seems to be the most effective, but there is a need for screening of existing lines/germplasm/cultivars against pea rust. Singh and Tripathi (2004) reported that KFP 106, DMR 11, HUP 8603, Type 163, and KPMR 22 showed a good level of resistance, which being conditioned by a number of genes. Xue and Warkentin (2002) observed that Tara and Century were the most resistant to both UF-1 and UF-3 isolates while Victoria and Topper were the most resistant to UF-2 only. Barilli et al. (2009) collected 2759 pea accessions and screened for resistance against pea rust. All accessions displayed a complete interaction (high infection type) both in adult plants under field condition and in seedlings under growth chamber conditions, but with varying levels of disease reduction and no complete resistance was observed. Khan et al. (2009) reported that among all tested pea cultivars only Climax gave the highest yield due to lowest disease severity and Meteor had the lowest yield due to maximum disease severity.

Induced resistance seems to a new approach for the control of pea rust. Walter and Murray (1992) observed that inoculation of the lowest two leaves of *Vicia faba* with urediospores of rust fungus *Uromyces viciae-fabae*. The resistance was seen as diminished infected areas on the leaves and as fewer urediospores for standard area for 29 days from challenge. The resistance was very high when first days separated the two inoculations but had disappeared when 12 days separated to two. In further experiments with plants at the same stage and using the same isolates of the rust fungus, Walter, and Murray (1992) found that treatment of the first two leaves with either 10 m^M tripotassium phosphate or 5m^M EDTA in place of rust inoculation also caused significant increase in resistance of the upper leaves to challenge inoculation with the rust fungus. In general resistant genotypes contained higher level of phenolics and susceptible ones had more sugar content.

13.8.4.3 CHEMICAL CONTROL

Chemical pesticides are backbone to control of rust disease in pulses and cereals crops till today. Spraying of fungicides alone or in combination has been considered necessary to provide adequate protection to the crop from rust incidence. Fungicides found effective against pea rust are Diathane M-45 (0.2v a.i.) and Calixin (0.2% a.i.). First spraying is done as soon as the disease appears in the field and three more sprays are given at ten days interval. Upadhyay and Gupta (1994) reported that Tridimefon, Maneb + Tridemorph were effective against rust disease under field conditions. Systemic sterol biosynthesis inhibiting fungicides were effective against *Uromyces viciae-fabae* (Fuzi, 1995). Khaled et al. (1995) reported that fungicides (benomyl) carboxin, metalaxyl, oxycarboxin, thiram, triadimefon, and triforine) alone or with Dithane M-45 (mencozeb) were effective against rust but Propiconazole (Tilt) gave the best control, reducing rust intensity and increased pod yield. Folicur and calixin were also effective against rust (Ayub et al., 1996). Gupta et al. (1998) observed that hexaconazole (0.1%) and difenoconazole (0.01%) were best against rust and increased yield. Mancozeb seed treatment was the most effective followed by carboxin and benomyl. Gupta and Shyam (2000) reported that seven ergostral biosynthesis inhibiting fungicides cyperconazole, flusilezole, propiconazole, and hexacanozole completely inhibited rust incidence and rust severity on leaves. Singh and Tripathi (2004) observed that Baycor

Diseases of Garden Peas (Pisum sativum L.)

(0.1%) prophylactic 2 to 3 sprays at 15 days interval was found most effective in reducing the disease severity and appreciable increase in grain yield. Individual fortnightly foliar sprays, commenced when the disease appeared with carbendazim, score, Tebuconazole, and hexaconazole among systemic and at 10 days interval of antracol, microsul, and shareamong the nonsystemic fungicides proved effective for combating the rust disease and in ameliorating the crop yield (Sugha et al., 1998).

13.9 ASCOCHYTA BLIGHT

Ascochyta blight was first reported from the North Western Province of India presently in Pakistan (Butler, 1918). Ascochyta disease complex has been reported from Poland, Germany, Chile, India, Austria, East Africa, Bulgaria, Scandinavia, and Netherland (Kaiser et al., 1998). *P. medicaginis* var. *pinodella* associated with Ascochyta disease complex of pea was isolated from pea cultivar Lincoln from Bajaura, in Himachal Pradesh by Sagar and Bhardwaj (1997). Srivastava and Gupta (1990) reported pea blight (*Ascochyta pisi*) from Sikkim, where it caused heavy losses from December to March. Under favorable weather conditions, it causes significant yield losses. The disease may cause yield losses up to 40%, but sometimes in blight phase alone the losses may go up to 60% (Bretag et al., 1995; Tivoli et al., 1996).

13.9.1 SYMPTOMS

Ascochyta blight is characterized by presence of brown to purplish, irregular areas on the foliage. Under high moisture condition for long period, the lesions become circular and larger in size. The small, brown to purplish irregular spots appear on the pods and enlarge to irregular, purplish, large area could become blotched with the coalescing of lesions. The early symptoms on stem appear as black to purplish streak which is more pronounced at the nodes and could enlarge into brown to purplish irregular areas on the stem. Pycnidia are usually darker in color and produced on lesions especially on leaves and pods in characteristic ring pattern. Fruiting bodies, the pycnidia, form concentric rings, which is the characteristic symptom of the disease. The lesions are circular on leaves and pods, whereas these are elongated on stem and branches. The apical twigs, branches, and stems often show girdling and plant parts above girdle portion are break-off even before drying. Lesions on pods are prominent and usually circular with dark margins. Pod infection often leads to seed infection through testa as well as cotyledons. Dark lesions with pycnidia in the concentric rings are formed even on the seed coat. In the field condition, disease appears in patches after 6–8 days of infection and rapidly spread to the entire field under congenial environmental conditions.

13.9.2   CAUSAL ORGANISM

Ascochyta blight disease complex consists of three pathogens, which include *Ascochyta pisi*, which causes spots on leaves and pods; *Ascochyta pinodes* (teleomorph *Mycosphaerella pinodes*) causes blight and *Phoma medicaginis* var. *pinodella* which causes foot rot. The fungus is homothallic. The ascostromata are globose with beaked ostioles. Asci are cylindrical-clavate with a wall made up of two membranes. The inner membrane is thickened at the tip and is provided with an apical pore. At the time of ascospores discharge the outer membrane is ruptured at the tip and the inner membrane stretches to approximately three times its length. The spores move towards the apex and when the membrane ruptures at the pore the spores are ejected and the stretched membrane contacts.

13.9.3   DISEASE CYCLE AND EPIDEMIOLOGY

Ascochyta blight is both externally and internally seed-borne. The fungus may be present on the seed surface, within seed coat, cotyledons, and embryo. The infested seeds are the main source carrying pathogens from one season to the next and one place to another. Infected seedling dies quite early and may serve as substrate for growth of the fungus and formation of pycnidia and conidia for secondary spread. In *M. pinodes* development of perithecia in crop refuse serves as another source of perennation. Development of pycnidia and perithecia of *M. pinodes* was studied on pea cultivar Solara under greenhouse and field conditions (Roger and Trivoli, 1996). Development and quantity of pycnidia were related to inoculums concentration and physiological state of the plants. Pycnidia were produced on both green and senescent organs while perithecia on senescent organs only. Spores trapping showed that both pycnispore dispersal and ascospore discharge were

Diseases of Garden Peas (Pisum sativum L.)
251

initiated by rainfall or dew. Pycniospores and ascospores were dispersed through the growing season, indicating that ascospores also play an important role in secondary infections. Waterlogging of pea already infected with *M. pinodes* may result in more severe infection and greater reductions in plant growth, cultivars more sensitive to waterlogging may suffer greater losses from disease (McDonald and Dean, 1996).

Spore germination of *A. pisi* was 85–87% in a drop of water or at 100% RH when temperature was 20–21C (Susuri, 1976). Singh (1987) reported that the infection did not occur below 80% RH, but it rapidly increased above 90%. Singh et al. (2005) reported the significant effect of temperature, moisture duration, and their interaction on *Ascochyta* blight development. Temperature ranging between 20–30°C with moisture duration of 12–24 hours resulted in severe disease development.

13.9.4 MANAGEMENT

13.9.4.1 CULTURAL PRACTICES

The use of disease-free or healthy seed is very important for managing Ascochyta blight disease. Disease-free seed can be produced if the crop is grown in low rainfall areas. The disease can be reduced by following long crop rotation and by reduction of the crop refuse by burning either in the field or after threshing (Bedlan, 1985).

13.9.4.2 HOST RESISTANCE

Kavasnikov and Krotova (1977) reported, out of 260 genotypes of pea only 13 were showed resistant against *Ascochyta pisi*. Sandhu and Dhillon (1984) reported from Ludhiana and Gurudaspur that pea cultivars Pleiofila and ML-21 were moderately resistant to *M. pinodella*. Iqbal et al. (2001) found three lines 89P117-5, 88P022-6-28 and 88P0-6-29 of pea as highly resistant to blight (*A. pisi*). Pea lines Bartel, Brite, Bodil, Borek, Karo, Meteor, Rondo, Zolty Pomorski, KM01, KM02, KM03, Solara, Bohatyr, and Lu15/92 were found to be resistant against *M. pinodes*. Lines K1632, K3055, K5072, K5117, K5513, K6391, K7354, and K8195 (Vladimirtseva et al., 1990). Pea cultivar viz., Oscar, Pony Express and Ru/53 were found resistant to *Ascochyta pisi* (Obradovic et al., 1994).

252   *Diseases of Fruits and Vegetable Crops*

13.9.4.3 BIOLOGICAL CONTROL

The effect of seed coating with antagonist *Pythium oligandrum* or fungicides (thiabendazole (TBZ) + fosetyl-aluminum + captan or TBZ + metaalxyl + thiram) under *in vitro* and *in vivo* conditions and found that *P. oligandrum* was an aggressive parasite of pathogens under *in vitro* condition but failed to control footrot under field conditions while fungicidal seed treatments significantly reduced foot rot under field condition (Bradshaw-Smith et al., 1991). Lacicowa and Pieta (1996) reported that seed dressing with *Trichoderma koningii* and *Gliocladium roseum* was found effective in protecting the seed from *Ascochyta pisi*. Pretorius et al. (2002) observed that crude *Eucomis autumnalis* bulb extract prevented the *M. pinnodes* spore infection of the leaves under *in vivo* conditions by inhibiting the spore germination and showed no phytotoxic reaction on the leaves. Four isolates were tested alone and in combinations for suppression of the disease and promotion of plant growth under field conditions. The mean of disease reduction with the most promising isolate 51 was 60% in foliar and 55% in the plant debris treatment. In addition to disease suppression, pseudomonads promoted plant growth in terms of increased plant height and grain yield. Moreover, pseudomonads were compatible with some fungicides at concentrations as high as 100 ppm in *vitro* condition.

13.9.4.4 CHEMICAL CONTROL

Seed treatment with fungicides is an effective measure to reduce the severity of disease. Seed treatment with fungicides viz., thiram, TBZs (Bretag, 1985) were found effective in reducing the seed-borne infection of Ascochyta spp. The seeds treated with cymoxanil + oxadixyl + carbendazim + thiram showed less rotting than the cymoxanil + oxadixyl treated seeds (Sanssene et al., 1998). Foliar application of fungicides such as TBZ (Bretag, 1985), copper oxychloride, chlorothalonil, and benomyl (Warkentin et al., 1996) have been found effective in reducing the severity of Ascochyta blight and increasing the yield and seed weight of pea. Single application of mancozeb at the early flowering stage was effective in reducing the disease severity and in increasing yield (Warkentin et al., 2000). Combined seed treatment with carbendazim and thiram and two foliar sprays each with mancozeb and dinocap reduced the severity of the disease and increased yield (Singh et al., 1992). EBI fungicides like Prochloraz were also found effective in reducing the disease and increasing yield (Nasir and Hoppe, 1997). Fungicides like

Diseases of Garden Peas (Pisum sativum L.) 253

chlorothalonil and benomyl are found effective in managing Ascochyta blight epidemics (Warkentin et al., 1996).

KEYWORDS

- **ascochyta blight**
- **disease cycle**
- **Food and Agriculture Organization**
- **neem bark methanol extract**
- **northeastern plain zone**

REFERENCES

Abo-El Seoud, M. A., Sarhan, M. M., Omar, A. E., & Helal, M. M., (2005). Biocides formulation of essential oils having antimicrobial activity. *Ar. of Phytopath and Pl. Protec., 38*, 175–184.

Anwar, S. A., Bhutta, A. R., Raut, C. A., & Khan, M. S. A., (1994). Seed borne fungi of pea and their role in poor germination of pea seed. *Pakistan J. Phytopathology, 6*, 135–139.

Ayub, A., Rahaman, M. Z., Ali, S., & Khatun, A., (1996). Fungicidal spray to control leaf rust of lentil. *Bangladesh J. Plant Pathol., 12*, 61–62.

Azmat, M. A., Nawab, N. N., Shahid, N., Abdul, R., Khalid, M., Khan, A. A., & Khan, S. H., (2010). Single gene recessive controls powdery mildew resistance in pea. *Inter. J. Veg. Sci., 16*, 17–24.

Bains, S. S., Dhiman, J. S., & Singh, H., (1993). Occurrence of *Peronospora pisi* with *Uromyces viciae fabae* on *leaves* on pea genotypes. *Indian Phytopathol., 48*, 365–366.

Barilli, E., Sillero, J. C., Aparicio, M. F., & Rubiales, D., (2009). Identification of resistance to uromyces pisi (Pers.) Wint. In: *Pisum spp. germplasm. Field Crop Research, 117*(2), 198–203.

Batra, L. R., & Stavely, R., (1994). Attraction of two spotted spider mites to bean rust uredinia. *Plant Dis., 78*, 282–284.

Bedlan, G., (1985). Pea leaf and pod spot. *Pflanzenschutz, 147*, 14–15.

Biddle, A. J., Knott, C. M., & Gent, G. P., (1988). *Pea Growing Handbook* (p. 264). Processors and Growers Research Organization, Peterborough, UK.

Bradshaw-Smith, R. P., Craig, G. D., & Biddle, A. J., (1991). Glasshouse and field studies using *Pythium oligandrum* to control fungal foot rot pathogens of pea. *Aspects App. Biol., 27*, 347–350.

Bretag, T. W., (1985). Chemical control of *Ascochyta* blight of field peas. *Aust. Plant Pathol., 14*, 42–43.

Bretag, T. W., Kearne, P. J., & Price, T. V., (1995). Effect of ascochyta blight on the grain yield of field peas (*Pisum sativum* L.) grown in southern Australia. *Aust. J. Expt. Agric., 35*, 531–536.

Brokenshire, T., (1980). Control of pea downy mildew with seed treatments and foliar sprays. In: *Test of Agrochemicals and Cultivars, Suppl. Annals of Applied Biology*, *64*(1), 34–35.

Buchheim, A., (1922). On the biology of *Uromyces pisi* (Pers.) Winter. Preliminary note. *Centralbl. fur Bakt. Ab.*, *2*, 507–508.

Butler, E. J., (1918). *Fungi and Diseases in Plants*. Thacker Spink and Co., Calcutta, India.

Campbell, L., (1935). *Downy Mildew of Peas Caused by Peronospora Pisi (De Bary) Syd* (Vol. 318, p. 42). Washington Agric. Exp. Sta. Bull.

Chupp, C., & Sherf, A. F., (1960). *Vegetable Disease and Their Control* (p. 693). The Ronald Press Company, New York.

Datar, V. V., (1983). Reaction of pea wilt (*Pisum sativum* L.) varieties to wilt caused by *Fusarium oxysporum* f. sp. *pisi. Indian J. Mycol. and Pl. Pathol.*, *13*, 88–89.

Deutelmoser, E., (1926). Plant proection measures in vegetable culture. I. Measures against fungus pests. *Obst.-U. Gemusebu I. XXII*, *19*, 291–292. (*Rev. Appl. Mycol.*, 1927, 137).

Devi, M., & Paul, Y. S., (2003/2005). Management of pea wilt/root rot complex by integrating plant extracts and biocontrol agents. *Integrated Plant Disease Management* (pp. 101–105). Challenging problems in horticultural and forest pathology, Solan, India.

Dixon, G. R., (1981). Downy mildews on peas and beans. In: Spencer, D. M., (ed.), *The Downy Mildews* (pp. 87–154, 636). Academic Press, New York, USA.

Fondevilla, S., Carver, T. L. W., Moreno, M. T., & Rubiales, D., (2006). Macroscopic and histological characterization of genes er1 and er2 for powdery mildew resistance in pea. *Euro. J. Plant Pathol.*, *115*, 309–321.

Fuzi, I., (1995). Fungicides against diseases of pea, *Pisum sativum* L., in Hungary. *Pesticides Science*, *45*(3), 292–295.

Gangopadhayay, S., & Kapoor, K. S., (1979). Wet seed treatment to control Fusariumwilt of pea. *Veg. Sci.*, *3*, 74–78.

Gritton, E. T., & Ebert, R. D., (1975). Interaction of planting date and powdery mildew on pea plant performance. *J. Am. Soc. Hortic. Sc.*, *100*, 137–142.

Gupta, P. C., Maheshwari, S. K., & Suhag, L. S., (1989). Efficacy of some seed dressing fungicides in controlling the wilts and root rot complex of pea. *Indian J. Mycol. Plant Pathol.*, *18*, 176.

Gupta, S. K., & Sharma, H. R., (2004). Efficacy of some EBI fungicides against pea powdery mildew. *Plant Dis. Res.*, *19*, 190–191.

Gupta, S. K., & Shyam, K. R., (1998). Control of powdery mildew and rust of pea by fungicide. *Indian Phytopathology*, *51*(2), 184–186.

Gupta, S. K., & Shyam, K. R., (2000). Post infection activity of ergosterol biosynthesis inhibiting fungicides against pea rust. *J. Mycol. Pl. Pathol.*, *30*(3), 414–415.

Gupta, V. K., & Mate, G. D., (2009). Conidial size of *Erysiphe polygoni* influenced by root reaction. *J. Plant Dis. Sci.*, *4*, 215–217.

Hagedorn, D. J., (1984). Root rot of peas. In: *Compendium of Pea Diseases* (pp. 8–11). American Phytopathological Society, Minnesota.

Hebblethwaite, P. D., (1983). *The Faba Bean* (p. 573). Butterworths London, U. K..

Iqbal, S. M., Jamali, A. R., Rauf, C. A., & Akram, A., (2001). Screening of pea (*Pisum sativum* L) germplasm against blight disease caused by *Ascochyta pisi. Pakistan J. Phytopathol.*, *13*, 64–66.

Jorstad, I., (1948). Erterust, *Uromyces pisi* (D. C.) Fuck., 1. Norge (Pea rust: *Uromyces pisi* (D. C.) Fuck., in Norway). *Nord, Jordbr-Forskn.*, *7–8*, 198–207. (*Rev. Appl. Mycol.*, 1950, 242–243).

Diseases of Garden Peas (Pisum sativum L.) 255

Joshi, A., & Tripathi, H. S., (2012). Studies on epidemiology of lentil rust (*Uromycesviciae fabae*). *Indian Phytopathology, 65*(1), 67–70.

Kaiser, W. J., Muehlbauer, F. J., Hannan, R. M., & Mihov, M., (1998). First report of natural infection of *Pisum sativum* sub species *elatius* by *Mycosphaerellapinodes* in Bulgaria. *Plant Dis., 82*, 830.

Kavasnikov, B., & Krotova, T., (1997). Resistance of varieties of garden pea to *Ascochyta* diseases. *Referativnyl Zhurnal., 1*, 14–24.

Khaled, A. A., EL-Moity, S., & Omar, S., (1995). Chemical control of some faba bean diseases with fungicides. *Egyptian J. Agri. Res., 73*(1), 45–56.

Khan, I. A., Khan, H., Ali, A., Raziq, F., Hussain, S., Ahmad, M., & Attauddin, (2009). Evaluation of various fungicides and cultivars for the control of pea rust under natural conditions. *Sarhad J. Agric., 25*, 261–268.

Koler, W., Allan, C. R., & Kolattukudy, P. E., (1982). Protection of *Pisum sativum* from *Fusarium solani* f. sp. pisi. *Phytopathology, 72*, 1425–1430.

Kosevkii, II, & Kirik, N. N., (1979). Features of the development of mycelium of *Peronospora pisi* Syd in pea tissue, Osobennosti razvitiya mitseliya *Peronospora pisi* Syd v tkanyak gorokha. *Mikilogiya I Fitopatologiya, 13*, 46–48.

Kumar, A., & Kohli, U. K., (2001). Evaluation of garden pea genotypes for horticultural traits and resistance to Fusarium wilt. *Haryana J. Sci., 30*, 217–219.

Kumar, A., (1983). *Studies on Resistance to Fusarium Wilt in Some Cultivars of Peas (Pisum sativum L)* (p. 71). M. Sc. Thesis, Dr. Y. S. Parmar Univ. Horti. Forestry, Solan, H.P.

Kushwaha, C. Chand, R., & Srivastava, C. P., (2006). Role of aeciospores in outbreak of pea (Pisvum sativum) rust (Uromyces fabae). *Eur. J. Plant Ptho., 115*, 323–330.

Laciocowa, B., & Pieta, D., (1996). The efficiency of microbiological dressing of pea seeds (*Pisum sativum* L) against pathogenic soil-borne fungi. *Roczniki-Nauk-Rolniczych. –Seria-E, -Ochrona-Roslin, 25*, 15–21.

Linford, M. B., (1928). A Fusarium wilt of peas in Wisconsin. *Bull. Wis. Agric. Exp. Sta., 85*, 1–44.

Maheshwari, S. K., Jhooty, U. S., & Gupta, J. S., (1981). Studies on the wilt and root rot of pea. Efficiency of various chemicals in controlling the wilt and root rot complex of pea. *Agri. Sci. Dig., 1*, 37–38.

Maheshwari, S. K., Jhooty, U. S., & Gupta, J. S., (1983). Survey of wilt and root rot complex of pea in northern India and assessment of losses. *Agric. Sci. Dig. India, 3*, 139–141.

Maheshwari, S. K., Nazir, A. Bhat, M. S. D., & Beig, M. A., (2008). Chemical control of lentil wilt caused by *Fusarium oxysporum* f. sp. *lentis. Ann. Pl. Protec., 16*(2), 419–421.

Mashita, N., & Fukaya, M., (2006). Race of *Fusarium* wilt in the *Pisum* occurred in Aichi and resistant variety to *Fusarium* wilt race 2. *Res. Bull. Aichiken Agric. Res. Center, 38*, 65–71.

Maurya, S., Singh, D. P., Srivastava, J. S., & Singh, U. P., (2004). Effect of some plant extracts on pea powdery mildew *(Erysiphe pisi)*. *Annals of Plant Protection Sciences, 12*, 296–300.

McDonald, G. K., & Dean, G., (1996). Effect of waterlogging on the severity of disease caused by *Mycosphaerella pinodes* in pea (*Pisum sativum* L) *Aust. J. Exp. Agric., 36*, 219–222.

Mence, M. J., & Pegg, G. F., (1971). The biology of *Peronospora viciae* on pea: Factors affecting the susceptibility of plants to local infection and systemic colonization. *Ann. Appl. Biol., 67*, 297–308.

Mittal, R. K., (1997). Effect of sowing dates and diseased development in lentil as sole and mixed crop with wheat. *J. Mycol. Pl. Pathol., 27*(2), 203–209.

Munjal, R. L., Chenulu, V. V., & Hora, T. S., (1963). Assessment of losses due to powdery mildew (*Erysiphe polygoni*, DC.) on pea. *Indian Phytopath., 16*, 268–270.

Nasir, M., & Hoppe, H. H., (1997). Evaluation of foliar fungicides for control of *Mycosphaerella pinodes* on peas. *Tests Agrochemicals and Cultivars No. 18, Ann. App. Biol., 130*, 12–13.

Neumann, S., & Xue, A. G., (2003). Reaction of field pea cultivars to four race of *Fusarium oxysporum* f. sp. *pisi. Canadian Journal of Plant Science, 83*, 377–379.

Neweigy, N. A., Eidin, I. F. G., Ziedan, M. I., Hanazy, E. A., & Meame, M. A. A., (1985). Effect of seed dressing of peas and soyabean seeds on their growth in soil infested with *Sclerotium rolfsii* or mixture of three pathogenic fungi (*Rhizoctonia solani, Fusarium solani, S. rolfsii). Ann. Agric. Sci. Moshtohor, 23*, 1109–1114.

Obradovic, A., Marinkovic, N., Mijatovic, M., & Dordevic, R., (1994). Resistance of pea genotypes to *Ascochyta pisi. Zestita-Bilja, 45*, 151–154.

Olofsson, J., (1966). Downy mildew of peas in western Europe. *Plant. Dis. Rep., 50*, 257–261.

Pandey, K. K., & Upadhyay, J. P., (1999). Comparative study of chemical, biological and integrated approach for management of Fusarium wilt of pigeonpea. *J. Mycol. Pl. Pathol., 29*(2), 214–216.

Patel, M. K., Kamath, M. N., & Bhide, V. P., (1949). Fungi of Bombay supplement I. *Indian Phytopath., 2*, 142–155.

Paul, Y. S., &. Kapoor, J. N., (1983). A revision of *Erysiphe polygoni* sensu Salmon from India. *Indian Phytopath., 36*, 247–250.

Peshne, N. L., (1966). A comparative study of morphology, physiology and pathogenicity of Nagpur and Poona isolates of *Fusarium oxysporum* f. sp. *pisi. Ann. Agril. Res. Abst. Grad. Res., W. K. 1960–1965, Agric. Coll. Mg. Space Res., 12*, 174–175.

Prasad, P., & Dewedi, S. N., (2007). Fungicidal management of field pea (*Pisum sativum* L) powdery mildew caused by *Erysiphe polygoni* DC. *Prog. Res., 2*, 116–118.

Prasada, R., & Verma, U. N., (1948). Studies on lentil rust, *Uromyces fabae. Indian Phytopathol., 1*, 142–146.

Pratibha, S., & Amin, K. S., (1991). Integrated disease management of powdery mildew of field peas (*Pisum sativum L.*) caused by *Erysiphe polygoni* D. C. *Legume Res., 14*, 59–63.

Pretorius, J. C., Craven, P., Watt, E., Vander, D. W. E. V., &Van der Wart, E., (2002). *In vivo* control of *Mycosphaerella pinodes* on pea leaves by a crude bulb extract of *Eucomis autumnalis. Ann. App. Biol., 141*, 125–131.

Ramphal, & Choudhary, B., (1978). Screening of garden peas for resistance to Fusariumwilt. *Indian J. Agric. Sci., 48*, 407–410.

Rana, D. P. S., Bhardwaj, P. K., Rao, M. V. B., & Chatterjee, D., (1991). Field evaluation of fenarimol and triadimefon for control of powdery mildew (*Erysiphe polygoni* D. C.) of pea. *Indian J. Plant Protec., 19*, 31–35.

Roger, C., & Tivoli, B., (1996). Spatio- temporal development of pycnidia and perithecia and dissemination of spores of *Mycosphaerellapinodes* on pea (*Pisum sativum*). *Plant Pathol., 45*, 518–528.

Sagar, V., & Bhardwaj, C. L., (1997). Identity of 'Mega' form of *Phomamedicaginis* var. *pinodella* on pea from Bajaura, Himachal Pradesh. *Indian J. Mycol. Plant Pathol., 27*, 195–198.

Sagar, V., & Sugha, S. K., (1998). Effect of soil type and available nutrients on Fusarial population and severity of pea root rot. *Indian J. Mycol. Plant Pathol., 28*, 294–299.

Sahani, R. K., & Saxena, A. R., (2009). Fungitoxic properties of medicinal and aromatic plants against *Fusarium oxysporum* f. sp. *pisi. Ann. Pl. Protec. Sci., 17*, 146–148.

Diseases of Garden Peas (Pisum sativum L.)

Saksirirat, W., Sirmung, K. S., Steinmetz, J., & Schoenbeck, F., (1994). Applicability of Gliocladium roseum Bain for biological control of fusarium wilt of pea. *Kaen kaset- khon Kaen Agri. J., 22*, 37–42.

Sandhu, K. S., & Dhillon, G. S., (1984). Reaction of pea varieties to *Ascochyta* blight. *J. Res. Punjab Agric. Univ., 21*, 391–395.

Sangar, R. B., & Singh, V. K., (1994). Effect of sowing dates and pea varieties on the, severity of rust, powdery mildew and yield. *Indian J. Pul. Res., 7*(1), 88–89.

Sanssene, J., Boulos, G., & Gascoin, B., (1998). Efficiency of seed treatment on root necrosis of peas (*Pisum sativum* L.). 3rd European conference on grain legumes. *Opportunities for High Quality, Healthy and Added-Value Crops to Meet European Demands* (p. 263). Valladolid, Spain.

Schroeder, W. T., & Walker, J. C., (1942). Influence of controlled environment and nutritive on the resistance of garden pea to Fusarium wilt. *J. Agric. Res., 65*, 221–248.

Shabeer, A., & Irfan, U. D., (2006). Field assay of promising phytobiocides vs. fungicides for control of powdery mildew of pea. *Sahad J. Agric., 22*, 303–305.

Sharma, P., Singh, S. D., & Rawal, P., (2003). Antifungal activity of some plant extracts oils against seed borne pathogen of pea. *Plant Disease Research, Ludhiana, 18*(1), 16–20.

Sharma, R. R. Rajiv, Sharma, K., & Munshi, A. D., (1998). Breeding for Fusarium wilt resistance in pea (*Pisum sativum*). *Ann. Pl. Protec. Sc., 6*(1), 1–10.

Sharma, S. K., & Sharma, A., (2003). Effect of time of planting, soil hydrothermal regimes and some meteorological parameters on the development of *Fusarium* wilt in autumn pea. *Plant Dis. Res., 18*, 127–130.

Shaw, C. G., (1981). Taxonomy and evolution. In: Spencer, D. M., (ed.), *The Downy Mildews* (pp. 17–29, 636). Academic Press, New York, USA.

Shroff, S., & Chand, R., (2010). Pre-infection Biology of aeciospores of *Uromyces fabae*. *Inter J. Curr. Trends Sci. Tech., 1*(2), 1–10.

Shukla, P., Singh, R. P., & Mishra, A. N., (1979). Efficacy of seed dressing fungicides for the control of pea wilt. *Pesticide, 13*, 40–47.

Singh, D., & Tripathi, H. S., (2004). Epidemiology and management of field pea rust. *J. Mycol. Pl. Pathol., 34*(2), 675–679.

Singh, H. B., & Singh, U. P., (1988). Powdery mildew of pea (*Pisum sativum* L.). *Int. J. Trop. Pl. Dis., 6*, 1–18.

Singh, P. P., Amit, W., & Sandhu, K. S., (2003). Influence of post-inoculation diurnal conditions on development of *Ascochyta blight* of peas. *Indian Phytopathology, 56*, 446–447.

Singh, P. P., Amit, W., & Sandhu, K. S., (2005). Effect of epidemiological parameters on the development of *Ascochyta* blight of peas. *Indian Phytopathology, 58*, 217–220.

Singh, R. S., (1987). *Disease of Vegetable Crop* (p. 362). Oxford and IBH Publishing Co. Pvt. Ltd. New Delhi.

Singh, R. S., (1996). *Plant Diseases* (p. 396). Oxford and IBH, New Delhi.

Singh, R. S., (1998). *Plant Diseases* (p. 512). Oxford and IBH, New Delhi.

Singh, S. K., Rahman, S. J., Gupta, B. R., & Kalha, C. S., (1992). An integrated approach to the management of the major diseases and insect pests in India. *Trop. Pest. Mgt., 38*, 265–267.

Sinha, V., & Upadhyay, R. S., (1990). Fungitoxicity of some pesticides against *Fusarium udum*, the wilt causing organism of pigeonpea. *Hindustan Antibiotic Bulletin, 32*, 1–2, 36–38.

Sitara, U., Niaz, I., Naseem, J., & Sultana, N., (2008). Antifungal effect of essential oils on *in vitro* growth of pathogenic fungi. *Pak. J. Bot., 40*, 409–414.

Srivastava, L. S., & Gupta, D. K., (1990). Occurrence of powdery mildew, rust and Ascochyta blight of pea in Sikkim. *Plant Dis. Res.*, *5*, 86.

Stegmark, R., (1988). Downy mildew resistance of vari-ous pea genotypes. *Acta Agric Scand.*, *38*, 373–379.

Sugha, S. K., Sharma, O. P., & Sharma, P. N., (1998). Performance of 'vegetable type' French bean genotypes against floury leaf spot disease. *Plant Dis. Res.*, *13*, 85–86.

Sukapure, R. S., Bhide, V. P., & Patel, M. K., (1957). Fusarium wilt of garden peas (*Pisum sativum* L.). In Bambay State. *Indian Phytopath.*, *10*, 11–17.

Surwase, A. G., Badgire, D. R., & Suryawanshi, A. P., (2009). Management of pea powdery mildew by fungicides, botanicals and bio-agents. *Ann. Plant Protec. Sci.*, *17*, 384–388.

Susuri, L., (1976). Contribution to the study of some factors of *A. pisi* Lib. Parasite of pea. *Zastata Bilja.*, *27*, 69–87.

Taylor, P. N., Lewis, B. G., & Matthews, P., (1990). Factors affecting systemic infection of *Pisum sativum* by *Peronospora viciae. Mycol. Res.*, *94*, 179–181.

Tommerup, I. C., (1981). Cytology and genetics of downy mildews. In: Spencer, D. M., (ed.), *The Downy Mildews* (pp. 121–142). Academic Press, New York, USA.

Trivoli, B., Beasses, C., Lemarchand, E., & Marsson, E., (1996). Effect of *Ascochyta* blight (Mycosphaerella) on yield components of single gene (*Pisum sativum*) plants under field conditions. *Ann. App. Bio.*, *129*, 207–216.

Upadhyay, A. L., & Gupta, R. P., (1994). Fungicidal evaluation against powdery mildew and rust of pea (*Pisum sativum* L.). *Ann. Agric. Res.*, *15*(1), 114–116.

Utikar, P. G., Gade, U. A., & More, B. B., (1978). Seed treatment to control of *Fusarium* wilt of pea. *Pesticides*, *12*, 29–30.

Van Der Gaag, D. J., & Frinking, H. D., (1997). Survival characteristics of oospore populations of *Peronospora viciae* f. sp. *Pisi* in soil. *Plant Pathol.*, *46*, 978–988.

Velikanov, L. L., Cukhonosenko, E. Y., Nikolaeva, S. I., & Zavlisko, I. A., (1994). Comparison of hyperparasitic and antibiotic activity of the genus *Trichoderma* Per: Fr and *Gliocladium virens* Miller Giddens et Foster isolates towards the pathogen causing root rot of pea. *Mikologiya-I- Fitopatologiya*, *28*, 52–56.

Verma, S., & Dohroo, N. P., (2002). Evaluation of fungicides against *Fusarium oxysporum* f. sp. *pisi* causing wilt of autumn pea in Himachal Pradesh. *Plant Disease Research*, *17*, 261–268.

Verma, S., & Dohroo, N. P., (2003). Comparative efficacy of biocontrol agents against *Fusarium* wilt of pea. Integrated plant disease management. *Challenging Problem in Horticultural and Forest Pathology* (pp. 93–99.). Solan, India.

Verma, S., & Dohroo, N. P., (2005a). Effect of sowing dates on development of Fusarium wilt of autumn pea caused by *Fusarium oxysporum* f. sp. *pisi, Plant Dis. Res.*, *20*, 177–179.

Verma, S., & Dohroo, N. P., (2005c). Novel approach for screening different antagonists against *Fusarium oxysporum f. sp. Pisi* causing fusarium wilt of autumn pea. *Plant Dis. Res.*, *20*, 58–61.

Virin, W. J., & Walker, J. C., (1940). Relation on near wilt fungus to the pea plant. *J. Agric. Res.*, *60*, 241–248.

Vladimirtseva, L. V., Guseva, N. N., & Ovchinnikova, A. M., (1990). Forms of pea resistant to *Ascochyta pisi. Sel. Semenovodstvo-Moskva. No. 3*, 30–31.

Walters, D. R., & Murray, D. C., (1992). Induction of systemic resistance to rust in *viciae faba* by phosphate and EDTA: Effects of calcium. *Plant Pathology*, *41*, 444–448.

Diseases of Garden Peas (Pisum sativum L.) 259

Warkentin, T. D., Rashid, K. Y., & Xue, A. G., (1996). Fungicidal control of *Ascochyta* blight of field pea. *Canadian Journal of Plant Science, 76*, 67–71.

Warkentin, T. D., Xue, A. G., & McAndrew, D. W., (2000). Effect of mancozeb on the control of *mycosphaerella* blight of field pea. *Can. J. Pl. Sci., 80*, 403–406.

Xue, A. G., & Warkentin, T. D., (2002). Reaction of fieldpea varieties to three isolates of *Uromyces fabae. Canadian J. Plant Sci., 82*(1), 253–255.

Yarwood, C. E., Sidky, S., Cohen, M., & Santilli, V., (1954). Temperature relations of powdery mildews. *Hilgradia, 22*, 603–622.

CHAPTER 14

Diseases of Ginger and Turmeric and Their Management

SUNIL KUMAR[1*] and GIREESH CHAND[2]

[1]*School of Agricultural Sciences and Rural Development, Nagaland University, Medziphema, Nagaland, India*

[2]*Department of Plant Pathology, Bihar Agricultural University, Sabour, Bhagalpur, Bihar, India*

Corresponding author. E-mail: drsunilk81@gmail.com

14.1 INTRODUCTION

In Indian food, ginger is a key ingredient, especially in thicker gravies, as well as in many other dishes, both vegetarian and non-vegetarian. Ginger also has a role in traditional Ayurvedic medicine. It is an ingredient in traditional Indian drinks, both cold and hot, including spiced *masala chai*. Fresh ginger is one of the main spices used for making pulse and lentil curries and other vegetable preparations, confectionery, pickles, medicinal, etc. The other commercial forms of ginger are raw ginger, dry ginger, ginger powder, ginger oil, ginger oleoresin, and ginger beer.

14.2 BACTERIAL WILT

14.2.1 SYMPTOMS

Bacterial wilt is one of the most important diseases of ginger, and their symptoms appeared in the month of July and August. The affected margins of the leaves are turned in to bronze color and curled to the backside. The severely affected plants appear as wilt and finally die. The base of the infected pseudostem and the rhizome produce a foul smell. A water-soaked linear streaks appear at the collar regions of the pseudostem. When the suspected

262   *Diseases of Fruits and Vegetable Crops*

pseudostem is cut and immersed in a glass of clean water, milky exudates will ooze out from the cut end. The typical symptom is the wilting observed during the afternoon in young seedlings.

14.2.2   *CAUSAL ORGANISM: RALSTONIA SOLANACEARUM*

The ginger strain of *Ralstonia solanacearum* survives at 34°C. This bacterium has two biotypes, biotype 3 is reported from India, and biotype 3 and 4 are from Queensland. The pathogen infects most of the solanaceous crops such as tomato, potato, capsicum, eggplant, groundnut, and tobacco.

14.2.3   *MANAGEMENT*

The contaminated seed is the major source of infection. Use only healthy rhizome for sowing. Treat the rhizome with Streptocyclin (20g/100 liter water) before planting. Destroy the affected clumps from field and drench the soil with solution of copper oxychloride @ 0.2%. Maintain a good drainage in the fields. Ishii and Aragaki (1963) revealed that soil fumigation with methyl bromide @ 1.362 kg/1.21 sq. m. control the disease significantly. Always obtain rhizome from areas where the disease is not reported. When the plant shows severe infection of the disease should be put under soybean cultivation for 3 years. The infected rhizomes removed from the field and bunt.

14.3   SOFT ROT

14.3.1   *SYMPTOMS*

It is a serious seed as well as soil-borne disease, and the symptoms can be seen from the month of July. Yellowing appears first on the lower leaves, and it proceeds to upper leaves. Roots arising from the affected rhizome become rotten and show brown discoloration of the rhizome tissue. As the disease advances, water-soaked discoloration appears at the basal portion of the plant. Laterally the shoot and pseudostem become soft, and it is easily pulled out. Soft rot also affects the rhizomes. The infected parts also attract other microorganisms like fungi and bacteria and insects, particularly the rhizome fly. During the rainy season, this disease spreads very fast from the infected field to a healthy field. The infected rhizomes fail to produce new rhizomes or roots. The infected rhizomes harvests and stores it further causes storage rots.

Diseases of Ginger and Turmeric and Their Management

14.3.2 CAUSAL ORGANISM: PYTHIUM APHANIDERMATUM

14.3.3 MANAGEMENT

Avoid waterlogging. The disease is easily managed by using healthy and disease-free rhizomes for planting. At the time of sowing, treat the rhizome with Bordeaux mixture (1%) and again with *Trichoderma* @ 8–10 gm/liter water. Rhizome treatment with Dithane M 45 (0.3%) for 30 minutes and soil drenching with the same fungicide with the same concentration have been recommended (NRCS, 1986). The disease can be managed by rhizome treatment with chlorothalonil @ 0.2% and soil drenching with difolatan @ 0.2% or copper oxychloride @ 0.4%. Treat the rhizome with Ridomil MZ 72 WP @ 2.5 g/l for and air dry before planting also effective for reducing the disease. Drenching of soil with Bordeaux mixture (1%) and rhizome treatment with same were found to be effective in managing this disease. Fungal biocontrol agents (BCAs) like *Trichoderma lignorum, T. harzianum, Gliocladium virens*, and VAM fungi viz. *Glomuscn strictum* and *G. mosseae* were found in reducing the incidence and intensity of rhizome rot. *Trichoderma harzianum* with a cfu 10^{11}/g is applied @ 50g per a raised bed size of 3 m². This is mixed with organic manures such as neem oil cake or decomposed farmyard manure (Anandraj et al., 2001). Solarization does not leave any toxic residue. It is a hydrothermal process dependent for success on moisture of the sample for maximum heat transfer. In more temperate regions soil is covered with clear plastic in order to trap solar radiation and raise the temperature sufficiently to suppress or eliminate soil-borne pathogens and pests (Katan, 1981; Katan and DeVay, 1991; Kumar et al., 2003). This is one of the most eco-friendly and energy-efficient methods available for rhizome treatment. Rhizome temperatures of 40 and 50°C were recorded after 1 and 2 hours of solarization from 9:00 a.m. to 11:00 a.m. on a bright sunny day (January to May in India) (Prasheena, 2003). The effectiveness of *Trichoderma harzianum* was reported from Sikkim (Rajan et al., 2002). A number of plant growth-promoting (PGP) endophytic bacteria (EB), *Pseudomonas* spp., has been isolated from surface-sterilized leaves, stems, and roots of tomato (Aino et al., 1997), *Solanumnigrum*, a wild solanaceous plant species (Hoang et al., 2010, 2011). They exerted with/without strong *in vitro* antagonism against *Ralstonia solanacearum* and significantly reduced the wilt incidence in tomato or tobacco. It is indicated that resistance in tobacco against *R. solanacearum* induced by endophytic *Pseudomonas* spp. is associated with the systemic induction of PR proteins in the SA-dependent pathway Hoang et al. (2011).

The crop rotation with non-hosts such as cereals and millets results in a reduction in the wilt incidence in the ensuing crop of ginger. Application of organic amendments like oil cakes of *Azadirachta indica*, *Callophyllum inophyllum*, *Pongamia glabra*, *Hibiscus sabdariffa* and *Brassica campestris* are found to be effective.

14.4 *PHYLLOSTICTA* LEAF SPOT

The disease was first reported from Godavari district of Andhra Pradesh (Ramkrishan, 1942). Latter, *Phyllosticta* leaf spot was reported from the Philippines (1966), Mauritius (1971), and Sarawak (1972). In India, again, it was reported from Himachal Pradesh (1973), Kerala (1974), and Maharashtra (1974).

14.4.1 *CAUSAL ORGANISM: PHYLLOSTICTA ZINGEBERI*

The pycnidium of *Phyllosticta* measures 78–150 μ in diameter and has definite ostiole. When it is mounted on slide characteristics worm-like mass of spores coming out of the ostiole can be seen under microscope. The spores are hyaline oblong measuring an average 4.3–1.6 μ the range being 3.7–7.4 × 1.2–2.5 μ. The pycnidia are light in color in the beginning but with age the color deepens and finally they turns light to deep brown. Each pycnidium has an ostiole and a very short neck. The pycnidia forms in cultures are much bigger than those in nature. Some are spherical but in most cases they are only subglobose.

14.4.2 SYMPTOMS

The spot appears on leaves as white and papery elongated or oval shape measuring 0.5 mm to 1–10 mm. The spots appear with dark center and it's surrounded by yellow halo. Minute dark color fruiting bodies of pycnidia are seen on mature lesions. Initially, the spots appear as small size and in advance stage the spots are coalesce and forms large spot. The affected leaf areas rolls and sometimes form very thin thread like structure. The extensive foliar destruction leads to reduction in yield. The plant shows a burnt appearance. In the case of severe infection the entire leaves dry up.

Diseases of Ginger and Turmeric and Their Management 265

14.4.3 MANAGEMENT

Spray of Bordeaux mixture (1%) 3–4 times at 15 days interval with the initiation of the disease. Good control is achieved by growing the crop under partial shade condition. Sood and Doharoo (2005) study the rhizome treatment with 14 fungicides for 1 hr followed by two protective foliar sprays starting from the second fortnight of July at 15 days intervals and one eradicating spray during last week of August. They found that the rhizome treatment as well as foliar sprays with Bordeaux mixture (1%), companion (0.2%), Indofil M 45 (0.25%), Unilax (0.2%) and Baycor (0.05%) were found effective in managing the disease severity.

14.5 DRY ROT

14.5.1 SYMPTOMS

The disease is caused by complexion of fungus and nematode. In comparison to rhizome rot, dry rot appears in field as small patches and spreads slowly. Stunted growth appears after the infection and also exhibits the yellowing of foliage. The dryness of leaves appears first on older leaves followed by younger leaves. In advanced stages of the disease the infected dry rot rhizome, when cut and open, we can see the brownish ring and it is mainly restricted to cortical region. The dry rot affected pseudostem does not move towards likewise soft rot. If stored in a humid condition the infected rhizomes may develop a surface growth of white mold. The dry rot infection reduced the market price of the rhizomes.

14.5.2 CAUSAL ORGANISM: FUSARIUM AND PRATYLENCHUS COMPLEX

This disease caused by the complexion of *Fusarium* spp. and nematode *Pratylenchus* spp.

14.5.3 MANAGEMENT

The basal application of mustard oil cake @ 40 kg/ha before sowing can reduce the nematodes infestation. Hot water treatment at 51°C for 10 min followed by seed treatment with Bordeaux mixture (1%) effectively checks the problem. Use healthy rhizomes for sowing.

14.6 *COLLETOTRICHUM* LEAF SPOT

This disease was first time reported from Godavari district of Andhra Pradesh (Sundraraman, 1922). Later it was reported from Tanganyika territory by Wallace and Wallace (1945). The disease causes conspicuous yield loss and also reduces the quality of the rhizomes.

14.6.1 SYMPTOMS

The symptoms appear as a small round to oval, light yellow color spots on leaves and leaf sheaths. These spots gradually increase in size and often coalesce to form large discolored areas. Such infected areas often dry up at the center, forming holes. In the case of a severe attack, the entire leaf dries up.

14.6.2 CAUSAL ORGANISM: COLLETOTRICHUM ZINGIBERI (SUNDAR)

Colletotrichum (sexual stage: *Glomerella*) is a symbionts to plant as endophyte or phytopathogens. The genus *Colletotrichum* belongs to the class sordariomycetes, order: Glomerella and family: Glomerellaceae.

14.6.3 MANAGEMENT

The best approach to management of the disease is use of resistant and healthy rhizomes for sowing. Affected plant should be removed from the field and burnt. Sprays of Bordeaux mixture @ 1.0% at the initiation of disease gave good control.

14.7 ROOT-KNOT NEMATODE

Ginger rhizomes severely infested with the root-knot nematode. The nematode affects the market quality of the crop very seriously but normally does not destroy the rhizomes.

14.7.1 SYMPTOMS

The disease does not form prominent surface galls in ginger rhizomes as it does in other plants and in ginger the external symptoms are usually masked. In severe infestations the cortex of the rhizomes appears somewhat lumpy and cracked. Apparently, when the female nematode reaches maturity, the surface galls break the epidermis of the rhizome and the rhizome surface becomes corky in appearance. When the cork layer is peeled off, small, circular, water-soaked, slightly brown lesions are observed. These water-soaked areas below the epidermis of the rhizomes are quite numerous in severely infested rhizomes.

14.7.2 CAUSAL ORGANISM: MELOIDOGYNE INCOGNITA

Mature females of the nematode are found in the lesions. The nematode lesions can serve as points of entry to common bacteria and fungi which are otherwise unable to invade uninjured tissues, and the rhizomes may be destroyed in storage by such organisms. Nematodes such as *Meloidogyne* sp. are known to invade immature tissues only. It is not known how infection of the ginger rhizome takes place.

14.7.3 MANAGEMENT

Infestation of the nematodes can be avoided by using the disease-free planting material. A very effective nematode management practice is soil solarization, but it does not eliminate completely the nematode inoculum. Solarization by covering the field with transparent polythene sheet for 6 hours during summer months caused 58% reduction of *Meloidogyne incognita* population density. Vadhera et al. (1998) reported that the maximum reduction of soil nematode and highest obtained with neem cake treated plots. They also reported the denematization of ginger rhizomes at 450C for three hours + summer plowing and covering of soil with polythene sheet from 15th to 30th May gave significantly higher yield and reduction in nematode population. Lopez-Perez et al. (2005) used crop residue of broccoli, melon, tomato, and chicken litter as a source of biofumigation to control *M. incognita* in tomato. Nico et al. (2004) reported the effective management of *M. incognita* by using the amendment of potting mixes with composted agro-industrial wastes such as dry cork, dry gapes residue after extraction of juice, dry rice husk, etc.

14.8 TURMERIC

Turmeric (*Curcuma longa*) is rhizomatous herbaceous perennial plant of the family Zingiberaceae. The origin place of the turmeric is southwest India. It is used all over the world as a part of daily cuisine. Turmeric has one active ingredient cucurmin which has distinctly earthy, slightly bitter, slightly hot peppery flavor and a mustardy smell. India plays a significant role in world turmeric production. It has antifungal and antibacterial properties. Turmeric is under study for its potential to affect human diseases, including kidney and cardiovascular diseases, arthritis, cancer, irritable bowel disease, Alzheimer's disease, diabetes, and other clinical disorders. It is also used widely in production of different cosmetic products due to its antifungal and antibacterial properties. Turmeric is considered auspicious and holy in India and has been used unvarious Hindu ceremonies of millennia. It remains popular in India for wedding and religious ceremonies. Turmeric has played a important role in Hindu spiritualism. The robes of the Hindu monks were traditionally colored with a yellow dye made of turmeric. The turmeric crop is affected by several fungal, bacterial, and viral diseases. Among them, fungal diseases are most important and cause enormous yield loss. Bacterial and viral diseases are of minor importance.

14.9 RHIZOME ROT

The disease appears all over the turmeric growing areas but in moist and humid condition noted as severe form. This disease is soil-borne and rhizome-borne. The appearances of disease as severe form have been noted from the month of June to September.

14.9.2 SYMPTOMS

The infection of the disease started at the collar portion of the pseudostem, and further progression appears upwards as well as downwards. Upward movement of the disease affects the pseudostem leaf sheaths, and downward movement affects the rhizome. Water-soaked lesions appear on the collar region of affected pseudostem, which start the rotting and spreads to the downwards resulted rhizome soft rot. Root infection also achieved in later stages. A light yellowing color appears at the tip of the lower leaves which gradually spreads to the leaf blades. At initiation or early stage of the disease, the middle lamella of the leaves remain green while the margins become yellow. Later, the yellowing spreads to all leaves of the plant from the lower region upwards and is followed by drooping, withering, and drying of pseudostems.

14.9.3 CAUSAL ORGANISM: PYTHIUM SPP.

Several workers reported the disease-causing organism is as *Pythium graminicolum* Subram (Ramkrishan and Sowmini, 1954), *Pythium aphanidermatum* (Edson) Fitz (Park, 1934) and *Pythiummyriotylum* Drechsler (Rathariah, 1982).

14.9.4 MANAGEMENT

Soil drenching with Ridomil and rhizome treatment with same fungicide is an effective measures for the management of this disease. Anusuya and Sathiyabama (2014) reported the chitosan played an important role in growth suppression of *Pythium aphanidermatum* infection in turmeric plants. They also studied the chitinase and chitosanase activity may play a role in enhanced resistance in turmeric plants against this disease. Maheswari and Sirchabai (2011) studied on biocontrol agents (BCAs) against turmeric rhizome rot pathogen *Pythium aphanidermatum* under lab conditions and revealed that fungal bioagent was inhibiting the growth of pathogen. *Trichoderma viride* was best inhibiting the colony growth of *Pythium aphanidermatum* statistically significant difference. Experts recommend that rhizome treatment with bio-fungicide like *Trichoderma* spp. can effectively control of this disease.

14.10 LEAF SPOT

Leaf spot is one of the most serious diseases of turmeric that occurs in the country. It has become a major constraint in the successful cultivation of turmeric. The disease has resulted in drastic reduction in rhizome yield and also affects the quality of rhizomes. Disease is soil-borne noticed on the leaves from July to October.

14.10.1 SYMPTOMS

Symptom appears as brown spots of various sizes on the upper surface of the young leaves with grey centers. Severely affected leaves dry and plant become wilt. The spots are irregular in shape. Later, spots may coalesce and form an irregular patch covering almost the whole leaf. The center of spots contains fruit head-shaped fruiting structures.

14.10.2 CAUSAL ORGANISM: COLLETOTRICHUM CAPSICI

14.10.3 MANAGEMENT

The use of resistant/tolerant variety is the cheapest and eco-friendly management of the disease. Narasimhudu and Balasubramanian (2002) reported that Topsin—M (0.1%) was found to be most effective in managing the disease followed by Indofil M-45 (0.25%) and Bavistin (0.1%). Singh et al. (2003) revealed the effectiveness of Score 25 EC (difenoconazole) @ 0.1% followed by Tilt 25 EC (propiconazole) @ 0.1%. Eco-Friendly management of leaf spot of turmeric by partial and heavy shading reported by Singh and Edison (2003). Rao et al. (2012) reported the management of leaf spot of turmeric through fungicides and found that rhizome treatment with Carbendazim + Mancozeb (0.1%) gave the best results for germination. They recorded the highest reduction of disease, and the maximum yield was obtained with the application of Propiconazole (0.1%).

KEYWORDS

- biocontrol agents
- endophytic bacteria
- ginger
- plant growth-promoting

REFERENCES

Aino, M., Maekawa, Y., Mayama, S., & Kato, H., (1997). Biocontrol of bacterial wilt of tomato by producing seedlings colonized with endophytic antagonistic pseudomonads. In: Ogoshi, A., Kobayashi, K., Homma, Y., Kodama, F., Kondo, N., & Akino, S., (eds.), *Plant Growth-Promoting Rhizobacteria: Present Status and Future Prospects* (pp. 120–123). OECD, Paris.

Diseases of Ginger and Turmeric and Their Management 271

Anandraj, M., Venugopal, M. N., Veena, S. S., Kumar, A., & Sarma, Y. R., (2001). Ecofriendly management of disease of species. *Indian Species, 38*, 28–31.

Anusuya, S., & Sathiyabama, M., (2014). Effect of chitosan on rhizome rot disease of turmeric caused by *Pythiumaphanidermatum*. *International Scholarly Research Biotechnology,*1–5.

Hoang, H. L., Furuya, N., Baldwin, I. T., & Tsuchiya, K., (2010). *Multitasking Bacterial Endophytes: Potential Biocontrol Agents of Bacterial Wilt of Solanaceous Plants.* The Phytopathologica Society of Japan (PSJ) Annual Meeting 2010, Kyoto, Japan.

Hoang, H. L., Furuya, N., Takeshita, M., & Tsuchiya, K., (2011). Biocontrol of bacterial wilt of tobacco via induced resistance by endophytic bacteria. *Phytopathology, 101*, S72. APS-IPPC Joint Meeting, Honolulu, Hawaii.

Ishii, M., & Aragaki, M., (1963). "Ginger wilt caused by *Pseudomonas solanacearum* E. F. Smith." *Plant Disease Reporter, 47*, 710–713.

Katan, J., & DeVay, (1991). *Soil Solarization.* CRC Press. Boca Ratan, FL.

Katan, J., (1981). Solar heating (solarization) of soil for control of soil borne pests. *Annual Review of Phytopathology, 19*, 211–236.

Kumar, A., Anandraj, M., & Sarma, Y. R., (2003). Rhizome solarization and microwave treatment: Eco-friendly methods for disinfecting ginger seed rhizomes. In: Allen, C. Hayward, A. C., & Prior, P., (eds.) *Bacterial Wilt and the Ralstonia Solanacearum Species Complex. American* Phyto-Pathological Society Press.

Lopez-Perez, J. A., Tatiana, R., & Antoon, P., (2005). Effect of three plant residue and chicken manure used as biofumigants at three temperature on meloidogyne incognita infection of tomato in greenhouse experiments. *Journal of Nematology, 37*, 489.

Maheswari, N. V., & Sirchabai, T. P., (2011). Effect of trichoderma species on *Pythium aphanidermatum* causing rhizome rot of turmeric. *Biosciences Biotechnology Research Asia, 8*(2), 723–728.

Narasimhudu, Y., & Balasubramanian, K. A., (2002). Fungicidal management of leaf spot of turmeric incited by *Colletotrichumcapsici*. *Indian Phytopathology, 55*(4), 527–528.

National Research Centre for Spices, (1986). *Annual Report for 1985.* NRCS, Calicut Kerala India.

Nico, A. I., Jimenez-Diaz, R. M., & Castillo, P., (2004). Control of root knot nematode by composted agro-industrial wastes in potting mixes. *Crop Protection, 23*(7), 581.

Park, M., (1934). *Report on the Work of the Mycology Division* (pp. 126–133). Admin. Rept. Dir Agric. Ceylon.

Prasheena, E., (2003). *Effect of Rhizome Solarization on Microbial Population of Ginger with Special Reference to Survival of Bacterial Wilt Pathogen Ralstonia Solanacearum.* Unpublished M. Sc. Thesis submitted to Bharathiar University, Coimbatore, India.

Rajan, P. P., Gupta, S. R., Sarma, Y. R., & Jackson, G. V. H., (2002). Diseases of ginger and their control with *Trichoderma harzianum*. *Indian Phytopathology, 55*(2), 173–177.

Rama, T. S., Krishnan, & Sowmini, C. K., (1954). Rhizome and root rot of turmeric caused by Pythiumgraminicolum. *Indian Phytopathology, 7*, 152–159.

Ramakrishnan, T. S., (1942). A leaf spot disease of *Zingiberofficinale* caused by *Phyllostictazingiber*. *Proc. Indian Acad. Sci., 15*, 167–171.

Rao, S. N., Kumar, K. R., & Anandraj, M., (2012). Management of leaf spot of turmeric (*Curcuma longa* L.) incited by Colletotrichumcapsici through fungicides. *Journal of Spices and Aromatic Crops, 21*(2), 151–154.

Rathaiah, (1982). Ridomil for control of rhizome rot turmeric. *Indian Phytopathology, 35*, 297–299.

Singh, A. K., & Edison, S., (2003). Eco-friendly management of leaf spot of turmeric under partial shade. *Indian Phytopathology, 56*(4), 479–480.

Singh, A., Basandrai, A. K., & Sharma, B. K., (2003). Fungicidal management of Taphrina leaf spot of turmeric. *Indian Phytopathology, 56*(1), 119–120.

Sood, R., & Dohroo, N. P., (2005). Epidemiology and management of leaf spot of ginger in Himachal Pradesh. *Indian Phytopathology, 58*(3), 282–288.

Sundararaman, S., (1922). A new ginger disease in Godavari district. *Mem. Dept. Agric. India* (*Bot. Ser.*), *11*, 209–217.

Vadhera, I., Tiwari, S. P., & Dave, G. S., (1998). Integrated management of root knot nematode, meloidogyne incognita in ginger. *Indian Phytopathology, 51*(2), 161–163.

CHAPTER 15

Diseases of Pointed Gourd Crops and Their Management

R. C. SHAKYWAR,[1*] M. PATHAK,[2] MUKUL KUMAR,[3] and R. B. VERMA[4]

[1]*Department of Plant Protection, College of Horticulture and Forestry, Central Agricultural University, Pasighat, Arunachal Pradesh, India*

[2]*KVK, East Siang, College of Horticulture and Forestry, Central Agricultural University, Pasighat, Arunachal Pradesh, India*

[3]*Department of Tree Improvement, College of Horticulture and Forestry, Central Agricultural University, Pasighat, Arunachal Pradesh, India*

[4]*Department of Horticulture (Veg. Sci. and Flr.), Bihar Agricultural University, Sabour, Bhagalpur, Bihar, India*

Corresponding author. E-mail: rcshakywar@gmail.com

15.1 INTRODUCTION

Pointed gourd (*Trichosanthes dioica* Roxb.), also called 'parwal' or 'patal,' is an important and highly accepted cucurbitaceous vegetable crop. It is called 'King of Gourds' because of its higher nutrient content than other cucurbits. Pointed gourd is extensively cultivated in Eastern Indian states particularly in Bihar, Eastern Uttar Pradesh, West Bengal, Assam, Tripura, and to some extent in Orissa (Khare, 2004). It is also grown in Madhya Pradesh, Maharashtra, and Gujarat. This crop is widely accepted and is available for nearly eight to ten months of a year. It is usually propagated through vine cuttings and root suckers. It is a dioecious (male and female plants) vine (creeper) plant with heart-shaped leaves (cordate) and is grown on a trellis. The fruits are green with white or no stripes. Size can vary from small and round to thick and long 2–6 inches (5–15 cm). It thrives well under a hot to moderately warm and humid climate. The plant remains dormant during the winter season and prefers a fertile, well-drained sandy loam soil due to its susceptibility to water-logging. The fruit is the edible part of the plant, which is cooked in various ways either alone or in combination with other vegetables or meats. Pointed gourd is rich in vitamin and contains 9.0 mg, 2.6 mg Na, 83.0 mg K, 1.1 mg Cu, and 17.0 mg S per 100 g edible part. It is purported that pointed gourd possesses the medicinal property of lowering total cholesterol and blood sugar (CSIR, 1998).

15.1.1 HEALTHY BENEFITS OF POINTED GOURD

Some of the healthy benefits and medicinal values of pointed gourd is given below:

- Pointed gourd is a good starting place of vitamins;
- Pointed gourd helps in improving digestion;
- Pointed gourd circle blood sugar levels;
- Pointed gourd controls cholesterol level and keeps heart healthy;
- Pointed gourd helps in treating;
- Pointed gourd helps in weight loss program;
- Pointed gourd is a blood purifier;
- Pointed gourd helps in reducing flu symptom;
- Pointed gourd fight with aging.

The management of diseases is one of the fundamental requirements for profitable pointed gourd production. However, the large number of organisms that can threaten a crop makes management challenging. In addition legislative requirements in relation to chemical control measures are complex and still changing. In practice, there is no substitute for knowledge and experience in identifying problems and choosing the most appropriate management technique for addressing them. The advice of a basis qualified agronomist can be invaluable in identifying threats to the crops and the most appropriate strategy for their management. In addition, a range of tools are available to help identify diseases and then to select the best plant protection product for the particular problem. The level of intervention a grower uses will very much depend on the particular circumstances of the farm and crop. At one end of the spectrum, certified organic growers will seek to avoid all chemical inputs; at the other end, some growers will routinely treat crops without paying much attention to the actual economic need for intervention. Most farmers lie somewhere between these two extremes. The most economically important diseases of pointed gourd in the country are given in Table 15.1.

Diseases of Pointed Gourd Crops and Their Management

TABLE 15.1 Important Diseases of Pointed Gourd Crop and Their Causal Organism

S. No.	Disease Name	Causal Organism
1.	**Bacterial**	
	Angular leaf spot	*Pseudomonas lachrymans* (E. F. Sm. and Bryan) Carsner
	Bacterial wilt	*Erwinia tracheiphila* (E. F. Smith)
	Bacterial leaf spot	*Xanthomonas campestris*
2.	**Fungal**	
	Anthracnose	*Colletotrichum lagenarium (Glomerellacingulataf sp. orbicularae)*
	Downy mildew	*Pseudoperenospora cubensis*
	Fruit rot	*Pythium aphanidernatum, P. butleri*
	Fruit and vine rot	*Phytophthora cinnamomi*
	Powdery mildew	*Erysiphae cichoracearum, Sphaerotheca fuliginea*
3.	**Viral (Major)**	
	CMV	*Cucumber mosaic virus*
	SqMV	*Squash mosaic virus*
	WMV-1	*Watermelon mosaic virus-1*
	WMV-2	*Watermelon mosaic virus-2*
	ZYMV	*Zucchini yellow mosaic virus*
	(Minor)	
	TRSV	*Tobacco ring spot virus*
	TmRSV	*Tomato ring spot virus*
	CYVV	*Clover yellow vein virus*
	Aster yellow	*Phytoplasma*
4.	**Nematodes**	
	Root knot	*Meloidogyne* spp.
	Sting	*Belonolaimus longicaudatus*
5.	**Non-Parasitic**	
	Low temperature injury	
	Guttation salt injury	
	Chemical and mechanical injury	
	Sunscald	
	Molybdenum deficiency	
6.	**Physiological Disorder**	
	Unfruitfulness	
	Blossom end rot	
	Hollow heart	
	Light belly color	

However, part of the reason for the high economic significance is that diseases are particular problems in the most widely grown crops. Other diseases can be equally, or even more, devastating in other crops. Crop walking is an essential part of any grower's management system. Only by careful and continual monitoring of the levels of diseases, pests, and weeds in the crop can an economically useful decision be made on the appropriate management strategies. It would not be economically or environmentally justifiable to try and totally control these natural processes. Instead, the grower needs to monitor activity and take action if and when it makes economic sense to do so. Crop rotations are also an important tool in the fight against diseases. Through growing a totally different type of crop, for one or two years, in many cases host specific diseases, in particular, will become exhausted. The details of all the diseases are given in systematic manner.

15.1.2 BACTERIAL DISEASES

15.1.2.1 ANGULAR LEAF SPOT

It occurs on cucumber in the field in most humid and semi-humid regions. While, it has not been reported on other cucurbits in the field, it has been found on honeydew melons in transit (Smith, 1946).

15.1.2.1.1 Symptoms

This disease appears on leaves, stems, and fruits as small water-soaked spots. On leaves, they enlarge up to about 3 mm in diameter, becoming tan on the upper surface and gummy or shiny on the lower surface and assuming an angular shape as the lesion is delimited by veins. The necrotic centers of leaf spots may drop out. On stems, petioles, and fruits, the water-soaked spots are converted with a white, crusty bacterial exudate (Smith and Bryan, 1915). Because, fruits begin to mature, brown lesions in the fleshy tissue beneath the membrane develop and the discoloration continues along the vascular system, which extends to the seeds. Bacterial soft rot commonly follows the disease on cucumber fruits. On honeydew melons slightly sunken, circular to oblong, water-soaked, greenish tan spots enlarge up to 6 mm and coalesce into brown to black patches (Gardner and Gilbert, 1921).

15.1.2.1.2 Causal Organism

Pseudomonas lachrymans (E. F. Sm. and Bryan) Carsner. The organism is a rod with one to five polar flagella, forming capsules and green fluorescent pigment in culture. Circular, smooth, glistening, transparent, white colonies form on beef peptone agar (Weber, 1929).

15.1.2.1.3 Disease Cycle

The bacterium is seed-borne and persists on infected crop refuse. The bacteria invade the seed coat through the vascular system from the fruit into funiculars. Cotyledons are not known to be invaded until after germination. The organism is disseminated locally chiefly by rain, and penetration occurs through stomata. The organisms develop intercellularly in the early stage of infection (Carsner, 1918).

15.1.2.1.4 Management

1. Rotation and sanitation are essential to eliminate inoculums in infected plant debris.
2. The organisms in the seed are eliminated in part by treatment with 1–1,000 mercuric chlorides for 5–10 minutes, followed by a rinse (Gardner and Gilbert, 1918).
3. No method for complete eradication is known.
4. The most effective means of eliminating inoculums in the seed is to grow crops for seed in arid regions such as interior California.

15.1.2.2 BACTERIAL WILT

Bacterial wilt is a common and often destructive disease on cucurbitaceous family. Squash and pumpkin are susceptible but are not affected so severely. A number of other cucurbits, including watermelon, have been infected artistically.

15.1.2.2.1 Symptoms

The first signs of wilt appear usually on individual leaves as dull green patches which become flaccid in sunny weather. As the disease progresses, more leaves wilt and eventually an entire branch is affected (Clayton, 1931). The wilting then becomes permanent, and the leaves and vines shrivel and die. Occasionally exudates on fruits are visible. When wilted stems are cross-sectioned, the viscid, sticky bacterial matrix exudes from the bundles and may be drawn out in strands 1 in crore more. This feature is used as a means of diagnosis (Harris, 1940).

15.1.2.2.2 The Causal Organism

Erwinia tracheiphila (E. F. Smith) Holland. The organism is a motile rod with four to eight peritrichous flagella. Capsules are formed on agar colonies are internally reticulated, small circular, smooth, glistening, white, and viscid (Harris, 1940).

15.1.2.2.3 Disease Cycle

The bacteria overwinter in the bodies of adult cucumber beetles of which there are two species: the striped beetle, *Acalymmavittata.* Fabricius and the spotted beetle, *A. duodecimpunctata* Oliver. Primary infection is produced when beetles deed upon young leaves or cotyledons (Rand, 1915). This is the only means of natural infection known. This pathogen is one of the unusual cases in which the organism is completely dependent upon of insect for its survival. After infection, the bacteria invade and progress within the tracheal vessels. The capsular material of the organisms provides a viscid matrix in the vessels and is believed to cause wilt by mechanical plugging of the xylem (Clayton, 1927).

15.1.2.2.4 Management

1. Some the pathogen is entirely dependent upon the beetle for its perpetuation and dissemination, control of wilt involves control of the insect.
2. It is not enough to reduce insect damage to a commercially acceptable degree, since insects will still feed on the plants in sufficient numbers to cause an epidemic (Wilson and Sleesman, 1947).
3. It essential therefore to apply insecticides early to forestall the colonization of overwintering adults and to continue applications frequently.

Diseases of Pointed Gourd Crops and Their Management

There are good indications that a fungicide applied with the insecticide enhances the degree of control (Rand and Enlowa, 1916).

4. There is a considerable difference between cucumber varieties in their susceptibility to wilt, but none are very highly resistant.

15.1.2.3 BACTERIAL LEAF SPOT

15.1.2.3.1 Symptoms

Dark, angular lesions on leaves; leaf lesions may coalesce and cause severely blighted foliage; water-soaked lesions which enlarge and develop into tan scabs, or blisters, on the fruit; blisters eventually flatten as they reach their full size. Disease can spread rapidly in a field. The bacterium can be introduced through contaminated seed.

15.1.2.3.2 Causal organism: Xanthomonas campestris

15.1.2.3.3 Management

1. Avoid overhead irrigation.
2. Crop rotation away from cucurbit species to prevent disease.
3. Use new seed each planting as saved seed is more likely to carry bacteria.
4. Apply appropriate protective fungicides; copper containing fungicides generally provide good management.

15.1.3 FUNGAL DISEASES

15.1.3.1 ANTHRACNOSE

Anthracnose of cucumber was first reported from Italy in 1867 and later from England in 1871. Now, the disease is distributed worldwide and is prevalent in Europe, America, Australia, and Asia. In India, the disease occurs in almost all the regions, wherever cucumber is grown and in some seasons it causes serious damage to the crop and yield loss. It attacks cucumbers, watermelon, muskmelon, gourds, vegetable marrow, chayote, and gherkin.

15.1.3.1.1 Symptoms

The fungus attacks all aerial parts of the plants. Typically, the young plants are not affected. However, if young plants are attacked at the soil level, the stems are girdled and the tissues shrink and the seedlings may collapse and die. Frequently, the disease occurs somewhat late in the growing season and spreads rapidly in wet weather. Initial symptoms appear on the leaves as small, pale-green, water-soaked spots, which soon become reddish-brown in color, with a yellow zone around. The spots enlarge and coalesce to form large bronzed areas. These areas later become necrotic, develop cracks and fall out, and soon the entire leaf dies. Severely affected crop presents a scorched appearance. On the upper surface of the leaf lesions, scattered, minute, pin point like dots representing the aceruli of the fungus are produced in large numbers. On leaf petioles and stem oval to linear, slightly sunken, water soaked, yellowish-brown lesions are formed. Sometimes, the stem lesions become very deep and girdle the stem completely; as a result the entire shoot may wither and die. Spots on the fruits are pale green, almost circular and sunken and they turn black later on. On these lesions, buff to pink colored, gelatinous masses of spores are produced abundantly. On ripe fruits, the black lesions may become white, crater-like cankers on which acervuli appear as black dots. From such seep lesions on the fruits, seed infection may occur.

15.1.3.1.2 Causal Organism: Colletotrichum Lagenarium (Glomerella Cingulata f. sp. Orbiculare)

The mycelium of the fungus is septate, hyaline when young, becoming darker with age, inter, and intracellular. Aceruli are produced from sub epidermal stromata. They are brown to black in color. The conidia are single-celled, thin-walled, hyaline, but pinkish in mass, oblong or ovate-oblong, slightly pointed at one end are 13.0–19.0 X 4.0–6.0 μ in size. Setae are numerous. They are 2–3 septate, brown, thick-walled, and 90–120 μ in length. The perfect stage is very rarely found in nature.

15.1.3.1.3 Mode of Survival, Spread, and Epidemiology

The fungus is mainly soil-borne. It can survive in the infected plant debris in the soil for prolonged periods and continues to produce conidia, which may cause primary infection. The disease may be carried through contaminated seed externally, if the infected fruit pulp remains adhered on the seed. The inoculums may come from the various alternate hosts of the fungus also. The disease may spread through irrigation or rainwater and through soil. The secondary spread is through conidia carried by wind or rain splash. The optimum temperature for the germination of conidia is 22–27°C. The high humidity of 87–95% and low temperatures of about 24°C are favorable for disease occurrence and development (Jones and Everett, 1966).

15.1.3.1.4   Management

1. Diseased crop debris should be collected and destroyed by burning.
2. Crop rotation may be followed.
3. Cultivation of susceptible cucurbits or varieties in endemic areas should be avoided.
4. Spraying with Copper oxychloride @ 2.5 gm or Mancozeb @ 2.0 gm or Carbendazim @ 1.0 gm or Chlorothalonil @ 1.5gm or Captafol @ 1.5/l of water at fortnightly intervals, commencing from the time the plants are well-established controls the disease.
5. Treating the seeds with Captan or thiram @ 4.0 gm/kg of seed, at least 24 hours prior to sowing, will eradicate the externally seed-borne inoculums and also protects the seedlings from early infection in the field.

15.1.3.2   DOWNY MILDEW

The disease, which was first recorded in Cuban in 1868, is now prevalent in all parts of the world. It occurs in most of the cultivated cucurbits, such as cucumber, muskmelon, watermelon, squash, pumpkin, gourd, and several other species.

15.1.3.2.1   Symptoms

The symptoms appear as pale yellow, angular patches on the upper surface of the leaves, which are usually limited by the veins. The color of the patches soon deepens to brownish- yellow, as they grow older. On the corresponding lower side of these patches, a fine, purplish, downy growth appears during periods of high humidity. The growth appears scanty and is seen only under close examination. Severely affected leaves become yellow completely, dry, and die. Mostly, the disease attack leaves in the middle portion of the plant and gradually spreads to the younger leaves. Fruits are rarely affected directly. Only a few small and misshapen fruits may be formed in the infected plants. Muskmelon fruits may be attacked and may be covered with the fungal growth (Ahamad et al., 2000).

15.1.3.2.2 Causal Organism: Pseudoperonospora Cubensis (Berk. and Curt.) Rostow.

The fungus causing the disease is an obligate parasite. The mycelium of downy mildew fungus was hyaline, coenocytic, and intercellular, developed abundantly in the mesophyll but also penetrated palisade tissues. Haustoria were small, ovate, intercellular sometimes with finger-like branches. Sporangiophores were 180–400 µm in length, dichotomously branched in their upper third, emerged in groups of 1–5 from stomata. The sporiferous tips, on which the sporangia were borne singly, were subacute. Sporangia were pale grayish to olivaceous-purple, ovoid to ellipsoidal, thin-walled, with a papilla at the distal end, measuring 20–40 x 14–25 µm. Sporangia germinated by the reproduction of flagellate zoospores, rarely by infection hyphae (Brunelli et al., 1989).

15.1.3.2.3 Mode of Survival, Spread, and Epidemiology

Because of the great diversity of climate, season, and cropping patterns in different parts of India and the presence of a number of wild hosts, the pathogen is able to survive from season to season and cause fresh infection. The pathogen may also perpetuate in the form of active mycelium on self-sown crops or cucurbits cultivated all through the year. In areas where oospores are formed, they may serve as an important source of primary inoculums (Khatua et al., 1981). Secondary spread of the disease is through sporangia dispersed by wind or rain splash. Some insects, such as the cucumber beetles, may also disperse the sporangia mechanically. Maximum sporangial production takes place in the early morning hours, under saturated atmospheric conditions and low temperatures between 18–22°C. Sporangial production almost ceases at temperatures above 35°C. A dry period of 6 hours, followed by a wet period of 6 hours, is highly favorable

Diseases of Pointed Gourd Crops and Their Management

for sporulation. Optimum temperature for germination of sporangia is 20C. Free water on the leaf surface is essential for the zoospores to cause infection. The disease usually appears in the latter part of the rainy season and may cause extensive damage to the crops (Bains and Jhooty, 1978).

15.1.3.2.4 *Management*

1. Alternate weed hosts in and around the fields should be eradicated.
2. Severely affected plants or plant parts may be removed and destroyed.
3. Resistant varieties may be grown.
4. Proper drainage facilities should be provided.
5. Protective fungicides, such as Copper oxycholoride, Dithiocarbamates, Captafol, Chlorothalonil, etc. are effective only in checking the disease before its appearance. These fungicides are not effective in controlling the disease after it has established.
6. Systemic fungicides, such as Metalaxyl and Fosetyl-Al have been found to be effective in managing the disease (Colucci and Holmes, 2010).
7. Spraying with Metalaxyl (Ridomil) @ 1.0 gm. + Mancozeb @2.0 gm or Fosetyl - Al (Aliette) @ 1.0 gm. + Moncozeb @ 2.0 g/l of water is very effective in managing the disease.

15.1.3.3 *FRUIT ROT*

Fruit rot is a very common disease of cucurbits in India. It occurs in almost all the regions, wherever cucurbits are cultivated. The occurrence of the disease is more in extended wet periods. Several species of cucurbits, including bottle gourd, sponge gourd, snake gourd, cucumber, bitter gourd, muskmelon, and watermelon, are attacked. In muskmelon and watermelon, sometimes 50% or more of the fruits undergo rotting due to this disease, especially after rains.

15.1.3.3.1 *Symptoms*

The symptom appears as dark green, water-soaked lesion on the skin of the fruit during prolonged wet periods. The fruits, which are directly in contact with the soil or slightly above the ground level, are more prone to attack by the fungi. On the water-soaked lesion on the surface of the fruit, a whitish, luxuriant weft of wooly mycelia growth results, while the interior fleshy portion is turned into a soft, pulpy watery, rotten mass called 'cottony leak.' The decaying fruit emits a bad and repulsive smell. The margin of the skin around the area covered by the wooly growth gradually becomes dark green and water-soaked, indicating the 'killing in advance' of the tissues below, followed by mycelia growth in that area also. Subsequently, the disease spreads to other fruits during storage and transit.

15.1.3.3.2 Causal Organism (Pythium aphanidermatum and P. butleri.)

Generally, both species (*Pythium aphanidermatum* and *P. butleri*) are responsible for causing the disease. For the morphology of *Pythium aphanidermatum*, refer 'Damping off seedling of eggplant. The morphology of *Pythium butleri* is almost similar to that of *P. aphanidermatum*. However, *P. butleri* produces more copious and robust mycelium, larger, more swollen and much branches sporangia, larger zoospores and larger oospores. Some species of *Fusarium, Rhizoctonia,* and *Phytophthora* also cause fruit rot of cucurbits.

15.1.3.3.3 Mode of Survival, Spread, and Epidemiology

The fungi *Pythium aphanidermatum* and *P. butleri* are soil inhabiting ones and survive on dead organic matter present in the soil and cause disease, when suitable hosts are available. Secondary spread of the disease may occur through sporangia carried by irrigation and rainwater and mechanically by some insects, such as cucurbit beetles. The oospores produced by the fungi may also remain in the soil in a viable state for long periods and cause primary infection. Any kind of injury or wound on the surface of the fruit caused by either mechanical means or by insect's predisporeses the fruit to infection. High soil moisture and high temperature of about 30°C favor disease occurrence and development.

15.1.3.3.4 Management

1. Proper drainage facilities should be provided to prevent water stagnation in field.

2. As far as possible, the fruits should not be allowed to come into contact with the soil.
3. During cultural operations and during storage and transit care should be taken to avoid causing injuries to the fruits.
4. The fruits should be stored in clean, moisture free and well aerated store houses.
5. Disease affected fruits should not be stored along with healthy fruits and should be removed and destroyed.

15.1.3.4 FRUIT AND VINE ROT

Fruit and vine rot is popularly known as Haja. It is a serious disease of pointed gourd that results in huge loss of crop in case of severe infection. In winter months, rise in atmospheric temperature accompanied by 2–3 showers favors the disease (Mondal et al., 2013) (Figure 15.1).

FIGURE 15.1 Vine rot of pointed gourd.

15.1.3.4.1 Symptoms

The disease starts with sign of drying of upper side of the nodes and internodes, while the lower portion remains unaffected. But with the advance of disease the whole vine get affected. In monsoon period the severity of disease became higher with the rotting of fruits and vines. Sudden rainfall after a long dry spell or water stagnation in the field helps in spreading of the disease (Khatua et al., 2013).

15.1.3.4.2 Causal Organism

Phytophthora cinnamomi: Sporangia are ovoid, obpyriform with an apical thickening, tapered or rounded at the base, and terminally borne. Sporangia, which release motile zoospores, are not readily produced in axenic culture. Chlamydospores are produced abundantly axenically and from infected tissue. They are borne from hyphae, and globose with thinner walls. Sizes range from 31 to 50 μm in diameter and are either terminal to intercalary in the mycelium. The fungus is heterothallic, requiring compatible types (A1 and A2) to sporulate sexually. Antheridia are amphigynous, averaging 19x17 μm. Oogonia are round with a tapered base, smooth, hyaline to yellow with size ranging from 21 to 58 μm. Oospores are hyaline to yellow and plerotic. Sizes range from 19 to 54 μm (Guharoy et al., 2006).

15.1.3.4.3 Disease Cycle and Epidemiology

The survival spores of *P. cinnamomi* may persist up to six years and occur in infected roots, crown, and infested soil (Coyier and Roane, 1986). *P. cinnamomi* grows and infects optimally at higher temperatures through the months of June and August, but is able to infect at temperatures as low as 15°C. Under optimal water and temperature conditions, chlamydospores germinate to produce sporangia, which releases infectious zoospores and (Erwin and Ribiero, 1996). High water tables and excess irrigation provide suitable conditions for increased zoospore inoculum levels and subsequent root infections. Thus, excessive soil water increases the incidence and severity of disease (Saha et al., 2004). Once a host is infected, the water flow through the xylem is reduced via wilt-inducing toxins such as ß-glucans and ß-glucan hydrolases. Unlike healthy plants, those infected with *P. cinnamomi* do not recover from the stresses of low soil moisture. Excessive use

Diseases of Pointed Gourd Crops and Their Management   287

of nitrogen-based fertilizers further increases susceptibility to disease due to the increased uptake of water from the soil matrix. Dispersal occurs via multiple avenues: groundwater, streams, and irrigation as well as infested potting soil, splash from pot to pot, infested pot bases on polyethylene, and diseased nursery stock (Reynolds et al., 1985).

15.1.3.4.4 Management

1. Continuously use strict sanitary measures.
2. Sustain good drainage facility.
3. Using straw mulching.
4. Alternate spraying of Metalaxyl + Mancozeb @ 2.5g/l of water and copper oxychloride @ 4g/l of water can effectively check the disease (Urech et al., 1977).

15.1.3.5 POWDERY MILDEW

The disease is distributed worldwide. In India, the disease occurs in the entire region almost every year and cause extensive damage to the crops and yield loss.

15.1.3.5.1 Symptoms

Under excessively moist conditions, the mildew appears as small, white or dirty gray patches on the leaves and stems but not on the fruits. These areas become powdery as they enlarge and soon cover the entire surface of leaves and stems. The effect of mildew, by covering the leaves and stems with its mycelia web arrest their growth, induce chlorosis and cause reduction in the size and quality of fruits.

15.1.3.5.2 The Causal Organism

Erysiphe cichoracearum and Sphaerotheca fuliginea. The pathogens causing the disease are obligate parasites.

1. ***Erysiphe cichoracearum:*** The fungus forms a tangled web of mycelium over the surface of leaves and stems, consisting of hyaline,

septate hyphae, which produce rather large spherical haustoria in the epidermal cells. The superficial mycelium gives rise to a great profusion of erect conidiophores, bearing conidia in long chains. Conidia are ellipsoid to barrel-shaped and measure 25–45 X 14–26 μ in size. They germinate and produce appressoria. Cleistothecia are of rare occurrence and found only on certain parts of the diseased leaves in the cold season. They are found scattered or in groups and are globose becoming depressed or irregular in shape. They measure 90–135 μ in diameter. Appendages on the cleistothecia are numerous, basally inserted, mycelioid, interwoven with the mycelium, hyaline to dark- brown, 1-time as long as the diameter of the cleistothecium and rarely branched. Each cleistothecium contains 10–15 ellipsoids, more or less stalked asci, measuring 60–90 X 25–50 μ. Each ascus contains 2, very rarely 3, hyaline, oval, or sub cylindrical ascospores, measuring 20–28 X 12–20 μ.

2. **Sphaerotheca fuliginea:** The mycelium of the pathogen is hyaline, becoming reddish-brown or brownish with age. The superficial mycelium produces spherical haustoria inside the epidermal cells. Erect conidiophores are produced in profusion from the superficial mycelium. Conidia are borne singly at the end of fibrosin bodies. They are hyaline, single-celled, ellipsoid to barrel-shaped, and measure 20–28 X 14–24 μ in size. The conidia on germination produce a characteristic forked germ tube, distinguishing from those of *Erysiphe cichoracearum* Cleistothecia are dark, spherical with many mycelioid appendages and measure 90–180 μ in diameter. Each cleistothecium usually contains only one ascus. The asci are broadly elliptic to sub- globose and 50–80 X 30–60 μ in size. In each ascus, there are 6–10, usually eight hyaline, continuous, single-called, ellipsoid ascospores, measuring 20–25 X 12–15 μ in size.

15.1.3.5.3 *Mode of Survival, Spread, and Epidemiology*

In areas where cleistothecia are formed, they may carry over the disease from one season to the next cleistothecia may be formed, when sexually compatible strains of the fungus establish union in certain parts of the diseased leaves. Cleistothecia may develop on diseased plant debris leftover in the field and may perpetuate the disease. It is possible that the pathogen may be harbored on some wild cucurbits during the offseason and produce conidia,

Diseases of Pointed Gourd Crops and Their Management

which may serve as primary inoculums and infect cultivated cucurbits. Secondary spread may occur through conidia carried by wind, rain splash, and by insects that attack cucurbit.

Heavy dew, cloudy, and damp weather, the moderate temperature range of 22–28°C and low light intensity favor the occurrence and development of the disease. However, even under dry condition, the conidia can germinate and cause the disease. At least 13 form species of *E. cichoracearum* and 84 f. sp. of *S. fuliginea* have been identified.

15.1.3.5.4 *Management*

1. Removal and destruction of diseased plant debris.
2. Destruction of weed hosts harboring the pathogen help in minimizing the occurrence of the disease.
3. Spraying the crop with Wettable Sulfur @ 4.0 gm or Dinocap (Karathane) @ 1.0 gm or Carbendazim @ 1.0 gm or Benomyl @ 1.0 gm or Thiophanate methyl (TopsinM) @ 1.0 gm or Tridemorph (Calixin) @ 1.0 g/l of water is found to be effective in managing the disease.
4. Resistant varieties of muskmelon only, such as Diguria and Huragola have been reported to be immune to powdery mildew.

15.1.4 *VIRAL AND PHYTOPLASMAL DISEASES*

15.1.4.1 *MAJOR VIRAL DISEASES*

Viruses are the most common causes of diseases affecting cucurbits in different parts of the country. These diseases result in losses through reduction in growth and yield are responsible for distortion and mottling of fruit and making the product unmarketable. A complex of viruses is able to infect cucurbits. A plant group includes cucumber, melon, squash, pumpkins, watermelon, and gourds. The most important viruses are cucumber mosaic (CMV), squash mosaic (SqMV), watermelon mosaic I (WMV-1), watermelon mosaic 2 (WMV-2), and zucchini yellow mosaic (ZYMV). With the exception of SqMV, which is seed-borne in melon and transmitted by beetles, the other major viruses are transmitted by several aphid species in a nonpersistent manner.

15.1.4.1.1 Cucumber Mosaic

The disease was first reported from the United States of America in 1934 on *Cucumber sativus* and now, the disease is distributed worldwide. In India, the disease occurs in all the regions, wherever cucumbers are cultivated on a large scale intensively. Cucumber mosaic virus (CMV) infects more different kinds of plants that any other virus. The host range includes several vegetable crops, ornamental plant, weeds, and other plants. The important crops affected by the virus are cucumbers, gladioli, melon, squash, pepper, spinach, celery, tomato, beet, beans, banana, and crucifers.

15.1.4.1.2 Symptoms

Young seedlings are usually not attacked in the field during the first few weeks. Infection starts, when the plants are about 6 weeks old and growing vigorously. The initial symptoms appear in the young developing leaves as small, greenish yellow areas, 1.0–2.0 mm in diameter and are limited by the veins. Mottling, distortion, and wrinkling of the leaves follow this and the edges of the leaves begin to curl downward. The growth of the affected plant is reduced markedly and the internodes and petioles become shortened. The leaf size is also reduced considerably. Such plants produce few runners few flowers and fruits. The older leaves of infected plants become chlorotic and necrotic areas are formed along the margins. The necrosis gradually spreads over the entire leaf area. Such leaves dry and remain hanging from the petioles or fall off, leading to defoliation of the vine. The flowers produced are also dwarfed. Fruits produced on infected plants show pale green or whitish patches, intermingled with dark-green raised areas, which later become rough and wart like. The fruits are also deformed and they have a bitter taste.

15.1.4.1.3 Causal Organism

Cucumber mosaic virus (CMV): The virus causing the disease belongs to the *Genus-Cucumovirus.* The viruses have tripartite genomes consisting of ssRNAs and are encapsidated in three types of protein covering. The three identical isometric particles are about 28 nm in diameter each. All the three particles are necessary for infection. The virus can easily be transmitted mechanically by sap inoculation and by insect vectors in a nonpersistent manner. The virus infects and multiplies in the phloem and parenchyma cells. Drying destroys the virus.

15.1.4.1.4   Mode of Survival and Spread

The virus can be transmitted to a healthy plant by touching it, after handling a disease infected plant. The aphid vectors, *Aphis craccivora* and transmit the virus in a nonpersistent manner. Because the virus has numerous hosts and *Myzus persicae*, the aphid vectors are also present throughout the year in various crops, the disease can be perpetuated easily.

15.1.4.1.5   Management

1. Alternate weed hosts belong to species of *Commolina, Physalis. Phytolacca,* etc. should be eradicated.
2. Continuous cropping of cucumbers in a particular area should be avoided.
3. After handling diseased plants, healthy plants should not be touched.
4. Resistant varieties may be grown.
5. The aphid vectors should be controlled by using insecticides, such as dimethoate, methyl demeton, or phosphamidon.

15.1.4.2   SQUASH MOSAIC VIRUS (SQMV)

It can cause an important disease of melons and squash in the country. The virus is seed-borne in muskmelon and is spread in nature principally by the spotted and striped cucumber beetles. The virus is carried within the seed and cannot be eliminated by hot water or chemical treatment with tri-sodium phosphate.

Symptoms consist of pronounced chlorotic mottle, green vein banding and distortion of leaves of young seedlings. On mature plants, leaves show intense dark green mosaic, blistering, and hardening suggestive of a hormonal herbicide effect. Infected fruit coming from such plants show a strong mottled pattern with a lack of netting on melons. Management includes selection of disease-free seed and cucumber beetle control.

Watermelon mosaic virus 1 (WMV-1) is aphid transmitted and infection is limited to cucurbits. This virus has been recovered in New York several times since it first occurred in epidemic proportions in 1969. This virus is capable of infecting all commercial cucurbit crops. The foliage of affected plants shows strong mosaic, distortion, and deep leaf serration. Fruits are also malformed with a knob by overgrowth.

Watermelon mosaic virus 2 (WMV-2) is the second most important cucurbit virus. This virus can infect and produce symptoms on all commercially grown cucurbits. This aphid transmitted virus causes milder symptoms on the foliage of most infected plants like squash, and growers have seen a lessening of foliar symptoms following fertilization. Fruit distortion and color breaking are still a problem on varieties like yellow straight-neck squash. The host range for WMV-2 is not limited to cucurbits, thus opening the possible overwintering of this virus in several leguminous species such as clover. Mixed infections of cucurbits with CMV and WMV-2 are common by the end of the season.

Zucchini yellow mosaic virus (ZYMV) is a recently described virus disease of cucurbits, first identified in Europe in 1981. It has since been reported from most southern and southwestern states and was found in New York State in 1983. The virus has characteristics very similar to WMV-1 and WMV-2 (nonpersistent aphid transmission, etc.), and like WMV-2, its host range is not limited to cucurbits. Currently, none of the genetic factors that confer resistance to WMV-1 or WMV-2 are able to control ZYMV, but other resistance sources have been identified. Muskmelon, watermelon, and squash are severely affected by ZYMV. Foliar symptoms consist of a prominent yellow mosaic, necrosis, distortion, and stunting. Fruits remain small, greatly malformed, and green mottled, including the fruit of the variety 'Multipik.' It is too early to tell which weed hosts may serve to overwinter this virus in New York.

15.1.4.3 MINOR VIRAL DISEASES

Tobacco ringspot virus (TRSV) is mainly transmitted by nematodes (*Xiphinema americanun*). Melons and cucumbers are most commonly affected by this virus. The virus has been known on rare occasions to be seed-borne in cucurbits. The newly infected leaves show a very bright mosaic with plant stunting. However, subsequent leaves are reduced in size and develop a dark green color.

Tomato ringspot virus (TmRSV) causes severe damage to summer and winter squash but shows only mild symptoms in the other cultivated cucurbits. Like TRSV, TmRSV is nematode transmitted and can be overwintered on many weed species without expressing symptoms.

Diseases of Pointed Gourd Crops and Their Management

Clover yellow vein virus **(CYVV)** is an aphid-transmitted virus that can infect summer squash and was previously considered to be the severe strain of bean yellow mosaic virus. The virus produces a yellow specking on the foliage of infected plants.

15.1.5 PHYTOPLASMAL DISEASES

15.1.5.1 ASTER YELLOWS

15.1.5.1.1 Symptoms

The disease symptoms are foliage turning yellow. Secondary shoots begin growing prolifically; stems take on a rigid, upright growth habit; leaves are often small in size and distorted, may appear thickened; flowers are often disfigured and possess conspicuous leafy bracts; fruits are small and pale in color. Disease is transmitted by leafhoppers and can cause huge losses in cucurbit crops.

15.1.5.1.2 Management

1. Remove any infected plants from the field to reduce spread.
2. Control weeds in and around the field that may act as a reservoir for the phytoplasma.
3. Protect plants from leaf hopper vectors with row covers.

15.1.6 NEMATODES DISEASES

Cucurbits are susceptible to attack from nematodes- microscopic round-worms that live in the soil. Nematodes can cause considerable damage and kill the plants if left untreated.

Root knot nematodes are the most damaging plant parasites. These pests, which are particularly problematic in sandy or sandy loam soil, attack cucurbit roots when soil temperatures rise above 25°C. They puncture the tissue with a specialized; needle like mouth part called a stylet and feed by extracting the contents of the plant cells. Sting nematodes are a type of ecto-parasitic nematode that can cause considerable damage to cucurbit crops by feeding on the outside of the roots (Khan and Verma, 2004).

15.1.6.1 NEMATODE LIFE CYCLE

Root knot nematodes usually only live for three or four weeks. Each female produces several hundred eggs that hatch almost immediately. The juvenile nematodes enter the roots and continue feeding, molting three more times before they become adults. Sting nematodes follow a similar life cycle but feed on the outside of the roots rather than from the inside.

15.1.6.2 SYMPTOMS OF ROOT KNOT NEMATODES (MELOIDOGYNE SPP.)

Root knot nematodes feed on the root systems of cucurbit plants and create abnormal, knotty growths on the roots called galls. The galls, which can grow to be 1 inch or more, make it difficult for the plant to transmit water and nutrients from the roots to the above-ground plant parts. Consequently, cucurbit plants infested with nematodes are often stunted. They may have yellow leaves and tend to wilt easily in warm weather but the symptoms do not improve when the plants are watered. Infested plants produce fewer leaves, flowers, and fruits than healthy plants and the fruit may be of poor quality (Khanna and Kumar, 2006).

15.1.6.3 SYMPTOMS OF STING NEMATODES (BELONOLAIMUS LONGICAUDATUS)

Cucurbits infested with sting nematodes have irregularly shaped roots. The parasites kill the root tips and cause them to stop growing, so the roots often look swollen and maybe in a tightly woven cluster (Verma and Ali, 1990). Sting nematodes are particularly damaging to young plants with underdeveloped root systems. Infested seedlings may emerge but stop growing. Severely infested plants may die (Mukhopadhyay et al., 2006).

15.1.6.3.1 Management

1. Nematode infestations can be managed by solarization of soil.
2. Intensify the soil temperature during the summer by covering it with a clear tarp. Leave it for four to six weeks.
3. The sun energy heats the first 12 to 18 inches of soil to temperatures up to 50–55°C killing the nematodes.
4. Deep tilling may also help control nematode populations.

15.1.7 NON-PARASITIC DISEASES

1. **Low Temperature Injury:** When cucumber fruits are held in storage or transits at low temperatures and low humidity, circular pits 3–10 mm in diameter and up to 1 mm deep appears, often coalescing to form large, irregular, sunken areas. Injury increases in severity from about 15°C down to 1°C. At any given temperature the injury deceases with rise in relative humidity (RH). Honeydew melons are subject to low temperature breakdown when stored at 0–2°C. For 2 weeks. Cantaloupes are not so affected. The rind first becomes water-soaked, and cell sap oozes onto the surface. The flesh and flavor of the melon remain unchanged. The fruits are storage at 5–8°C (Morris and Platenius, 1939).
2. **Guttation Salt Injury:** The injury from increasing concentration of salts and their deposit leads to necrosis of the leaf margins and some-times the entire leaf may die (Curtis, 1943; Ivanoff, 1944).
3. **Chemical and Mechanical Injuries:** When cucurbits fruits are shipped, it is not unusual for bruises to occur which develop into watersoaked blemishes which detract from their market value. Other injuries may be due to fertilizers or lime on the floors of cars. In view of the sensitivity of melons to mechanical and chemical injuries, it is important to clean shipping cars and trucks thoroughly and provide adequate excelsior bedding at the base of the load (Ramsey, 1938).
4. **Sunscald:** The honeydew melon is susceptible to the sun's rays, especially when the foliage has been reduced by *Alternaria* leaf blight (Leclerg, 1931). The disease is described by the Rocky Ford area of Colorado. Small brown spots appear on the exposed side of the melon, enlarging up to 12 cm. in diameter. They are roughly circular and sunken, becoming dark with a white or gray chlorotic margin. Ripe melons are more susceptible than immature ones.
5. **Molybdenum Deficiency:** On cantaloupe is described in Australia by Wilson as pale chlorosis of leaves and stunting.

15.1.8 PHYSIOLOGICAL DISORDERS

15.1.8.1 UNFRUITFULNESS IN POINTED GOURD

Pointed gourd is a dioecious cucurbit. So, male and female plants are separate. Female plants produce the fruit, whereas male plants act as a pollen donor. So, the required number of male plants should be there in the population of female plants to ensure adequate pollination, fertilization, and fruit set. A common problem is met with where pistillate flowers in female plants are shed due to lack of pollination and fertilization. In some cases, ovary of the unfertilized flower may flow a bit due to parathenocarpic stimulation which also abscises after a few days (Figures 15.2–15.4).

FIGURE 15.2   Wilt of pointed gourd.

15.1.8.1.1 *Management*

Male plants must be grown in the field along with the female plants at the rates of 10–12 male plants per 100 female plants to ensure adequate pollination and fruit set. Hand pollination may be done successfully to achieve fruit set. Hand pollination to the female flowers should be done in the early morning hours because stigma receptivity decreases with an advancement of the day.

FIGURE 15.3 Powdery mildew of pointed gourd.

15.1.8.2 BLOSSOM END ROT

15.1.8.2.1 Symptoms

The blossom end of the fruit develops a dark leathery appearance. Symptoms may progress until the entire end of the fruit turns black and rots.

15.1.8.2.2 Conditions for Disease Development

This disorder is associated with insufficient calcium uptake and alternating periods of wet and dry soil. Damage to the root system may also account for decreased calcium uptake and the development of blossom end rot.

FIGURE 15.4 Downy mildew of pointed gourd.

15.1.8.2.3 Management

1. The disorder is minimized by mulching to maintain constant soil moisture.
2. Applying calcium fertilizers and avoiding high levels of nitrogen.
3. Drip irrigate crops to manage water management.

Diseases of Pointed Gourd Crops and Their Management

15.1.8.3 HOLLOW HEART

15.1.8.3.1 Symptoms

Cracks in internal watermelon fruit flesh can occur due to accelerated growth in response to ideal growing conditions.

15.1.8.3.2 Conditions for Disease Development

There is a genetic component to this disorder, but growing conditions can account for much of the variation observed. It appears to be associated with conditions that result in poor pollination (enough pollination to set the fruit but not enough to fertilize a high percentage of the ovules) followed by rapid fruit growing conditions (too much fertility, water, and high temperatures).

15.1.8.3.3 Management

Avoid watermelon varieties with a tendency to exhibit a hollow heart. Implement best practices for irrigation and fertilization programs.

15.1.8.4 LIGHT BELLY COLOR

15.1.8.4.1 Symptoms

This disorder is characterized by the undersurface of cucumber fruit remaining light in color instead of turning dark green.

15.1.8.4.2 Conditions for Disease Development

Commonly occurs on fruit lying on cool, moist soil.

15.1.8.4.3 Management

1. Avoid excessive nitrogen.
2. It is partially controlled by avoiding luxuriant vine growth.

KEYWORDS

- **cucumber mosaic virus**
- **non-parasitic diseases**
- **physiological disorder**
- **squash mosaic**
- **squash mosaic virus**
- **zucchini yellow mosaic**

REFERENCES

Ahamad, S., Narain, U., Prajapati, R. K., & Chhote, L., (2000). Management of downy mildew of cucumber. *Annals of Plant Protection Sciences*, 8(2), 254–255.

Bains, S. S., & Jhooty, J. S., (1978). Epidemiological studies on downy mildew of muskmelon caused by *Pseudoperonos poracubensis*. *Indian Phytopathology*, 31, 42–46.

Brunelli, A., Davi, R., Finelli, F., & Emiliani, G., (1989). Control trials against cucurbit downy mildew (*Pseudoperonos poracubensis* (Berck. & Kurt.) Rostow) on melon. Difesadelle Piante, 12(1&2), 265–273.

Carsner, F., (1918). Angular leaf spot of cucumber: Dissemination, overwintering and control. *Jour. Agr. Res.*, 15, 201–220.

Clayton, E. E., (1922). Comparative susceptibility of European and American varieties of cucumbers to bacterial wilt. *Ibid.*, 12, 143–146.

Clayton, E. E., (1927). Effect of early spray and dust applications on later incidence of cucumber wilt and mosaic disease. *Phytopathology*, 17, 473–481.

Clayton, E. E., (1931). Cucumber disease investigations on Long Island. *N. Y. State Agr. Expt. Sta Bul.*, 590.

Colucci, S. J., & Holmes, G. J., (2010). Downy mildew of cucurbits. *The Plant Health Instructor*. doi: 10.1094/PHI-I-2010-0825-01.

CSIR., (1998). *A Dictionary of Indian Medicinal Plant's Raw Material and Industrial Products* (pp. 289, 290). New Delhi: The Wealth of India.

Curtis, L. C., (1943). Deleterious effects of guttated fluids on foliage. *Amer. Jour. Bot.*, 30, 778–781.

Gardner, M. W., & Gilbert, W. W., (1921). Field tests with cucumber angular leaf spot and anthracnose. *Phytopathology*, 11, 298–299.

Gilbert, W. W., & Gardner, M. W., (1918). Seed treatment control and overwintering of cucumber angular leaf spot. *Ibid.* 8, 229–233.

Guharoy, S. Bhattacharyya, S. Mukherjee, S. Mandal, N., & Khatua, D. C., (2006). *Phytophthora melonis* associated with Fruit and vine rot disease of pointed gourd in India as revealed by RFLP and sequencing of ITS region. *Journal of Phytopathology*, 154, 612–615.

Harris, H. A., (1940). Comparative wilt induction by *Erwinia trachciphila and Phytomons stewarti*. *I Bid.*, 30, 625–638.

Ivanoff, S., (1944). Guttation salt injury on leaves of cantaloupe, pepper and onion. *Phytopathology, 34,* 437.

Jones, J. P., & Everett, P. H., (1966). Control of anthracnose, downy mildew and soil rot of cucumbers. *Plant Disease Reporters, 50,* 340–344.

Khan, M. N., & Verma, A. C., (2004). Biological control of *Meloidogyne incognita* in pointed gourd (*Trichosanthes dioica* Roxb.). *Ann. Pl. Protec. Sci., 12*(1), 115–117.

Khanna, A. S., & Kumar, S., (2006). *In vitro* evaluation of neem-based nematicides against *Meloidogyne incognita. Nematol. Medit., 34*(1), 49–54.

Khare, C. P., (2004). *Encyclopedia of Indian Medicinal Plants* (p. 458). Berlin, Heidelberg, New York: Springer Verlag.

Khatua, D. C. Mondal, B., & Jana, M., (2013). Host range, medium for isolation and technique for bioassay of fungicides against *Phytophthora nicotianae*: The causal pathogen of leaf rot of betelvine. *Research on Crops, 14*(2), 592–595.

Khatua, D. C., Das, A., Ghanti, P., & Sen, C., (1981). Downy mildew of cucurbits in West Bengal and preliminary field assessment of fungicides against downy mildew of cucumber. *Pestology, 5*(2), 30–31.

Mondal, B. Saha, G., & Khatua, D. C., (2013). Fruit and vine rot of pointed gourd in West Bengal. *Research Journal of Agricultural Sciences, 4*(1), 44–47.

Mukhopadhyay, A. K., Roy, K., & Panda, T. R., (2006). Evaluation of nematicides and neem cake for managing root-knot nematodes in pointed gourd, *Trichosanthes dioica. Nematol. Medit., 34*(1), 27–31.

Saha, G. Das, S., & Khatua, D. C., (2004). Fruit and vine rot of pointed gourd. *Journal of Mycopathological Research, 42,* 73–81.

Urech, P. A., Schwinn, F. J., & Staub, A. J., (1977). A novel fungicide for the control of late blight, downy mildew and related soil borne disease. *Proceedings of British Crop Protection Conference, Brighton,* 623–632.

Verma, A. C., & Ali, A., (1990). Twig galls incited by *Meloidogyne incognita* on *Trichosanthes dioica* R. *Indian J. Nematol., 20,* 215.

CHAPTER 16

Diseases of Brinjals (Eggplant) and Their Management

NARENDER KUMAR,[1*] AJAY KUMAR,[1] GIREESH CHAND,[2] and S. K. BISWAS[1]

[1]*Department of Plant Pathology, C.S. Azad University of Agriculture and Technology, Kanpur, Uttar Pradesh, India*

[2]*Department of Plant Pathology, Bihar Agricultural University, Sabour, Bhagalpur, Bihar, India*

Corresponding author. E-mail: kumar.narendra6887@gmail.com

16.1 INTRODUCTION

Eggplant has many properties that make it an integral part of a healthy diet. Eggplant is rich in nutrients and supplies vital vitamins, minerals, and dietary fiber to the human diet, especially in the rainy season, when other vegetables are in short supply for the rural and urban poor. It also has low fat, zero cholesterol, and very low-calorie content. In Indian cuisines, eggplant is used in a variety of ways ranging from curries to chutneys. Baigunbhurta is the most popular eggplant dish in India. Aside from being used as food, eggplant is also used in medicines in China, Southeast Asia, and the Philippines. China is the leading producer of eggplant in the world, contributing around 55% of the world's output. Other major producers are India (28% of the world's production), Egypt, Turkey, and Japan (Raabe et al., 1981).

16.2 BRINJAL DISEASES

There are a number of plant diseases that can limit eggplant production, including Damping-off, Phomopsis blight, Verticillium wilt, little leaf, and early blight.

16.2.1 DAMPING OFF

Damping-off is a term often given to the sudden death of seedlings. It is an important disease of solanaceous crops. It is usually associated with the fungi *Pythium, Rhizoctonia,* or *Phytophthora.* The disease causes severe damage to the nursery. Damping-off generally occurs under cold, wet conditions.

High soil moisture and moderate temperature, along with high humidity, especially in the rainy season, leads to the development of the disease. In damping-off two types of symptoms are observed. They are described in the subsections.

16.2.1.1 PRE-EMERGENCE DAMPING-OFF

The pre-emergence damping off results in seed and seedling rot before these emerge out of the soil. Poor and weak seed germination is the result of pre-emergence damping off. In pre-emergence damping off, the seed fails to emerge after sowing. They become soft, mushy, turn brown, and decompose as a consequence of seed infection.

16.2.1.2 POST-EMERGENCE DAMPING-OFF

After seed germination young plants infected by pathogens. At the ground level, the infected tissues become soft and show water-soaked symptoms. The affected portions (roots, hypocotyls, and crown of the plant) are pale brown, soft, water-soaked, and thinner than non-affected tissues. The infected stems collapse. The collar portion rots, and ultimately the seedlings collapse and die.

16.2.1.3 CAUSAL ORGANISMS

The damping-off is caused by several fungal pathogens (*Pythium, Phytophthora, Phomopsis, Rhizoctonia* and *Sclerotium,* etc.), but most common are *Pythium* spp. such as *P. debaryanum, P. aphanidermatum, P. ultimum* and *P. arrhenomanes.* Fungi (*Phytophthora parasitica, Phomopsis vexans, Rhizoctonia solani,* and *Sclerotium rolfsii*) other than *Pythium* spp. and bacteria can also cause similar symptoms. Pythium produces white mycelium on which sporangia are developed. The hypha is 2.8 to 7.5 µm in diameter. The sporangia either germinate directly by germ tube or by producing a balloon-like secondary sporangium known as a vesicle. In this vesicle, over 100

Diseases of Brinjals (Eggplant) and Their Management

zoospores are formed, and they swim after their release and then rounded off to form a cyst. Encysted zoospores are 6.7–12.0 μm in size. The zoospores create new infections after germination. The sexual stage develops as oogonia and antheridia. Oogonia are spherical and 1–2 antheridia attached with each oogonium. After fertilization, oogonium develops a thick wall and known as oospore. The oospores can survive in adverse conditions, and this is the resting stage. The oospores germinate in the same way as sporangia.

16.2.1.4 DISEASE CYCLE AND EPIDEMIOLOGY

Causal organisms of damping-off are either seed or soil-borne. Off-season propagules of *Pythium* survive in plant debris (mycelium) or soil (oospore). Under favorable conditions resting structures (mycelium and oospores) germinate and produce zoosporangia. The zoosporangium produces several zoospores that are encysted and cause infection to the germinating plant seedling below the soil surface or at the collar region of the seedlings adjoining to the soil surface. Spore germ tube directly penetrates the seed or young seedlings when it comes in contact. Proteolytic enzymes break the protoplast, and in some cases, the cell wall is disintegrated by cellulolytic enzymes causing the death of seeds or young seedlings.

If crop planted in heavily infested soil, overwatering with poor drainage, overcrowding of the seedling with poor ventilation, excessive doses of nitrogen, stressful environmental conditions such as cloudy, wet weather that results in etiolated plants, prolonged soil moisture, low light that prevents drying, or pathogenic nematodes are present then seedlings will be susceptible during the initial three weeks after sowing. The movement of the infection is very fast and can quickly rot even the fleshy vegetables or fruits in the field, in storage, in transit, and even in the market when healthy, and infected fruits come in contact. Clay soil, which poorly aerated and drained soil, favored the most damping-off diseases incidence caused by *Pythium aphanidermatum* and minimum in sandy soil.

16.2.1.5 DISEASE MANAGEMENT

There is a range of fungicides registered to control of damping-off, but an integrated approach is required for damping-off management. The incidence is significantly reduced by planting under good conditions when practical, careful irrigation management and the use of appropriate fungicides when required.

306   *Diseases of Fruits and Vegetable Crops*

- Avoid nursery sowing in the same bed year after year with frequent heavy irrigation.
- Soil should be solarizing for 30 days.
- Healthy seed should be selected for sowing.
- Application of native bio-control agent *Trichodermaviride* in soil @ 1.2 kg/ha is also found effective to control damping-off to a considerable extent.
- The topsoil of the nursery should be treated with Thiram @ 5g/m^2 area of the soil, and nursery should be drenched with the same chemical @ 2 g/liter of water at the fortnightly interval.
- The seed should be treated with Thiram @ 2 g/kg of seed before sowing.

Prevention is the key to managing damping off. The following cultural practices will help reduce the incidence of this disease:

- Use sterilized soil when planting seeds in pots or flats.
- Disinfest tools, potting containers, and workbenches used in seeding/ transplanting operations.
- Use seed treated with a fungicide to protect against seed decay.
- Purchase seed that has already been treated.
- Do not overwater.
- Plant seed in the garden or field after the soil has warmed.
- Plant in well-drained soils.

16.2.2   *PHOMOPSIS BLIGHT*

It is a serious disease of brinjal infecting the foliage and the fruits caused by fungal pathogens. This disease was reported for the first time from Belgaon district of Bombay, India (Uppal et al., 1935; Pawar and Patel, 1957). Damage the fruits partly or completely in the field as well as during transit. The disease is favored by warm, wet weather and is spread by splashing water. Yield loss is 10–20% due to fruit rot.

16.2.2.1   *SYMPTOMS*

Infection may occur when rain or overhead irrigation splash inoculum to foliage and stems. The fungus infects aboveground plant parts at all stages of

Diseases of Brinjals (Eggplant) and Their Management

development, and in the nursery, it fungus causing damping-off symptoms. Generally, spots first appear on seedlings shortly after they emerge. When the leaves are infected, small circular spots appear, which become grey to brown with irregular blackish margins and ultimately die. Lesions may also develop on petiole and stem, causing blighting of the affected portion of the plant. Symptoms on the infected fruits pale dull, sunken spots, which later merge and enlarge to form rotten areas and produce concentric rings with yellowish and brown zones on the entire fruit, which later merge to form rotten areas. Affected fruit becomes soft and watery at first; decay may penetrate rapidly throughout the fruit. Phomopsis blight is generally favored by hot, wet weather.

16.2.2.2   CAUSAL ORGANISM

The disease is caused by *Phomopsis vexans* (Sacc, and Syd.) Harter. The fungus can survive from season to season in plant debris in the soil as well as in or on the seed. The fungus can survive for more than a year in fields where a diseased crop was grown. The mycelium of fungus made up of hyaline and septate hyphae (2.7 to 3.9 µm in diameter). The beak may or may not be a presence in the pycnidium. The conidiophores (10 to 16 µm in length) are hyaline, simple or branched, sometimes septate. Pycnidium produces two types of conidia *viz.*, alpha, and beta.

16.2.2.3   DISEASE CYCLE AND EPIDEMIOLOGY

Phomopsis vexans perennates both as mycelium and spores during the offseason in plant debris in the soil. The pathogen also seed-borne and pycnidia of the fungus have also been observed in the seed. Spores of the fungus are released from the pycnidia. The major means of spread is by rain splashing. Wind dispersal is usually considered to be of minor importance. Disease is favored by hot and wet weather. The optimum temperature for fungal growth is 29°C and it grows well up to 32°C.

16.2.2.4   DISEASE MANAGEMENT

- Adopting good field sanitation, destruction of infected plant material and crop rotation (3 year) to break the disease cycle. Seeds obtained from disease plants should be used for planting.

308 Diseases of Fruits and Vegetable Crops

- Sow high-quality seed to help produce pathogen-free transplants.
- Remove and destroy all infected plant material.
- Mulch and furrow irrigate to help reduce splashing of water and soil.
- Seed treatment with Thiram (2 g/kg seed) protects the seedling in the nursery stage. Spraying with Dithane Z-78 (0.2%) or Bordeaux mixture (1%) effectively controls the disease in the field.
- Grow resistant varieties.
- Spray various systemic and non-systemic fungicides on crops have been found effectively controls the disease in the field conditions.

16.2.3   LEAF SPOT

16.2.3.1   SYMPTOM

The disease symptoms are characterized by chlorotic lesion, angular to irregular in shape, later turning grayish-brown. Severely infected leaves drop off prematurely, resulting in reduced fruit yield.

Symptoms appear on the older, lowest leaves first, and if unchecked can move upwards and infect young leaves and stems. Severely infected leaves curl and prematurely drop off the plant, often causing a reduction in yield (Chupp, 2006). The pathogen does not infect eggplant fruits but does reduce the growth of plants and thus reduce yield. Where the disease is severe, it can be fatal to plants.

Symptoms appear on the leaves, petioles, and stems of eggplant. Initially, small, circular to oval chlorotic spots appear that may develop angular or irregular shapes. The spots on leaves can easily be confused with spots caused by a bacterial disease. On closer inspection, however, the spots due to *Cercospora* show distinguishing feature. They are generally circular with light to dark tan centers, and the stomata have black spots that can be readily observed using a hand-held magnifying lens. Elliptical to oval lesions may occur on the leaf blades, veins, and petioles. Bacterial leaf spots, conversely, are irregularly shaped or circular spots with clear stomata (Windels et al., 2003).

16.2.3.2   CASUAL ORGANISM

The leaf spots on eggplant caused by *C. melongenae* are 4–10 mm in diameter and visible on both leaf surfaces, though they may appear more abundantly on the lower surface. They are brown to steel-gray on the upper surface and

Diseases of Brinjals (Eggplant) and Their Management 309

light brown on the lower surface (Anonymous, 2011). During the later stages of disease, lesions become grayish brown, with excessive sporulation at their centers. Concentric rings of diseased tissue may appear as the lesions gradually expand in size. As a lesion dries, the tissue in the center may crack and drop out.

A plant-pathogenic fungus, *Cercospora melongenae*, causes Cercospora leaf spot of eggplant. The genus *Cercospora* is a hyphomycete fungus comprised of many plant-pathogenic species. They produce leaf spot diseases on a wide range of agriculturally important plants. These diseases are major problems for large-scale growers and backyard gardeners (Chupp, 1953; Farr et al., 1989; Kranz and Werner, 1978; Gonsalves et al., 1994). There are about 1,200 known species in the genus *Cercospora*, with over 69 species reported in Hawaii (Raabe et al., 1981). Conidiophores (fungal structures that bear the spores/conidia) appear in fascicles (clusters) of 3 to 12. They are pale to medium brown, paler towards the apex, occasionally septate, and unbranched, and they bear hyaline, mildly curved conidia.

16.2.3.3   DISEASE CYCLE AND EPIDEMIOLOGY

Cercospora melongenae can survive for at least one year in plant debris or in soil. The disease process begins when the fungal spores are dispersed to susceptible plants by rain, irrigation water, or wind, or on agriculture equipment or by people (Nelson, 2008). Leaf wetness and high relative humidity (RH) favor infection and disease development.

16.2.3.4   DISEASE MANAGEMENT

- Removal and destruction of affected plant parts.
- Use disease-free transplants.
- Control weeds to reduce RH in the eggplant canopy.
- Irrigate in the morning to reduce humid and damp conditions overnight.
- Avoid over-irrigation to reduce RH.
- Avoid overhead sprinkler irrigation in order to minimize leaf wetness and spread of the pathogen in splashing water droplets.
- Do not work with wet plants or move through a field of wet plants, as such movement can disperse fungal spores among plants.
- Increase the spacing between plants to improve aeration and drying of wet foliage.

310 *Diseases of Fruits and Vegetable Crops*

- Keep plants adequately fertilized.
- Intercrop eggplant with other vegetables to interrupt pathogen transmission between plants.
- Grow eggplant undercover (i.e., in greenhouses or under shade cloth) where possible to minimize leaf wetness.
- A calendar-based spray program using a protectant fungicide, combined with cultural practices, can reduce losses from the Cercospora leaf spot on eggplant.
- Spraying the affected plants with Bavistin (0.1%) or Chlorothalonil (2 g/liter of water) is useful for disease control.

16.2.4 ALTERNARIA LEAF SPOTS

Alternaria leaf spots also known as early blight. Early blight is more commonly known as a leaf spotting or foliar blight disease. Early blight can cause serious defoliation of tomato crops. It is often associated with septoria leaf spot, and these two fungal diseases, either separately or together, are responsible for most of the defoliation caused by diseases in field tomato crops in Canada. The pathogen also infects potato and solanaceous weeds. Pepper and eggplant are rarely affected. This disease is responsible for damaging the leaves and fruit.

16.2.4.1 SYMPTOMS

Early blight can affect all the above-ground plant parts throughout the growing season. The disease causes characteristic leaf spots with concentric rings (target spots). Spots first appear on the older leaves. The spots are mostly irregular and coalesce to cover large areas of the leaf blade. Severely affected leaves drop off. The symptoms on the affected fruits are in the form of large deep-seated spots. The infected fruits turn yellow and drop off prematurely. Leaf spots, which typically form on older leaves first, begin as small, dark, irregular spots. On eggplant, these lesions are lighter in color, and the concentric ring pattern may not be as noticeable as that seen on tomato or potato. When spots are numerous, leaves die prematurely and drop, which exposes the fruit and makes it more likely to be damaged by sun-scald. Lesions may also develop on the fruit. These spots are dark, leathery, and sunken and usually have distinct concentric rings. Spots on the fruit slowly expand until they decay much of the surface area and the internal flesh of the fruit.

Diseases of Brinjals (Eggplant) and Their Management   311

16.2.4.2 CASUAL ORGANISM

Early blight also known as Alternaria leaf spot is most often caused by the fungus *Alternaria solani*. In addition to *Alternaria solani* is often isolated from typical early blight lesions, especially on fruit. Conidia of the two species differ in morphology and length.

16.2.4.3 DISEASE CYCLE AND EPIDEMIOLOGY

The fungus survives from season to season on crop debris in the soil and grows well in warm, wet conditions. Spores are spread by wind and splashing water. Spores can be carried for several kilometers by the wind. The early blight fungus also can be seed-borne. Its relatively short disease cycle allows for numerous, repeated infections, resulting in rapid defoliation under favorable conditions. Plant susceptibility increases with age, heavy fruit load, and inadequate nutrition. The fungus is able to grow through exposed epidermal layers, resulting in small black lesions. In time, even during cold storage, these lesions may coalesce and cover large areas of the fruit, especially on the shoulders.

16.2.4.4 DISEASE MANAGEMENT

- Control can be achieved by extending crop rotations to three or four years break between Solanaceous crops.
- Using disease-free transplants, minimizing plant injury, and maintaining plant vigor. When irrigation is required, morning watering will allow the leaves to dry before a new dew period begins in the evening.
- Removal and destruction of affected plant parts.
- Spraying the affected plants with Bavistin (0.1%) is useful for disease control.
- Properly timed foliar fungicides are effective in reducing losses caused by early blight.
- Destroy a diseased crop immediately after harvest to reduce the chance of the fungus overwintering.

16.2.5 VERTICILLIUM WILT

The disease attacks the young plants as well as mature plants. The infected young plants show dwarfing and stunting due to the shortening of the internodes.

Such plants do not flower and fruit. Infection after the flowering stage results in development of distorted floral buds and fruits. The affected fruits finally drop off. The infected leaves show the presence of irregularly scattered necrotic pale yellow spots over the leaf lamina. Later on, these spots coalesce resulting in complete wilting of the leaves. The roots of the affected plants are split open longitudinally, a characteristic dark brown discoloration if the xylem vessels are observed.

Although seedlings can be infected, symptoms usually are not observed until plants are older. In eggplant, symptoms of Verticillium Wilt infection progress slowly. A characteristic symptom of infection is a V-shaped lesion that develops on older leaf tips that later expand to cover the leaf. As the disease progresses in eggplant, stunting, and chlorosis become severe with diurnal wilting. Disease causes a brown discoloration in the stem and root. Fruit that form are small and deformed with internal discoloration.

16.2.5.1 SYMPTOMS

Symptoms can appear any time but commonly appear around anthesis and include unilateral wilting, chlorosis, and defoliation. The first signs of this disease are wilting of the older leaves, then yellowing, and finally death. The leaves will often have a V-shape of yellow. The whole plant may wilt and die. As with Fusarium wilt, there are no external markings on the stem, but if the stems of affected plants are split lengthwise, a brown discoloration of the vascular tissue is seen. This disease is spread by contaminated soil, with the fungus entering the plant through the roots to the vascular system. Cold weather favors the disease, which survives in the soil for long periods as sclerotia. Leaves wilt starting at the bottom of plants and turn yellow before they wither and fall off. Disease causes a brown discoloration in the stem and root.

16.2.5.2 CAUSAL ORGANISM

Verticilium wilt in brinjal caused by *Verticilium dahliae*. This pathogen also affected other crop plants, i.e., tomato, pepper, and many other hosts. Hyphae of this fungus are hyaline and white to grayish after a week. *V. dahliae* produces microsclerotia, which are dark brown to black in color and each microsclerotium (18–100 µm) arises from single hypha by repeated budding. Root-knot nematodes and lesion nematodes in combination

Diseases of Brinjals (Eggplant) and Their Management

with Verticillium wilt on eggplant, pepper, potato, and tomato have been reported to have a synergistic effect. *Verticillium dahliae* produces one-celled, colorless conidia that are short-lived. *V. dahliae* also produces minute, black, resting structures called microsclerotia. After the host dies or the growing season ends, *Verticillium* survives as mycelia overwintering in dead plant parts that have fallen to the ground; the fungi may also live saprophytically in the soil whether or not a host is available. Survival can occur in roots of non-host species in which systemic infection does not occur. *Verticillium* is known to naturally colonize soils where susceptible hosts have never been grown.

16.2.5.3 DISEASE CYCLE AND EPIDEMIOLOGY

These soil-borne pathogens cause a wilt by blocking the vascular system of the plant. *Verticillium* have an extremely wide host range and can survive in soil and plant debris for several years. *V. dahliae* prefers slightly higher temperatures (25°C–28°C) for disease development and is somewhat more common in warmer regions. *Verticillium* enters plants through root wounds caused by cultivation, secondary root formation and nematode feeding. Symptomatic plants may be few and restricted to one area, or occur throughout an entire field or greenhouse. Disease development is favored in heavy clay soils.

Agricultural soils may contain up to 100 or more microsclerotia per gram. Six to 50 microsclerotia per gram are sufficient to generate 100% infection in such susceptible eggplant or other vegetable crops like pepper, potato, and tomato. *Verticillium* invade the root systems of host plants through wounds or by direct penetration. Once within a root, the fungi invade the xylem of the host and then spread upward through the plant. The disease is spread by fungal spores (conidia) being transported upward in the sap stream where they may become lodged, germinate, and affect new plant parts. Wilt symptoms typically are not observed until the fungus has colonized the roots, stems, or trunks of trees and shrubs. Wounds on the trunk, branches, and twigs of trees may also serve as sites for infection by insect transmission of the fungus.

The microsclerotia are capable of long-term survival (up to 15 years) without contact with a host plant. They may be introduced into uncolonized areas by wind or water where they serve as new inoculum. In addition, mycelia, and microsclerotia may be transported by normal tillage operations, hand tools, or farm machinery. Once contact is made with a new host, the fungi again infect the root system, progress upward, and the cycle is repeated.

16.2.5.4 DISEASE MANAGEMENT

- Crop rotation with bhendi, tomato, potato should be avoided.
- Soil fumigation and solarization reduce disease incidence.
- Soil application and foliar application with Benlate (0.1%) is effective in reducing the wilt disease.

16.2.6 BACTERIAL WILT

Bacterial wilt disease causes a severe problem in brinjal cultivation. Bacterial wilt affects chiefly tomato crops in the southern United States, but some Canadian growers have experienced losses after using infected imported transplants. This disease can also affect pepper and eggplant. The pathogen is capable of attacking more than 200 species of plants in 33 families.

16.2.6.1 SYMPTOMS

The characteristic symptoms of the disease are wilting of the foliage followed by the collapse of the entire plant. The wilting is characterized by gradual, sometimes sudden, yellowing, withering, and drying of the entire plant or some of its branches. Infected plants exhibit dark vascular browning that extends into the cortical or pith regions and sometimes deep into the below-ground part of the stem. Most infected transplants die within two weeks, beyond which no further loss occurs. When infected stems or roots are cut crosswise and squeezed firmly, a gray to yellowish ooze appears. This disease can be distinguished from other wilts by placing an infected stem section in a glass of water. If bacterial wilt is present, a milky stream flows from the cut surface within five minutes.

16.2.6.2 CAUSAL ORGANISM

Bacterial Wilt is caused by *Pseudomonas solanacearum*, which is an aerobic, Gram-negative, rod-shaped bacterium with one to four polar flagella. All

Diseases of Brinjals (Eggplant) and Their Management

strains except those from banana and other musaceous hosts produce a dark brown diffusible pigment on a variety of agar media containing tyrosine. It is a non-fluorescent pseudomonad. The pathogen is negative for levan production, starch hydrolysis, indole production, hydrogen sulfide, and aesculin hydrolysis.

16.2.6.3 DISEASE CYCLE AND EPIDEMIOLOGY

Pseudomonas solanacearum occurs throughout the world in areas with warm climates. Infection and disease development are favored by high temperatures (optimum 30 to 35°C) and high moisture. It survives in infested soil and crop residues, is seed-borne and can be found in numerous weed hosts. The bacteria do not survive in the field in Canada, but may persist in greenhouses. The disease quickly spreads within the cortex and pith of an infected plant, eventually causing death. In Canada, bacterial wilt has not been observed to spread in the field beyond initially infected plants.

16.2.6.4 DISEASE MANAGEMENT

- Removal and destruction of the affected plant parts.
- Using disease-resistant varieties helps to reduce disease incidence.
- Crop rotation with bhendi, tomato, potato should be avoided.
- Before sowing, the seeds should be dipped in a solution of Streptocycline (1 g/40 liters of water) for 30 minutes.

16.2.7 LITTLE LEAF OF BRINJAL

This disease is common in all parts of India cause drastic reduction in leaf size and excessive growth of axillary shoots, giving the plants a bushy appearance. The plants remain sterile and the floral parts turn green. The disease agent transmitted by *Hishmonus physcitis* and *Empoasca devastans* (Thomas and Krishnaswamy, 1939).

This is a phytoplasmal disease of brinjal. In India, this disease was first reported in 1938. The disease is transmitted by leaf hopper. The yield and growth of the plants are highly affected when plants are infected during early growth stage, i.e., before 35 and 55 days after sowing.

316   *Diseases of Fruits and Vegetable Crops*

16.2.7.1   SYMPTOMS

In early stage, the leaves of the infected plants are light yellow in color. The leaves and shoots show a reduction in size and are malformed. Disease affected plant are generally shorter in stature bearing a large number of branches, roots, and leaves than healthy plants. The petioles get shorter considerably, many buds appear in the axil of leaves and internodes get shortened thus giving the plants a bushy appearance. Flowering and fruiting is rare in affected plants. The growth of the fruits after infection is restricted or stopped and the fruits become hard and fail to mature. Flower parts are deformed leading the plants to be sterile. Infected plants do not bear any fruit. However, if any fruit is formed it becomes hard and tough and fails to mature.

16.2.7.2   CAUSAL ORGANISM

The disease is caused by a *Candidatus* phytoplasma. Phytoplasma bodies (230–770 nm) are found associated with the phloem cells of roots of the eggplant affected with little leaf. Each phytoplasma body contained ribosomes and strands of nuclear material surrounded by 16.5 µm wide triple layer unit membrane. Presence of Phytoplasma in the phloem cells of roots further strengthen the Phytoplasmal etiology of the disease and indicates uniform distribution of phytoplasmas in all parts of the diseased plants.

16.2.7.3   DISEASE CYCLE AND EPIDEMIOLOGY

The host range of little leaf of brinjal pathogen is very wide. About 30 genera of 5 families with 60 species infected by this pathogen. Phytoplasma transmitted by vector *Hishimonus phycitis* from infected plants to healthy plant. Continuous cropping of the susceptible varieties increased disease incidence, as did the presence of weeds, sweet potato or tomato crops in the vicinity, which helps to increase in vector populations. However, there is no appreciable effect of NPK application on the incidence of the disease in the field.

16.2.7.4   DISEASE MANAGEMENT

- The sowing time can be adjusted to avoid the main flights of the beet leafhopper. Early sown crops (June sown) escape the disease compared to July and August sown crops due to differences in vector populations.

Diseases of Brinjals (Eggplant) and Their Management   317

- Adopting sanitary measures including the eradication of susceptible volunteer crop plants from a previous planting can reduce the damage. Rogue of affected plants, destruction of weeds hosts from in and around the fields, avoid planting of crops like sweet potato and tomato nearby the eggplant fields.
- Use resistant cultivars/lines for cultivation.
- Dip Pre-transplant seedling in carbofuran (0.2%) for 24 h.
- Apply tetracycline to seedlings by girdling delayed the symptoms development by 4 months.
- Dipping the seedlings in tetracycline solution before transplanting followed by soil application of phorate.
- Spraying Malathion (2ml/liter of water) starting with the appearance of the leaf hoppers controls their population.

16.2.8   ROOT KNOT DISEASE

The crop happens to be highly susceptible to infection by root knot nematodes of the genus *Meloidogyne*, such as *M. incognita* and *M. javanica* (Khan et al., 1984). The root knot nematodes, causing root galls affecting almost all agricultural crops including Brinjal in all parts of the world. Root-knot nematode (*Meloidogyne* spp.) is an endoparasite. Root knot disease is found in most of the countries on various vegetable crops such as potato, tomato, brinjal, chili, okra, bhendi, cucurbit, etc. The losses caused by this disease are from 20–75% in tomato and 17–81% in brinjal.

16.2.8.1   SYMPTOMS

The affected tomato and brinjal plants show unthrifty development and stunted growth if the infection occurs during early stages of plant growth. Leaves become yellowish green to yellow and sometimes show scorching along the margins. The roots show galls of various sizes and shapes. Presence of root galls is the most characteristic symptom of the disease.

16.2.8.2   CASUAL ORGANISM

The disease is caused by the nematode pathogens of the genus *Meloidogyne*. The common species are *M. incognita, M. javanica,* and *M. arenaria,* which commonly attack brinjal and other vegetables. The male nematodes are vermiform and 1.2 to 1.5 mm × 30–36μm in size. The females are pear shaped and measure 0.15–0.25 mm wide at the base. Up to 600 eggs are laid by each female.

16.2.8.3 DISEASE CYCLE AND EPIDEMIOLOGY

A large number of females are found in the root galls which survive in the soil on plant debris. The second stage larvae are liberated in the soil. Sandy light soil is best for their movement. These larvae can be killed at higher temperature of 40–50°C and excess of soil moisture. The larvae are attracted by root exudates and thus come in contact of the roots. They gathered in a mass around roots and cause infection by penetrating roots of healthy plants. The male and female nematodes develop depending on the number of larvae per unit area of the host tissues. If they are more crowded mostly male develops but if only few larvae are there than females will develop. The reproduction is mostly parthenogenetic. A temperature of 25–28°C is best for infection and gall formation by nematodes.

16.2.8.4 DISEASE MANAGEMENT

- Cultural practices such as crop rotation, fallow soil, sanitation, deep plowing during summer, soil solarization, and flooding of soil are important to avoid infection by the nematodes.
- Resistant varieties such as Vijay, black beauty, T-3, S-149, etc. should be used.
- Among numerous organisms that have shown antagonism against root-knot nematodes, *P. chlamydosporia* (Kerry, 2000; Siddiqui, 2009), *P. lilacinus* (Jatala, 1985; Khan et al., 2005) and *T. harzianum* (Bokhari, 2009; Siddiqui and Shaukat, 2004) have been found to be highly suppressive to plant nematodes, especially under greenhouse conditions (Khan, 2007).
- *Paecilomyces lilacinus* increased growth of Brinjal and reduced root galling, egg masses and number of eggs per egg mass of *Meloidogyne incognita* (Sharma and Trivedi, 1989).
- Neem derivatives such as azadirachtin, nimbin, cake extract, leaf extract have been effectively employed for seed treatment of Brinjal against *Meloidogyne incognita*.

Diseases of Brinjals (Eggplant) and Their Management 319

- 10% neem cake extract (at 20 ml/pot) was used with spores of *Paecilomyces lilacinus* (at 1×106 spores/ml). This gave maximum reduction in root galling (87.53%) and egg masses (71.20%) thereby increased yield (64.36%) (Zareena and Das, 2014).
- Root dip treatment of Brinjal seedlings in neem cake extract with *Paecilomyces lilacinus* (at 5×106 spores/ml) for 20 min resulted in 84.23% reduction in root galls and 70.90% reduction in the egg (Zareena and Das, 2014).
- Application of caster cake extract with *Trichodermaharzianum* (at 500ml/m^2 containing 9.9×103 spores/ml) was proved the least effective treatment where reduction in root galls was 70.30% and reduction in egg masses was 59.26% (Zareena and Das, 2014).

16.2.9   FRUIT ROT

High humidity favors the development of the disease. The symptoms first appear as small water-soaked lesions on the fruit, which later enlarges in size considerably. Skin of infected fruit turns brown and develops white cottony growth. This disease is caused by *Phytophthora nicotianae.*

16.2.9.1   DISEASE MANAGEMENT

Removal and destruction of the affected fruits and spraying the crop with Difolatan (0.3%) thrice at an interval of 10 days effectively controls the disease.

16.2.10   MOSAIC

This is a viral disease caused by potato virus Y (PVY) and transmitted by aphids (*Aphis gossypi* and *Myzus persicae*). The important symptoms of the disease are mosaic mottling of the leaves and stunting of plants. The leaves of infected plants are deformed, small, and leathery. Plants show a stunted growth when infected in the early stages.

16.2.10.1   DISEASE MANAGEMENT

- The disease incidence can be minimized by reducing the population of aphids, removal.

320 *Diseases of Fruits and Vegetable Crops*

- In the nursery, aphids can be controlled by application of Carbofuran (1 kg *a.i.*/ha) in the nursery bed at the time of sowing seeds followed by 2–3 foliar sprays of Phosphamidon (0.05%) at an interval of 10 days.
- Destruction of infected plants and eradication of susceptible weed hosts.
- Spraying Phosphamidon (0.05%) at 10 days interval starting from 15–20 days after transplanting effectively controls the aphids in the field.

KEYWORDS

- *Alternaria* leaf spots
- bacterial wilt
- brinjal
- leaf spot
- root knot disease
- *Verticillium* wilt

REFERENCES

Akhtar, M., (1997). Biological control of plant parasitic nematodes by neem products in agricultural soils. *Applied Soil Ecology, 7*, 219–223.

Akthar, M., & Mahmood, M. M., (1994). Control of root knot nematode by bare root-dip in undecomposed and decomposed extracts of neem cake and leaf. *Nematologia Mediterranea, 22*, 55–57.

Anonymous, (2011). *Small Fruit and Vegetable IPM Advisory–Integrated Pest Management.* Utah State University Cooperative Extension Web portal. http://utahpests.usu.edu/ipm/htm/advisories/small-fruitand-vegetable-advisory/articleID=13472–leafspot (Accessed on 18 November 2019).

Bhatti, D. S., & Jain, R. K., (1977). Estimation of loss in okra, tomato and Brinjal yield due to *Meloidogyne incognita. Indian Journal of Nematology, 7*, 37–41.

Bokhari, F. A. M., (2009). Efficacy of some *Trichoderma* species in the control of *Rotylenchulus reniformis* and *Meloidogyne javanica. Archives of Phytopathology and Plant Protection, 42*, 361–369.

Chupp, C., (1953). *A Monograph of the Fungus Genus Cercospora.* Cornell Univ., Ithaca, New York.

Chupp, C., (2006). *Manual of Vegetable Plant Diseases.* Discovery Publishing House, India.

Diseases of Brinjals (Eggplant) and Their Management

Dhawan, S. C., & Sethi, C. L., (1977). Inter-relationship between root-knot nematode, *Meloidogyne incognita* and little leaf of Brinjal. *Indian Phytopathology, 30*, 55–63.

Farr, D. F., Bills, G. F., Chamuris, G. P., & Rossman, A. Y., (1989). *Fungi on Plants and Plant Products in the United States*. APS Press, St. Paul, Minnesota.

Gonsalves, A. K., & Ferreira, S., (1994). *Cercospora* primer. http://www.extento.hawaii.edu/kbase/crop/Type/cer_prim.htm-DISEASES (Accessed on 18 November 2019).

Hui, Y. H., (2006). *Handbook of Food Science* (pp. 13–20). Technology and Engineering. CRC Press.

Jatala, P., (1985). Biological control of nematodes. In: Sasser, J. N., & Carter, C. C., (eds.), *An Advanced Treatise on Meloidogyne: Biology and Control* (pp. 303–308). North Carolina State University Graphics.

Jatala, P., (1986). Biological control of nematodes. *Annual Review of Phytopathology, 24*, 453–489.

Kerry, B. R., (2000). Rhizosphere interactions and the exploitation of microbial agents for the biological control of plant parasitic nematodes. *Annual Review of Phytopathology, 38*, 423–432.

Khan, M. R., (2007). Prospects of microbial control of root-knot nematodes infecting vegetable crops. In: Singh, N. S., (ed.), *Biotechnology: Plant Health Management* (pp. 643–665). International Book Distributing Co.

Khan, M. R., Khan, S. M., & Mohiddin, F. A., (2005). Root-knot nematode problem of some winter ornamental plants and its biomanagement. *Journal of Nematology, 37*, 198–206.

Khan, M. W., Khan, M. R., & Khan, A. A., (1984). Identity of root-knot nematodes on certain vegetables of Aligarh district in northern India. *International Nematological Network News, 1*, 6–7.

Kranz, J. S., & Werner, K., (1978). *Diseases, Pests and Weeds in Tropical Crops* (pp. 194–195). Wiley Publications.

Nagesh, M., Parvatha, R. P., & Rao, M. S., (1997). Integrated management of *Meloidogyne incognita* on tuberose using *Paecilomyceslilacinus* in combination with plant extracts. *Nematologia Mediterranea, 25*, 3–7.

Nelson, S., & Bushe, B., (2012). *Hawaii Host-Pathogen Database*. http://hawaiiplantdisease.net/qd/index.php?path=HawaiiHostPathogenDatabase (Accessed on 18 November 2019).

Pawar, V. H., & Patel, M. K., (1957). Phomopsis blight and fruit rot of Brinjal. *Indian Phytopath., 10*(2), 115–120.

Raabe, R. D., Conners, I. L., & Martinez, A. P., (1981). *Checklist of Plant Diseases in Hawaii*. Hawaii Institute of Agriculture and Human Resources, College of Tropical Agriculture and Human Resources, University of Hawaii.

Sen, K., & Dasgupta, M. K., (1977). Additional hosts of the root-knot nematode, *Meloidogyne* spp. from India. *Indian Journal of Nematology, 7*, 74.

Siddiqui, I. A., & Shaukat, S. S., (2004). *Trichoderma harzianum* enhances the production of nematicidal compounds in vitro and improves biocontrol of *Meloidogynejavanica* by *Pseudomonas fluorescens* in tomato. *Letters in Applied Microbiology, 38*, 169–175.

Siddiqui, I. A., Atkins, S. D., & Kerry, B. R., (2009). Relationship between saprotrophic growth in soil of different biotypes of *Pochoniachlamydosporia* and the infection of nematode eggs. *Annals of Applied Biology, 155*, 131–14.

Staples, G. W., & Herbst, D. R., (2005). *A Tropical Garden Flora, Plants Cultivated in the Hawaiian Islands and Other Tropical Places*. Bishop Museum Press, Honolulu, Hawaii.

Thomas, K. M., & Krishnaswamy, C. S., (1939). *Proc. Indian Acad. Sci., B10*, 201–212.

Uppal, B. N., Patel, M. K., & Kamat, M. N., (1935). *The Fungi of Bombay Bull., 178,* 56.

Varma, A., Chenulu, V. V., Raychaudhuri, S. P., Prakash, N., & Rao, P. S., (1969). Mycoplasma like bodies in tissue infected with sandal spike and brinjal little leaf. *Indian Phytopath., 22,* 289–291.

Windels, C. E., Bradley, C. A., & Khan, M. F. R., (2003). *Cercospora and Bacterial Leaf Spots on Sugarbeet.* http://www.ag.ndsu.edu/pubs/plantsci/rowcrops/pp1244.pdf (Accessed on 18 November 2019).

Zareena, S. K., & Vanita Das, V. V., (2014). Root knot disease and its management in brinjal. *Global Journal of Bio-Science and Biotechnology, 3*(1), 126–127.

CHAPTER 17

Diseases of Carrots, Radishes, and Knol Khol (Kholrabi) and Their Management

SHAILBALA

Junior Research Officer, Plant Pathology, Sugarcane Research Center, G.B. Pant University Agriculture and Technology, Uttarakhand, India, E-mail: shailbalasharma10@gmail.com

17.1 INTRODUCTION

Carrot is a root vegetable rich in carotene (pro-vitamin A) and consumed either fresh, as a salad crop, or cooked. It is grown all over India. Its large quantities are also processed either alone or in mixtures with other vegetables by canning, freezing or dehydration. Carrot root exists in a different color, i.e., orange, white with the orange or orange-red colors being by far the most popular today. Judging by seed usage, it is among the top ten vegetable crops on an area basis. Green carrot leaves are highly nutritive rich in protein, minerals, and vitamins and used as fodder and also for preparation of poultry feed. Radish is one of the choicest root crop grown all over the world. It can be eaten raw and also cooked. It has some medicinal value. Its leaves are rich source of vitamin A and can also be used as green leafy vegetable. It is a quick growing vegetable thus it occupies field for a very short period. It is equally suitable for growing in large scale as well as in kitchen or nutrition gardens. Being a quick growing crop, it can be easily planted as companion crop or intercrop between the rows of the other vegetables. It can also be planted on ridges, separating one plot from another.

Knol Khol is one of the lesser known vegetables. It is quite similar in looks to that of a cabbage. It is greenish in color with white patches. The main difference with the cabbage is that there are no leaves covering this vegetable. There may be some light green colored extrusion which is mostly shoots. It is one of those few vegetables which originated in Europe. It has a very cool and succulent stem which can be consumed. It is rich source of

324   *Diseases of Fruits and Vegetable Crops*

vital vitamins for human and has low fat content, almost 0% cholesterol, protein, dietary fibers and minerals.

17.1.1   CARROT (Daucus carota L.)

The carrot (*Daucus carota* L.) belongs to the family Apiaceae/Umbelliferae. The carrot originated in Asia. Carrot is a cool season vegetable that is cultivated in temperate and subtropical regions of the world. Carrot is a commercially important root vegetable cultivated for human consumption for over ten centuries and valued for its dietary and health benefits (Simon, 1990; Rubatzsky et al., 1999). The plant is a biennial, i.e., it grows vegetatively in the first season and produces seed in the second season. Wherever carrots are grown, a variety of diseases reduces both the yield and the market value of the roots. Yet many bacteria, fungi, etc., cause lesions that reduce their value. Phytoplasma also cause damage to carrots both in the form of malformed roots and direct yield losses of plants. Because of the nature of the carrot root, damage caused by various nematodes is an important limiting factor in carrot production. Disease management strategies, including cultural practices in addition to chemical disease control, are used to limit economic losses to diseases.

17.1.1.1   BACTERIAL DISEASES

The main problems associated with carrot production are diseases caused by bacteria. Bacteria are microscopic, single-celled, rod-shaped organisms that multiply quickly in the plant, which causes enormous yield loss in carrot. Often they occur as a secretion on the surface of diseased plants. These bacteria are spread to other plants by splashing water, rain, insects, man, etc.

17.1.1.1.1   Bacterial Leaf Blight (Xanthomonas hortorum pv. carotae, formerly Xanthomonas campestris pv. carotae)

Bacterial leaf blight is an explosive disease that develops rapidly under hot, wet, and windy conditions. The presence of a trace level of infection in the field requires action. First symptoms are yellow spots on the tips of leaf segments. Bacterial blight lesions are initially small roughly circular lesions that are light brown to tan and shiny on the lower leaf surface, as coated with

Diseases of Carrots, Radishes, and Knol Khol (Kholrabi)

varnish. As the infection progresses, the lesions become elongated in shape on the leaf blade, turn darker brown to black in color and develop water-soaked edges that are yellow and chlorotic. A yellowish halo often subtends the black center of the lesion. On the leaf margin, the lesions are crescent-shaped while on the carrot leaflets, the lesions can appear more 'V' shaped. Entire leaf segments or leaflets may be killed and lower leaves die and dry up as the disease advances.

In severe infections, long, dark-brown, water-soaked lesions develop on the petioles and main stem. A gummy bacterial substance frequently collects on them. It can also cause 100% crop losses in the area with warm and rainy weather. The affected roots of carrots may show small, water-soaked, greasy flecks or scab-like lesions at any point on the surface. They first appear as brown or maroon spots, which may become raised pustules or sunken craters. A grayish ooze may cover the surface of the lesions. The bacteria persist in the soil and are commonly carried with seed. The bacterium is spread by splashing rain, irrigation water, insects, animals, and machinery.

Pathogen may carry in the seed so seed treatment with hot water at 50°C for 30 minutes is effective. Always use certified and healthy seeds for planting. To reduce the soil-borne inoculum, crop rotation for 2 to 3 years is very important. After harvesting of the crop, plow under infected crop debris to hasten the decomposition. Water management is important, so always avoid overhead irrigation. Application of copper bactericides can slow disease development, especially if the application begins when plants are young (Reddy, 2010).

17.1.1.1.2 Bacterial Soft Rot (Erwina carotovora subsp. carotovora (Jones) Bergey et al.) (syn. = Pectobacterium carotovorum subsp. carotovorum)

This is a disease of storage and transit. The disease is characterized by a watery, smelly, soft decay of storage tissue. The infected tissue softens and become watery or slimy and water extrusion become more evident as rot progresses. Disease can easily distinguish by foul odor from it. The bacteria live in the soil and in the decaying refuse. It enters the root principally through cultivation wounds, harvest bruises, freezing injury and insect openings. After infection, high humidity is essential for progress of the disease. When soft rot occurs in the field, it usually follows a period of water logging in low areas following excessive rain or irrigation.

The bacteria must have relatively high moisture and high temperature for rapid development. A storage temperature just above the freezing point and a relative humidity (RH) below 90% check the development of the disease. The pathogenicity of the soft rot bacteria depends on the production of an enzyme, protopectinase which moves through the tissues ahead of the bacteria, loosening, and destroying cells as it does so. Byproducts of bacterial growth cause the cell contents to flow into the intercellular spaces where they are a nutrient medium for the bacteria. Thus the rot is watery.

In the field, maintain good drainage and avoid practices that could wound the plant roots. Avoid prolonged irrigation of mature carrots during warm months of the year. Follow crop rotation of cereals and fodder grasses. Always grow carrots in well-drained soil. Minimize mechanical damage during harvesting and in the packing shed. Discard affected carrots before transport and storage. Care in harvesting to avoid bruises is important. Grading as well as sorting of carrot root before storage is helpful to lower down the infection. Keep dry the root surface and store in well-ventilated place at low temperature with 90% RH. Spraying fungicide Carbendazim @ 2 g/l of water at an interval of 8–10 days is effective for the management of the disease (Reddy, 2010).

17.1.1.1.3 Carrot Scab (Streptomyces scabies (Thaxt.) Waksman and Henrici

This pathogen can attack several other species of root crop vegetables, including beet, potato, turnip, radish, and parsnip. Scab lesions are formed by the abnormal growth of the host cells, resulting in corky tissue that is usually darker than healthy tissue. Lesions are sometimes sunken below or raised above the surface of the healthy skin. Many single lesions may join to form continuous scabby areas. The disease tends to be more severe in dry, alkaline soils. In soils with a pH above 5.5, Streptomyces scabies is usually responsible for causing scab lesions.

Diseases reduction can be achieved by avoiding carrot production in alkaline soil. Maintain soil pH levels between 5.0 and 5.2 by using gypsum or acid tending fertilizers such as ammonium sulfate or sulfur. Local farm advisors can provide information on amounts of fertilizer that are appropriate for your soil conditions. Irrigation management is very important. Growing carrot in soils with good moisture-holding capacity or irrigating to maintain an even water supply may reduce scab. Growers should avoid planting carrot in fields used for potato production because the potato is more prone to this

Diseases of Carrots, Radishes, and Knol Khol (Kholrabi) 327

disease. Long rotations with small grains, grasses, or corn may also help to reduce scab severity.

17.1.1.2 FUNGAL DISEASES

Disease management in carrot production is necessary to produce high yields of high-quality carrots. The major concern is the production of a disease-free and cleans carrot root. In this context, diseases caused by fungus also play an important role in carrot production. Fungus produces countless number of spores which are carried by wind or other means to their hosts. Under the favorable environmental condition, they germinate and infect healthy carrot plants. Sclerotinia rot, Alternaria leaf blight, Cercospora leaf blight, powdery mildew are important diseases of carrot that causes yield losses if not manage properly.

17.1.1.2.1 *Sclerotinia Rot (Sclerotinia sclerotiorum (Lib.) de Bary)*

Sclerotinia rot of carrot, also referred to as watery soft rot or cottony rot was first reported on field carrots in Belgium by M. E. Coemans in 1860, while on stored carrots it was first described by E. Rostrup in 1871 (Mukala, 1957). It is an economical important disease of carrot occurring in the field and storage. Pathogen affect both above and below ground part of the carrot plant. The disease first appears on leaves and petioles as water soaked soft, olive green lesions associated with collapsed tissues which enlarge into a watery rotten mass of tissues that is covered by white silvery appearance. Eventually, characteristic black sclerotia (2 to 20 mm) form, embedded in the mycelium covering infected tissues is characteristic of *Sclerotinia* rot in cool and humid weather (Kora et al., 2008). Affected crop lose their turgidity and fail to throw flowering shoots. The pathogen survives from year to year as sclerotia in soil or plant debris.

Disease in storage develops mainly from infected roots introduced from field and can substantially reduce yield. Symptoms on infected roots appear as soft, watery rot lesions characterized by a rapidly spreading white mycelium and superficial black sclerotia. Mycelium arising from an infected carrot can spread to adjacent roots forming enlarging pockets of infection (McDonald, 1994). Colonized carrots are usually held together in large clumped by the extensive mycelial growth. In addition, secondary pathogens may gain entrance into infected areas and may contribute to further disintegration of the roots (Kora et al., 2008). The fungus is a necrotrophic and soil-borne in nature. *S. sclerotiorum* spends about 90% of its life cycle in soil as dormant sclerotia which develop primarily from mycelia on diseased tissues (Adams and Ayres, 1979). Mature sclerotia can survive in soil for 1 to 5 years, depending on the interaction of various physical and biological factors (Adams and Ayres, 1979; Cook et al., 1975).

1. **Epidemiology:** To understand the epidemiology of disease, it is essential to know the life cycle of pathogen *S. sclerotiorum* in relation to carrot development. Carrot is a biennial crop and mainly grown for root production. The presence of senescing leaves is a prerequisite for infection of carrot and determines the susceptible stage of the crop. However, the contact of senescing leaves with soil after they collapse is the most important phenological event for initiation of disease (Kora et al., 2005b). The susceptible stage of carrot crop is relatively long, extending from the collapse of first senescing leaves on the soil until harvest. Accumulation of collapsed senescing leaves and lodging healthy leaves in furrow increases the potential infection site.

2. **Pre-Harvest Epidemics:** Ascopspores originating from apothecia within the crop appear to be the most important primary inoculum to initiate foliar infection and subsequent epidemics in the field (Couper, 2001; Kora et al., 2005b). Airborne ascospores from apothecia in surrounding fields can contribute to development of disease when inoculum coincides with the susceptible stage of the crop (Kora et al., 2005b). Mycelium from sclerotia on or near the soil surface can also initiate disease through direct colonization of leaf and stem. Root infection results from infected foliage and occur via the crown. Some factors, i.e., presence of senescing leaves, inoculum, and free surface moisture are require for foliar epidemic of disease. Air temperature of 12° to 18°C and 12 to 24 hours of leaf wetness are optimum conditions for infection of roots (Kora, 2003).

3. **Post Harvest Epidemics:** Infection in storage is initiated mainly by mycelium arising from the crowns of diseased roots introduced from the field. Infected foliar debris near harvested roots show an additional source of contamination in carrot in storehouse. Mycelium of the pathogen persisting on the surface of infested wooden bins used for storage can also initiate new infection on stored carrots (Subbarao, 2002). However, this source of inoculum is unlikely to be important if propagules on bins previously used to store diseased crops at low frequency (Kora, 2003).

Diseases of Carrots, Radishes, and Knol Khol (Kholrabi)

Disease on carrot roots can develop at temperature as low as 0°C but progresses faster when temperature of surrounding air increases. The optimum temperature for disease development in storage is 13°C to 18°C (McDonald, 1994). Once infection is established, infected tissues usually provide sufficient moisture for further disease development. Post-harvest epidemics occur sporadically in that pre-harvest epidemics do not always result in epidemic in storage; however, it appears to be connection between high disease incidence at harvest and high disease severity in storage (Kora et al., 2005b).

Cultural practices such as deep burial of crop refuse in soil, reduced irrigation frequencies (Ferraz et al., 1999), weed management, soil solarization and grass mulching of the soil helps to reduce population of sclerotia in soil. Surface drip irrigation, instead of surface irrigation can provide good control of apothecia production by maintaining dry soil conditions in the top 5 to 8 cm of planting bed (Subbarao, 2002; Davis, 2004). Crop rotations with the plants that are not susceptible to *S. sclerotiorum* help to manage the disease.

2–3 year rotation with beets, onion, spinach, cereals, and corn manage the disease (Anonymous, 2001, 2004b). A bio-fungicide containing *Coniothyrium minitans* reduced apothecial production by up to 100% and the proportion of Sclerotinia infected roots by up to 36% (Couper, 2001). *C. minitans* can use as biological alternative and replace fungicides (Weber, 2003). Post-harvest losses are minimized by reducing mechanical damage during harvesting by rapid cooling and storing carrots at 0°C. Foliar applications of fungicide Carbendazim or Benomyl or Thiophanate Methyl @ 0.1% reduced infections of carrot in the field.

17.1.1.2.2 *Alternaria Leaf Blight (Alternaria Dauci (Kuhn) Groves and Skolko)*

Alternaría blight appears first as irregular dark brown to spots with yellowish centers near the margins and tips of the leaves and yellowish areas surrounding the spots. The decayed tissue is greenish black to jet black due to presence of masses of black spores. Under favorable conditions, the lesions become numerous and continue to expand until they ultimately coalesce giving the leaf tissue a blighted (burned) appearance. The leaflet tip or entire leaflet may be invaded and turn brown.

Large lesions can also develop on the petioles and may girdle and kill the leaves. Under severe infections, entire fields may be bronzed as if scorched by heat. The disease affects roots in the field as well as in storage. This fungus is seed and residue-borne. Complete loss of foliage can take place during periods of prolonged wet and humid weather. The pathogen requires the presence of moisture for infection. Heavy dews are almost as favorable for infection as are the rains. The optimum temperature for growth and infection by Alternaria is 82F (Chupp and Sherf, 1960).

Good crop nutrition always helps to control the disease. Never use excessive nitrogen fertilizer as lush green top growth of plant make difficult to control the disease. Long Crop rotation of 3 to 4 years and destruction of infected plant material in the field will minimize infection. Prompt incorporation of carrot residues after harvest speeds decomposition of the crop debris that harbors the pathogen. Control seed borne Alternaria by treating seed with hot water at 50°C for 20 minutes is effective to manage the disease. Seed treatment with fungicide Thiram @ 2 g/kg of seed for 24 hours also reduces the infection. Carrot seed can be dusted with fungicide Thiram @ 5 g/kg seed. Spraying of Foltaf or Copper oxychloride @ 0.2% effective to control Alternaria leaf blight (Reddy, 2010).

17.1.1.2.3 Cercospora Leaf Blight (Cercosporacarotae (Passerini) Solheim.)

Cercospora leaf blight lesions are initially small necrotic flecks that develop into cream to grey colored lesions with dark colored definitive margins. These lesions are circular in shape and more elongate along the leaf margin as the lesions expand, they can coalesce and lead to leaflet death. Petiole lesions are circular to elliptical in shape and have a lighter colored center. The infected leaflet shows lateral curling. The lesions are more numerous on the younger as well as on older leaves. Under very moist conditions, lesions of Cercospora may be quite extensive and dark-colored, closely resembling Alternaría. Disease appears on the youngest leaves first.

The fungus is active under moderately warm conditions. The fungus overwinters on/in the seed and in diseased plant refuse in the soil or on foliage placed in storage and discarded in the spring. Minute black stromata are present in old diseased leaves and produce conidiophore and conidia. It is disseminated by wind, water, and moving objects. When they are blown to a wet, warm leaf, they germinate and enter the stomata within 2 to 3 days. The fungus pathogen can grow in the culture at 45°F to 98°F temperatures with the optimum between 66° and 82°F (Chupp and Sherf, 1960).

Crop rotation and destruction of infected plant material in the field will minimize infection. Plow under crop residues will hasten the decomposition. Hot water treatment of carrot seeds at 50°C for 20 minutes is effective to

Diseases of Carrots, Radishes, and Knol Khol (Kholrabi) 331

manage the disease. Seed treatment with fungicide Thiram @ 2 g/kg of seed for 24 hours also reduces the infection. Fungicidal sprays with Foltaf @ 0.2% or Copper oxychloride @ 0.3% (Reddy, 2010) or Mancozeb @ 0.25% or Zineb @ 0.25% at 7–10 days interval is effective. Start the spraying program when the first sign of blight appears.

17.1.1.2.4 *Powdery Mildew (Erysipheheraclei)*

The disease is characterized as white powdery mass of ectophytic mycelium on the upper surface of leaves as well as on petiole. The underneath or lower surface of the leaf gives brown or purplish colored growth of fungus. Heavily infected leaves will senesce and die. Older leaves are typically infected first and then the pathogen spreads to the younger leaves. Powdery mildew is favored by hot and dry conditions and, unlike many other fungal pathogens, the spores do not require free water to germinate.

The fungus also affects the other members of the Apiaceae family. Source of the pathogen likely is wind-dispersed spores from crops or weeds. These spores can be moved long distances, thus the source could be from another region. The pathogen (*Erysiphe heraclei*) can produce cleistothecia (structure containing spores that can survive in the absence of living host plant tissue), but it is not known if they have a potential role as a source of inoculum from affected plants of previous season or as contamination on seed. Glawe et al., (2005) reported first occurrence of powdery mildew of carrot in Washington State.

Isolate new carrot fields from established infected fields. Management practices include providing adequate fertilizer and water to maintain good crop growth and avoid drought stress. Mulching can also minimize drought stress. Spray Wettable Sulphur @ 0.2% 4 weeks after sowing and repeat the spray at 7 and 10 weeks after sowing (Reddy, 2010).

17.1.1.3 *VIRAL DISEASES*

A variety of diseases reduce yield and market value of roots where ever carrots are grown. Bunching carrot must have damage-free top as well as root, but the foliage is attacked by large number of pathogen. The presence of heavy infection causes insufficient harvesting and yield loss. Mycoplasma also causes damage to carrot root both in form of malformed roots and direct yield losses of plant. They are transmitted by insects.

17.1.1.3.1 Yellows (Aster Yellow Mycoplasma)

Diseases are caused by aster yellow mycoplasma which is transmitted to carrot by aster leafhopper. Its first symptom is a yellowing of young leaves as they emerge from the crown. The disease symptoms on the leaves become yellow accompanied by vein clearing. Witch broom appearance is noticed from the dormant buds in the crown. As the disease develops, the entire cluster of shoots may become a sickly yellow color and the older outside leaves may become bronzed or reddened and twisted. By late season, the crown often becomes dead and blackened.

The internal texture, color, and flavor show marked damage causing reduction in the value of carrots for fresh market as well as for processing. Roots of infected plants have bitter and astringent flavor. The virus is mainly transmitted by leafhopper, *Macrosteles divisus*. H. H. P. Severin, of the University of California, was first to show that the aster yellows virus was the cause of yellows in carrot. L. O. Kunkel, of the Rockefeller Institute, had earlier shown that the six-spotted leaf hopper *(Macrosteles divisus)* was the vector of the disease on aster.

The most important method of control is to eradicate all perennial or biennial weeds on which pathogen can survive. Infected plants and weeds should be removed to eliminate the source of the mycoplasma and minimize spread. Spray with insecticide Carbofuran, Dimethoate @ 0.04% for controlling the insect vectors. Also, spray the adjacent area of the field to destroy the insects on the weed. Immediate after harvesting, destroy or plow all crop residues on which the disease and the leafhopper can continue to thrive.

17.1.1.4 NEMATODES

The root knot nematode is the major limiting factor in successful carrot cultivation. Nematodes develop characteristic forking of the root in carrot which are responsible for reduction in market value. Considerable losses can be experienced with carrots because of attacks by root-knot nematodes (*Meloidogyne* spp). The symptoms are nodular thickenings on the taproot and particularly on the finer lateral roots. Splitting and forking of roots can occur. Nematodes are often more prevalent on lighter soils. Various soil fumigants may be used before planting to control nematodes. Because of nature of the carrot roots, damage caused by root knot nematode becomes a limiting factor in carrot production.

Diseases of Carrots, Radishes, and Knol Khol (Kholrabi) 333

17.1.1.4.1 *Root Knot Nematodes (Meloidogynehapla Chitwood)*

When the density of nematodes in soil is high, there may be areas within fields with missing or stunted plants. Leaves may be smaller and lighter colored. A reddish tinge may appear on the back of leaves while they are still green. Older leaves often turn yellow and dry prematurely. Infected plants usually senesce early in the season. A few weeks after planting, small swellings and branches may be visible on the lateral roots, even before the tap roots start to size.

Tap root development is delayed and mature roots are deformed, short, and branched or knobby. Secondary roots are often abnormally branched and hairy. There may be numerous root swellings, from which small rootlets originate. Marketable yields are reduced considerably because of the poor appearance of tap roots or tubers, rather than by a direct weight loss.

- **Life History:** Nematodes are attracted by root secretions and migrate toward roots soon after seed germination and root elongation. Second-stage juveniles penetrate the root tips. They position themselves with their head in the vascular tissue and induce the formation of giant cells upon which they feed. The juveniles enlarge considerably, undergoing three molts. Migration of these parasites through the cortex and the establishment of feeding sites in the vascular tissue cause changes in root morphology.

The root tissue increases in size through hypertrophy and enlargement (hyperplasia) of vascular parenchyma cells, resulting in small swellings, knobs or galls. At each gall and especially at the root tips, nematode development causes the roots to branch, giving them a matted, bushy appearance. Females become so large that they often protrude from the gall. At soil temperatures around 20°C, several hundred eggs are produced by each female within a few weeks. The eggs are laid at the surface of the gall in dark brown, gelatinous egg masses the size of a small pin head, which can be seen with the naked eye.

Infective second-stage juveniles develop in approximately two weeks. They can re-infect newly formed roots and form additional galls. *Meloidogyne hapla* reproduces quickly and by mid-season, medium to high densities of juveniles in soil or eggs on roots usually develop. Low to medium densities of nematode before planting generally mean that susceptible vegetable crops will suffer some damage.

Crop rotation with non-hosts such as cereals helps to reduce populations of root-knot nematodes in soil. Combination of spinach-marigold-carrot and spinach-radish-carrot can decrease the nematode population (Hasan and Jain, 1998). Sudan grass is a non-host to this nematode and when incorporated as a green manure will further suppress the soil population of this nematode.

The use of cover crops grown between the main crops may provide an alternative management strategy. Rye grain, barley, oats, Sudan grass, annual rye grass, and wheat have been shown to be non- or poor hosts to this nematode.

In small plantings areas inter-planting with marigolds *(Tagetespatula* L. and *T. erecta* L.), solarization, and fumigation also are effective. Use a bare fallow between susceptible crops is important to lower down the nematode population. The oil cakes (neem, ground nut, mustard, and castor) were effective in reducing parasitic nematode population (Reddy, 2008). Carrot cultivar Arka Suraj is reported to be tolerant to root knot nematode (Swamy and Sadashiva, 2004).

Application of *Paecilomyces lilacinus* (containing 1×10^6 spores at 10 g/2.5 sq. m.) recorded least gall index and reduced nematode population by 45.23% and increased yield by 147.6% (Sivakumar, 1998). Seed treatment with *Trichoderma harzianum* at 10 g per kg (with 1×10^6 cfu/g) and *Pseudomonas fluorescens* at 10 g per kg (with 1×10^9 cfu/g) and subsequent field application of 5 tonnes enriched FYM with *T. harzianum* (with 1×10^6 cfu/g) and *P. fluorescens* (with 1×10^9 cfu/g) per hac significantly reduced root knot nematode in carrot root (Reddy, 2008).

17.1.1.5 PHYSIOLOGICAL DISORDERS

Physiological disorders are caused by environmental conditions that seriously affect normal growth of plant. Rocky, stony or heavy soil may also responsible for forked carrot root. Carrot plants with full foliage top and small or limited roots can result from seeding too close in the field and no thinning. Excessive nitrogen fertilizer and inappropriate/poor environmental condition are responsible for disorder in carrot root. Heat canker, cavity spot, splitting of carrot root are major problem in many carrot growing areas.

17.1.1.5.1 Heat Canker

Heat canker results when the carrot root tissue at or near the soil surface is injured and killed by high temperatures. Heat canker can cause considerable losses in carrot seedlings and may also cause injury at later stages of growth. In carrot seedling, heat will cause the tissue at or near the soil surface to collapse and die. The top of the plant often falls over or breaks off and the seedling dies. If the plant is larger when affected, only the cells of the cortex are killed. These shrivel, discolor, and form a constriction near the top of the root.

Diseases of Carrots, Radishes, and Knol Khol (Kholrabi)

Irrigation management is very important to control of heat canker. Using overhead irrigation to cool the soil surface and provide moisture to the seedlings is also helpful. Broadcast seeding of a cover crop such as barley or spinach to shade the soil and reduce wind erosion has been successful.

17.1.1.5.2 Cavity Spot

Calcium deficiency leads the disorder known as cavity spot (Maynard et al., 1963). In this, cavity appears in the cortex and in most cases the subtending epidermis collapses to form a pitted lesion. It may be induced by excess potassium uptake during the ontogeny of the plant. Increasing calcium accumulation reduced the incidence of cavity spots in carrots (Bhat, 2016).

Calcium application may control this problem. Increased calcium level in the growing medium results in increase calcium accumulation in the plant and reduce the incidence of cavity spot. Always apply balanced dose of nitrogenous fertilizer. Irrigation management is essential and do not allow the carrot to grow in the field for longer period of time in water scarcity.

17.1.1.5.3 Splitting

This is a major problem in carrot growing areas in which the carrot roots crack. Some factors like heavy doses of nitrogen fertilizer in early stages as well as boron deficiency are responsible for this problem. According to Boss and Some (1986) high nitrogen and chloride levels in the soil may induce this problem. This may also be due to fluctuating water supply in area where there is heavy rainfall after a long spell of drought. In this case, inner flesh of carrot expands faster than the toughened skin, causing the skin to fissures, sometimes root splits often exposing core (Bhat, 2016).

Balanced dose of nitrogen fertilizer and irrigation management is important factor to reduce the problem. Never allow the carrot to grow in the areas for longer periods of time without water.

17.1.2 RADISH (RAPHANUS SATIVUS LINN.)

Radish *(Raphanus sativus* L.) is a member of the family Brassicaceae and is grown for its large succulent, slightly spicy, crunchy tap root. It is a hardy, cool-season vegetable that can produce many crops each season due to its rapid days to maturity. It is mainly used for salad, pickling, and other processing purposes. Radish roots are low in calories and are usually eaten raw. It is usually grown as annuals and are harvested before they flower. The history of this crop is vague. This crop is probably originated in China and presently cultivated all over the world. In India, it seems to have been cultivated from ancient times. There are two distinct genetical groups in radish. The Asiatic varieties, which are primarily for tropical climates produce edible roots in the first season and seed in second season as a biennial crop. The exotic or European varieties produce roots in the plains of tropical and subtropical climate and seed in the hills of temperate climate.

17.1.2.1   BACTERIAL DISEASES

Radish develops so quickly and the entire period of culture, even of a seed crop, is so much short that the crop matures before any slow developing disease can show its serious effects in plants. But this crop is attacked by bacteria which cause qualitative as well as quantitative losses. Some of the important bacterial diseases of radish are root rot, bacterial soft rot and black rot recognized as the major limiting factor responsible for yield loss.

17.1.2.1.1   Root Rot of Radish (Erwinia rhapontici)

It is an important bacterial disease of radish crop. Symptoms appear as rotting of pith tissues resulting in cavity formation and wilting of plants. The disease is more severe in seed crop grown from cut roots. The disease spreads when the roots are transplanted for seed production and through implements, irrigation water, insects, etc. This disease was first reported from Bangalore, Karnataka during 1983 and its incidence ranged from 15 to 50% (Kishun, 1983b).

Use only certified and disease-free seed. Clean all farm equipment to avoid the further spread of disease. Water management is essential with proper drainage facility. Always apply balanced and optimum dose of fertilizers which helps to reduce the favorable conditions for disease development. Dipping of the seeds in a solution of Agrimycin-100 (100 ppm) at the time of sowing is effective in checking the disease (Kishun, 1983b).

Diseases of Carrots, Radishes, and Knol Khol (Kholrabi)

17.1.2.1.2 Bacterial Soft Rot (Pectobacterium carotovorum subsp. carotovorum (syn. Erwinia carotovora var. carotovora)

The pathogen causes a soft, watery, slime rot. Plant tissue at first appears as water-soaked and then rapidly breaks down into a soft, mushy rot. Secondary invaders may also attack. An offensive odor usually is present. The soft-rot bacterium enters through natural openings or wounds caused by insects and equipments. Any plant part can be affected but the most serious economic loss is from storage root infections. Prolonged moisture from rain or irrigation and mild temperatures encourage disease development. It is a serious transit and storage problem if affected roots are not discarded. These bacteria can survive in soil and plant debris.

Crop rotation must be practiced so that infected crop resides have time to break down. Plant to plant distance must be maintained which will allow good air drainage. Intercultural operations should be done carefully to minimize injury in the plants. Water management is essential along with good drainage. Always avoid stagnant water sources. In store houses, humidity (90 to 95%) and temperature (32°F to 39°F) should be maintained.

17.1.2.1.3 Black Rot (Xanthomonas campestris)

The plant may be affected at any time during its growth from the youngest seedling until it mature. On young seedlings, the cotyledons are affected at the margins which show blackening and such shoots die. Later, infection of leaves occurs as water pores at the margins. The infected tissues turn yellow and chlorosis occurs. The veins show a brown or black discoloration. It is a seed borne disease.

Seed treatment with Agromycin @ 0.01% or Streptocycline @ 0.01% will check the seed borne infection. One of the recommendations is to give hot water treatment at 50°C for 30 minutes can also be effective. Application of 10–12 kg/ha stable bleaching powder as soil drench can also be used. The two-year crop rotation must be practice to check the soil borne infection as pathogen lives only for one year in the field.

17.1.2.2 FUNGAL DISEASES

Radish is one of the common root vegetable grown all over India. Fungal diseases, i.e., Alternaria blight, white rust and black root of radish can be

338   *Diseases of Fruits and Vegetable Crops*

limiting factors in sustainable and profitable production wherever they are grown. In the absence of disease management practices, these diseases can cause extensive crop damage.

17.1.2.2.1 *Alternaria Blight (Alternaria raphani Grover and Skolko)*

Leaf spotting is the major symptom associated with *Alternaria* infection. Pre and post-emergence damping-off and damage to the inflorescence of seed crops and to seed can also occur. Pinpoint spots on leaves enlarge to become circular lesions with target-like concentric rings. Lesions are initially yellow-brown and later turn brown to black. The center of the ring often dries out and drops, leaving the leaves with a shot-hole appearance. Complete leaf drop may occur. The pathogen also attacks the stems and pods. The entire pod may be so infected that the styler end becomes black and shriveled.

The fungus penetrates in pod tissues, ultimately infecting the seeds. The infected seed fails to germinate. The pathogen also cause post harvest disease in Chinese radish in storage (Su et al., 2005). Pathogen can survive in susceptible weeds and perennial crops; diseased residue is a major source of inoculum and a means of perennial perpetuation of pathogens. *Alternaria raphani* can be readily isolated from diseased tissue using routine procedures. *A. raphani* does not sporulate abundantly and forms a white, cottony mycelium that also develops many irregular forms, varying from thick, dark, heavily septate mycelium to one with characteristic chlamydospores.

Pathogen Alternaria survives saprophytically outside the host with diseased crop debris as the primary site of survival. Resting spores (chlamydospores, micro-sclerotia) also develop. Temperature range from 15° to 25° and 90% RH at least for 12 hours favor the disease. The fungi sporulate profusely and are spread throughout fields by wind, splashing water, equipment, workers, etc. The introduction of pathogen into new areas is on infected seed.

The use of certified, disease-free seed and resistant varieties is always recommended. Long crop rotation with no cruciferous crop, incorporation of diseased crop residues into the soil, elimination of cull piles, eradication of cruciferous weeds, is always essential. Minimize the length of leaf wetness periods by reducing plant density, orienting rows with prevailing winds and irrigating in the morning help to lower down disease development. Avoidance of overhead irrigation during head development will reduce inoculum levels.

Diseases of Carrots, Radishes, and Knol Khol (Kholrabi) 339

Irrigate the field in the morning to allow the foliage to dry. Seed beds and successive crops of crucifers should be located away or upwind from existing cruciferous crops to avoid wind-borne inoculum. Also, seed beds should be kept disease-free to prevent the spread of disease. Hot water treatment of seed at 50°C reduces or eliminates both internal infection and external infestation of seed by this fungus. Regular spraying with Difolatan @ 0.3% or Mancozeb @ 0.2% or Ridomil @ 0.1% beginning in midseason particularly if conditions are warm and wet will arrest disease development

17.1.2.2.2 White Rust (Albugo candida (Pers ex. Chev) Kunze)

The first symptoms may appear as small, light green spots which later turn white and finally result in blister-like raised white pustules usually on the lower leaf surface. Pustules can develop on the upper or lower leaf surfaces or on stems and consist of masses of sporangia. The infected leaves get devitalized. Disease attacks the leaves and flowering shoots. Affected flowering shoots get deformed and bear only malformed flowers. Systemic infections can occur, causing the above-ground portion of young seedlings to be distorted and appear abnormally shaped. Infections of the flower parts may cause a bizarre outgrowth on the seed stalk. Seed pedicels may terminate and form staghead without seeds developing.

The pathogen overwinters as oospores in staghead (galls formed on infected seed heads) or plant debris as well as mycelium in infected hosts. Pieces of staghead are often found as seed contaminants and can be the primary source of infection in a field. The oospores can germinate and infect young plants, leading to pustules on leaves. Sporangia are produced in pustules, can be moved by wind, rain or insects and can germinate by giving rise to zoospores. Zoospores are motile in nature. They swim for a short distance and then invade the plant by germinating through the stomata. Sporangia require some drying in order to germinate but disease development is favored by moist conditions and temperatures between 10°C and 25°C. Dew, fog or periods of extended rainfall and cool temperatures are ideal for zoospore activity. This disease thrives in dry condition and is spread by the wind.

Clean cultivation and use of resistant varieties help to prevent the disease. Plow under infected plants or volunteers and incorporate plant debris into the soil soon after harvest. Always select fields away from the established radish fields are also helpful. Three-year rotation helps to reduce the disease.

Destruction of infected plant debris, weeds in which pathogen easily harbor is essential. Spraying with Mancozeb or phytoalexin/Macimin @ 0.3% (Reddy, 2010) or Dithane Z-78 @ 0.2% as soon as disease incidence is noticed gives effective control of disease.

17.1.2.2.3 Black Root (Aphanomyces raphani Kendrick)

Black root is characterized by stunting, severe blackening and deformity of the fleshy roots tissues and commonly regarded by commercial growers as indicative of boron deficiency in soil. Lesions on younger roots may restrict growth in that area causing pronounced girdling and deformity as diseased tissues fail to keep pace with root growth. In the early stages, the darkened areas are small, superficial, and located in the immediate vicinity of the point of emergence of secondary roots.

The dark lesions may grow together, resulting in constricted rings around the root and reducing growth. By this time, the pathogen has usually penetrated deeply causing the tissues to be partially blackened. Dark-colored strands of infected tissue frequently extend irregularly through neighboring healthy white tissues. In the advanced stages, the internal root tissues are uniformly blackened and collapsed (Wenham, 1960). It is a minor soil-borne fungal disease.

The disease is favored by high temperatures with infection of radish seedlings occurring at 16° to 32°C with a maximum at 27°C. High soil moisture or free water in the soil is necessary for penetration by the motile zoospores. Oospore germination in soil is stimulated by the presence of radish seedlings. Oospores are produced in large numbers in the secondary roots of many crucifers and in host residues or soil particles that accompany seed. They allow the fungus to persist for long periods and subsequently to infect hosts. Mycelium and zoospores are incapable of prolonged survival in soil but the fungus may subsist as mycelium in volunteer or seed plants. Dissemination is by splashing water or by movement of surface water carrying oospores or zoospores to other plants or neighboring fields. Inoculum dispersal is also possible by infected plants, wind-blown soil or host residue and by tools, agricultural machinery and workers.

Four-year crop rotation with non-cruciferous crops may reduce inoculum level of pathogen. Complete destruction of cruciferous seed is essential. Complete destruction of diseased crop residues helps to reduce inoculum. Good soil drainage should be provided.

Diseases of Carrots, Radishes, and Knol Khol (Kholrabi) 341

17.1.2.3 VIRAL DISEASES

Among the non-cellular or meso-biotic causes of infectious plant diseases, virus, and mycoplasma are the only agents. They are more dangerous than fungi and bacteria as it is much more difficult to control them because of their chemical nature and parasitism. Radish mosaic and radish phyllody are important viral diseases of radish.

17.1.2.3.1 Radish Mosaic (Radish Mosaic Virus)

The symptoms first appear as small, circular to irregular, chlorotic lesion in between and adjacent to the veins. Little or no leaf distortion is noticed and stunting or abnormal formation rarely occurs. The affected plants are conspicuous with stunting, reduced leaf lamina and root size. Mosaic pattern is also seen on the stem, petiole, and pods. Seeds from diseased plants show poor germination. The virus is readily sap transmitted. The virus is not transmitted through seeds in radish. Aphid also transmits the virus in non persistent manner. The disease was first reported by Thakur and Dube (1981), later by Ahlawat and Chenulu (1984). The disease incidence ranges from 7 to 63% (Ahlawat and Chenulu, 1984). The virus causes losses ranging from 3.3% to 100% in root weight and up to 95.8% in seed weight depending upon the time of infection (Reddy, 2010).

Complete destruction of diseased plant is essential to reduce the inoculum. Timely inspection of field is essential. Quarantine law along with certification regulation help to prevent the spread of disease. The disease can be effectively checked by controlling aphids with 2–3 foliar sprays of insecticide either Dimecron @ 0.05% or Monocrotophos @ 0.05% at 10 days interval. The disease spread can be minimized by soil application of Carbofuran at 1.5 kg ai/ ha at the time of sowing of seeds, followed by 2–3 foliar sprays of Dimethoate or Monocrotophos both at 0.05% at 10 days interval (Reddy, 2010).

17.1.2.3.2 Radish Phyllody (Mycoplasma Like Organism)

The disease is transmitted by jassid (*Orosiusalbicinctus*). The diseased plant assumes a dull grey to light violet coloration. The symptoms of the disease appear at the time of flowering when all the floral parts become green violet and leafy. The sepals and petals become green thick knob headed leaves. The sepal, petals, and carpels of affected flowers show phyllody and the stamens become sapaloid. Generally, the whole plants show symptoms of the disease. If the infection occurs at an early stage of growth in the nursery then the whole plant is affected but delayed infection results in partial phyllody.

The jassid requires minimum of 4 hr of feeding on a diseased plant to acquire pathogen in sufficient quantity. The jassids become capable of transmitting the pathogen in 2 to 3 weeks after acquiring the mycoplasma. If an infective jassids feeds on a healthy plant for more than 30 minutes, it injects sufficient quantity of mycoplasma to make the plant diseased. About two month after injection, the radish plants show symptoms. Both male and female can transmit the pathogen. The nymph cannot inherit the disease from their respective parents but they acquire the pathogen by feeding on the diseased plants (Reddy, 2010)

One or two sprays of insecticide Monocrotophos (0.05%) or Phospham-idon (0.05%) or Oxydemetan Methyl (0.02%) is done to eradicate the jassids, the vector of the virus. Soil application of Thimet 10 G (1.5 kg a.i./ha) is also recommended. The application of Thimet should be followed by irrigation (Reddy, 2010).

17.1.2.4 NEMATODE

Damage to radish from nematodes can have a substantial economic impact on radish production. The important nematode of economic importance for radish is stunt nematode which causes decrease in yield.

17.1.2.4.1 Stunt Nematodes (Tylenchorhynchus spp.)

Disease is characterized as reduction in fresh weight of plants and suppressed root growth, resulting in reduction in water absorption. The optimum soil temperature, i.e., 30°C and soil moisture (25–30%) are requiring for reproduction. Further, high temperature (35°C–40°C) and low moisture are highly unfavorable for survival of nematode. However, at moderate temperature and moisture conditions, the nematode is capable of surviving even up to 240 days (Reddy, 2008).

The mustard (*rabi*), radish (summer), sesame (*kharif*) rotation sequences considerably reduced the population of *Tylenchorhynchus* spp. (Haque and Gaur, 1985). Aldicarb, Carbofuran, Ethoprophos, and Phenamiphos each at 1 to 2 kg ai per hac were found effective in reducing the stunt nematode population (Ahuja, 1983).

Diseases of Carrots, Radishes, and Knol Khol (Kholrabi) 343

17.1.2.5 PHYSIOLOGICAL DISORDERS

Radishes are also susceptible to freezing injury (Bhat, 2016). Ryall and Lipton (1972) stated that freezing followed by thawing causes the injured tissue to appear translucent. Where in severe cases, root softens loose moisture rapidly and shrivels. Parsons and Day (1970) reported that in red radishes the pigment oozes out of the roots with moisture, leaving them yellowish and bleached.

17.1.3 KNOL KHOL (BRASSICA OLERACEA VAR. GONGYLODES L., SYN= BRASSICCAULORAPA)

Knol Khol also known as stem turnip, colinabo, gunthgobhi, kholrabi, kholrabi greens, Novalkoletc, and belongs to family Brassicaceae (syn: Cruciferae) but it is the most commonly known as Kholrabi which is the German name for cabbage turnip resembles an above ground turnip. It is very widely used in the Northern state of Kashmir in India. It is also used popularity in many other states of country. It is one of those few vegetables which originated in Europe. The whole plant is edible; however, it is mostly used for its bulged swollen stem. It is excellent vegetable if used at its early stage before it becomes tough and fibrous. Formation of knob which arises from thickening of the stem tissue above the cotyledons. The fleshy edible portion is an enlargement of the stem which develops entirely above ground and is used as vegetable. The edible portion is globular to slightly flattened stem. This crop can withstand extreme cold and frost better than other cool season crop. Knol Khol is a good source of potassium and has rich contents of anti-oxidants.

17.1.3.1 BACTERIAL DISEASES

Black rot is an important bacterial disease of Knol Khol and can become major limiting factor in successful cultivation of Knol Khol crop.

17.1.3.1.1 Black Rot (Xanthomonas campestris pv campestris (Pam.) Dowson)

The pathogen is seed borne and also survives in the infected plant debris as well as cruciferous weed hosts. It is one of the most destructive vascular disease of crucifers and occurs in all parts of the temperate and subtropical zones of the world where rainfall or heavy dews are plentiful and average temperatures are between 25° and 30°C (Gupta et al., 2013). Initially, the marginal chlorotic spots appear on the leaves followed by darkening or blackening of midrib and veins. The pathogen moves systemically through the vascular bundles turning them black.

The pathogen colonizes the vascular system after its entry into the plant and produces plentiful extracellular polysaccharide called xanthan, which along with the bacterial cells plug the xylem vessels, restricting the water flow and resulting in the characteristic V-shaped chlorotic lesions originating from the margins of the leaves. As the bacterium moves throughout the plant, the vascular tissues darken, resulting in blackening of the veins. Systemic infection during storage often renders the product unmarketable. The infection of this pathogen may be followed by attacks of soft rot bacterium *Erwinia carotovora* or fungus *Sclerotinia sclerotiorum* (Agrios, 1997). The disease can cause significant yield losses when warm, humid conditions follow periods of rainy weather during early crop development.

Select the land which has been free of undecayed cruciferous debris for over 12 months, use of pathogen free seed or resistant cultivars if possible, control of weeds and volunteer plants and removal as well as destruction of diseased plants. This disease can be managed by hot water seed treatment for 30 minutes at 50°C. Onsando (1988) advocated the use of grass mulch reduces the extent of splashing of infested soils and hence secondary spread of the pathogen. Crop rotation with non-cruciferous crops should be followed for 3 to 5 years.

Irrigation should be performed with care, to avoid over watering the crop. Fields should be deeply plowed after harvest to kill bacterium and speed the decay of plant debris. *Terminalia chebula* extract was found effective in reducing the disease incidence (Bora and Bhattacharyya, 2000). Cultural practices can be conditioned for their effective use for eco-friendly disease management in the field either alone or as a component of integrated disease management (Kashyap and Dhiman, 2010).

17.1.3.2 FUNGAL DISEASES

Plant diseases caused by a fungal pathogen is often recognized from the particular plant organ infected and the type of the symptoms produced. So fungus attacks the Knol Khol crop and cause damping off, Alternaria leaf

Diseases of Carrots, Radishes, and Knol Khol (Kholrabi) 345

spot, club root and downy mildew diseases which must be controlled to obtain desired quality and good yield.

17.1.3.2.1 Damping Off (Pythium aphanidermatum (Eds) Fitz. and Rhizoctonia solani)

Pathogen is a soil-borne fungus that attacks germinated seedlings that have not yet emerged or have just emerged. Cool, wet weather promotes infection by *Pythium* fungus where as cool to moderate weather promotes *Rhizoctonia* infection. Fields that have poor drainage, compacted soil and/or high green organic matter are the most susceptible to damping-off. Young seedlings develop a sunken, necrotic lesion near the soil line. If the lesion girdles the stem, the seedling collapses and soon dies. Also, seeds may simply not emerge.

Damage usually occurs at soil level, leaving lesions in the stem tissue. The tissue becomes dark and withered, the weak support causes the seedling to collapse and die. Beneath the soil a general rot is observed. *Pythium* can also attack the seedling's roots, causing them to turn brown and rotten. Seedlings that are attacked by *Rhizoctonia* but continue to grow will develop wire stem. In some cases, the seedling may continue to grow even though the lesion girdles the stem. The girdled seedling eventually dies. In 1948, the first occurrence of this disease in Knol Khol in India was recorded (Mahmud, 1950).

Complete destruction of residues from the previous crop through plowing is very important to practice to manage disease. Water management through overhead or sprinkler irrigation is the best methods for promoting rapid germination. Fields should be properly drained to avoid water accumulation. Disease can be prevent by practicing long crop rotation with cereals may reduce pathogen population in the soil and follow sanitation practices by decomposing crop debris. Bio-control agent *Trichoderma virens* or *Glocladium virens* can manage *Pythium* and *Rhizoctonia* induced damping-off through antagonism process.

17.1.3.2.2 Alternaria Leaf Spot (Alternaria brassicae)

Initially small dark spot appear on the leaf. As the disease progresses, concentric rings will develop around the spot creating a bulls-eye pattern. Eventually, velvety, brown spore-bearing growths develop within the spots. As disease progress, pathogen will eventually defoliate a plant. Pathogen can survive for long periods of time on plant debris or within infected seeds.

The disease occurs during a wet winter. Spores can survive in weed hosts and plant debris for long periods of time. The fungal spores are wind and rain dispersed and requires free water for germination.

Always use disease free seed because pathogen can pass to the next generation in infected seed. Cole crop should be planted once in 4 years in same piece of land to avoid disease carry over. Destruction of infected plant debris, cruciferous weed is very important to prevent the transmission of pathogen from infected debris and weed to Knol Khol. It is important to clean equipment between uses in different fields to prevent contamination of an uninfected field. Neem oil can also be used for disease management. Foliar spray of fungicide Maneb @ 0.25% as a prophylactic spray to control the disease.

17.1.3.2.3 Club Root (Plasmodiophora bassicae)

P. brassicae is a soil-borne fungus that forms no mycelia (numerous threads like branches). It is an obligate parasite that means it can develop and multiply only in living host cells. Clubroot may develop extensively on plant roots before the first sign (an abnormal wilting and yellowing of leaves, especially on warm days) is noticed aboveground. If the soil is moist, these symptoms may not become apparent until water stress occurs. When infection occurs at an early stage of growth, young plants are stunted and may die, whereas plants infected in a later stage fail to make marketable heads or growth.

When diseased plants are pulled from the soil, the roots are usually swollen and distorted. The name clubroot is derived from these symptoms. Root malformation may vary in size from very small swellings on the tap and lateral roots to large club-shaped roots, depending on when the plants became infected. In addition to reducing the plant's ability to take up water, the clubbed tissue fails to develop a protective outer layer and thus is susceptible to invasion by soft rotting bacteria. The fungus produces resting spores that can remain alive for years in some soil types in the absence of susceptible crops.

Pathogen *P. brassicae* is capable of surviving in the soil for 7–10 years or longer as resting spores. The resting spores of the fungus can be spread from field to field by infested soil, contaminated water supplies, infected transplants, infested soil on farm machinery, and even by roving animals such as cattle. In the presence of susceptible roots, resting spores of the fungus germinate to produce motile zoospores that swim in free water and penetrate the surface of the root hairs. The fungus develops into a sporangial

Diseases of Carrots, Radishes, and Knol Khol (Kholrabi)

plasmodium and subsequently divides into multinucleate portions that develop into zoosporangia. These are released from the host through pores dissolved in the epidermal cell wall, and four to eight secondary zoospores are liberated from each zoosporangium. Some of these zoospores fuse in pairs before germinating and infecting the host. Plasmodia in the roots cause cells to enlarge abnormally and divide repeatedly, resulting in gall formation. Infected cells are distributed in small groups among healthy cells within the gall. As the plasmodia mature, they transform into masses of resting spores that are released back into the soil when secondary organisms decay the galls. The plasmodium-infected galls use nutrients required by the plant and interfere with the absorption and translocation of nutrients and water in the roots, causing stunting and wilting of the plant. Decay of the galls releases toxic substances that are partly responsible for the wilt symptoms. The fungus invades young root hairs readily but wounds are necessary for infection of thickened roots and underground stems, although stems may also be invaded through leaf scars.

Clubroot is a disease that prefers warmer temperatures and high soil moisture conditions. Ideal conditions for the proliferation of this disease would be a soil temperature between 20°–24°C and a pH less than 6.5. Therefore, this disease tends to be prominent in lower fields where water tends to collect.

Always use certified and diseases free seed and it is advisable to reject the entire seed lot if trace of clubroot is present on few plants. Never contaminate the area by using irrigation water from sources that may be contaminated. Removal of diseased plant along with cabbage family weed (mustard) is recommended because pathogen can reside in the soil for many years. This disease is more severe in acidic soil, so add lime in acidic soil can help to control the disease. Farm equipments must thoroughly clean after working in a field suspected of having the club root fungus. Six to seven years of crop rotation may be effective to avoid infestation.

In small areas, soil fumigation can be another way to decrease the buildup of pathogen. Good soil drainage and the maintenance of a high pH by regular application of lime help to reduce disease incidence. The movement of soil or plant material from infested areas into clean fields must be avoided and manure from animals fed infected cull plants or pastured in infected crops should not be used. Transplants should be produced on non-infested soil and the field site should be well drained and have no history of club root. Avoid use of infested fields for crucifer crops for about seven years. Weeds of the mustard family will maintain or increase the level of infestation of club root in a field so destruction of weeds is essential.

17.1.3.2.4 Downy Mildew (Peronospora parasitica (Persoon: Fries))

Initially infection begins with the growth of grey-white fungus on the lower leaf surface. Damage occurs on both leaf surfaces, beginning with chlorotic lesions that later turn purple and eventually brown. Young leaves sometimes dry and drop off, while older leaves generally remain on the plant and develop a papery texture. Downy mildew thrives in mild, humid weather which promotes spore formation, spore dispersal and plant infection. A wet surface is required for spore germination. The pathogen *P. parasitica* infects knol kohl through its leaves and then grows between the leaves cells.

When conditions are favorable, the pathogen can spread rapidly. The fungus also produces resting spores which can survive in the soil or crop residue until the following season. The pathogen is spread by wind, rain, infected seed and infected transplants. Severe infections of mature knol kohl can result in decreased photosynthesis, stunted plants and reduced yield. Downy mildew is a systemic disease that can also result in darkened areas and/or black streaks in the stem. This damage to the leaf tissue as well as on stem makes the plant more susceptible to secondary infections.

Destruction of cruciferous weeds that can act as a host for downy mildew and crop rotation with non-cole crop helps to reduce the infection. Overhead irrigation should be avoided as it aids in spread of pathogen *P. parasitica*. Fields should be plowed under the following harvest to promote the decomposition of infected plant debris. Neem oil is also registered for the control of downy mildew. Spreading compost on the soil is sometimes used for the control pathogens. Fungicide Maneb as preventive measures is important for chemical control of downy mildew.

17.1.3.3 PHYSIOLOGICAL DISORDER

Hollow stem is an important physiological disorder that is related to rapid growth of the plant that Knol Khol can suffer and affect the crop yield.

17.1.3.3.1 Hollow Stem

In this problem, the inner tissue of the stem and the pith will crack and collapse often leaving the inner stem hollow. When hollow stem is caused by a boron deficiency, the cracked tissue is also dark in color. High temperatures along with high levels of nitrogen, large stem diameters, wide plant

Diseases of Carrots, Radishes, and Knol Khol (Kholrabi)   349

spacing and boron deficiencies are responsible for quick growth of the plant and cause the hollow stem problem.

The best way to manage this problem is to maintain optimum nutrients availability and prevent rapid stem growth.

Carrot, radish, and Knol Khol are grown intensively in certain areas and there is a large scope for improving the farming practices. The return from these crops being relatively more and the possibility of taking plant protection measures are great. But work on diseases of these crops is a little. The various foliar diseases occurring on these crops need immediate attention. The extensive damage caused by soil borne pathogens, bacteria, root knot nematodes is menace in their cultivation. The virus infection also causes great crop losses. There is urgent need to replace the susceptible varieties with resistant or tolerant varieties. The seed certification program in these crops must be introduced and strengthened. Generally, we focus on the fungal diseases and have little knowledge about the viral diseases of these crops to work out the proper control measures. So there is needed to do research work and concentrated work on viral diseases, etc. A well-planned breeding program to develop high yielding, disease resistance cultivars along with better nutritional and agronomic qualities is urgently required. We have to pay due attention in diseases of these crops along with eco-friendly management.

KEYWORDS

- club root
- downy mildew
- hollow stem
- Knol Khol
- radish phyllody
- stunt nematodes
- kholrabi

REFERENCES

Adams, P. B., & Ayres, W. A., (1979). Ecology of *Sclerotinia* species. *Phytopathology, 69,* 896–899.
Agrios, G. N., (1997). *Plant Pathology* (p. 803). Academic Press INC. London.

Ahlawat, Y. S., & Chenulu, V. V., (1984). Radish mosaic-A new disease caused by turnip mosaic in India. *Trop. Agric. (Trinidad)., 61,* 188–192.

Ahuja, S., (1983). *Studies on the Variability in Infection of Root Knot Nematode Meloidogyne incognita on Potato* (p. 4). Third Nematology Symposium, Himachal Pradesh Agricultural University, Solan.

Anonymous, (2001). Carrots. In: *Vegetable Production Guide for Commercial Growers, 2001/2002 Edition* (pp. 71–77). British Columbia. Ministry of Agriculture, Food and Fisheries, BC, Canada.

Anonymous, (2004b). Carrots. In: *Vegetables Production Recommendations, 2004–2005* (pp. 70–73). Publication 363. Ontario. Ministry of Agriculture and Food, ON, Canada.

Bhat, K. L., (2016). Physiological disorders of other vegetable crops. In: *Physiological Disorders of Vegetable Crops* (p. 92). Daya Publishing House. New Delhi.

Bora, L. C., & Bhattacharyya, A. K., (2000). Integrated management of black rot of cabbage caused by *Xanthomonas campestris* (Pammel) Dowson. *J. Agric. Sci. Soc. NE India, 13,* 229–233.

Boss, T. K., & Some, M. G., (1986). *Vegetable Crops in India.* NayaProkash, Calcutta, India.

Chupp, C., & Sherf, A. F., (1960). *Vegetables Diseases and Their Control* (p. 693). The Ronald Press Company, New York.

Cook, G. E., Steadman, J. R., & Boosalis, M. G., (1975). Survival of *Whetzelinia sclerotiorum* and initial infection of dry edible beans in Western Nebraska. *Phytopathology, 65,* 250–255.

Couper, G., (2001). *The Biology, Epidemiology and Control of Sclerotinias clerotiorum on Carrots in North East Scotland.* PhD thesis, University of Aberdeen, Aberdeen, Scotland, UK.

Davis, R. M., (2004). Carrot diseases and their management. In: Naqvi, S. A. M. H., (ed.), *Diseases of Fruits and Vegetables* (Vol. 1, pp. 397–439). Kluwer Academic Publishers, The Netherlands.

Ferraz, L. C. L., Café, F. A. C., Nasser, L. C. B., & Azevedo, J., (1999). Effect of soil moisture, organic matter and grass mulching on the carpogenic germination of sclerotia and infection of bean by *Sclerotinia sclerotiorum. Plant Pathology, 48,* 77–82.

Glawe, D. A., Pelter, G. Q., & Toit, L. J., (2005). First report of powdery mildew of carrot and parsley caused by *Erysiphe heraclei* in Washington State. *Plant Health Progress,* 1–3.

Gupta, M., Vikram, A., & Bharat, N., (2013). Black rot - A devastating disease of crucifers: A review. *Agri. Reviews, 34*(4), 269–278.

Haque, M. M., & Gaur, H. S., (1985). Effect of multiple cropping sequences on the dynamics of nematode population and crop performance. *Indian Journal of Nematology, 15,* 262–263.

Hasan, N., & Jain, R. K., (1998). Nematological research in Uttar Pradesh: An overview. In: Trivedi, P. C., (ed.), *Phytonematology in India* (pp. 150–169). C. B. S. Publishers and Distributors, New Delhi.

Kashyap, P. L., & Dhiman, S. J., (2010). Eco-friendly strategies to suppress the development of Alternaria blight and black rot of cauliflower. *World Appl. Sci. J., 9,* 345–350.

Kishun, R., (1983b). Soft rot of radish caused by *Erwinia* spp. and its control (Abstr.). *Indian J. Mycol. Pl. Pathol., 13,* 3.

Kora, C., (2003). Etiology, epidemiology and management of sclerotinia rot caused by *Sclerotinia sclerotiorum* (Lib.) de Bary. *PhD Thesis,* University of Guelph, Canada.

Kora, C., McDonald, M. R., & Boland, G. J., (2003). Sclerotinia rot of carrot, an example of phenological adaptation and bicyclic development by *Sclerotinia sclerotiorum. Plant Disease, 87*(5), 456–470.

Diseases of Carrots, Radishes, and Knol Khol (Kholrabi)

Kora, C., McDonald, M. R., & Boland, G. J., (2005b). Epidemiology of sclerotinia rot of carrot caused by *Sclerotinia sclerotiorum*. *Canadian J. of Plant Pathology, 27*, 245–258.

Kora, C., McDonald, M. R., & Boland, G. J., (2008). New progress in the integrated management of Sclerotinia rot of carrot. In: Ciancio, & Mukerji, (eds.), *Integrated Management of Diseases Caused by Fungi: Phytoplasma and Bacteria* (pp. 243–270). Springer Science.

Mahmud, K. A., (1950). Damping-off of cabbage, cauliflower and knolkhol caused by *Pythium aphanidermatum*. Fitz. *Current Sci., 19*(2), 67–68.

Maynard, D. M., Gersten, B., Vlach, E. F., & Vernell, H. F., (1963). The influence of plant maturity and calcium level on the occurrence of carrot cavity spot. *Proc. Am. Soc. Hort. Sci., 83*, 506.

McDonald, M. R., (1994). Sclerotinia rot (white mold) of carrot. In: Diseases and pests of vegetable crop in Canada. In: Howard et al., (eds.), *The Canadian Phytopathological Society and Entomological Society of Canada* (pp. 72–73). Ottawa, Canada.

Mukula, J., (1957). On the decay of stored carrots in Finland. *Acta Agric. Scand. Suppl., 2*, 2.

Onsando, J. M., (1988). Management of black rot of cabbage caused by *Xanthomonas campestris* pv. *campestris* in Kenya. *Acta Horticulturae, 218*, 311–314.

Parsons, C. S., & Day, R. H., (1970). Freezing injury of root crops: Beets, carrot, parsnip, radishes and turnips. U.S. Department. *Agric. Marketing Res. Rep.*, p. 866.

Reddy, P. P., (2008). *Diseases of Horticultural Crops: Nematode Problem and Their Management* (p. 379). Scientific Publishers, Jodhpur, India.

Reddy, P. P., (2010). *Bacterial and Viral Diseases and Their Management* (Vol. 3, p. 288). In plant protection in horticulture. Scientific Publishers, Jodhpur, India.

Reddy, P. P., (2010). *Plant Protection in Horticulture: Fungal Diseases and Their Management* (Vol. 2, p. 359). Scientific Publishers, Jodhpur, India.

Rubatzky, V. E., Quiros, C. F., & Simon, P. W., (1999). *Carrot and Related Vegetables Umbelliferae* (p. 294). CABI Publishing, New York.

Ryall, A. L., & Lipton, W. J., (1972). *Handling, Transportation and Storage of Fruits and Vegetables* (Vol. 1.). Vegetables and Melons, AVI Pub. Co. Westport, CT.

Simon, P. W., (1990). Carrots and other horticultural crop as a source of pro-vitamin A carotenes. *Hort. Science, 25*, 1495–1499.

Sivakumar, C. V., (1998). Bacterial antagonists for suppression of plant parasitic nematodes. In: Singh, & Hussaini, (eds.), *Biological Suppression of Plant Diseases, Phyto-Parasitic Nematodes and Weeds* (pp. 128–137). Project Directorate of Biological Control, Banglore.

Su, X. J., H. Yu., T. Zhou., X. Z. Li., & Gong, J., (2005). First report of *Alternaria raphani* causing black patches on Chinese radish during post harvest storage in Canada. *Plant Disease, 89*(9), 1015.

Subbarao, K. V., (2002). Cottony rot/pink rot. In: Davis, R. M., & Raid, R. N., (eds.), *Compendium of Umbelliferous Crop Diseases* (pp. 29, 30). APS Press, St. Paul, MN.

Swamy, K. R. M., & Sadashiva, A. T., (2004). *Vegetable Varieties/Hybrids of IIHR* (p. 30). Tech. Bull. 2. Indian Institute of Horticulture Research, Bangalore.

Thakur, M. S., & Dube, G. S., (1981). Studies on mosaic disease of radish (*Rhaphanus sativus*) (Abstr.). *Proc. 3rd Int. Symp. Pl. Path.*, New Delhi.

Weber, Z., (2003). Efficacy of biological and chemical protection of winter oilseeds rape against white mould. *Bulletin of the Polish Academy of Science and Biological Sciences, 51*, 149–152.

Wenham, H. T., (1960). Black root disease of radish caused by *Aphanomyces raphani* Kendr. *New Zealand Journal of Agricultural Research, 3*, 179–184.

CHAPTER 18

Major Diseases of Chili and Their Management

C. S. AZAD,* PANKAJ RAUTELA, SUPRIYA GUPTA, and R. P. SINGH

*Department of Plant Pathology, College of Agriculture,
G.B. Pant University of Agriculture and Technology, Pantnagar,
Uttarakhand, India*

Corresponding author. E-mail: azadbau81@gmail.com

18.1 INTRODUCTION

The introduction of chili in India is believed to be the 17th century through Portuguese. India is the major producer and exporter of chili, followed by China. India is one of the major chili producing countries in the world. In India, chili is cultivated over an area of 775 thousand hectares with an annual production of 1492 thousand tones and productivity of 1.9 metric tonnes per hectare. The area and production under this crop indicate that there is an ample scope to increase per hectare tonnage, but the attack of pest and diseases is one of the major constraints that attribute low yield. About 750 pathogens of different origins have been reported from different parts of the world, but only a few pathogens cause considerable damage to this crop in a particular region. Among the fungal diseases, damping off, leaf spot, anthracnose, dieback, powdery mildew, and fruit rot important in India, whereas among the bacterial diseases bacterial canker, bacterial spot, bacterial wilt, and seedling blight and leaf spot. In the case of noninfectious disorder, blossom end rot, cracking, stip, and sunscald are important.

18.1.2 FUNGAL DISEASES

18.1.2.1 DAMPING OFF AND ROOT ROT

Damping-off and root rot is caused by *Pythium* spp. *Rhizoctonia solani* and *Fusarium.*

354 *Diseases of Fruits and Vegetable Crops*

18.1.2.1.1 Symptom

Pre-emergence damping off is a result of the death of young seedling after germination but prior to emergence above the soil, which results in poor seed germination. Post-emergence damping-off is characterized by the development of disease after seedlings have emerged out of soil surface, but before the stems are lignified lesion formed at collar region. Infected areas appear brown, and water-soaked plants shrivel and collapse as a result of softening of tissues infected stems become hard, thin (Wire stem symptoms), and infected seedlings topple. Disease appears in patches both in nursery and field beds.

18.1.2.1.2 Conditions for Disease Development

Pre -emergence damping-off, which consists of a decay of the germinating seed or death of the seedling is caused by *Pythium* spp. and *Phytophthora* spp. while *Rhizoctonia* spp. is mostly responsible for post-emergence damping-off which occurs after the seedlings have emerged from the soil but while small and tender. These fungi are soil borne and exudates derived from host plants stimulate growth of these fungi. Excess soil moisture accelerates the disease development. Infection is favored by heavy soils, poor aeration, low pH, heavy seedlings resulting in dense planting, low light and presence of weeds. The fungi survive for long periods in soil and in host residue in the form of oospores and sporangia. Spread of damping-off fungi depends primarily on the mechanical transfer of mycelia, sclerotia or resting spores in infested soil particles or infected plant tissue.

18.1.2.1.3 Management

1. Deep plowing and seed bed management should be used.
2. Solarization of soil in nursery beds in warm weather for 4–6 weeks before sowing of seeds is beneficial.
3. Thinning of seedlings in seedbeds to permit good air circulation.
4. Adopt sanitary practices in the nursery and follow rotation of nursery bed site every year.
5. Light and frequent irrigation and stagnation of water should be avoided.
6. Treat the seed with captan or thiram @ 2.5g/Kg or carbendazim @ 1 g/Kg of seed, before sowing or treat the seed with bioagents pant

Major Diseases of Chili and Their Management 355

bioagent-1 or pant bioagent-2 @ 10g/Kg at least 24 hrs before sowing the seeds.

7. Drench the nursery bed with mancozeb (0.25%) or carendazim (0.05%) or suspension of bioagents (1.0%) when damping off appears.

18.1.2.2 DIE-BACK AND FRUIT ROT

18.1.2.2.1 Causal Organism

Colletotrichum capsici, C. gloeosporioides, C. coccodes, and *C. acutatum.*

18.1.2.2.2 Symptoms

Anthracnose affects all the above-ground parts of peppers during any stage of growth. Seedling infection may be confined to cotyledons and not spread. Small, circular to irregular, brownish-black scattered spots appear on leaves and stems. The infection of growing tips leads to necrosis of branches from tip backward. Fruit lesions are the most economically important aspect of this disease. Ripe fruits are more liable for attack than the green ones small, circular, yellowish to pinkish sunken spots appear on fruits. Spots increase along fruit length attaining elliptical shape. Severe infection results in the shriveling and drying of fruits. Such fruits become white or grayish in color and lose their pungency. Depending on the *Colletotrichum* species present, black or brown filamentous structures may be visible in the lesion. Anthracnose can affect both green and ripe fruit, but symptoms are usually not visible until fruit ripen and turn red.

18.1.2.2.3 Conditions for Disease Development

Humid weather, wet weather with rainfall at frequent intervals generally favors infection and development of symptoms. Depending on the species of *Colletotrichum* present, optimal temperatures 28°C and relative humidity (RH) more than 97%. Free moisture is necessary for infection. Fog and dew are conducive to disease development. Rain disseminates the pathogen's spores and often leads to severe losses, especially if fruit is wounded. These fungi can survive in infected seed and persist in leaf or stem lesions in plant debris for long periods of time.

18.1.2.2.4 Management

Use disease healthy and free seeds from a certified agency. Removal and destruction of Solanaceous weed hosts and infected plant debris. This disease can be managed by sowing good quality seed, rotating out of solanaceous crops for two to three years, removing weeds and infected debris, and choosing fields that drain well. Minimize fruit wounds by controlling insects. Copper fungicides are available, but have limited economical value for controlling this disease. Spray thrice with captan@1.5% or mancozeb@0.25%. Just before flowering, at fruit formation stage and 15 days after second spray.

18.1.2.3 FUSARIUM WILT

18.1.2.3.1 Causal Agent: Fusarium Oxysporum f. sp. Capsici

18.1.2.3.2 Symptoms

All vascular wilts, regardless of the pathogen, have some symptoms in common. The leaves or other parts of the infected plants lose turgidity, turn lighter green to greenish-yellow, droop, and wilt, turn yellow then brown, and die. In cross-sections of infected stems, discolored brown areas appear as a complete or interrupted ring of discolored vascular tissues. Fusarium wilts can affect many vegetables that include pepper, tomato, potato, eggplant, crucifers (radish, cauliflower, cabbage), cucurbits (cucumber, squash, melon), beans, and peas. In chili, symptoms begin as slight vein clearing on outer leaflets and drooping of leaf petioles. Later the lower leaves wilt, turn yellow and die and the entire plant may be killed often before the maturity of the plant. In many cases, a single shoot wilts before the rest of the plant shows symptoms or one side of the plant is affected first. If the main stem is cut, dark, chocolate-brown streaks may be seen running lengthwise through the stem. This discoloration extends upward for some distance and is especially evident at the point where the petiole joins the stem.

18.1.2.3.3 Conditions for Disease Development

The fungus survives in the soil for several years and is spread by farm equipment, irrigation water, and infected plant debris. Warm soil temperatures (33°C; 92°F) and high soil moisture generally favor rapid disease development.

Major Diseases of Chili and Their Management 357

18.1.2.3.4 Management

Plant on raised beds to help promote soil water drainage away from roots. Thoroughly disinfect equipment before moving from infested to clean fields.

18.1.2.4 POWDERY MILDEW

18.1.2.4.1 Causal Agent: Leveillulataurica

18.1.2.4.2 Symptoms

Powdery mildews are characterized by spots or patches of white to grayish powdery growth. White, powdery fungal growth develops on leaf surfaces, petioles, and stems. It usually develops first on crown leaves, on shaded lower leaves, and on leaf undersurfaces. Yellow spots may form on upper leaf surfaces opposite powdery mildew colonies. Older plants are affected first. Infected leaves usually wither and die. The whole stem of the infected plant is blackened in a severe attack. Severely infected plants may reduce yield, shortened production times, and fruit that has little flavor.

18.1.2.4.3 Conditions for Disease Development

Powdery mildew is caused by two different fungi, one is caused by *Leveillullataurica* (anamorph: *Oidiopsis taurica*) which grows endophytically and the other is caused by *Oidium lycopersici* which produce epiphytic conidiophore bearing mycelium that grows superficially on host surface powdery mildew fungi are host specific, obligate parasites and cannot survive in the absence of living host plants, except as cleistothecia. They survive the winter as dormant mycelium on perennial plants or as spores in thick-walled fruiting structures.

Powdery mildew develops quickly under favorable conditions because the length of time between infection and the appearance of symptoms is usually only 3–7 days. Powdery mildew fungi thrive under conditions of high RH, warm temperatures, low light, high fertility, and succulent plant growth. Free moisture on leaf surfaces inhibits infection by these pathogens. The minimum and maximum temperatures for conidial formation and host penetration are 10°C and 32°C, respectively, and the optimum being 26–28°C. Spores and fungal growth are sensitive to extreme heat and direct sunlight. Mature foliage is most readily infected; very young leaves are nearly immune.

18.1.2.4.4 Management

1. Adopt field sanitation practices like cleaning and burning of diseased crop debris.
2. Use cultural practices that avoid excessive succulence, overcrowding, shading, overwatering, or excess fertilization especially with nitrogen.
3. Spray dinocap @ 0.05% or wettable sulfur @0.2% or calixin @ 0.1%. Sulfur dust may be applied @ 20–25 Kg/ha but not when temperature is above 30°C.
4. Plant resistant/ tolerant cultivars.

18.1.2.5 CERCOSPORA LEAF SPOT (FROGEYE)

18.1.2.5.1 Causal Agent: Cercospora Capsici, C. melongenae

18.1.2.5.2 Symptoms

This disease affects the leaves, petioles, stems, and peduncles of pepper and eggplant. Symptoms first appear as small, circular to oblong chlorotic lesions. Lesions later turn necrotic with a sporulating light-gray center and a dark-brown margin. Concentric rings may be observed as individual lesions expand. These lesions often resemble frog eyes, giving this disease its common name. As the lesions dry, the center crack and drop out. When the disease is severe, defoliation, and reduction in fruit size occur.

18.1.2.5.3 Conditions for Disease Development

These fungi can survive for at least one year in infected plant debris. Wet, warm weather conditions favor disease development. Spores are spread by wind, rain, and irrigation water or mechanically by equipment and people.

18.1.2.5.4 Management

A calendar-based protectant fungicide spray program combined with cultural practices can help reduce losses from Cercospora Leaf Spot. Turn under or remove all plant debris and rotate to non-host crops to lower field inoculum levels. Mulch and furrow or drip irrigates to help reduce spread of the pathogen from splashing water.

Major Diseases of Chili and Their Management 359

18.1.2.6 ANTHRACNOSE FRUIT ROT (RIPE FRUIT ROT)

18.1.2.6.1 Causal Agent: Colletotrichum Capsici, C. Coccodes, C. Gloeosporioides, C. Acutatum

18.1.2.6.2 Symptoms

These fungi may infect the epidermis of immature fruit and remain latent until harvest. Symptoms usually develop on ripe fruit, giving this disease its common name: "Ripe Fruit Rot." Fruit lesions first appear as small, indefinite, slightly sunken, water-soaked spots that may enlarge rapidly and coalesce. Later, fruiting bodies form in concentric circles to cover the surface of lesions. The lesions appear tan or brown and are covered with salmon to orange gelatinous spores. If the fruit rot extends to the seed cavity, it may infest and infect the seed.

18.1.2.7 FUNGAL FRUIT ROTS: PHYTOPHTHORA CAPSICI, P. NICOTIANAE VAR. PARASITICA

18.1.2.7.1 Symptoms

Phytophthora rot occurs when fruit are in contact with the soil or mycelia grow through the peduncle into the fruit. Infected fruit tissue is water-soaked and dark-green at first; later, white mycelium and sporangia develop on the surface of the affected area and, within several days, consume the entire fruit. In contrast to infected chili fruit, no concentric rings develop. Fruit affected by these fungi dry rapidly and shrivel, but do not drop.

18.1.2.8 RHIZOPUS ROT: RHIZOPUS STOLONIFER

18.1.2.8.1 Symptoms

Contamination and wounding of fruit during the packing process is the primary means of infection. Symptoms first appear as soft, water-soaked lesions that are not discolored. Lesions develop from wounds, the stem end or inner walls, and quickly enlarge to engulf the entire fruit. When the epidermis ruptures, liquefied tissue is released. Under high humidity, profuse, coarse mycelia cover lesions. Later, white sporangia develop that turn black as they mature, giving a peppery, speckled appearance to the mycelia. In storage, these fungi penetrate directly from nests of infected fruit into adjacent healthy fruit.

18.1.2.8.2 Conditions for Disease Development

Rain splashes overwintering spores from soil and crop debris onto developing fruit. Symptom development is generally favored by high humidity. Botrytis fruit rot occurs during periods of cool, wet weather. The remaining four fruit rots occur during warm, wet weather.

18.1.2.8.3 Management

Fruit injury during harvest and packing should be avoided. Improved sanitation in the field and in the packing house is effective at helping to reduce losses due to fruit rots. All harvest equipment, the packing line and packing boxes should be sanitized daily. Dump tank water and packing-line washers should maintain a minimum available chlorine concentration of 150 ppm at pH 6.0–7.5. Culling infected and injured fruit during packing help reduce losses due to post-harvest decays. Wet surfaces should be dried promptly before packing, and fruit should be cooled quickly to 10°C (50°F).

18.1.2.9 CHOANEPHORA BLIGHT (WET ROT): CHOANEPHORA CUCURBITARUM

18.1.2.9.1 Symptoms

Symptoms are visible on apical growing points, flowers, and fruits. Initially, water-soaked areas develop on leaves, and apical growing points become blighted. Later, the fungus grows rapidly downward, causing dieback. Darkgray fungal growth can be seen on some lesions. Close inspection will reveal silvery, spine-like fungal structures and dark spores. In seedlings, symptoms may be confused with phytophthora blight. A black soft rot can also develop in fruit.

Major Diseases of Chili and Their Management 361

18.1.2.9.2 Conditions for Disease Development

The fungus is found throughout the tropics on many crops including beans, peas, squash, cucumber, eggplant, and pepper. Extended periods of rain, high humidity and high temperatures generally favor fungal sporulation and disease development. The fungus is generally spread via wind and splashing water, and on clothing, tools, and cultivation equipment.

18.1.2.9.3 Management

There are few management techniques available; fungicide sprays may help reduce disease damage.

18.1.2.10 GRAY LEAF SPOT: STEMPHYLIUM SOLANI, S. LYCOPERSICI DISTRIBUTION

18.1.2.10.1 Symptoms

Small spots develop on pepper leaves, petioles, stems, peduncles, and calyxes. Although mature plants can be infected, young seedlings are most susceptible. Infection begins as small red to brown spots that later expand into lesions with white to gray centers and red to brown margins. When numerous lesions develop, leaves turn yellow and drop. Gray leaf spot does not affect fruit.

18.1.2.10.2 Conditions for Disease Development

These fungi survive in soil and on plant debris from one year to the next. In addition, volunteer pepper and tomato and solanaceous weeds can serve as sources of inoculum. Fungal spores are spread from the surface of infected tissues by wind and splashing water. Warm and humid or wet weather generally favor disease development. The disease also can be a problem in arid climates when dew periods are long.

18.1.2.10.3 Management

Remove plant debris, provide adequate ventilation for seedling beds and treat with fungicides to help reduce losses from this disease.

18.1.2.11 LEAF SPOTS

Causal Agent: *Alternaria spp., Septoria melongenae, Cercospora spp.*

18.1.2.11.1 Symptoms

Septoria leaf spot is a disease of the foliage and stems. Although the symptoms may appear on the leaves and stems at any stage of plant development, they usually become evident after plants have begun to set fruit. Small circular spots are first observed as water-soaked areas on the undersurface of the lower leaves (Figure 18.1). As the spots enlarge, they develop dark brown margins and sunken, white grey centers (Figure 18.2). Yellow haloes often surround the spots. The number and size of the spots depends on susceptibility of the host. In susceptible cultivars the spots may be enlarged whereas in less susceptible cultivars there may be numerous spots of smaller size. The centers of the spots on the upper surface of the leaf show minute black fruiting bodies (Figure 18.3). Severely infected leaves die and drop off. The spots may also develop on floral calyx and on the stem but very rarely on the fruits. Severe leaf spotting and defoliation is common in severe infection. Favorable weather permits infection to move up the stem, causing a progressive loss of foliage from the bottom of the plant upward. Plants appear to wither from the bottom up. Loss of foliage causes a decrease in the size of the fruits and exposes fruit to sunscald. Spotting of the stem and blossoms may also occur (Figures 18.4 and 18.5).

18.1.2.11.2 Cause and Condition for Development

Septoria lycopersici is responsible for causing leaf spot in chili. The fungus lives in the soil on infested debris of chili and weeds. Spores formed on crop debris splash onto foliage and start the disease. Lower, weaker leaves catch infection from the soil. Wind and rain spread spores produced in the dark bodies formed on leaf spots to adjacent uninfected leaves. Seeds and transplants may also carry the fungus.

Free water or dew on the leaf surface is essential for infection. The disease is favored by moderate temperatures and abundant rainfall. Two to three day's cloudy weather or drizzle provides favorable conditions. Optimum temperature for growth of the fungus and disease development is 25°C with minimum of 15 and maximum 27°C.

Major Diseases of Chili and Their Management 363

FIGURE 18.1   Leaf spot of chili.

FIGURE 18.2   Wilt of chili.

FIGURE 18.3   Anthracnose rot of chili.

FIGURE 18.4 Die back of chili.

FIGURE 18.5 Fruit rot of chili.

18.1.2.11.3 *Management*

1. Seed has been implicated as a source therefore use of disease free seed is necessary.
2. Rotate chili with cereals, corn, or legumes for four years where disease has been severe.
3. Apply proper sanitation practices like deep plowing, to bury all plant refuse and control weeds. Fallen leaves should be removed from the field or burnt on the spot.
4. Treat the seeds in hot water for 25 minutes at 50°C or thiram @ 2.5 g/kg seed.

Spray mancozeb (0.25%) or chlorothalonil (0.2%) 3–4 sprays at 10–15 days interval.

18.1.2.12 PHOMOPSIS BLIGHT

Causal Agent: *Phomopsis vexans.*

18.1.2.12.1 Symptoms

This fungus attacks seedlings soon after emergence. Dark-brown lesions develop on the stem above the soil line. Eventually, a dry rot or canker girdles the stem and the seedling collapses and dies. When older plants are infected, circular or irregular gray to brown lesions develop on lower leaves and stems. Lesions enlarge and coalesce, causing complete yellowing of foliage and severe defoliation. Cankers on stems can cause wilting and death of the upper plant. Fruit lesions start as soft, light-brown, sunken oval areas. Later, fruit lesions deepen, enlarge, and coalesce to develop a soft, spongy rot. In dry weather, fruit may shrink and mummify. A diagnostic characteristic is the minute black fruiting bodies (pycnidia) that develop in a circular pattern in the center of mature lesions. Pycnidia are the inoculum source for later infections.

18.1.2.12.2 Conditions for Disease Development

This fungus can survive in plant debris or in mummified fruit in the soil. Infection may occur when rain or overhead irrigation splash inoculum to foliage and stems. Seed produced on plants grown in affected fields can be infested with fungal spores and may initiate disease on seedlings. Phomopsis blight is generally favored by hot, wet weather.

18.1.2.12.3 Management

Sow high-quality seed to help produce pathogen-free transplants. Remove and destroy all infected plant material, and establish a crop rotation to break the disease cycle. Mulch and furrow irrigate to help reduce splashing of water and soil. A regular schedule of protectant fungicide sprays may reduce damage in areas where the disease is known to occur.

366   *Diseases of Fruits and Vegetable Crops*

18.1.3   BACTERIAL DISEASES

18.1.3.1   BACTERIAL CANKER

Causal Agent: *Clavibacter Michiganensis* sub sp. *Michiganensis.*

18.1.3.1.1   Symptoms

Infected seedlings may be killed, stunted or malformed, or may show no signs until they are transplanted. In larger plants, initial signs are scorched or 'firing' markings on leaflets and wilting of lower leaves. Symptoms on leaves may be unilateral, the leaflets on one side of the rachis being wilted while those on the opposite side appear healthy As infection advances through the vascular system, wilting progresses until the whole plant collapses. Parts of the pith may collapse and become hollow. Internal browning of vascular tissue and pith cavities can be seen by snapping off a leaf at a node (the point where the leaf joins the stem). In wet conditions, brown, raised cankers may form on the stems and fruit. The fruit cankers have pale halos and are called 'bird's-eye' spots.

18.1.3.1.2   Conditions for Disease Development

Bacterial canker is caused by *Clavibacter michiganensis* subsp. *michiganensis.* The bacteria can live from season to season in or on the seed, diseased plant debris in soil, Infected seed or seedlings are the primary source of this pathogen. Within the field dissemination is from open cankers mainly through raindrop splashes. Seeds are infected through systemic invasion of fruits.

18.1.3.1.3   Management

1.   2–3-year crop rotation with nonhost crop should be followed
2.   Soil solarization for 6weeks in summer eliminates the bacterial population in soil
3.   Use healthy and disease-free seed and transplants
4.   Seed should be disinfected by submersion in water at 52°C for 20 minutes and with copper compounds. Treatment of chili plants with beta-amino butyric acid (BABA) offers protection against bacterial canker.

Major Diseases of Chili and Their Management 367

18.1.3.2 BACTERIAL STEM AND PEDUNCLE CANKER (SOFT ROT)

Causal Agent: *Pectobacteriumcarotovora, P. atrosepticum, Dickeyachrysanthemi.*

18.1.3.2.1 Symptoms

This disease affects pepper stems and fruit. Internal discoloration appears on the stem, followed by hollowing-out of the pith and wilting. As lesions expand along the stem, branches break. Foliar chlorosis and necrosis may also develop. Symptoms of post-harvest decay start as sunken, water-soaked areas around the edge of wounds or on the stem end next to the peduncle. These areas may be light or dark and become soft as they rapidly expand. Often, the epidermis splits open, releasing watery, macerated tissue.

18.1.3.2.2 Conditions for Disease Development

Soft rot bacteria are common inhabitants of soils. Under warm, humid conditions, infection through wounds or cut stems occurs. Splashing rain and irrigation water spread bacteria to foliage and fruit.

18.1.3.2.3 Management

In greenhouse operations, provide adequate air circulation to help reduce RH. Avoid injuries to plants during the growing season and on fruit during harvest. Improved sanitation in the field and in packing houses is effective in reducing losses. All harvest equipment, the packing line and packing boxes should be sanitized frequently. Dump tank water and packing line washers should maintain a minimum available chlorine concentration of 150 ppm at a pH of 6.0 to 7.5. Wet fruit should be dried promptly before packing and then cooled quickly to below 10°C (50°F).

18.1.3.3 SYRINGAE SEEDLING BLIGHT AND LEAF SPOT

Causal Agent: *Pseudomonas syringae* pv. *Syringae.*

18.1.3.3.1 Symptoms

Affected leaves or cotyledons develop irregular, water-soaked lesions that later become necrotic, turning dark-brown with a light center. Lesions may coalesce to form relatively large necrotic areas. Lesions with chlorotic halos are rare. However, under heavy disease pressure, large areas of the leaf may be affected and the whole leaf may prematurely turn yellow and drop. Infected fruit develop brownish-black, watery lesions that expand and rot. Symptoms of syringae seedling blight can be confused with those caused by Bacterial Spot. However, the lower temperatures at which Syringae Seedling Blight occurs can help differentiate these two diseases.

18.1.3.3.2 Conditions for Disease Development

Temperatures between 16°C and 24°C (60° and 75°F) and high humidity favor syringae seedling blight. Bacteria are generally spread by splashing water that enters the plant through natural openings or wounds.

18.1.3.3.3 Management

Avoid low temperature and high humidity conditions in nurseries. Inspect seedlings for symptoms before transplanting to avoid introducing the disease to the field. Avoid overhead irrigation whenever possible.

18.1.3.4 BACTERIAL SPOT

Causal Agent: *Xanthomonas euvesicatoria (synonym: Xanthomonas campestris pv. vesicatoria) X. vesicatoria, X. gardneri.*

18.1.3.4.1 Symptoms

On leaves, small, irregular areas with a grassy appearance develop first. These areas dry out and form slightly raised dry spots that are grayish brown, particularly at the center. The bacteria often ooze from these spots and, when dry, form a glistening, cream-colored film on and around the lesions. Where infection is severe, the dried-out parts may combine and kill large areas of leaf. Marginal and tip burns on leaves have often been noted. On flowers,

Major Diseases of Chili and Their Management 369

water-soaked dark-brown-to-black areas develop, which later dry out and turn grey. Flower infection often causes blossoms and young fruit to wither and fall. Main stems are occasionally attacked. Spots are often elongated but retain the same grayish, scab-like characteristics as other affected parts. Small, water-soaked areas develop on green fruit. These dry out and form slightly raised and wrinkled, brown-to-grey, scab-like bodies, making the fruit unmarketable and susceptible to secondary rots.

18.1.3.4.2 Causes and Conditions for Disease Development

Xanthomonas campestris pv. *vesicatoria,* causes bacterial spot in chili, the bacterium is strictly aerobic, gram negative, rod shaped and posses a single polar flagellum. It is seed borne and overwinters on surface of seeds, in infected debris, in reservoir hosts and in soil. Disease is favored by warm (24–30°C) humid conditions.

The bacterium is disseminated within the field by water droplets, clipping of transplants and aerosols; it penetrates through stomata, or enters through insect puncture or mechanical injury. The bacteria, carried in water droplets, can form new spots on the leaves, stems or fruit where the droplet comes to rest. In wet weather, particularly if strong winds are blowing, the disease may spread rapidly through a crop from a few affected plants. Once the bacterium is established in the soil it may persist for 2 or 3 years.

18.1.3.4.3 Management

1. Use disease-free seed or transplants seed that has been hot water treated.
2. At least 3-year crop rotation with non-host crops should be followed.
3. Spray plants with streptomycin before transplanting. After transplanting, apply a mixture of mancozeb plus copper before the occurrence of disease. Protection is most needed during early flowering and fruit setting periods.

18.1.3.5 BACTERIAL WILT

Causal Agent: *Ralstonia solanacearum* (synonym = *Burkholderia solanacearum, Pseudomonas solanacearum).*

18.1.3.5.1 Symptoms

Characteristic symptoms of bacterial wilt are the rapid and complete wilting of normal grown up plants. Lower leaves may drop before wilting. The disease starts with stunting, yellowing of the foliage, wilting, and finally the entire plant collapse. Pathogen is mostly confined to vascular region; in advance cases, it may invade the cortex and pith and cause yellow-brown discoloration of tissues. Vascular bundles of the stem seen to be brown and this browning is often visible from the surface of the stems of infected plants as dark patches or streaks. Infected plant parts when cut and immerse in clear water, a white streak of bacterial ooze is seen coming out from cut ends. Development of adventitious roots from the stem is enhanced. In humid weather and high-temperature sudden drooping of the leaves without yellowing and rotting of the stem occurs in chili. The roots appear healthy and are well-developed however, brown discoloration is present inside.

18.1.3.5.2 Conditions for Disease Development

Bacterial brown rot or wilt in most of the vegetables is caused by *Ralstonia solanacearum* (formerly *Pseudomonas solanacearum* and more recently, *Burkholderia solanacearum*). *R. Solanacearum* can survive through soil, seed tubers and cultivated hosts. Soil is considered most important source of primary inoculum. Optimum temperature for growth of the bacterium is 35°C–37°C, with maximum of 41 and minimum of 10°C. Soil moisture more than 50% favors the disease development. Organic manures promote activity of the bacterium while inorganic fertilizers decrease its activity.

18.1.3.5.3 Management

1. Follow crop rotations for three years with wheat, maize, oat, barley, sun hemp, finger millet and vegetables like cabbage, onion, and garlic.
2. The field should be plowed to expose the soil to summer heat in month of May to June.
3. Uproot the diseased plants as soon as the disease appears.
4. Treat the seedlings of chili with 0.02% streptocycline for 30 minutes.
5. Drench the soil with streptocycline @ 0.01% or copper oxychloride @ 0.3% at early stage of plant showing symptoms.

Major Diseases of Chili and Their Management 371

18.1.4 VIRAL DISEASES

18.1.4.1 CHILI VEINAL MOTTLE

Causal Agent: Chiliveinal mottle virus (ChiVMV), Vector: aphids.

18.1.4.1.1 Symptoms

Typically, leaves of infected plants develop a mottle or mosaic with dark-green vein banding. Plants infected when young usually are stunted, with dark-green streaks on stems. Most flowers drop before fruit set. Affected fruit may be mottled and deformed. Symptom severity is dependent upon the variety, infecting strain and age of the host at the time of infection. Eggplant can be infected, but remains asymptomatic.

18.1.4.1.2 Conditions for Disease Development

Peppers, tobacco, tomatoes, and weeds such as *Physalis* spp. are hosts of ChiVMV. Tropical climates support the continuous presence of ChiVMV and its vectors. ChiVMV is transmitted by several species of aphids in a non-persistent manner and can also be transmitted mechanically through pruning and grafting. There is no evidence of seed transmission.

18.1.4.1.2 Management

Use resistant varieties and virus-free transplants. Manage aphid populations using reflective mulches, stylet oil sprays and insecticides. Controlling the aphid vector population with chemical treatment is very difficult and generally provides limited control. In mature plants, it is difficult to achieve complete insecticide coverage of leaves to effectively eradicate all aphids.

18.1.4.2 CUCUMBER MOSAIC

Cucumber mosaic virus (CMV), Vector: Many species of aphids.

18.1.4.2.1 Symptoms

Symptoms can vary greatly depending upon the affected variety, age of the plant at the time of infection, and the strain of the virus. Leaves may become narrow, distorted, and mottled. In peppers, defoliation may occur when mature plants are infected. Tip dieback and leaf discoloration in an "oak-leaf" pattern may develop. Infected plants are usually stunted, and the fruits distorted with occasional concentric rings. The infection of young plants results in unmarketable fruit and severe yield losses.

18.1.4.2.2 Conditions for Disease Development

The host range of CMV includes as many as 800plant species. The virus is acquired by aphids and is transmitted from plant to plant in a non-persistent manner. CMV often remains in infected alternate hosts near agricultural areas and is transmitted to peppers and eggplant when environmental conditions support disease development. This virus is also mechanically transmitted.

18.1.4.2.3 Management

Control aphids, and rogue-infected plants to reduce the incidence of CMV in greenhouse crops. In field-grown peppers and eggplant, eliminate adjacent weeds and ornamentals, and use reflective mulches to deter aphids and a combination of stylet oil and insecticide sprays to reduce losses caused by this virus. Controlling the aphid vector population with chemical treatment is very difficult and generally provides limited control. In mature plants, it is difficult to achieve complete insecticide coverage of leaves to effectively eradicate all aphids. Resistant varieties are available in pepper.

18.1.4.3 BEET CURLY TOP

Beet curly top virus (BCTV), Vector: The beet leafhopper (*Circulifer tenellus*).

18.1.4.3.1 Symptoms

When seedlings are infected, leaves turn yellow, twist, and curl upward, and thicken to become stiff and crisp. Petioles may curl downward. The fruit set

Major Diseases of Chili and Their Management 373

is reduced. The fruit appears dull and wrinkled, and tend to ripen prematurely. This virus is not mechanically transmitted.

18.1.4.3.2 Conditions for Disease Development

This virus has a wide host range, affecting more than 300 species. Common hosts are tomatoes, beets, peppers, squash, beans, cucurbits, spinach, potatoes, cabbage, and alfalfa. The beet leafhopper transmits BCTV in a persistent manner. Warm temperatures and dense leafhopper populations are conducive to the spread of BCTV. Viruliferous leafhoppers migrate seasonally and can be moved long distances by wind.

18.1.4.3.3 Management

Transplant virus-free seedlings. Rogue infected plants to help avoid transmission in the field. Control weeds near pepper fields to reduce vector and virus reservoirs. Transplant early or late to escape leafhopper infestations, and increase plant density to compensate for losses due to BCTV. Insecticides are generally not effective in controlling beet curly top.

18.1.4.4 PEPPER YELLOW MOSAIC

Causal agent: Pepper yellow mosaic virus (PepYMV), Vector: Many species of aphids.

18.1.4.4.1 Symptoms

Typical symptoms of pepper yellow mosaic include vein banding, blistering, and a bright-yellow mosaic. Leaves also may be distorted and develop epinasty. Generally, plants are stunted. Fruit develops a mosaic and may be distorted.

18.1.4.4.2 Conditions for Disease Development

Aphids transmit this virus in a non-persistent manner. The widespread use of potyvirus Y resistant cultivars in commercial vegetable production in

374 *Diseases of Fruits and Vegetable Crops*

the 1970s may have contributed to the emergence of PepYMV as a serious pepper pathogen in Brazil. Pepper yellow mosaic also affects chili.

18.1.4.4.3 Management

Plant PepYMV-resistant pepper varieties. The use of insecticides to control the disease is generally not efficient due to the short times for acquisition and transmission of this virus by aphids. Cultural practices that may delay infection include the use of reflective mulches to deter aphids and weed control to help remove viruses and vector reservoirs.

18.1.4.5 PEPPER MOTTLE

Causal Agent: Pepper mottle virus (PepMoV), Vector: Many species of aphids.

18.1.4.5.1 Symptoms

Symptoms vary depending on the pepper variety infected, isolate of the virus present, age of plants when infected, and environmental conditions. Affected field plants may develop a systemic mottle, distortion, and may be stunted. Infected greenhouse plants first develop vein-clearing, followed by chlorotic mottle in newer leaves. Fruit may be distorted, mottled, and small. Field plants often are infected with more than one virus. Multiple infections usually result in symptoms different from those caused by PepMoV alone and may appear more severe.

18.1.4.5.2 Conditions for Disease Development

Like TEV, PVY, and CMV, PepMoV can be transmitted by many species of aphids. PepMoV is transmitted on the mouthparts of aphids in a non-persistent manner. Weeds, such as *Datura,* and other solanaceous plants can harbor the virus and vector. Aphids may spread infection within a crop, or the virus may be transmitted mechanically through cultural practices such as staking, pruning, or handling of infected plants.

Major Diseases of Chili and Their Management 375

18.1.4.5.3 Management

Remove crop residues and weeds that serve as a reservoir for both virus and vector. Use reflective mulches to deter aphids, and combine the use of stylet oils and insecticide sprays to reduce losses in young plants. Controlling the aphid vector population with chemical treatment is very difficult and generally provides limited control. In mature plants, it is difficult to achieve complete insecticide coverage of leaves to effectively eradicate all aphids. Resistant varieties are available in both hot and sweet peppers. However, resistance in available commercial varieties may not be effective against all isolates of PepMoV found.

18.1.4.6 TOBAMOVIRUSES

Causal Agent: Tobacco mosaic virus (TMV), Tomato mosaic virus (ToMV), and Pepper mild mottle virus (PMMV). Vector: Mechanically transmitted with no known insect vectors.

18.1.4.6.1 Symptoms

Symptoms of infection by TMV and ToMV in peppers and eggplant can vary greatly with the strain of virus, temperature, light intensity, day length, age of the plant when infected, and cultivar. Foliar symptoms include chlorotic mosaic, distortion, and, at times, systemic necrosis and defoliation. Plants infected as seedlings can be stunted and are generally chlorotic. Infected plants produce disfigured fruit that are usually small with distinct chlorotic and/or necrotic areas. Foliar symptoms of PMMV in peppers are also variable but are generally mild. Plants infected as seedlings remain stunted. Leaves develop a subtle mosaic, can be crinkled, and remain small. Symptoms may first appear on fruit. Fruit can be mottled and necrotic, are usually small and distorted, and have a rough or wrinkled appearance.

18.1.4.6.2 Conditions for Disease Development

TMV and ToMV have very wide host ranges and infect over 200 plant species, including varieties of peppers, tomatoes, eggplant, and tobacco. PMMV can infect all species of peppers, but not tomatoes, tobacco, or eggplant. Infected transplants, seed, and debris are common sources of inoculum. These viruses can be found on and under the seed coat and in the endosperm. Tobamoviruses are very stable, and extensive spread can occur through handling, tools, trays, pots, stakes, twine, and clothing, as well as pollination, pruning, and other cultural practices. Tobamoviruses can remain viable for several years in plant debris, but generally, lose their ability to infect as debris decomposes.

18.1.4.6.3 Management

Enforce strict sanitation practices during production and harvest to minimize infection and prevent spread. Restrict access to the crop, wash hands and equipment with a soap solution between plants or rows of plant sand before entering a greenhouse. There are reports of successful prevention of bamovirus spread by coating hands, plants, and equipment with a solution of powdered non-fat milk. Rogue symptomatic and adjacent plants, and rotate to non-solanaceous crops to manage the disease. Use seed tested and treated for bamoviruses. Many hot pepper varieties contain hypersensitive resistance to TMV and ToMV. Use resistant varieties in greenhouse production where to bamoviruses are a problem. Some strains of PMMV may overcome the commercial resistance.

KEYWORDS

- beet curly top virus
- beta-amino butyric acid
- cucumber mosaic virus
- pepper mild mottle virus
- tobacco mosaic virus
- tomato mosaic virus

REFERENCES

Agrios, G. N., (2005). *Plant Pathology* (p. 902). Academic Press, San Diego, CA.
Ahmed, S. S., (1982). Studies on seed borne aspects of anthracnose of chilies caused by *Colletotrichum capsici* (Sydow.) Butler and Bisby. *M.Sc. (Agri.) Thesis, Univ. Agric. Sci.,* Bangalore.

Major Diseases of Chili and Their Management 377

Alam, S., Banu, M. S., Ali, M. F., Akhter, N., Islam, M. R., & Alam, M. S., (2002). In vitro inhibition of conidial germination of *Colletotrichum gloeosporioides* Penz. by fungicides, plant extracts and phytohormons. *Pakistan J. Biological Sciences, 5*, 303–306.

Anonymous, (1991). *Efficacy Test, Protocols* (p. 5). Anthracnose of Capsicum sp. FAO/APIO21.

Anonymous, (2011). *India Horticultural Database–2011*. National Horticulture Board, Ministry of Agriculture, Govt. of India, Gurgaon.

Anonymous, (2015). *National Horticulture Board Database* (pp. 177–185). NHB Publication. New Delhi.

Asalmol, M. N., Kale, V. P., & Ingle, S. T., (2001). Seed borne fungi of chili, incidence and effect on seed germination. *Seed Res., 29*(1), 76–79.

Bagri, R. K., Choudhary, S. L., & Rai, P. K., (2004). Management of fruit rot of chili with different plant products. *Indian Phytopathology, 57*(1), 107–109.

Bailey, J. A., & Jeger, M. J., (1992). *Colletotrichum: Biology, Pathology and Control* (p. 388). Wallingford: Commonwealth Mycological Institute.

Khodke, S. W., & Gahukar, K. B., (1995). Fruit rot of chili caused by *Colletotrichum gloeosporioides* Penz. In Amravati district. *P. K. V. Research J., 19*(1), 98–99.

Manandhar, J. B., Hartman, G. L., & Wang, T. C., (1995). *Plant Dis., 79*, 380–383.

Mesta, R. K., (1996). Studies on fruit rot of chili caused by *Colletotrichum capsici* (Sydow.) Butler and Bisby. *M.Sc. (Agri.) Thesis*, Univ. Agric. Sci., Dharwad.

Mishra, D., (1988). Fungicidal control of anthracnose and fruit root (*Colletotrichum capsici*) of chili (*Capsicum annuum*). *Indian J. Agricultural Science, 58*(2), 147–149.

Mridha, M. A. U., & Choudhary, M. A. H., (1990). Efficacy of some selected fungicides against seed borne infection of chili fruit rot fungi. *Seed Research, 18*(1), 98–99.

Pakdeevaraporn, P., Wasee, S., Taylor, P. W. J., & Mongkolporn, O., (2005a). *Plant Breeding, 124*(2), 206–208.

Pakdeevaraporn, P., Wasee, S., Taylor, P. W. J., & Mongkolporn, O. M., (2005b). Inheritance of resistance to anthracnose caused by *Colletotrichum capsici* in Capsicum. *Plant Breeding, 124*, 206–208.

Panagopoulos, C. G., (2000). Diseases of vegetable crops. In: *Vegetable Disease* (pp. 15–189). Stamoulis, Athens.

Persley, D., Cooke, T., & House, S., (2010). *Diseases of Vegetable Crops in Australia*. CSIRO Publishing: Collingwood, Victoria.

Prabhavathy, K. G., & Reddy, S. R., (1995). Post-harvest fungal disease of chili (*C. annuum*) from Andhra Pradesh. *Indian Phytopathology, 48*(4), 492.

Sherf, A. F., & MacNab, A. A., (1986). *Vegetable Diseases and Their Control* (2nd edn., p. 728). John Wiley & Sons, NY.

Singh, R. S., (2005). *Plant Diseases* (p. 720). Oxford and IBH Publishing Co. Pvt Ltd., New Delhi.

Than, P. P., Jeewon, R., Hyde, K. D., Pongsupasamit, S., Mongkolporn, O., & Taylor, P. W. J., (2008). *Plant Pathology, 57*(3), 562–572.

Thind, T. S., & Jhooty, J. S., (1985). Relative prevalence of fungal diseases of chili fruits in Punjab. *Indian J. Mycol. Pl. Path., 15*, 305–307.

Ushakiran, L., Chetry, G. K. N., & Singh, N. I., (2006). Fruit rot disease of chili and their management in agro-climatic conditions of Manipur. *J. Mycopathol. Res., 44*(2), 257–262.

Whitelaw-Weckert, M. A., Curtin, S. J., Huang Steel, R. C. C., Blanchard, C. L., & Rxoffey, P. E., (2007). *Plant Pathology, 56*(3), 448–463.

CHAPTER 19

Diseases of Cucurbits and Their Management: Integrated Approaches

GIREESH CHAND

Department of Plant Pathology, Bihar Agricultural University, Sabour, Bhagalpur, India, E-mail: gireesh_76@rediffmail.com

19.1 INTRODUCTION

The cucurbit crops suffer from a number of fungal, bacterial, viral, nematodes, and nutrients deficiency diseases. Among these fungal diseases (powdery mildew, downy mildew, anthracnose fruit rot, damping off and fusarial wilt), bacterial diseases (bacterial wilt, bacterial fruit rot and angular leaf spot), viral diseases (cucumber mosaic virus (CMV) and pumpkin mosaic virus), nematodes diseases (root knot of cucurbits and Reniform nematode) and nutritional disorders are major to reduce production of cucurbits. Some of the known diseases and their causes are given in Table 19.1.

TABLE 19.1 Disease of Cucurbits and Their Causal Organisms

S. No.	Name of the Disease	Causal Organisms
A.	**Fungal Diseases**	
1.	Anthracnose	*Colletotrichum orbiculare, C. lagenarium*
2.	Powdery mildew	*Erysiphe cichoracearum, Sphaerotheca fuliginea*
3.	Downy mildew	*Pseudoperonospora cubensis*
4.	Septorialeaf spot	*Septoria cucurbitacearum*
5.	Alternaria blight	*Alternaria cucumerina, Alternaria alternate*
6.	Fusarial wilt	*Fusarium oxysporum f. sp. Melonis*
7.	Damping off	*Pythium aphanidermatum, P. debaryanum, P. butleri, Phytophthora, Fusariumspp.*
8.	Fruit rot or Cottony leak	*Pythium aphanidermatum, P. debaryanum, P. butleri, Phytophthora spp., Fusarium spp., Rhizoctonia spp.*

TABLE 19.1 *(Continued)*

S. No.	Name of the Disease	Causal Organisms
9.	Cercospora leaf spot	*Cercosporacitrullina, C. trichosanthes, C. memordicae, C. lagenariae, C. cucurbiticola, Corynespora melonis*
10.	Fusarial root rot	*Fusarium solanif. sp. cucurbitae*
11.	Gummy stem blight	*Didymella bryoniae*
12.	Target spot	*Corynespora cassiicola*
13.	Charcoal rot	*Macrophomin aphaseolina*
14.	Blossom end rot	*Choanephora cucurbitarum*
15.	Verticillium wilt	*Verticillium alboatrum, V. dahlia*
B.	**Bacterial Diseases**	
16.	Bacterial wilt	*Erwinia traccheiphila*
17.	Bacterial fruit blotch	
18.	Angular leaf spot	*Pseudomonas syringaepv. Lachrymans*
19.	Bacterial leaf spot	*Xanthomonas campestris pv. Cucurbitae*
20.	Bacterial soft rot	*Erwinia carotovora*
C.	**Viral Diseases**	
21.	Cucumber mosaic virus	*Cucumber mosaic virus Cucumber green mottle mosaic virus, Water melon mosaic virus*
22.	Pumpkin mosaic virus	*Cucumo virus group*
23.	Pumpkin leaf curl	*Poty virus group*
D.	**Nematode Diseases**	
24.	Root knot of cucurbits	*Meloidogyne incognita, M. javanica*
25.	Reniform nematode	*Rotylenchulus reniformis*
E.	**Nutritional Disorders**	
26.	Yellowing, curling, deformation, blossom end and necrosis of leaves	Nitrogen, Potassium, Calcium, Iron, Manganese, Molybdenum, and Zinc deficiency

19.1.1 HOST RANGE

The cucurbit crops suffering from a number of fungal, bacterial, viral, nematodes, and nutrients diseases. Cucurbits are warm weather crops which are sown, grown, and harvested over spring, summer, and autumn. Sponge gourd, ridge gourd, muskmelon, watermelon, round melon, pumpkin, vegetable marrow cucumber, snake gourd, pointed gourd, kheera, bottlegourd, etc., reduces quality and quantity of the produce or crop yield. Some of the known pathogens and their host crops are given in Table 19.2.

Diseases of Cucurbits and Their Management

TABLE 19.2 Pathogens of Cucurbits and Their Host-Range

S. No.	Name of the Disease	Host Crops
A.	**Fungal Diseases**	
1.	*Colletotrichum orbiculare, C. lagenarium*	Sponge gourd, ridgegourd, muskmelon, watermelon, roundmelon, pumpkin, cucumber, snakegourd, kheera, bottlegourd, and bittergourd etc.
2.	*Erysiphe cichoracearum, Sphaerothecafuliginea*	Sponge gourd, ridge gourd, muskmelon, watermelon, round melon, pumpkin, vegetable marrow cucumber, snake gourd, pointed gourd, kheera, bottlegourd, bitter gourd, potato, tobacco, lettuce, sunflower, mango, castor, etc.
3.	*Pseudoperonosporacubensis*	Sponge gourd, ridge gourd, muskmelon, watermelon, round melon, pumpkin, vegetable marrow cucumber, snake gourd, pointed gourd, kheera, bottlegourd, bitter gourd, etc.
4.	Gummy stem blight	
5.	*Alternaria cucumerina*	Sponge gourd, ridge gourd, muskmelon, watermelon, round melon, cucumber, snake gourd, kheera, bottlegourd, etc.
6.	*Fusarium oxysporum f. sp. niveum, melonis*	Sponge gourd, ridge gourd, muskmelon, watermelon, round melon, pumpkin, vegetable Marrow cucumber, snake gourd, pointed gourd, kheera, bottlegourd, bitter gourd, etc.
7.	*Pythium aphanidermatum, P. debaryanum, P. butleri, Phytophthoraspp.*	Sponge gourd, ridge gourd, muskmelon, watermelon, round melon, pumpkin, vegetable marrow cucumber, snake gourd, kheera, bottlegourd, bitter gourd, etc.
8.	*Pythium aphanidermatum, P. debaryanum, P. butleri, Phytophthora, Fusarium spp., Rhizoctonia spp.*	Sponge gourd, snake gourd, pointed gourd, kheera, bottle gourd, bitter gourd, etc.
9.	*Cercosporacitrullina, C. trichosanthes, C. emordicae, C. lagenariae, C. cucurbiticola, Corynespora melonis*	Sponge gourd, ridge gourd, muskmelon, watermelon, round melon, pumpkin, vegetable marrow cucumber, snake gourd, pointed gourd, kheera, bottlegourd, bitter gourd, etc.
10.	*Fusarium solani f. sp. cucurbitae*	Sponge gourd, ridge gourd, muskmelon, watermelon, round melon, pumpkin, vegetable marrow cucumber, snake gourd, pointed gourd, kheera, bottlegourd, bitter gourd, etc.
11.	*Didymella bryoniae*	Cucumber, snake gourd, pointed gourd, kheera, bottlegourd, bitter gourd, etc.
12.	*Corynespora cassiicola*	Sponge gourd, snake gourd, pointed gourd, kheera, bottle gourd, bitter gourd, etc.
13.	*Macrophomina phaseolina*	Cucumber, snake gourd, pointed gourd, kheera, bottlegourd, bitter gourd, etc.
14.	*Choanephora cucurbitarum*	Sponge gourd, snake gourd, pointed gourd, kheera, bottle gourd, bitter gourd, etc.

TABLE 19.2   *(Continued)*

S. No.	Name of the Disease	Host Crops
15.	*Verticillium alboatrum, V. dahliae*	Cucumber, snake gourd, pointed gourd, kheera, bottlegourd, bitter gourd, etc.
B.	**Bacterial Diseases**	
16.	*Erwinia traccheiphila*	Sponge gourd, ridge gourd, muskmelon, watermelon, round melon, pumpkin, vegetable marrow cucumber, snake gourd, pointed gourd, kheera, bottlegourd, bitter gourd, etc.
17.	Bacterial fruit blotch	—
18.	*Pseudomonas syringaepv. lachrymans*	Muskmelon, watermelon, pumpkin, cucumber, kheera, bottlegourd, etc.
19.	*Xanthomonas campestris pv. cucurbitae*	Muskmelon, watermelon, pumpkin, cucumber, kheera, bottlegourd, etc.
20.	*Erwinia carotovora*	Sponge gourd, ridge gourd, muskmelon, watermelon, round melon, pumpkin, vegetable marrow cucumber, snake gourd, kheera, bottlegourd, etc.
C.	**Viral Diseases**	
21.	*Cucumber mosaic virus Cucumber green mottle mosaic virus, watermelon mosaic virus*	Sponge gourd, ridge gourd, muskmelon, watermelon, round melon, pumpkin, vegetable marrow cucumber, snake gourd, pointed gourd, kheera, bottlegourd, bitter gourd, Potato, tobacco, lettuce, sunflower, mango, castor, etc.
22.	*Pumpkin mosaic virus*	Muskmelon, watermelon, pumpkin, cucumber, kheera, bottlegourd, etc.
23.	*Pumpkin leaf curl*	Muskmelon, watermelon, pumpkin, cucumber, kheera, bottlegourd, etc.
D.	**Nematode Diseases**	
24.	*Meloidogyne incognita, M. javanica*	Sponge gourd, ridge gourd, muskmelon, watermelon, round melon, pumpkin, vegetable marrow cucumber, snake gourd, pointed gourd, kheera, bottlegourd, bitter gourd, Potato, tobacco, lettuce, sunflower, mango, castor, etc.
25.	*Rotylenchulusreniformis*	Pumpkin, cucumber, brinjal, okra, bean, carrot, cauliflower, peas, radish, tomato, onion, etc.
E.	**Nutrient Deficiency**	
26.	Nitrogen, Potassium, Calcium, Iron, Manganese, Molybdenum, and Zinc deficiency	Muskmelon, watermelon, pumpkin, cucumber, kheera, bottlegourd, etc.

Diseases of Cucurbits and Their Management 383

19.1.2 DAMPING-OFF OF SEEDLINGS AND FUNGAL ROOT ROTS

19.1.2.1 DISEASE OCCURRENCE AND DISTRIBUTION-WORLDWIDE; COMMON OCCURRENCE IN NURSERY BEDS AND YOUNG SEEDLINGS

19.1.2.1.1 Diagnostic Symptoms

The first symptoms are water-soaked lesions occurring at soil level. This leads to wilting and seedling death. Often, plants that have survived damping-off might show symptoms of root rot. Roots can have a watery grey appearance, particularly the finer feeder roots. These fungi are common soil inhabitants that might also infect weeds or survive on decaying plant material. Can be spread by water and soil. Survival: Cool temperatures, high soil moisture and poor aeration. Root rots of older plants are also common in moist hot conditions, especially under plastic mulches.

19.1.2.2 ALTERNARIA LEAF SPOT

19.1.2.2.1 Disease Occurrence and Distribution

Several species of *Alternaria* can attack the foliage of cucurbits causing leaf spots and blight and fruits of musk melon usually causing a dry rot.

19.1.2.2.2 Symptoms

Small spots develop on the upper surface of leaves which could develop into larger coalescing lesions. These lesions might have concentric rings.

19.1.2.2.3 Disease Cycle

The fungi survive in soil on plant debris and seeds may also be the source of new infection. Favored by moist conditions, such as rain, fog, and heavy dews and therefore more common in districts of higher rainfall.

384
Diseases of Fruits and Vegetable Crops

19.1.2.3 ANTHRACNOSE

19.1.2.3.1 Disease Occurrence and Distribution

The first time reported in 1867 on gourds from Italy; occurs mainly in all humid regions of the world. It rarely affects pumpkin. In temperate regions heavy losses occur on fruits of watermelon.

19.1.2.3.2 Diagnostic Symptoms

Brown to black spots develops on leaves (Figure 19.4); long dark spots develop on stems and round sunken spots develop on fruit. Fruit symptoms might develop in transit.

19.1.2.3.3 Biology and Molecular Characterization

The anthracnose of cucurbits is caused by the fungus *Colletotrichum lagenarium* (Pass). Ellis and Halsted. The mycelium is septate, hyaline when young and dark when old. Stomata (acervuli) are brown to black and variable in size. Setae are brown, thick-walled, 2–3 septate, and 90–120 μm long. Conidia are produced one at a time at the tip of the conidiophores and accumulate in a slimy, pinkish mass. Individually the conidia are hyaline, oblong to ovate oblong, 1-celled, and measure 13–19 x 4–5 μ microns. They germinate which, on contact with a hard surface by a germ tube.

19.1.2.3.4 Disease Cycle

Infected crop trash and infected seeds are sources of fresh infection. The disease spreads by windblown rain, people, animals, and machinery moving through the crop in wet or moist conditions.

19.1.2.4 DOWNY MILDEW

19.1.2.4.1 Disease Occurrence and Distribution

First reported from Cuba in 1868; prevalent in the warm temperate and tropical regions like North America, Europe, and Asia. In India, it is present all over the country except in temperate zone in high altitude of the Himalayas.

Diseases of Cucurbits and Their Management

19.1.2.4.2 Diagnostic Symptoms

The initial green to yellowish and dark green areas like mosaic, later turn into yellow colored angular spots on upper surface of the leaves (Figure 19.5). Their number and size increase. Finally, the leaves become chlorotic, brown, and shrivel. Leaves have a mottled appearance. Leaf spots turn pale yellow in color, enlarge, and dry out. C These may be confused with the bacterial disease angular leaf spot. The 5–15 days old leaves are more susceptible. The whole vines are wilted. The infection directly affects fruits of musk melon; reduces number and size of fruits and prevents fruit maturation.

19.1.2.4.3 Biology and Molecular Characterization

Pseudoperonospora cubensis (Berk. and Curt.) Rostow. The upper third of the sporangiophore is branched either dichotomously or intermediately between dichotomous and monochotomous branching habit. The spore-bearing tips are sub-acute. The sporangia are grayish to olivaceous purple, ovoid to ellipsoidal, thin-walled, and with a papilla at the distal end. They measure 21–39x14–23 μm. The germination of sporangia occurs by production of biflagellate zoospores which are 10–13 μm in diameter when in resting state. Oospores are not common in the species. However, in India presence of oospores on certain cucurbits has been reported from Madhya Pradesh, Punjab, and Rajasthan. The oospores are spherical, rarely obovoid to ellipsoid, light yellow, and smooth walled. They measure 19–22 μ microns in diameter. The smooth wall is 1.5–3.5 μm thick.

19.1.2.4.4 Disease Cycle

The exact disease cycle in different regions of the country is not known. However, in Punjab, when winter temperatures are too low for the growth of cucurbits, the fungus perpetuates in the form of active mycelium on self-sown or cultivated sponge gourd growing in sheltered places during severe winters and also in open during milder winters. In areas where oospores has been detected these may be an important source of perpetuation of the pathogen.

Other cucurbit crops affected by downy mildew and old infected crop trashes are sources of infection; spreads by spores in wind, air currents, workers, and machinery. The sporangia are produced before midnight, get matured by 3 AM., dispersed between 6 and 10 AM., and germinate within one hour. The day temperature of 25–30°C, night temperature of 15–21°C and RH> 75% help the disease development. The pathogen survives as both mycelia and sporangia on a wide range of wild host.

Transmission: by insects and other invertebrates.

Fore casting: A prediction model based on duration of temperature of 13–30°C and leaf wetness has been developed.

19.1.2.5   POWDERY MILDEW

19.1.2.5.1   Disease Occurrence and Distribution

In India, the disease is prevalent in almost all the states particularly in the warm and dry areas where moisture is present as dew. Powdery mildew affects cucumber, muskmelon, bottlegourd, squash, pumpkin, and watermelon

19.1.2.5.2   Diagnostic Symptoms

First symptoms are small, white to dirty gray spots (sometimes with reddish brown tinge) on lower surface followed by upper surface of leaves. The superficial powdery growth finally covers entire host surface (Figure 19.3). Leaves turn pale yellow and later brown, shrivel, dry, and fall off. Fruits are undersized, ripe prematurely, lack flavor and sweet taste.

19.1.2.5.3   Biology and Molecular Characterization

Erysiphe cichoracearum and *Sphaerotheca fuliginea* are the fungi belong to Kingdom-Fungi, Phyllum-Ascomycota, Subdivision-Ascomycotina, Class-Pyrenomycetes, Order-Erysiphales, and Family-Erysiphaceae. *S. fuliginea* has been found to be mainly responsible for causing the disease on majority of the cultivated cucurbits except watermelon. In addition to these two fungi, *Leveillula taurica* has also been found to be associated with powdery mildew symptoms. Perithecia of both *E. cichoracearum* and *S. fuliginea* have been observed on greenhouse cucumber in Germany. Perithecia of L. taurica are

Diseases of Cucurbits and Their Management

immersed in the dense mycelium. Perithecia of all three pathogens have indeterminate mycelial appendages.

E. cichoracearum: Conidia singe-celled, hyaline, barrel-shaped, in chains, measuring 20.25–44.00x11.25–16.87 µ m in size; perithecia dark spherical, with mycelioid appendages, 100–180 µ m in diameter; asci 8–18 in number per perithecium, 36.00–65.75x22.50–40.50 µ m in size; ascospores 2 rarely 3 per ascus, single-celled hyaline and 18.00–29.25x12.37–18.00 µ m in size.

S. fuliginea: Conidia single-celled, hyaline, borne singly and measuring 20.25–27.90x 13.50–18.00 µ m; perithecia contain a single ascus measuring 50.75–74.50x49.50–67.50 µ m; ascospores single-celled, hyaline, spherical, 6–10 per ascus and measure 11.12–20.25x 9.00 µ m in size

19.1.2.5.4 Disease Cycle

There are several possible ways in which the two fungi can live from one season to the next. Where cleistothecia are formed, they can explain the mode of perpetuation from one crop season to the next. In India, these sexual fruiting bodies develop on leftover cucurbit crops during winter in isolated areas. These may initiate the disease in the local hosts, and from there, the primary inoculums in the form of conidia might be wind to the main crop in the plains. However, as in the downy mildew, the main source of primary inoculums seems to be the existence of wild and cultivated cucurbits in one or the other locality of the country from where the conidia are blown to the new crop. The conidia germinate and cause direct penetration of the epidermal cells in which they produce haustoria. Incubation period about 3–7 days.

19.1.2.5.5 Epidemiology

The fungi are influenced greatly by age of the host plant and air humidity and temperatures. 16–23 days old leaves are highly susceptible while very young leaves are almost immune. The fungi can sporulate and cause infection in a very dry as well as wet atmosphere but infection increase as the atmospheric humidity increase, heavy dew deposits favoring the penetration by germ tubes most. The minimum and maximum temperatures for conidial formation and host penetration are 10°C and 32°C, respectively, the optimum being about 26 to 28°C.

19.1.2.6 FRUIT ROT OR COTTONY LEAK

19.1.2.6.1 Disease Occurrence and Distribution

Fruit rot is common disease of cucurbits in India. It occurs in almost every locality during the rainy season. It is not only field disease but market and transit disease also.

19.1.2.6.2 Symptoms

The fruit in contact with the soil suffer from the disease. The skin of the fruit shows soft, dark green water-soaked lesions that develop into a watery soft rot. On this rotting portion the cottony mycelial growth develops abundantly during humid atmosphere. The affected fruits look as if wrapped in absorbent cotton. In watermelons, the decay frequently starts at the blossom end. On the margin of the cottony growth the skin of the fruit looks dark green and water-soaked.

19.1.2.6.3 Disease Cycle

The main pathogens are warm weather fungi. They persist in soil as oospores and if moisture, temperature, and other suitable conditions are present they cause infection of the fruits on or near the soil. Infection is always facilitated by bruises and wounds on the fruit. This is common when the fruit are rubbing against soil particles or when there are insect bites. Once the cottony growth has appeared zoospores produced by this growth cause secondary spread of the disease. The fungi are present in the soil living in a saprophytic manner on dead organic matter.

19.1.2.7 CERCOSPORA LEAF SPOT

19.1.2.7.1 Disease Occurrence and Distribution

In the tropics and subtropics regions throughout the word, this disease is prevalent. It may become a serious disease if weather conditions are favorable. In Europe, this is a destructive disease of greenhouse crops.

Diseases of Cucurbits and Their Management 389

19.1.2.7.2 Diagnostic Symptoms

The characteristic symptoms are the appearance of the water-soaked areas on the leaf lamina. These spots enlarge rapidly to become circular to irregular spots with pale brown, tan or white centers and wide purple to almost black margins (Figure 19.2). Many coalesce to from blotches. The leaf many dry and finally die.

19.1.2.7.3 Disease Cycle

The fungi perpetuate on the perennial weeds and on the disease crop debris.

19.1.2.8 FUSARIUM WILT

19.1.2.8.1 Disease Occurrence and Distribution

This disease is known for the last more than 95 years. In India, it was reported in 1955 from Maharashtra, it now occurs in many states of the country. Yield losses up to 80% have been reported in the worst affected areas.

19.1.2.8.2 Diagnostic Symptoms

Plants of all stages of growth are attacked by the disease. The germinating seeds may rot under the soil. In young seedlings, cotyledons droop and wilt. The hypocotyls are girdled by a watery soft rot. The characteristic wilt symptoms appear on the older plants when the leaves droop during hot period of day. In wet weather, the dead stems show a white or pinkish mass of fungal spores.

The fungus is seed and soil borne as well as present in the soil.

19.1.2.9 SCAB OR GUMMOSIS

19.1.2.9.1 Diagnostic Symptoms

Can affect leaves, petioles, stems, and fruits. Water-soaked spots occur on leaves and runners. These spots eventually turn grey to white. The center of the spots could then drop out to give a 'shot-hole' appearance. Lesions on fruit are often confused with anthracnose. These spots are 3–4 mm in diameter and might ooze a gummy substance. The spots could then be invaded by secondary rotting bacteria which cause the spots to smell.

19.1.2.9.2 Disease Cycle

The fungi can be seed-borne, but also survive in soil on undecomposed plant material. Spread by wind in moist conditions. Cool, wet weather, including rain, dew, and fogs are most favorable for development of the disease.

19.1.2.10 SEPTORIA LEAF SPOT

19.1.2.10.1 Diagnostic Symptoms

Leaf spots are brown with small, black, fruiting bodies. Raised spots develop on fruit, often with star-shaped cracks.

19.1.2.10.2 Disease Cycle

The fungi can survive on crop debris, especially from the previous season. It can be spread by rain splash and windborne rain and favored by cool, rainy weather. Mainly found on pumpkins.

19.1.2.11 GUMMY STEM BLIGHT

19.1.2.11.1 Diagnostic Symptoms

Stems, leaves, and fruit could all be affected, with leaves displaying brown to black spots. Stems near the crown might have a bleached appearance and exude a brownish gum; however, similar symptoms appear with charcoal rot and fusarium wilt.

19.1.2.11.2 Disease Cycle

Infected crop trash, soil, weeds, and infected seed are source of infection. Spread by spores carried in wind and air currents. Moist conditions and high

Diseases of Cucurbits and Their Management   391

relative humidity (RH) are most favorable for the development of disease incidence.

19.1.2.12   CHARCOAL ROT

19.1.2.12.1   Diagnostic Symptoms

Bleaching of stems and leaf death near the crown of the plant. The stems could also have gum exuding from these areas as the bleached areas turn drier. Symptoms may be very similar to gummy stem blight. Symptoms often appear late in the season.

19.1.2.12.2   Disease Cycle

The fungus is a common soil-borne organism. It has a wide host range. Crop trash also provides a source of further infection. The disease is favored by a common soil-borne fungus that is favored by warm to hot conditions and spread by trash and soil.

19.1.2.13   ANGULAR LEAF SPOT

19.1.2.13.1   Disease Occurrence and Distribution

This disease is common on pumpkin, muskmelon, cucumber, and allied plant species in regions of cold climate.

19.1.2.13.2   Diagnostic Symptoms

On leaves, the disease first appears as small, water-soaked spots which enlarge to about 3 mm in diameter (Figure 19.1). The spots become tan on the upper surface and gummy or shiny on the lower surface, due to bacterial ooze which dries out and turns white. Round lesions occur on fruit. Could be confused with anthracnose. Mainly found on cucumbers, but also found on rockmelon, honeydew, watermelon, and squash. Seed and infected crop refuse and favored by warm and humid conditions. Spread by rain, irrigation water, hands, and clothing of workers.

19.1.2.14   BACTERIAL LEAF SPOT

19.1.2.14.1   Disease Occurrence and Distribution

This disease is distinct from angular leaf spot and causes maximum damage to cucumber. It was reported from India in 1931.

19.1.2.14.2   Diagnostic Symptoms

Spots first appear on squash and pumpkin leaves as small water-soaked or greasy areas on the underside of leaves, and as indefinite yellow areas on the upper side of leaves. In about five days the spots become round to angular with thin, brown, translucent centers and a wide, yellow halo. The spots enlarge up to about 7 mm diameter. Occasionally young stems and petioles are attacked. Young fruits may also be affected. Fruits may produce light brown ooze from small, water-soaked areas, which can extend into the seed cavity, causing seed infection.

19.1.2.14.3   Disease Cycle

Disease can be transmitted by seed and infected crop residue, as well as rain splash and the movement of people and machinery. Cool, damp weather favored disease development.

19.1.2.15   BACTERIAL SOFT ROT

Soft rot found in many vegetables is common in cucurbits. It is mainly a fruit disease and occurs due to injury to fruits and poor transit and storage conditions.

19.1.2.15.1   Mosaic

19.1.2.15.2   Disease Occurrence and Distribution

Different members of family Cucurbitaceae are attacked by viruses belonging to cucumovirus, poty virus and to bamovirus groups. All these viruses produce varied types of mosaic symptoms resulting in considerable losses.

19.1.2.15.3 Diagnostic Symptoms

Light and dark green mottling of the leaves. Distortion of leaves and stunting of the plant might occur (Figure 19.6). Marrow and summer squash fruit might show sunken concentric circles or a raised marbled pattern. Papaya ring spot virus may cause lumpy distorted fruit on zucchini. Viruses may also affect fruit set. All commercially grown cucurbits are susceptible. Pumpkin, squash, rockmelon, and zucchini crops are most commonly affected (Figures 19.1–19.6).

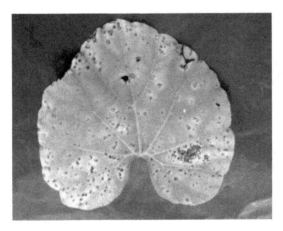

FIGURE 19.1 Angular leaf spot of bottlegourd.

FIGURE 19.2 Cercospora leaf spot of sponge gourd.

FIGURE 19.3 Powdery mildew of bottlegourd.

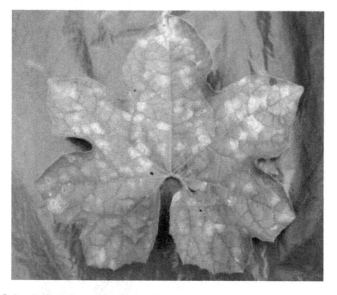

FIGURE 19.4 Anthracnose of bottlegourd.

Diseases of Cucurbits and Their Management

FIGURE 19.5 Downy mildew of bitter gourd.

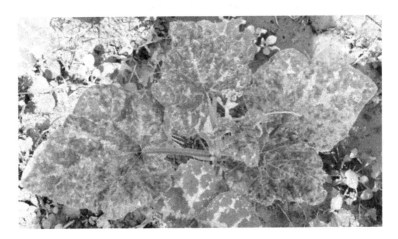

FIGURE 19.6 Mosaic of pumpkin.

19.1.2.15.4 Disease Cycle

Other mosaic-affected cucurbit crops and weeds. Spread by aphids that only need a very short period of feeding to transmit the virus.

19.1.2.16 ROOT KNOT

19.1.2.16.1 Disease Occurrence and Distribution

Root knot is one of the most important groups of phytonematodes and was first recorded in greenhouse vegetables in England in the year 1855. In India it was reported from southern parts on tea plants. Although more than 65 species of *Meloidogyne* have been reported from different parts of the world but four, namely, *M. incognita, M. javanica, M. arenaria,* and *M. hapla* are important (Johnson and Fassuliotis, 1984). The average crop yield losses due to root knot nematode are thought to be about 5%. India losses varying from 20 to 80% have been reported

19.1.2.16.2 Diagnostic Symptoms

The infected plants show unthrifty development and the increased susceptibility of roots to other root pathogens. Formation of gall on host root system is the primary symptoms associated with this nematode. The size and form of the galls vary and depends on the nematode species and their number in the tissues and type as well as age of host plants.

19.1.2.16.3 Disease Cycle

Dissemination of the nematodes takes place when juveniles or eggs are transported from infested to uninfected areas. Besides, irrigation water and soil adhering to animals and implements also spread them. Wind borne dissemination of root knot nematode has also been reported in the regions having windstorms. Dispersion over great distances occurs by the moment of infested plants.

19.1.2.17 NUTRITIONAL DISORDERS

1. **Nitrogen:** Its deficiency causes general yellowing of the plants, beginning at the lower leaves, affected leaves gradually turn yellow and dry up. Cucumber fruits become more slender and pinched at the blossom.

   Application of nitrogen to the soil (65 to 185 kg/ha) based on tests is useful in correcting the deficiency.

Diseases of Cucurbits and Their Management

2. **Potassium:** Its deficiency generally occurs in light soil or when the crop is grown continuously, applying only nitrogen and phosphorus without application of potassium.

   Deficient plants may have cupped young leaves. Cucumber fruit turns brown or spotted and may develop a club shape.

   Applying potassium to the soil based on tests effectively controls the deficiency.

3. **Calcium:** Deficiency occurs mostly in alkaline soils or when water containing high residual sodium carbonate is used for irrigation. Leaf margins stop expanding and leaves cup down.

   Application of single super phosphate or gypsum corrects calcium deficiency.

4. **Iron:** Deficiency symptoms occur first on developed leaf tissues. Younger leaves become chlorotic. Larger veins remain green in the early stages of the deficiency, but later the entire leaf becomes uniformly chlorotic. In later stages, necrotic spots develop in the leaves.

   Soil application of ferrous sulfate (5 kg/ha or 0.2% foliar spray) corrects iron deficiency.

5. **Manganese:** Symptoms of manganese deficiency occur on the new growth. The interveinal area of younger leaves becomes chlorotic while veins remain green. As the deficiency continues, older leaves become chlorotic.

   Manganese deficiency can be corrected by soil application of 10 kg manganous sulfate per hectare or by 0.2% foliar spray.

6. **Molybdenum:** Its deficiency results in interveinal chlorosis in old leaves. Later leaf margins, turn brown and the plants remain stunted.

7. **Zinc:** Symptoms of zinc deficiency occur as yellowing, curling; deformation and necrosis of leaves are the main symptoms. This can be controlled by spraying mixture of lime and zinc sulfate.

19.1.2.17.1 Management

- **Cultural Practices:**
  1. Deep plowing during summer to expose the nematodes to solar radiation and also the resting stages of the pests.
  2. Crop rotation with paddy in low-lying areas is very effective, against downy and powdery, mildew.

398   *Diseases of Fruits and Vegetable Crops*

3. Preventing the crops from overlapping and adequate spacing to minimize the length of leaf wetness periods, for downy mildew disease management.
4. Late planting helps in delayed development of powdery mildew.

- **Mechanical:**
  1. Eradication of disease infected plants.
  2. Removal of alternate host from the field particularly for virus disease management.
  3. Destruction of wide cucurbit plants and crop residues reduces to incidence of root knot and reniform nematodes and anthracnose and powdery mildew respectively.
  4. Removal of damage fruits and burnt.
  5. Careful cultivation and harvesting avoiding wounds, helps in reducing infection particularly to gummy stem blight disease.
  6. Keeping of fruits away from the soil contact.
  7. Hygienic transportation and storage of the fruits to avoid the incidence of the disease.

- **Chemical Control:**
  1. Foliar spray with mancozeb or zineb 1.5 to 2.0 kg per 1000 liter of water per hacter or Metalaxyl (0.05%) mixture or Blue Copper (0.4%) or Blitox (0.4%) at 10 days intervals helps in controlling the anthracnose disease.
  2. Spraying with dinocap 1 ml per liter of water gives good control over powdery mildew.
  3. Minimizing of spraying of fungicide during flowering stage.
  4. Spraying of metasystox or dimecron @ 1 ml per liter of water at 10–12 days interval for the viral diseases.
  5. Two sprays of streptomycin @ 200 ppm at 15 days interval for the bacterial diseases.

- **Host Resistance:** Cultivars/Genotypes as a source of resistance for diseases of cucurbits are the following:
  1. Cucumber cultivar GY-5937-587 is reported as resistant to Race 3 of *M. incognita* and cv. improved Long Green to *M. javanica*. Cultivar S-445 of muskmelon is resistant to *M. javanica*.
  2. Cucumber resistant variety against Powdery mildew-Ashley and Palmettu, Cherokee, Poinsette, Yomaki, Saparton Salad and PI-197083; Downy mildew—Yomaki, Saparton Salad Wautoma,

Diseases of Cucurbits and Their Management 399

Simoisrad, Topolek, Summer Prolific, Autumn Green Breso, Avamgard 234, Syaotsy-gua, and Poinsetta, Anthracnose, and Angular leaf spot- Poisetta Yomaki, Saparton Salad and Mosaic-Tokoyo Long Green, Chinensis Long, Winscrimson, Brasil and Table Green.

3. Muskmelon resistant variety against Wilt and Anthracnose—Analai fruit, IIHR190, Durgapura, Madhu, and Punjab Sunehri, Tavrichanka, UC Top Mark; Powdery Mildew-Arka Rajhans, PMR-45, RM43, Punjab Rasila, IIHR352, Home Garden, Diguria, and Hurragola; Downy Mildew- Punjab Rasila, IIHR352, Home Garden Rasila and Arka Rajhans; Mosaic virus-PI161375.

4. Watermelon resistant variety against Powdery Mildew, Downy Mildew, and Anthracnose-ArkaManik.

5. Pumpkin resistant variety against Anthracnose-*Cucurbita ecuadorensis* and *C. lunbelliana.*

- **Biological Control:**
  1. Seed treatment with *Trichoderma harzianum* @ 2 g per 100 g of seed.
  2. Application of *T. harzianum* @ 1 kg in 25 kg FYM during early stages of the crops along with the rows, which helps in inhibiting soil-borne fungus infection.

- **Integrated Approach:**

Stage	Disease	Practices
1. Pre-sowing	Resting stages of pathogens	Deep summer plowing. Application of 200 kg neem cake/ha.
2. Seed- seedling	Collar rot and virus-infected plants	Seed treatment with *T. harzianum* @ 2 g/100 g seeds; apply 1 kg *T. harzianum* in 25 kg FYM, along the rows and removal of the alternate host from the field.
3. Vegetative	Downy and powdery mildew leaf spot	Optimum irrigation. Spray dinocap 1 ml per liter of water and removal of the alternate host from the field.
4. Reproductive	Downy and powdery mildew, virus-infected plants	Field sanitation, optimum irrigation, and removal alternate host from the field.

KEYWORDS

- angular leaf spot
- bacterial soft rot
- charcoal rot
- diagnostic symptoms
- gummy stem blight
- mosaic

REFERENCES

Agrios, G. N., (1988). *Plant Pathology* (p. 803). Academic Press, San Diego, New York.

Anonymous, (1999). *Integrated Pest Management Package for Cucurbitaceous Vegetables, Directorate of Plant protection, Quarantine and Storage*. Faridabad, Haryana, India.

Arya, A., (2009). *Diseases of Fruit Trees*, International Book Distributing Co., Lucknow.

Chakrabarti, D. K., Kumar, S., & Chand, G., (2010). *A Guide Book for Diseases of Horticultural Crops (Diseases of Fruits, Ornamentals, Plantation, Spices, Medicinal, Forest and Vegetable Crops)* (pp. 1–152). Narendra Publishing House, Maliwara, Delhi–110006 (India).

Fageria, M. S., Arya, P. S., & Choudhary, A. K., (2005). *Vegetable Crops: Breeding and Seed Production* (pp. 150–187). Kalyani Publishers, Ludhiana, New Delhi, Noida, Hyderabad, Chennai, Kolkata and Cuttack.

Gangwane, L. V., & Khilare, V. C., (2008). *Crop Disease Identification and Management* (p. 244). Daya Publishing House, Delhi.

Gautam, P. L., Ram, H. H., & Singh, H. P., (2006). Cucurbits breeding and production technology. In: *Proceedings of National Seminar on Cucurbits, Held on 22–23 September, 2005*. G.B. Pant University of Agriculture and Technology, Pantnagar, Uttaranchal, India.

Gour, T. B., Babu, T. S. Sriramumlu, M., Reddy, D. D. R., Rao, K. C., Babu, R. R., & Reddy, P. N., (2000). *Pictorial Crop Based Identification Mannuals on Crucifers, Cucurbits, Insect, Pests, Diseases and Nutritional Disorders* (pp. 64–88). Agricultural Information and Communication Center, Acharya, N.G. Rangy Agricultural University, Rajendranagar, Hyderabad.

Gupta, V. K., & Paul, Y. S., (2008). *Diseases of Vegetable Crops* (pp. 100–120). Kalyani Publishers, Ludhiana.

IIHR, (1987). *Diseases of Fruit and Vegetable Crops and their Control* (p. 37). IIHR, Bangalore.

Mehrotra, R. S., & Aggrawal, A., (2008). *Plant Pathology Tata McGraw-Hill Publishing Co. Pvt. Ltd* (p. 846). New Delhi – 110008.

Ram, H. H., (2002). *Vegetable Breeding: Principles and Practices* (pp. 222–274). Kalyani Publishers, Ludhiana, New Delhi, Noida, Hyderabad, Chennai, Kolkata and Cuttack.

Saha, L. R., (1990). *Hand Book of Plant Protection* (pp. 655–661). Kalyani Publishers, Ludhiana.

Singh, R. S., (1999). *Diseases of Vegetable Crops* (pp. 207–236). Oxford & IBH Publishing Co. Pvt. Ltd. New Delhi – 110001, Calcutta.

CHAPTER 20

Diseases of Beans Crop and Their Management

SHAILBALA

Junior Research Officer, Plant Pathology, Sugarcane Research Center, G.B. Pant University Agriculture and Technology, Uttarakhand, India, E-mail: shailbalasharma10@gmail.com

20.1 INTRODUCTION

Cluster bean is grown for green vegetable, dry seeds, and also as a forage crop. It is also used as a green manuring crop. Broad bean is a hardy plant and is the only bean which can withstand sufficiently in cold temperature. An illness sometimes fatal and known as favism is caused. The winged bean also known as four-angled bean and Goa bean is a tropical legume of exceptionally high value. It produces edible pods, seeds, leaves, flowers, tuberous roots and edible seed oil.

Soybean is an important bean crop and has been cultivated in many parts of country. Soybean produce a lot of product like oil, soya milk, soya cake, soya peat, cattle feed, etc. Other beans which are cultivated and used as green pods though on a very limited scale are scarlet runner or multi-flora bean, tepary bean, velvet bean, sword bean, etc. The major production constraints of beans include moisture stress, diseases, insect pest, weeds, poor soil fertility, and lack of improved seeds (Kidane, 1987; Ayele, 1991). Of which, diseases are known to be the major factors which threatened the productivity of beans in all growing area (Fininsa, 2001; Abiy et al., 2006; Tadesse et al., 2009). So disease management is very important to make viable cultivation and to ensure productivity of the bean crop.

20.2 BACTERIAL DISEASES

Among many diseases affecting bean crop, common bacterial blight and halo blight are the most destructive diseases of beans worldwide (Ariyaratne

402   *Diseases of Fruits and Vegetable Crops*

et al., 1998; Dursun et al., 2002). Among which, common bacterial blight is recognized as economically important disease of bean (Popovic et al., 2012). There are four major bacterial diseases of beans, i.e., common bacterial blight (*Xanthomonas campestris* pv. *phaseoli*), halo blight (*Pseudomonas syringae* pv. *phaseolicola*), bacterial brown spot (*Pseudomonas syringae* pv. *syringae*) and bacterial wilt (*Corynebacterium flaccumfaciens)* which can limit the bean production and needs proper management practices.

20.2.1   COMMON BACTERIAL BLIGHT (XANTHOMONAS CAMPESTRIS PV. PHASEOLI) (SYN: X. AXONOPODIS PV. PHASEOLI)

The disease can attack leaves, stems, pods, seeds, and reported to causes considerable yield loss (Fininsa, 2001; Fourie, 2002). Common bacterial blight affects bean foliage, pods, and seedlings. Initially, symptoms appear as small, angular, light green, water-soaked or translucent spots on foliage. During warm and wet conditions, these lesions rapidly enlarge and merge. Gradually, the centers of the lesions become dry and turn brown in color and are surrounded by a distinct, narrow zone of yellow tissue. In highly susceptible varieties, the lesions continue to expand until the leaves appear scorched, ragged, and torn by wind and rain. Pod symptoms consist of generally circular, slightly sunken and dark red-brown water-soaked lesions. In severe infection, entire pods may shrivel and die.

Common bacterial blight can survive on or in the seed/contaminated seed is the primary source of the pathogen. Bean plant residues can also be a source of the bacterium. The disease is favored by warm temperature (i.e., 28°C) and high humidity. The pathogen survives better in residues on the soil surface than residues that have been worked into the soil. Once introduced, secondary spread can occur by wind-blown rain, overhead irrigation, contaminated equipment or peoples and animals. Yield is affected most, if blight develops before the pod-fill stage.

20.2.2   HALO BLIGHT (PSEUDOMONAS SYRINGAE PV. PHASEOLICOLA)

Halo blight affects bean foliage, pods, and seedlings and causes significant losses in both yield and seed quality. Symptoms appear as water-soaked spots on the lower surface of leaf. A zone of yellow-green tissue (halo) appears around each water-soaked spots. The spots maybe 3 to 6 mm in diameter with halo up to 2.5 cm in diameter. Plants develop generalized

Diseases of Beans Crop and Their Management   403

systemic chlorosis in case of severe leaf infection. Pod symptoms generally consist of red or brown lesions which may also appear water-soaked. As pods mature and turn yellow, pod lesions may remain green and may exhibit crusty bacterial ooze on the surface. Developing seed may be shriveled or discolored if lesions expand to involve the pod suture. Both leaf and pod symptoms often coalesce. The upper foliage of diseased plants develops a characteristic yellow color infected seed. The infected seed may be smaller in size with wrinkled seed coat.

Halo blight Infected bean seed is the most important source of the halo blight bacterium. The pathogen can survive more than 4 years in bean seed. The disease development is favored by humid, cloudy conditions. Halo expression does not develop or tends to disappear if the temperature is above 21° to 23°C. The systemic chlorosis is particularly pronounced at 18° to 23°C. The pathogen can also survive in bean residues from previous seasons. The disease can be spread by wind-driven rain, overhead irrigation, equipment, or people and animals. Severe outbreaks of halo blight often occur after heavy rainstorms.

20.2.3 *BACTERIAL BROWN SPOT (PSEUDOMONAS SYRINGAE PV. SYRINGAE)*

Bacterial brown spot affects bean foliage and pods. The lesions are generally circular, brown, and necrotic and are often surrounded by a blight yellow zone. Bacterial ooze and water soaking are rarely observed prior to necrosis. Even water soaking is not as pronounced as it is in common and halo blight. The leaf tissue around the lesion may be puckered. Lesion coalesces and their center fall off. In severe cases, much of the foliage may be killed. Stem lesions are occasionally observed when the pathogen develops systemically. Lesions on pods are circular and initially water-soaked. They become brown and necrotic. Infected pods may be twisted or bent where lesions develop. Occasionally, ring sports of lesions occur around a central lesion.

Source of primary infection are usually weed hosts on which bacteria survives as an epiphyte on the leaf surface. It can survive in plant debris for one year. Pathogen can spread either by wind-blown rain or overhead sprinkler irrigation. Overcast, cloudy, humid weather favors the disease especially if such condition immediately follows rain or irrigation. Such a condition favors multiplication of bacteria on the leaf surface. Plant injured by high winds, hail or blowing sand is susceptible to infection.

20.2.4 BACTERIAL WILT (CORYNEBACTERIUM FLACCUMFACIENS)

Bacterial wilt kills young seedlings by plugging the water-conducting vascular tissue in the stems. Initially, plant shows temporary wilt during warm part of the day and come in normal appearance in cool periods. Eventually a gradual wilting persists for longer period of time and plant turn straw in color and dies. A systemic dark brown to black color discoloration is apparent in vascular bundle inside root and lower hypocotyle. Larger plants that become infected may survive in the entire season and produce seed.

Infection can occur on pod sutures. Infected seed may be bright yellow, orange or purple depending on the strain of the infecting bacterium. The bacteria can survive on seed for many years. It can overwinter in plant debris and on weed. The disease is favored by warm temperature and dry weathers. The wilt bacteria are not spread as easily by rain or contact with wet foliage as compared with other bean diseased bacteria.

20.2.5 DISEASE CYCLE OF BACTERIAL DISEASES

The bacteria that cause common bacterial blight, halo blight, bacterial brown spot and bacterial wilt survive from season to season in infected seed and infested bean residue. Bean plants grown from infected seed often serve as the initial source of bacteria. Infected plants may be volunteers from the previous bean crop or plants grown from infected seeds used to plant the current crop. The brown spot bacteria can also spread from some weed species (hairy vetch). The bacteria can spread from plant to plant and field to field in many ways including wind-driven and splashing rains, overhead irrigation, surface-drainage water, and farm machinery. The bacteria may survive for 6 to 18 months in plant refuse (on or above the soil surface and under dry conditions), in bean cull piles within or near fields, on volunteer plants from a previous crop and even on the surface of weeds. Bacteria enter plants through natural openings or injuries caused by insects or even accompany other diseases such as rust. There are two types of field infection. Primary infection involves the invasion of bean seedlings as they emerge from the soil. This occurs when the seed is infected or when an emerging seedlings comes into contact with infected plant material. Secondary infection is due to spread from primary source to other growing plants often involving only one or two isolated plants or it may cover an entire field or area. In severely diseased fields, nearly all infection is secondary (Reddy, 2010a).

Diseases of Beans Crop and Their Management 405

When an infected seed sprouts and the seedlings emerges, bacteria ooze to the surface of the diseased cotyledons, are splashed to a neighboring plant, enter the stomata and infect. Stomata or hydathode entry also occurs from bacteria that survives on surface debris and are rain splashed to the leaves and stems. Once inside the plant, the bacteria may move systematically by way of the vascular bundles to the leaves, stems, pods, and into the seed. The bacteria are carried by rain splashes to neighboring leaves and pod surface. They invade growing pods largely along the suture and enter the seed through the vascular tissue. The bacteria remain on or just below the surface of seed. The bacterial pathogens have been recovered from 3 to 14 years old seed. Common blight and bacterial brown spot are favored by cloudy, damp weather and high air temperatures of 28° to 32°F, whereas halo blight thrives under damp, cooler conditions and 18° to 22°C temperatures. Bacterial wilt requires more than 27°C temperature for disease development. Under favorable environmental conditions, bacteria continue to spread and the disease cycle repeats.

20.2.6 MANAGEMENT OF BACTERIAL DISEASES

Use of resistant varieties supplemented with proper cultural practices and chemical seed treatment could be the best alternative options in managing bacterial diseases of bean and avoiding yield losses (Belachew et al., 2015). Planting of bean cultivars resistant to *Xanthomonas axonopodis* is economically and technically the most practical method for effective management of common bacterial blight (Popovic et al., 2012). Bacterial diseases of bean are managed the most effectively by integrating several control practices.

20.2.6.1 CROP ROTATION

Bacteria do not survive well in the absence of bean plants. As infested bean residue decomposes, bacteria are exposed to the soil environment and quickly die. Crop rotation with non-host crop at least for two years or longer will provide sufficient time for residue decomposition and also will reduce the number of volunteer beans present during each succeeding season.

20.2.6.2 DISEASE FREE AND CERTIFIED SEED

Disease free and certified seed should always be planted to minimize the risk of introducing bacterial pathogens with the seed.

20.2.6.3 SEED TREATMENT

Seed treatment with antibiotic streptomycin will help to eliminate bacteria present on the surface of the seed. Thus, seed treatment helps to reduce contamination that may be present in a seed lot.

20.2.6.4 SANITATION

Because volunteer plants are important sources of bacteria for initiating disease outbreaks, volunteer bean plants within or near bean fields should be destroyed as soon as they are discovered. Incorporation of bean residue into the soil will reduce the amount of inoculum available to initiate disease development.

20.2.6.5 IRRIGATION MANAGEMENT

Overhead irrigation increases the risk of disease by providing the moisture and splashing water necessary for bacterial spread and disease development. So irrigation management by avoiding overhead irrigation is very important.

20.2.6.6 AVOID MOVEMENT THROUGH THE FIELD

Bacterial cells are easily spread between plants and between fields by machinery, people, and animals. Spread is more likely to occur when plants are wet. Consequently, it is important to minimize or avoid movement through bean fields when plants are wet.

20.2.6.7 BACTERICIDES

Foliar sprays with a fixed copper compounds, applied weekly starting four days after seedling emergence or at the first sign of disease have given fair to good control when continued to harvest (Reddy, 2010b). Thorough

Diseases of Beans Crop and Their Management 407

coverage of the plants is required. Properly applied aerial sprays are superior to applications by ground equipment.

20.2.6.8   OTHER PRACTICES

Field subjected to wind and water erosion should be planted to a non-susceptible cover crop before winter. Planters, harvesters, and other equipment should be sanitized by spraying with a disinfectant such as chlorine dioxide, sodium hypochloride or a quaternary ammonium compound before moving from an infected to a blight free field. Storage areas should also be sanitized (Reddy, 2010a).

20.3   FUNGAL DISEASES

Although various fungal diseases are reported in literature to be associated with bean crop, some important fungal diseases of bean i.e., bean rust, Alternaria leaf spot, angular leaf spot, white mold, Aschochyta leaf and pod spot, anthracnose, ashy stem blight, Fusarium root rot, Pythium root rot, Rhizoctonia root rot, Sclerotinia root rot, Cercospora leaf spot which may become major limiting factors in the successful production of bean crop if not managed properly.

20.3.1   RUST (UROMYCES PHASEOLI)

This disease occurs worldwide and cause 13 to 100% yield loss. Pre-flowering and flowering are crucial stages for fungus attack. The symptoms appear as tiny white raised pustules on the under surface of the leaves. Gradually these spots develop, enlarge, and turn into reddish brown pustules which eventually erupt to release the rusty spore mass (Hagedorn and Inglis, 1986). Disease occurs late in the growing season resulting premature leaf drop which lower down the seed quality.

20.3.1.1   DISEASE CYCLE

The fungus survives on crop debris, stubble, and self-sown bean plant. The teliospore can infect volunteer bean plants without need for an alternate host. Infection of volunteer bean plant can play important role in early

408 *Diseases of Fruits and Vegetable Crops*

development of rust epidemic. Rust spore from stubble, crop debris, volunteer plants can move to new crop through winds and infect the new plant. The new spore with rusty spore mass in the form of rust pustules will appear on infected plants. Secondary spread of disease occurs when these rust spores become airborne and spread to other plants. Extended periods (18 hours or more) of humidity, more than 95% and moderate temperature (17° to 27°C) are required for spore germination and infection.

20.3.1.2 MANAGEMENT

Cultural operations like good sanitation, destruction of infected plant material, two to three years crop rotation, suitable plant spacing and removal of weeds to lower down relative humidity (RH) in the crop are very important in minimizing the disease incidence. Frequent crop monitoring during blossom and early pod development is very important. Sprinkler systems and frequent rains provide favorable conditions for rust development and may wash protectant fungicides off leaves before they can inhibit rust spore germination, so avoid these conditions. Fungicidal sprays with fungicide Mancozeb @ 0.2% or Bayleton @ 0.05% are effective for disease control (Reddy, 2010b). Three sprays of fungicides Tridimefon @ 0.1% or Tridemorph at 45, 60, and 75 days after sowing are reported to give the best control of disease and the highest bean yields (Singh, 2003).

20.3.2 ALTERNARIA LEAF SPOT (ALTERNARIA ALTERNATA)

Initially, small brown irregular spots appear on the leaf, later develop into large grey brown oval shape spots with concentric rings. When many diseased spots come together, large leaf area become necrotic. Some time these necrotic areas fall out and give shot hole appearance. Seed produced on diseased plants are easily disseminated by wind, rain, insect, and seed. Cool and wet weather (leaves remain wet for 24 hours or more) are conducive for spore germination and infection. Alternaria leaf spot develops late in the season as the plant start to mature. The fungus survives on crop debris and other host plants.

20.3.2.1 MANAGEMENT

Use certified and disease-free seed material for planting is essential to manage the disease. Plant density should be maintained in the field. Plant

Diseases of Beans Crop and Their Management 409

extract of *Allium sativum* and *Allium cepa* lower down the disease intensity in soybean crop (Bhosale et al., 2014). Fungicides Cyprodinil + Fludioxonil, Copper hydroxide, Chlorothalonil, Azoxystrobin, and Pyraclostrobin significantly reduced disease severity caused by *A. alternata* on snap bean (Dillard and Cobb, 2008).

20.3.3   ANGULAR LEAF SPOT (PHAEOISARIOPSIS GRISEOLA (SACC.) FERRARIS, OR ISARIOPSIS GRISICOLA)

Fungus attacks the foliage and pods of bean during the growing season. The disease is spread through wind-blown spore. Initially, lesions appear as grey or brown irregular spots having a chlorotic halo in the leaves. Later, these lesions turn brown and necrotic and appear as an angular shape, which is a characteristic of this disease. Later symptoms also appear on pods, which results in defoliation. Primary inoculum comes from seed or infested crop residue. Humid condition favors the disease development. The optimum temperature for symptom development is 24°C, but it will develop over the wide range of temperature, i.e., 16°C to 28°C (Strausbaugh and Forster, 2003).

20.3.3.1   MANAGEMENT

Always use disease-free and certified seed for planting. Two to three years crop rotation is helpful to manage the disease. Deep plowing of field to bury the infected crop debris. Approved fungicides should be applied when the disease first appears, and conditions are favorable for disease development. Choose cultivars that have resistance against disease. On the appearance of disease, spray fungicide Chlorothalanil or Dithionon or Propineb or Quantaf or Carbendazim @ 0.1% will be effective to manage the disease (Reddy, 2010a).

20.3.4   WHITE MOLD OR SCLEROTINIA STEM ROT (SCLEROTINIA SCLEROTIORUM)

Since the first report of white mold in beans in the United States in 1915, it has become widespread and destructive throughout the most of bean production areas in the world (Strausbaugh and Forster, 2003). The fungus may invade the stem near soil line causing a rapid wilting and death of the entire plant or it may invade pods or branches particularly if they are in contact with the soil. The initial symptoms are water-soaked areas of indefinite size and shape on any aerial part of the plant. Lesions expand rapidly under moist conditions and the affected parts become a watery and rotten mass covered by white fungal growth.

After several days, the fungal growth on external plant surfaces forms a cushion shaped structure called a sclerotium. Sclerotia are typically 1/8 to 1/4 inch in diameter and are made of compact masses of fungal threads or hypae. It can be relatively hard particularly when dry. The white moldy growth and black sclerotia are characteristic of this disease. If climatic conditions within crop canopy become dry after infection, sclerotia, and mycelia on plant surfaces may be sparse or absent. However, infected tissue will appear off white due to bleaching by oxalic acid produced by the fungus and the epidermis will be rubbed off easily.

Under moderately cool and moist condition, sclerotia present in and on moist soil for several months can produce a fungal mushroom like fruiting bodies known as apothecium. This fruiting bodies (1/8 to 3/8 inch in diameter) produce millions of spores which can disperse through air. Severe outbreak of fungus is favored by average temperature (21°C) or less in combination with high humidity. Free moisture is requiring spreading the fungus on the plant surface. During harvest, sclerotia formed on or in plants may fall to the soil surface, remain in crop debris or be moved with seed.

Sclerotia can survive for three or more years but only sclerotia located within 2 inches of the soil surface form apothecia. Apothecia formation is optimum at soil moisture level of 50% field capacity and temperature of 60 to 65°F (Hogedorn and Inglis, 1986). The pathogen is most active between 60° and 70°F and ceases to grow when the temperature reaches 86°F (Strausbaugh and Forster, 2003).

20.3.4.1 MANAGEMENT

Always use sclerotia free seed of bean crop for planting. Four-year crop rotation with non-host plants (small grains, barley, wheat, corn, etc.) can be helpful but do not use potato, peas, sunflower in rotation because these crops are very susceptible to this fungus. Irrigation management is very important. Use the varieties having upright growing habit. Overdose of nitrogen stimulate vegetative growth; induce dense plant canopies which favor white mold development so excessive use of nitrogen fertilizer should be avoided.

Diseases of Beans Crop and Their Management 411

Proper row to row distance will reduce the disease severity by providing proper air circulation between plants. If available, use the tolerant varieties of bean cultivars for planting.

20.3.5 ASCHOCHYTA LEAF AND POD SPOT (ASCHOCHYTA BOLTSHAUSERI SACC. AND ASCHOCHYTA PHASEOLORUM SACC.)

Initially, small, circular, dark brown spots appear on the leaf. As the disease progress, lesion expands and turn to dark grey in color with concentric rings, irregular in shape often zonate appearance. These lesions coalesce together and cover most of the leaf surface. Leaf tissues may become black and necrotic. Under moist conditions, numerous pinhead size fruiting bodies known as pycnidia develop in diseased tissues. On the stem, lesions are more elongated, shunken, darker, and are covered with scattered pycnidia on the stem. Pod lesion is dark brown, slightly sunken, and zonate. These pod lesions commonly have pycnidia. Severe infection may cause premature defoliations and lowering of plant vigor. Stem girdling can cause plant collapse.

20.3.5.1 DISEASE CYCLE

The fungal pathogen can be seed-borne and can spread up to long distances. Local spread is accomplished by wind and rain borne pycnidiospores, the pathogen can survive on crop debris, volunteers plants, and infected seed. The disease often becomes established when fungal spores produce on old bean trash are carried by the wind into new crop and infect young crop. Disease is favored by cool temperature of 16° to 24°C (Hagedorn and Inglis, 1986). At temperature more than 30°C, the fungus will become inactive. Rainy, overcast weather favor the disease development. Under favorable conditions, fungus spreads from infected o healthy plants by rain splash and wind-borne water droplets. The disease is more severe early in the season and in wet years.

20.3.5.2 MANAGEMENT

Always use disease-free seed. Crop rotation with nonhost crops is also important. Use disease-resistant cultivars for planting. Seed treatment with fungicide Carbendazim @ 0.1% effectively manages the disease.

20.3.6 ANTHRACNOSE (COLLETOTRICHUM LINDEMUTHIANUM)

It is an important disease of beans, and under favorable conditions, it can cause 100% yield loss. Other commonly grown legumes susceptible to this disease include scarlet runner bean (*Phaseolus coccineus*), cowpea (*Vigna unguinculata*), broad or faba bean (*Vicia faba*) and lima bean (*Phaseolus lunatus*). The most characteristic symptoms appear on immature pods, leaves, and cotyledons. Stems are also infected. Initially, symptoms appear as small, dark brown to black lesions on cotyledons. Seedling stems (hypocotyls) may have rust-colored flecks or 1/4-inch-long elliptical, sunken lesions that weaken the stems, causing stunting or girdle the stems finally seedling death.

Circular to irregular, sunken lesions on bean pods are the most noticeable symptom of anthracnose. Pod lesions are up to 1/2-inch in diameter and are tan to rust-colored with a brown or purple border. Inside the pod, the seed coat may have brown to black lesions. During severe outbreaks of this disease, pods may dry and fail to fill. The lesion center is pale in color. It may contain gelatinous mass of pale salmon pink conidia under low temperature and high moisture condition. Later on, it starts drying and turn grey brown or black in color. Under favorable condition fungus invade the plant with high severity and young pod shrivel and dry up.

20.3.6.1 DISEASE CYCLE

The fungus survives primarily in bean seed, but inoculum can also survive in infested crop debris. Air currents, water, contaminated garden tools, or insects can spread the fungus from one place to another. Survival in soil or in plant residue varies greatly, depending on environmental conditions. Cool to moderate temperatures and prolonged periods of high humidity or free water on the foliage and young pods promote anthracnose development. Moisture is required for development, spread, and germination of the spores as well as for infection of the plant. A prolonged wet period is necessary for the fungus to establish its infection.

Lesions begin producing fruiting bodies (acervuli) and conidia within 2–4 days of infection. Frequent rainy weather increases disease occurrence and severity. The fungal spores can germinate and begin the infection process in as little as 6 hours when environmental conditions are favorable. Moist conditions favor sporulation and infection. Periods of wet weather combined with wind that carries spores to new infection sites, can result in serious outbreaks of this disease. Anthracnose is favored by cool temperature (16°)

Diseases of Beans Crop and Their Management 413

and wet conditions. The heavy and frequent rains with moderate temperature (19 to 25°C) and high RH (more than 70%) favor the progress of the disease in terms of vertical and horizontal spread (Kumar et al., 1999).

20.3.6.2 IMPORTANT FACTORS FAVOR THE DISEASE DEVELOPMENT

- Infected seeds and infected plant debris left in the field after harvest
- Moist and warm weather during the reproductive stages. Plants are most susceptible during the flowering stage from bloom to post-harvest
- Poorly drained soils
- Wet periods for about 12 hours or more favors the occurrence of infection
- Improper/inadequate nutrients to the plant

20.3.6.3 MANAGEMENT

Always use disease-free seeds for planting. Never save and use the seed from legume infected with anthracnose. Crop rotation and destruction of infected crop debris must be practiced because pathogen can survive in infected plant debris for more than 2 years. If possible, restrict the activity, entry, and movement of workers/farmers and farm implements in the field when the foliage is wet from rain or dew just to prevent the transport of fungal inoculum to new areas. Water management is very important, so avoid overhead irrigation, which will wet and liberate the spore on foliage.

Adequate plant to plant distance will promote foliar drying and inhibit fungal infection as well as sporulation. Weed management is also an important factor that will help proper air circulations and reduce moisture in the foliage. A hot-water seed treatment has been reported to kill the fungus in infested seed without reducing seed germination. Treat seed with fungicide Carbendazim @ 2g/kg seed before sowing. In disease-prone areas, spray right from the beginning fungicides Chlorothalanil @ 0.2% or Zineb @.02% or Mancozeb @ 0.1% (Reddy, 2010b).

20.3.7 POWDERY MILDEW (ERYSIPHE POLYGONI)

This disease occurs worldwide and occasionally causes significant yield loss. Initially, symptoms appear as faint dark areas on leaf which develop in small white powdery spots. Under favorable condition, these spots enlarge, coalesce, and cover the entire leaf surface. If infection occurs early in the growing season, leaf may become dwarf, yellow, and fall off. On pods small spots develop into white powdery mass of pathogen mycelium and spore. Fungus produces the spore easily transmitted from one area to other through rain, wind, and insects. It is seed-borne in nature. Infection and disease development is easily favored by moderate temperature (21°C) and low humidities (65% RH) (Hogedorn and Inglis, 1986).

20.3.7.1 MANAGEMENT

Infested plant debris must be burnt after harvesting the crops. Spray fungicides Dinocap or Karathane or Carbendazim or Baycor @ 0.1% and repeat the spray after 10–15 days if necessary.

20.3.8 ASHY STEM BLIGHT (MACROPHOMINA PHASEOLINA)

Initially, symptoms appear as black, sunken, elongate lesions just above the soil line of stem of bean crop. The infection may extend upward and downward. As a disease progress, the entire stem may infected and become diseased which can cause wilting, premature defoliation and death of the plant. Concentric rings may be seen within stem cankers. Numerous black sclerotial bodies and/or minute black pustule like pycnidia may form in diseased tissue.

The pinpoint pycnidia on gery ten or ash-colored lesion is the important characteristic of this disease, thus the name ashy stem blight. The fungus may be seed-borne and spread long distances due to this. The spread is by airborne conidia or pycniospores and by movement of sclerotia or pycniospore in infected crop debris or infested soil. The fungus is a warm temperature fungus and requires 24 to 27°C temperatures for disease development. Jabdel-Kader et al. (2010) also gave the first report of ashy stem blight on another host plant, i.e., *Aeonium canariese* in Egypt.

20.3.8.1 MANAGEMENT

Sanitations practices like deep plowing of field to bury the infected plant debris can be helpful. Seed treatment with fungicide is also important. Four to five years of crop rotation with nonhost or nonsusceptible crop should be practiced. Always use disease-free seed and resistant cultivars for planting.

Diseases of Beans Crop and Their Management 415

20.3.9 *ROOT ROT DISEASES*

Root rots are widespread diseases in the world and are often considered as a major constraint to bean production, reducing both yield and quality (Abawi and Widmer, 2000). Depending on the pathogen(s) involved in the development of the disease, general root rot symptoms might include any combination of various traits such as poor seedling establishment, damping-off, uneven growth, leaf chlorosis, premature defoliation, death of severely infected plants, and lower yield (Abawi et al., 2006; Schwartz et al., 2007). Fusarium root rot, Pythium root rot, Rhizoctonia root rot, and Sclerotinia root rot are important root rot diseases which reduce significantly the bean yield worldwide. Root rot problem is caused by soil-borne fungus.

20.3.9.1 *FUSARIUM ROOT ROT (FUSARIUM SOLANI [MART.] SACC. F. SP. PHASEOLI [BURK.] W.C. SNYDER AND H.M. HANS)*

Initially, symptoms appear as small, elongated, tan-red lesions in lower hypocotyle and upper tap root. Lesion increases in size and covers the entire root system and lower hypocotyle showing reddish-brown necrosis. The dry lesions can penetrate the root and hypocotyle tissue to develop deep elongated fissures leading to collapse of the plant under slight pressure. Diseased plants remain stunted in growth. The severely infected plant may die. Fusarium fungus is soil-borne in nature and can spread by irrigation or drainage water or by any mean which moves infested soil from one field to another field.

The disease is favored by 22 to 32°C temperatures, high soil moisture and by acidic soils (Hogedorn and Inglis, 1986). Compact soil also helps to increase disease severity. The fungus survives in the soil as thick-walled resting spores (chlamydospores). The fungus is able to live indefinitely in infested fields by germinating and reproducing on organic matter and roots and seeds of other plants in the soil until susceptible hosts become available. The spores germinate in response to nutrients exuded by germinating seeds and roots. Hyphae produced by the fungus penetrate the bean plant through wounds and natural openings.

20.3.9.2 *MANAGEMENT*

The best prevention strategy is to plant beans after soils have warmed up at a depth of 1/2 inch in a coarse, well-drained soil that has been optimally fertilized. A well-prepared seedbed promotes rapid seedling growth and minimizes root rot. Soil compaction should be minimized and hard pans should be broken up if they exist. The bean refuse should always be hauled where beans probably will not be grown for 6 or more years. Follow long crop rotation to check the disease. Always avoid spread of infected crop debris and infested soil from one field to another. Do not over irrigate the field. Break the soil compaction or deep dig the compact soil. If available, use disease-resistant cultivars in the field.

20.3.9.3 PYTHIUM ROOT ROT (PYTHIUM SPP.)

Pythium is a complex genus containing over 200 described species with a broad host range (Dick, 2001). The presence of the pathogens responsible for producing root rot and the severity of the disease are associated with intensification of land use, inappropriate crop rotations and/or reduced fallow periods, leading to a decline in soil fertility and a build-up of soil pathogen inoculum (Abawi et al., 2006). The genus *Pythium* belongs to the family Pythiaceae, order Pythiales, class Oomycetes, Phylum Oomycota, and kingdom Chromista (Kirk et al., 2008). Over the last 20 years, there has been an increase in the importance of *Pythium* bean root rots in several countries (Otsyula et al., 2003).

Pythium is a soil fungi causes huge losses under favorable conditions. The first symptoms appear as elongated, water-soaked area on the hypocotyls and roots. These areas slightly become shunken, tannish brown lesions which coalesce giving the entire root system and lower stem a collapsed, shunken, tan brown appearance because of wet soft rot. Rotting takes place both in primary and secondary root system. The total root system is destroyed, and the affected plant becomes stunted, wilts, and dies. High soil moisture is important for disease development.

20.3.9.4 DISEASE CYCLE

Pythium species can be found in various ecological areas such as soils in arable land, pastures, forests, nurseries, and marshes, and in water (Van der Plaats-Niterink, 1981). Fungus overwinters as resting spores (oospores) during periods of adequate soil moisture and temperature; oospores germinate to form structures called sporangia. In saturated soils, small swimming spores (called zoospores) are released. They swim to dry bean roots where they

Diseases of Beans Crop and Their Management

attach, germinate, invade the root and grow. Wet soils favor damping off. Disease is more severe in poorly-drained or saturated soils and where there is soil compaction. High amounts of organic matter and high planting densities can contribute to more severe disease. In general, soil temperature can affect spore germination, germ tube growth and zoospore discharge (Tedla and Stanghellini, 1992). However, each *Pythium* species has its specific optimal development conditions. For example, *P. ultimum* and *P. dissotocum* inhabit cool (10–15°C) and wet soil as saprophytes on crop residues while *P. aphanidermatum*, *P. irregulare*, *Pythium sylvaticum* Campbell and *P. myriotylum* occur in warm (25–36°C) and wet soil (Owen-Going et al., 2008).

20.3.9.5 MANAGEMENT

Always avoid over-irrigation in early stages of the crop development. Crop rotation with grain crops will always helpful to lower down the soil-borne inoculum. Soil should be well-drained, and proper spacing between plants should be maintained. Fungus may persist for many years through resistance structures such as oospores, zoospores, and sporangia (Onokpise, 1999). So fungicide Benomyl, Captafol, Captan, Carboxin, Metalaxyl, Propamocarb hydrochloride and Etridiazole will help to control *Pythium* root rot diseases on beans. Soil fumigants such as Methyl bromide, Chloropicrin, and Vorlex are highly effective biocides that kill *Pythium* agents (Abawi et al., 2006). Isolates of *Trichoderma* spp. and *Gliocladium* spp. are antagonists of *Pythium*-induced soil-borne diseases and several strains are already commercially available for the biological control of *Pythium* root rots (Howell et al., 1993; Fravel, 2005).

20.3.9.6 RHIZOCTONIA ROOT ROT (RHIZOCTONIA SOLANI KUHN) (TELEOMORPH: THANATEPHOROUS CUCUMERIS (FRANK) DONK.)

The fungus causes damping-off and attacks the stem below and above the soil surface. The young succulent plants die soon after infection. On older plants, as the stem becomes woodier, reddish-brown cankers extend longitudinally along the stem near the soil surface. The plants at this stage may show little indication of disease, except the yields may be reduced considerably. The fungus may enter the pith where it causes a brick-red discoloration.

Under moist conditions, the fungus mycelium can be seen as a brownish tuft on infected plant parts, small (0.3 to 0.5 mm), brown or tan sclerotia may develop later in diseased tissue. The fungus may attack pods in contact with soil, a rapid transit rot with off-white fungal growth can develop in these pods. Fungus can spread in any manner by which infested soil is disseminated. Optimum soil temperature, i.e., 18°C and normal moisture regimes are adequate for disease development (Hagedorn and Inglis, 1986). The fungus survives as mycelium or sclerotia in the soil or on infested crop debris. It can survive in or on bean seed. Irrigation water and soil movement spreads propagules of *Rhizoctonia* and moderate to high soil moisture and low soil temperature favor the disease.

20.3.9.7 MANAGEMENT

Crop rotation with grain crops is very important. Avoid growing bean in the land that has been cropped with potatoes, cole vegetables in last six months. If the crop is to be picked by hand or harvested as dry seed, hill up the soil around the base to encourage the new root growth above the diseased area. Irrigation management is also important to lower down the disease development. Shallow seeding and cultivation (soil not placed against the seedling) reduce the severity of the disease. Chemical seed treatments and in-furrow treatments reduce this disease. On disease appearance spray fungicide Carbendazim @ 0.1%. Soil application of *Trichoderma* spp. along with FYM before sowing the seeds, helps to control soil-borne problems. Integration of *Glomus mosseae* with *Trichoderma viride* gave total protection (Reddy, 2010a).

20.3.9.8 SCLEROTIUM ROOT ROT (SCLEROTIUM ROLFSII) OF CLUSTER BEAN

Sclerotium is a ubiquitous fungus with a broad host range. Initially wet rotting of bark is observed. Entire bark of the plant near collar region rots. The characteristic symptoms include white, cottony fungus growth is observed on affected portion as well as on parts in contact with soil. Gradually this haypal mat is converted into small, mustard like sclerotia that survive in the soil.

20.3.9.9 MANAGEMENT

Summer ploughing followed by irrigation helps in reducing the initial inoculum of the fungus. Field sanitation by destruction and burning all plant

Diseases of Beans Crop and Their Management 419

debris is effective. Crop rotation with cereals, etc., is effective in reducing the initial inoculum. Follow proper spacing between rows and plants. Avoid excessive irrigation from pre-flowering stage onwards. Soil drenching with fungicide Copper oxychloride @ 0.3% followed by Carbendazim @ 0.1% near the collar region in evening is recommended. Soil application of *Trichoderma* spp. supplemented with organic matter and green manuring are beneficial (Reddy, 2010b).

20.3.10 CERCOSPORA LEAF SPOT OF COWPEA (MYCOSPHAERELLA CRUENTA)

Initially symptoms appears as small, light yellow color spot in leaf, later they turn to bronze to dark grey, circular to elongated up to 10 mm across. The fungus produce airborne spore on the lower surface of leaf gives the spot as a grey to dark powdery appearance. When held up to the light the older leaf spots are darker, more reddish and often with a distinct ring. Dead tissues fall down and give a shot hole appearance. When spots expand and come together, leaf withers, die, and fall off. The fungus also infects the pods and stems.

The disease is carried over from one season to the next growing season on left over planting material, crop debris, alternative hosts, etc., The fungus also attack other bean i.e., French bean, soy bean, etc., These host plants can carry over infections to the next growing season. The disease spread through airborne spore.

20.3.10.1 MANAGEMENT

Destruction of crop remains, crop debris and removal of alternative hosts to prevent the carryover of fungus should be encouraged. Intercropping of cowpea with nonleguminous crop will check the spread of disease in field. Spray of fungicide Mancozeb @ 0.25% after flowering and pod initiation with maximum 2–3 sprays per cropping season will help to manage the disease.

20.4 VIRUS

Common mosaic, golden mosaic, yellow mosaic, phyllody, and curly top have been considered as widespread and serious viral diseases of bean crop. In the absence of proper disease management practices, it may cause extensive damage to the bean crop.

20.4.1 COMMON MOSAIC (COMMON MOSAIC VIRUS)

This is the common disease of bean worldwide. Initially irregular shaped, light-yellow, and dark green areas in mosaic like pattern appear in trifoliate leaves. These leaves may show considerably puckering, stunting, malformation, and downward curling. Infected leaves may be narrower than normal ones. Early infected plant is generally yellow and dwarfed. Some cultivars show systemic vein necrosis in leaves, stems, roots, pods or localized necrotic leaf lesions. Infected pods may be chlorotic and shorten in size. Infected plant produce very few, small, and abnormal fruits.

Black root is characterized by local necrotic lesions which extend into the vein causing systemic necrosis in the vascular system, this symptoms only occur in cultivars possessing the dominant resistance gene 1. This necrosis can extend into roots, stem, and meristem and may result in plant death if the plant is infected at an early stage. The virus overwinters in weed hosts or in infected seed. Long distance spread is by infected seed distributions. The virus is transmitted by mechanical sap inoculation, pollen, and infected seed. Several species of Aphid vectors, i.e., *Aphis gossypii*, *A. medicaginis*, *A. rumicis*, *Myzus persicae*, etc., transmit the virus under natural conditions.

20.4.2 MANAGEMENT

Always use virus free seed for planting. If available use resistant cultivars of bean. Removal and destruction of infected plants and weed hosts from bean field are important. Planting dates must be adjust to minimize the exposure to the aphids. The environment of bean plants can be altered by intercropping and this can result in a shift in host pathogen interaction which may affect disease incidence. Intercropping of bean with other non-host crops can reduce the disease incidence. Application of insecticide Carbofuran @ 1.0 kg ai/ha at the time of sowing followed by 3 to 4 sprays of Dimethoate or Monocrotophos both @ 0.1% at 10 days interval effectively control the insect vector (Reddy, 2010a).

20.4.3 GOLDEN MOSAIC (GOLDEN MOSAIC VIRUS)

Golden mosaic is a serious disease of beans. This disease is first reported by Capoor and Verma (1948) from India. Disease is characterized as systemic,

Diseases of Beans Crop and Their Management 421

bright yellow or golden mosaic pattern in leaves. The chlorophyll of the leaves is partially or completely destroyed. Infected plant remains dwarf. Infected pods may be malformed, stunted, and show mosaic pattern. Seeds remain discolored and small in size. Pod production is generally reduced. Disease is transmitted by white fly *Bemisia tabaci*. The virus is readily transmitted by bud grafting. Successful transmission of this virus require unusually warm temperature, i.e., 25° to 30°C. Likewise whitefly transmission is optimum at warm temperature i.e., 26.5°C (Hagedorn and Inglis, 1986).

20.4.5   MANAGEMENT

Follow crop rotation in the field. Destruction of infected plants from field is very important. It is essential to isolate field from virus reservoir plants and from insect vector sources plants, i.e., soybean, tobacco, tomato, cotton. Soil application of insecticide Carbafuron or Fensulfothion or Disulfoton or Disyston or Phorate @ 1.5 kg ai/ha at the time of sowing followed by 3–4 sprays of Dimethoate or Monocrotophos both @ 0.05% at 10 days interval significantly controlled the white fly population and disease spread. Spraying of 2% mineral oil has proved very effective in checking the spread of the disease (Singh, 1981).

20.4.6   YELLOW MOSAIC (YELLOW MOSAIC VIRUS)

Drooping of leaflet, malformed, and distorted leaves, intense yellow and green mottling of leaves is an important symptoms of this disease. The plant remains stunted in growth. Sometime puckering of lower leaves and veinal, stem or petiole necrosis or local lesions on leaves can be seen. Infected pods may be distorted or malformed. Virus is transmitted mechanically and by aphid vectors. Moisture and temperature is important which favor the spread of aphid vectors.

These viruses do not go through a life cycle like many other pathogens do. Pathogen cannot survive for long outside the plant and they rely solely on aphids to spread them from one live material to other live material. They can survive for long periods of time in seed but do not have stages that allow them to survive in the soil or stubble as sources of infection for future crops. The common sources of viruses are infected seed, other infected crops, and infected weed or pasture legumes.

422   *Diseases of Fruits and Vegetable Crops*

20.4.7   MEANS OF VIRUS SPREAD

These viruses are only spread by aphids in a non-persistent manner. Aphids acquire virus on their mouthparts when probing and are only infective for a maximum period of a few hours. Aphids spread or transmit the virus when they move to another plant and probe. After two or three probes, the virus is usually gone from mouthparts, so the period of infectivity may be short. Spread of this type of virus is usually fairly local (up to 200 m) due to the relatively short infectivity period.

20.4.8   MANAGEMENT

Destruction of alternate hosts from bean field or from adjoining areas is very important. Use of systemic insecticide Dimethoate or Monocrotophos @ 0.05% at 10 days interval control the aphid populations. Adjust the planting dates to minimize the exposure of crop to the aphid populations.

20.4.9   PHYLLODY (MYCOLPASMA-LIKE ORGANISM)

The disease is characterized by transformation of floral parts into green leafy structures. The infected plants are stunted. The pods proliferate too much and there is an excessive proliferation of branches which gives witches broom appearance to the plant. Leaves are drastically reduced in size. A few of the flowers on an infected plant look yellow in color but do not open and at maturity give rise to abnormal seedless pods (Reddy, 2010b). The disease is transmitted by grafting.

20.4.10   MANAGEMENT

Removal and destruction of infected plant are important. The disease symptoms can be suppressed temporarily by foliar application of 500 ppm Oxytetracycline hydrochloride solution at weekly intervals (Singh, 1991).

20.4.11   CURLY TOP (CURLY TOP VIRUS)

The prominent symptom is stunting. Leaf puckering, turning down, curling, yellowing are important. Finally, death of the plant takes place. Plant may

Diseases of Beans Crop and Their Management   423

drop their blossom. All the leaves become thicker than normal and being very brittle break off easily at the stem. Severely infected plant becomes dwarfed. Infected plants yield few or no pods. Pods are also stunted. The virus is transmitted to bean crop from various perennials and annuals plant by leaf hopper (*Circulifer tenellus*). Dry conditions during winters and spring favor the movement of insect vectors.

20.4.12   MANAGEMENT

Destruction of infected crop debris is important. If available, use curly top resistant cultivars for planting.

20.5   NEMATODE

Nematodes are microscopic, parasitic, and soil inhibiting in nature causes damage to the crop. Root-knot nematode, reniform nematode and cyst nematode attack the bean crops and become important factor for lower down the successful production of bean crops.

20.5.1   ROOT-KNOT NEMATODE (MELOIDOGYNE SPP.)

Meloidogyne incognita and *M. javanica* appear to be the most common root-knot species of beans and have been causing damage in tropics and sub-tropics. Lal and Ansari (1960) reported *M. arenaria* for the first time on beans from Bihar, India. *M. incognita* was responsible for 19.38 to 43.48% loss in pod yield of French bean (Das, 1994; Patel et al., 2001; Reddy and Singh, 1981) while *M. javanica* caused 30–40% loss in yield (Sharma et al., 2002). Severe infestation can result in 50% to 90% yield loss.

Aboveground symptoms usually involve stunting, chlorosis, or yellowing of lower leaves and yield reductions that often worsen over time. Plants may wilt during the heat of the day, especially under dry conditions or in sandier soils. These species are more troublesome in well-drained sandy soil with an average soil temperature 25 to 30°C. Root symptoms consist of root galls up to 1/2 inch (12 mm) in diameter in primary and secondary roots. Diseased roots are shortened and thickened, and there are few lateral roots, resulting in poor root system. Root-knot galls may vary in size and maybe nearly undetectable. On heavily infested plants, galls tend to fuse together so that large areas or the entire root may be swollen. The only visible symptoms on the roots are the presence of large egg masses (Reddy, 2008). *M. incognita* in association with soil-borne fungi such as *R. solani* and *F. solani* were responsible for root rot/wilt disease complex in French bean. In the presence of nematode, the root rot due to *R. solani* and *F. solani* increased considerably (Reddy et al., 1979; Sharma et al., 1981).

20.5.2 DISEASE DEVELOPMENT

Female root-knot nematodes deposit eggs in a gelatinous mass at or near the root surface. A worm-shaped larva hatches and then migrates either into the soil or to a different location in the root. The larva penetrates a suitable root by repeatedly thrusting its feeding structure (stylet) into cells at the root-surface. Within a few days, the larva becomes settled with its head embedded in the developing vascular system and it begins feeding. Enzymes secreted into cells at the feeding site cause an increase in cell size and number. As the nematode matures, the male reverts to the worm-shape (vermiform) and the female begins laying eggs. While root-knot nematodes move through soil slowly, anything that moves soil particles (equipment, shoes or boots, etc.) can carry nematodes to new locations. The rate of population increase and length of the life cycle depend on a number of factors including soil temperature, host susceptibility and soil type. Warmer soil temperatures and a suitable host will encourage the nematode to complete its life cycle considerably faster. Sandy, organic muck and peat soils are more favorable for population build up than are the heavier clay soils.

20.5.3 MANAGEMENT

Soil solarization for 4–8 weeks in summer can be effectively managing the root-knot nematode. Intercropping of bean with finger millet, chili, and groundnut helps in management of *M. incognita* race 2 (Rammappa, 1988). Crop rotation must be practiced. Spot application of neem cake @ 16.6 g/spot, 15 days before sowing was found effective in reducing root galls, number of egg masses and final nematode population (253 in neem cake treatment compared to 437 in control) and in increasing plant growth and yield (Ahmed and Choudhury, 2004). Seed treatment with Phenamiphos, Aldicarb, and Carbofuran all @ 1% concentration found to be effective in controlling root-knot nematode infecting French bean (Reddy, 2008).

Diseases of Beans Crop and Their Management

20.5.4 RENIFORM NEMATODE (ROTYLENCHULUS RENIFORMIS)

The amount and type of damage incurred by *R. reniformis* often depends on the host species and/or cultivar as well as the nematode population. General symptoms include reduced root systems, leaf chlorosis, overall stunting of host plants, reduced yields and finally affect plant longevity. Vadhera et al. (1995) demonstrated that interaction of root rot was maximum in simultaneous inoculation of *Rotylenchulus reniformis* and *F. solani* on beans.

20.5.5 LIFE CYCLE

Only the female reniform nematode parasitizes plant roots. An immature female imbeds her head into root tissue while the tail end remains in the soil. As she feeds and grows, her head end enlarges. Under favorable environmental conditions, a female will deposit approximately 50 eggs into the soil, surrounded by a gelatinous matrix, seven to nine days after infecting the root. A nematode goes through four molts before becoming an adult. The first molt occurs within the egg. After the eggs hatch, the larvae develop to the pre-adult stage without feeding or growing. Nematodes differentiate into adult males and females after the fourth molt. Nematode takes 24 to 29 days to complete cycle at optimum conditions.

Nematodes require at least a film of water in order to move through the soil, thus soil water content is a primary ecological factor. *R. reniformis* moves slowly through the soil under its own power. The overall distance travelled by a nematode probably does not exceed one meter per season. Nematodes move faster when soil pores are lined with a thin (a few micrometers) film of water than when soil is waterlogged. In addition, nematodes can be easily transported by anything that moves or carries particles of soil. Farm equipments, irrigation, flood or drainage water, animals (including humans) and dust storms spread nematodes in local areas, while over long distances nematodes are spread primarily with farm produce and nursery plants (Agrios, 1997).

20.5.6 MANAGEMENT

Soil solarization always helps to lower down the nematode population when soil is covered with polyethylene during summer months because it raises the soil temperature so that it becomes unfavorable to reniform nematodes (Heald

426 *Diseases of Fruits and Vegetable Crops*

and Robinson, 1987; Sharma and Nene, 1990). The fungus *Paecilomyces lilacinus*, a parasite of nematode eggs, with insecticide Carbofuran (Furadan 10G), Phenamiphos 15G @ 6.73 kg a.i./ha and Phenamiphos 15G @ 13.46 kg a.i./ha. gave better control than untreated plots (Vincente et al., 1991).

Furrow application of Carbofuran @ 2.5 kg ai/ha was also found effective (Reddy, 2008). Soil amendments have also been shown to aid in nematode reduction. Siddiqui and Alam (1990) reported neem (*Azadirachta indica* A. Juss) and mango (*Mangifer indica* L.) saw dust reduce nematode development and reproduction. Neem oil and karanj oil @ 2 tonnes/ ha were effective in reducing *R. reniformis* infecting French bean (Reddy, 2008).

20.5.7 *PIGEON PEA CYST NEMATODE (HETERODERA CAJANI)*

In many parts of our country, cyst nematode has been found associated with cowpea. The nematode retard emergence of leaves and reduce the number of flowering buds, flower, growing pods and yield (Reddy, 2008).

20.5.8 *MANAGEMENT*

Crop rotation of cowpea with paddy rice reduces the cyst nematode population. Cowpea rotated with paddy rice may be less affected by cyst nematode because of the negative effect of flooding on nematode densities (Reddy, 2008). If available, use resistant cultivars of crop for planting.

20.6 PHYSIOLOGICAL DISORDERS

Disturbance in normal functioning of the plant due to external, physical, chemical or mechanical agents lead to physiological disorders. Bald head and sneak head, sunscald, and ozone injury are considered as important physiological disorders of bean crop.

20.6.1 *BALD HEAD AND SNEAK HEAD*

Bald head is wide spread and common in occurrence. Severely stunted and malformed bean seedlings are the typical symptoms. This is the result of

Diseases of Beans Crop and Their Management 427

mechanical injury to bean and lima bean. Seed during threshing, seed with severely mutilated coats are removed in the milling and cleaning processes while those which have been injured only internally are ones which produce bald head and sneak head plant. Normally because of the lack of the normal growing tip, such plants are designated as baldhead plants. Sneak head, a malady brought about by chewing of the germinating seedlings by seed corn maggot has symptoms which develop those of bald head (Bhat, 2016). Seed with low moisture content are more prone to mechanical injury resulting in bald head.

20.6.2 MANAGEMENT

Take extreme care in all of the mechanical aspects of seed harvesting, cleaning, handling, and planting. The moisture content of seed should not be less than 14–15%.

20.6.3 SUNSCALD

It probably occurs almost everywhere beans are grown but rarely cause significant yield loss. Initially very tiny brown or reddish spots upon the upper or outer valve away from the center of the plant. These spots gradually increase in size until they appear as short streaks running backwards and downward from the vertical towards the dorsal suture. Sun scalding is caused by the concentration of the sun's heat on leaf tissue. Intense sunlight primarily causes sunscald but high temperature may also be responsible for this problem (Bhat, 2016).

20.6.4 MANAGEMENT

Optimum use of fertilizers and irrigation management are important to lower this problem.

20.6.5 OZONE INJURY

This problem is caused by air pollution by exposure to ozone. Ozone pollution originates in large cities and power plants. Polluted air can travel long distances at high altitudes and follow downdrafts into bean fields. Leaf symptoms may be tin white spot either or both surface of leaf. When these spots come together and gives white appearance in the leaf but the vein remains the green. The white spots are more common on the bean leaves but not exposed to direct sunlight. A common appearance of ozone injury in the field is bronzing, i.e., brownish purple discoloration on upper surface of the leaves. Severe injury can occur due to high warm temperature, high light intensity, high soil moisture and high plant nutrition.

20.6.6 MANAGEMENT

In possible, use tolerant cultivars of bean. Benomyl and other anti-oxidant chemicals have provided protection against ozone injury (Hagedorn and Inglis, 1986).

Bean crops are widely cultivated legumes in the world. Diseases caused by bacteria, fungus, virus, nematodes, etc., create a challenging problem in commercial agriculture and pose real economic threats to the farming system. The extensive damage caused by bacteria *Xanthomonas campestris* pv *phaseoli*, *Pseudomonas syringae* pv *phaseolicola*, *P. syringae* pv *syringae*, and *Corynebacterium flaccumfaciens* are menace to bean cultivation. The trend for management of bacterial diseases has shifted from antibiotics to eco-friendly technologies and lots of data need to be generated on efficacy of herbal extracts, neem, phylloplane bacteria and fluorescent pseudomonad. Biological management of bacterial diseases require a better understanding of the environmental factors which effect the survival, colonization, and spread of the agent. So there is need of much progress in this aspect. Fungal diseases cause huge loss in bean crops if not managed properly. Virus infection in bean crops also causes great yield loss. So there is urgent requirement to develop resistant as well as tolerant varieties against fungal and viral diseases of bean. Root-knot and reniform nematodes have been recognized as pathogenic causing economic damage to bean crop. We still have a long way to take a challenge against this destructive and insidious plant nematode. Continuous research efforts for working out integrated control of disease problem are essentially required due to rapid changes in the present cropping pattern and crop cultivation conditions. There is need to do a lot of research work and to popularize the integrated disease management practices against bean diseases.

KEYWORDS

- *Heterodera cajani*
- nematode
- ozone injury
- *Rotylenchulus reniformis*
- sunscald
- yellow mosaic

REFERENCES

Abawi, G. S., & Widmer, T. L., (2000). Impact of soil health management practices on soil-borne pathogens, nematodes and root diseases of vegetable crops. *Appl. Soil Ecol., 15,* 37–47.

Abawi, G. S., Ludwig, J. W., & Gugino, B. K., (2006). Bean root rot evaluation protocols currently used in New York. *Annu. Rep. Bean Improv. Cooperative, 49,* 83–84.

Abiy, T., Fekede, A., & Chemeda, F., (2006). Low land pulse diseases research in Ethiopia. In: Ali, et al., (eds.), *Proceedings of Food and Forage Legumes of Ethiopia: Progress and Prospects* (pp. 228–237). The workshop on food and forage legume, Addis Ababa, Ethiopia.

Agrios, G. N., (1997). *Plant Pathology* (4th edn., p. 803). Academic Press INC. London.

Ahmed, J. A., & Choudhury, B. N., (2004). Management of root-knot nematode, *Meloidogyne incognita* using organic amendment in French bean. *Indian J. of Nematology, 24,* 137–139.

Ariyaratne, H. M., Coyne, D. P., Vidaver, A. K., & Eskridge, K. M., (1998). Selection for common bacterial blight resistance in common bean: Effects of multiple leaf inoculation and detached pod inoculation test. *J. Amer. Soc. Hort. Sci., 123*(5), 864–867.

Ayele, H., (1991). Importance of haricot bean export to the Ethiopian economy. In: *Research on Haricot Bean in Ethiopia, an Assessment of Status, Progress, Priorities and Strategies* (pp. 31–34). Addis Ababa, Ethiopia.

Baudoin, J. P., Camarena, F., Lobo, M., & Mergeai, G., (2001). Breeding *Phaseolus* for intercrop combinations in Andean highlands. In: Cooper et al., (eds.), *Broadening the Genetic Bases of Crop* (pp. 373–384). Oxford, UK, CABI Publishing.

Belachew, K. M., Gebremariam, M., & Alemu, K., (2015). Integrated management of common bacterial blight (*Xanthomonas axonopodis* pv. *phaseoli*) of common bean (*Phaseolus vulgaris*) in Kaffa, Southwest Ethiopia. *Malays. J. Med. Biol. Res., 2*(2), 147–152.

Bhat, K. L., (2016). Physiological disorders of other vegetable crops. In: *Physiological Disorders of Vegetable Crops* (p. 92). Daya Publishing House. New Delhi.

Bhosale, S. B., Jadav, D. S., Patil, B. Y., & Chavan, A. M., (2014). Fungal disease of *Alternaria alternata* associated with soybean (*Glycine max (L.) Merr.*) and its biological control by efficient plant extract. *Indian Journal of Applied Research, 11*(4), 79–81.

Capoor, S. P., & Verma, P. M., (1948). Yellow mosaic of *Phaseolus lunatus* L. *Current Science, 17,* 152–153.

430   *Diseases of Fruits and Vegetable Crops*

Das, A. K., (1994). Assessment of yield loss due to *Meloidogyne incognita* on french bean and its management. *M.Sc. (Agriculture), Thesis*. Assam Agri. Univ. Jorhat.

Dick, M. W., (2001). The peronosporomycetes. In: McLaughlin et al., (eds.), *The Mycota VII. Part A. Systematic and Evolution* (pp. 39–72). Berlin, Germany: Springer-Verlag.

Dillard, H. R., & Cobb, A. C., (2008). *Alternaria alternata* and *Plectosporium tabacinum* on snap beans: Pathogenicity, cultivar reaction and fungicide efficacy. *Online: Plant Health Progress* doi: 10.1094/PHP-2008-1212-01-RS.

Dursun, A., Donmez, M. F., & Sahin, F., (2002). Identification of resistance to common bacterial blight disease on bean genotypes grown in Turkey. *Europ. J. Plant Pathol.*, *108*, 811–813.

Ferreira, S. A., & Boley, R. A., (1991). *Rotylenchulus reniformis* (p. 2). Crop Knowledge Master.

Fininsa, C., (2001). Epidemiology of bean common bacterial blight and maize rust in intercropping. *PhD Thesis*, Swedish University of Agricultural Sciences, Uppsal, Sweden.

Fourie, D., (2002). Distribution and severity of bacterial diseases on dry beans (*Phaseolus vulgaris* L.) in South Africa. *J. Phytopathol.*, *150*, 220–226.

Fravel, D. R., (2005). Commercialization and implementation of bio-control. *Ann. Rev. Phytopathol.*, *43*, 337–359.

Hagedorn, D. J., & Inglis, D. A., (1986). *Handbook of Bean Diseases* (p. 28). A3374. Madison, Wisconsin.

Heald, C. M., & Robinson, A. F., (1987). Effects of soil solarization on *Rotylenchulus reniformis* in the lower Rio Grande Valley of Texas. *J. Nematology.*, *19*(1), 93–103.

Hillocks, R. J., (2006). *Phaseolus* bean improvement in Tanzania, 1959–2005. *Euphytica.*, *150*, 215–231.

Howell, C. R., Stipanovic, R. D., & Lumsden, R. D., (1993). Antibiotic production by strains of *Gliocladium virens* and its relation to the bio-control of cotton seedlings diseases. *Biocontrol. Sci. Technol.*, *3*, 435–441.

Jabdel-Kader, M. M., El-Mougy, N. S., Aly, M. D. E. H., & Lashin, S. M., (2010). First report of ashy stem blight caused by *Macrophomina phaseolina* on *Aeonium canariense* in Egypt. *J. Plant Pathol. Microbiol.*, *1*, 101. doi: 10.4172/2157–7471.1000101.

Kidane, G., (1987). A review of bean agronomy research in semi-arid regions of Ethiopia. In: *Proceedings of Bean Research in Eastern Africa* (pp. 174–180). Mukono, Uganda.

Kirk, P. M., Cannon, P. F., Minter, D. W., & Stalpers, J. A., (2008). *Ainsworth & Bisby's Dictionary of the Fungi* (10th edn.). Wallingford, UK, CAB International.

Kumar, A., Sharma, P. N., Sharma, O. P., & Tyagi, P. D., (1999). Epidemiology of bean anthracnose *Colletotrichum lindemuthianum* under sub-humid mid hills zone of Himanchal Pradesh. *Indian Phytopath.*, *52*(4), 393–397.

Lal, B. S., & Ansari, M. N. A., (1960). Field studies on the root-knot nematodes (*Meloidogyne* spp.) (Nematoda: Heteroderidae). *Science and Culture*, *26*, 279–281.

Onokpise, O. U., (1999). Evaluation of macabo cocoyam germplasm in Cameroon. In: Janick, J., (ed.), *Perspectives on New Crops and New Uses* (pp. 394–396). Alexandria, VA, USA: ASHS Press.

Otsyula, R. M., Buruchara, R. A., Mahuku, G., & Rubaihayo, P., (2003). Inheritance and transfer of root rots (*Pythium*) resistance to bean genotypes. *Afr. Crop Sci. Soc.*, *6*, 295–298.

Owen-Going, T. N., Beninger, C. W., Sutton, J. C., & Hall, J. C., (2008). Accumulation of phenolic compounds in plants and nutrient solution of hydroponically-grown peppers inoculated with *Pythium aphanidermatum*. *Can. J. Plant Pathol.*, *30*, 214–225.

Patel, B. A., Patel, R. G., Patel, H. R., Vyas, R. V., & Patel, B. N., (2001). Management of root-knot of turmeric. *Natl. Cong. on Centenary of Nematol. in India-Appraisal and Future Plans, Indian Agri. Res. Inst* (p. 165). New Delhi.

Diseases of Beans Crop and Their Management

Popovic, T., Starovic, M., Aleksic, G., Zivkovic, S., Josic, D., Ignjatov, M., & Milovanovic, P., (2012). Response of different beans against common bacterial blight disease caused by *Xanthomonas axonopodis* pv. *phaseoli*. *Bulg. J. Agric. Sci., 18*, 701–707.

Ramappa, H. K., (1988). *Studies on Nematode Parasites of French Bean and Their Control.* M.Sc. (Agri), Thesis, University of Agricultural Science, Bangalore.

Reddy, P. P., & Singh, D. B., (1981). Assessment of avoidable yield loss in okra, brinjal, French bean and cowpea due to root-knot nematodes. *Third International Symposium on Plant Pathology* (pp. 93, 94), New Delhi.

Reddy, P. P., (2008). *Diseases of Horticultural Crops: Nematode Problem and Their Management* (p. 379). Scientific Publishers, Jodhpur, India.

Reddy, P. P., (2010a). Bacterial and viral diseases and their management. In: *Plant Protection in Horticulture* (Vol. 3, p. 288). Scientific Publishers, Jodhpur, India.

Reddy, P. P., (2010b). *Plant Protection in Horticulture: Fungal Diseases and Their Management* (Vol. 2, p. 359). Scientific Publishers, Jodhpur, India.

Reddy, P. P., Singh, D. B., & Ram, K., (1979). Effect of root-knot nematodes on the susceptibility of pusa purple cluster brinjal to bacterial wilt. *Current Science, 48*, 915–916.

Schwartz, H. F., Gent, D. H., Gary, D. F., & Harveson, R. M., (2007). Dry bean, *Pythium* wilt and root rots. *High Plains IPM Guide, a Cooperative Effort of the University of Wyoming.* University of Nebraska, Colorado State University and Montana State University.

Schwartz, H. F., Steadman, J. R., & Harveson, R. M., (2011). Rust of dry beans. *Crop Series/Diseases* (p. 2). Fact Sheet No. 2.936. Colorado State University.

Sharma, N. K., Thapa, C. D., & Nath, A., (1981). Pathogenicity and identity of myceliophagus nematode infesting *Agaricus bisporus* (Lange) Sing. in Himachal Pradesh (India). *Indian J. of Nematology, 11*, 230–231.

Sharma, S. B., & Nene, Y. L., (1990). Effects of soil solarization on nematodes parasitic to chickpea and pigeonpea. *J. Nematology, 22*(45), 658–664.

Sharma, S. B., Sharma, H. K., & Pankaj, (2002). Nematode problem in India. In: Prasad, D., & Puri, S. N., (eds.), *Crop Pest and Disease Management, Challenges for Millennium* (pp. 267–275). Jyoti Publishers, New Delhi, India.

Siddiqui, M. A., & Alam, M. M., (1990). Saw dusts as soil amendments for control of nematodes infesting some vegetables. *Biological Wastes, 33*, 123–129.

Singh, R. S., (2003). *Plant Diseases* (p. 686). Oxford and IBH publishing Co. Pvt. Ltd, New Delhi.

Singh, S. J., (1981). Relationship of pumpkin mosaic virus with its aphid vector, *Aphis gossypii* Glov. *J. Turkish Phytopath., 10*, 93–109.

Singh, S. J., (1991). Association of mycoplasma-like organism with french bean phyllody. *Ann. Rept.* Indian Institute of Horticultural Research, Bangalore.

Strausbaugh, C. A., & Forster, R. L., (2003). *Management of White Mold of Bean* (p. 4). A Pacific Northwest Extension Publication PNW 568. University of Idaho.

Tadesse, T., Ahmed, S., Gorfu, D., Beshir, T., Fininsa, C., Abraham, A., Ayalew, M., Tilahun, A., Abebe, F., & Meles, K., (2009). Review of research on diseases food legumes. In: Tadesse, A., (ed.), *Increasing Crop Production Through Improved Plant Protection, Proceeding of the 14th Annual Conference of the Plant Protection Society of Ethiopia (PPSE)* (Vol. 1, p. 598). Addis Ababa, Ethiopia. PPSE and EIAR, Addis Ababa Ethiopia.

Tedla, T., & Stanghellini, M. E., (1992). Bacterial population dynamics and interactions with *Pythium aphanidermatum* in intact rhizosphere soil. *Phytopathology, 82*, 652–656.

Vadhera, I., Sheila, B. N., & Bhat, J., (1995). Interaction between reniform nematode and *Fusarium solani* causing root rot of french bean. *Indian J. of Agriculture Science, 65,* 774–777.

Van der Plaats-Niterink, J., (1981). *Monograph of the Genus Pythium. spp in Mycology 21* (pp. 25–39). Baarn, The Netherlands: Centraalbureau voor Schimmelcultures.

Vincente, N. E., Sanchez, L. A., & Acosta, N., (1991). Effect of granular nematicides and the fungus *Paecilomyces lilacinus* in nematode control in watermelon. *J. Agric. Univ. P. R., 75*(3), 307–309.

Watson, A., (2009). Soil-borne diseases of beans. *Primefact, 586,* p. 6.

PART III
Other Food Crops

CHAPTER 21

Economically Important Diseases of Tea (*Camellia* sp.) and Their Management

KISHOR CHAND KUMHAR[*] and AZARIAH BABU

Tea Research Association, North Bengal Regional Research and Development Center, Nagrakata, Jalpaiguri, West Bengal, India

[*]*Corresponding author. E-mail: kishorkumarc786@gmail.com*

21.1 INTRODUCTION

Tea (*Camellia* spp.) belonging to family *Theaceae*, is an herbaceous, dicotyledonous, and perennial plantation crop. Globally, it is the second most commonly consumed non-alcoholic beverage next to water due to having numerous health benefits.

Globally, it is cultivated in China, India, Sri Lanka, Kenya, and Indonesia for domestic consumption as well as export purpose. In India, it is mainly cultivated in Assam, West Bengal, Himachal Pradesh, Uttar Pradesh (now Uttrakhand), Karnataka, Kerala, and Tamil Nadu states (Mohankumar and Radhakrishnan, 2013), covering approx. 5.63 lakhs hectare with an annual production of about 1208.78 million kg (Anonymous, 2015). In India, the CTC, green, and orthodox types teas are being manufactured. The first type of tea (CTC) is primarily for domestic consumption, and the other two types of teas are meant for export.

Tea favors moderate rainfall, high humidity, and partial shady environmental conditions for its luxuriant growth and production. However, such conditions are also conducive for the development of various air-borne as well as soil-borne fungal diseases, which are responsible for heavy crop loss in terms of quantity and quality. Indian tea plantation is subjected to attack of various diseases (Sarmah, 1960; Muraleedharan and Chen, 1999). Several important tea diseases are reported from Srilanka (Gadd, 1949), East Africa (Goodchild, 1952; Hansford, 1943), Mauritius (Ramlogun, 1971), China (Tzong and Shinfunchen, 1982), Japan (Hamaya, 1981) and Kenya (Sudoi and Langat, 1992).

436
Diseases of Fruits and Vegetable Crops

The fungi are the dominant ones among the different tea pathogens in India and more than hundred fungal species are reported to cause diseases of tea in India (Agnihothrudu, 1964). However, apart from fungi, bacteria, viruses, and algae also incite tea diseases in the world. Out of the several fungal diseases reported so far from India, in this article, our focus has been restricted to the economically important ones namely, blister blight, die back, grey blight, brown blight, branch canker, charcoal stump rot and brown root rot. The best possible efforts have been made to provide the latest information with respect to pathogen, its taxonomic position, disease symptoms, distribution, disease cycle, alternate host plants and also their management approaches.

List of Reported Tea Disease in India

Disease	Causal Organism
Root Diseases	
Brown root rot	*Fomeslamaoensis* (Murr.) Sacc. and Trott. = *Fomesnoxius* Corner.
Charcoal stump rot	*Ustulinazonata* (Lev.) Sacc. = Ustulinadeusta (Fr.) Petrak
Red root rot	*Poriahypolateritia* (Berk.) Cooke
Black root rot	*Roselliniaarcuata* Petch
Tarry root rot	*Hypoxylonasarcodes* (Theiss.) Mill.
Purple root rot	*Helicobasidium compactum* Boedijn
Violet root rot	*Sphaerostilberepens* B. Br.
Diplodia disease	*Botryodiplodia theobromae* Pat.
Rhizoctonia root rot	*Rhizoctonia bataticola* (Taub.) Butler = *Macrophomina phaseoli* (Maubl.) Ashby
Thorny blight	*Aglaospora* sp.
Poria canker	*Poriahypo brunnea* Petch.
Ganoderma	*Ganoderma lucidum* (Leys ex Fr.) Karst = (*Fomeslucidus*)
Ganoderma	*Ganoderma applanatum* (Pers. Ex Wallr.) Pat. = *Fomes Applanatus* (Pers) Wallr., and *Fomeslignosus* Klotzsch
Stem Diseases	
Nectria	*Nectria*spp. (N. cinnabarina (Tode ex Fr.) Fr. and *Nectria*sp.)
Branch canker	*Poriahypo brunnea* Petch.
Thread Blight (epiphytic)	*Marasmiuspulcher* (B. & Br.) Petch
Pink Disease	*Pelliculariasalmonicolor* (B. & Br.) Rogers = *Corticium Salmonicolor* B. & Br.
Thorny blight	*Aglaospora*sp.
Macrophoma	*Macrophomatheicola* Petch (= *Physalosporaneglecta* Petch)
Jew's ear fungus	*Auricularia auricular* (Hooke) Underwood.
Velvet blights	*Septobasdiumbogoriense* Pat., *S. tuberculatum* Boed. and Stein., *S. pilosum*
Aschersonia	*Aschersonia* sp.

Disease	Causal Organism
Leaf Diseases	
Black rot	*Corticiuminvisum* Petch and *C. theae* Bernard.
Blister blight	*Exobasidiumvexans* Massee
Botrytis	*Botrytis* sp.
Brown blight	*Colletotrichum camelliae* Mass. = *Glomerellacingulata* (Stonem) S. & v. S.
Grey blight	*Pestalozziatheae* Sawada.
Sooty molds	*Limacinulatheae* Syd. & Butl., *Capnodium* sp., *Meliola* sp.
Seedling Diseases	
Collar rot	*Phomopsis* sp.
Damping off	*Pythium* sp.
Diseases Caused by Other Than Fungi	
Red rust	*Cephaleurosparasiticus* Karst.

Source: Sarmah, 1960.

21.1.1 BLISTER BLIGHT

Pathogen: *Exobasidium vexans* Massee, 1898 Bull. Misc. Inf., Kew 111 (1898).

21.1.1.1 TAXONOMIC POSITION

Exobasidiaceae, Exobasidiales, Exobasidiomycetidae, Exobasidiomycetes, Ustilaginomycotina, Exobasidiomycota, Fungi.

The blister blight disease has been reported in India since 1855 (Venkata Ram, 1964). The fungus, *Exobasidiumvexans* Masee as a pathogen of this crop was reported in 1895 in upper Assam (McRae, 1910). The outbreak of blister blight on tea in the Darjeeling district of West Bengal was witnessed in 1908.

In North India, especially in the Himalayan region there are six species (*Exobasidium vexans* Massee, *E. vaccinii* (Fuck.) Wor, *E. butleri* Syd, *E. nilagiricum, E. celtidis,* and *E. trisepticum*) occurring on six different host plants including tea (Butler and Bisby, 1931).

This disease caused two major epiphytotics in N.E. India (Venkata Ram, 1964). Its third outbreak occurred in 1946 in South India and disease spread rapidly over most of the tea estates in Central Travancore, The Anamalias, and The Nilgiris due to exceptionally wet and prolonged rainy season. Later on, epiphytotics occurred in Ceylon, now Srilanka (Tubbs, 1947), Sumatra in 1949 and Jawa in 1951 (De Weille, 1959). From 1946, this disease laid to a crop loss estimating 180 million pounds of tea in South India from all tea districts. The huge crop loss due to its incidence was recorded in Indonesia as well (De Weille, 1959; Ordish, 1952). In Darjeeling region, generally, attains its peak infection during July to September and resulting in huge crop loss.

21.1.1.2 SYMPTOMS

The pathogen attacks the pluckable tender shoots, which results in an enormous yield loss up to 40% and quality deterioration is also evident even below 35% disease threshold level (Gulati et al., 1993). Its infection on the succulent leaves leads to development of tiny translucent spots within 3–10 days which later on enlarges. These leaf spots are depressed on the upper surface of the leaves. Simultaneously, the underside of the leaf becomes convex to form the typical blister lesions which turn necrotic and lead to dieback of the leaf ultimately. In addition to all these morphological changes, several physiological changes are reported in response to infection by the fungus (Rajalakshmi and Ramarethinam, 2000). It produces enzymes that degrade the polysaccharides of the cell wall and thereby gain entry into the cell (Albersheim et al., 1969).

21.1.1.3 ALTERNATE HOST

The pathogen is of bio-trophic nature and hence survives solely on tea plants (*Camellia* spp.) and no alternate host has been reported till date and it completes its life cycle exclusively on tea plants (Subba, 1946).

21.1.1.4 MANAGEMENT

Monitoring during active season (April onwards) is the most effective means to detect and manage it effectively. If the dull, humid, cloudy, and cool weather conditions prolong for longer time, there are chances of disease appearance. The plant protection code (PPC) has already been implemented with the aim to ensure the safe use of pesticides on this crop by the Tea Board of India (www.teaboard.in). As and when, its incidence is noticed (even if one or two blister spots), immediately spray the crop with hexaconazole (0.1%) or propiconazole (0.1%) and repeat the spray at an interval of two weeks, as prophylactic measure.

In the field, where disease has already been infected the crop, spraying of copper oxychloride (0.25%) or hexaconazole (0.1%) at an interval of 7–15 days depending upon its severity has been recommended. The shade trees plays a crucial role in regulation of disease, therefore, ensuring optimum shade is a prerequisite for the efficient management of this disease.

The fungicide application should also be done on nursery plants, pruned young tea, medium pruned tea, light pruned tea as well as un-pruned tea to minimize the inoculum load. The restriction on the movement of workers from infected area to healthy area will also be helpful in managing the disease to a considerable extent.

Application of the biological control agents like *Trichoderma viride, T. harzianum, Gliocladiumvirens, Serratia marcescens, Pseudomonas fluorescens* and *Bacillus subtilis* (Premkumar, 2001–2003; Sarmah et al., 2005; Phukan et al., 2006; Balasubramanian et al., 2006) have been utilized for the management of disease, when its incidence is low.

The indigenous isolates of *B. subtilis* and *Pseudomonas fluorescens* (Baby et al., 2004) were found to be more effective as these bacteria impart induce systemic resistance (ISR) in to plant system by producing higher amount of defense enzymes such as peroxidase (PO), polyphenol oxidase (PPO), phenylalanine ammonia lyase (PAL), chitinase, β-1,3-glucanase and phenolics (Saravanakumar et al., 2007). Certain plants species found growing in and around the tea fields like *Clerodendron infortunatum* and *Acorus calamus* were reported to possess very good fungicidal properties against some of the tea pathogens and control blister blight in Darjeeling and Upper Assam. Foliar application of aqueous extracts of *Equisetum arvense* also minimizes the blister blight incidence. Aqueous extracts of local herbs viz., *Clerodendron infortunatum, Polygonum hydropiper, Adhatodavasica, Acorus calamus, Xanthium strumarium* also found effective against various fungal diseases of tea when applied in two rounds at fifteen days intervals (Dutta et al., 2005).

21.1.2 DIEBACK: THE PATHOGEN: FUSARIUM SOLANI

21.1.2.1 TAXONOMIC POSITION

Nectriaceae, Hypocreales, Hypocreomycetidae, Sordariomycetes, Pezizomycotina, Ascomycota, and Fungi.

21.1.2.2 THE PATHOGEN PROFILE

The genus is one of the most important phytopathogens, mainly causing wilt, root rot and decline disease of several economically important crop plants. Its hyphae are hyaline, branched, and septate. The conidia are formed in slimy, effuse sporodochia or sometimes scattered on the mycelia. The conidiophores are quite short, simple or branched, septate, bearing a terminal phialide. The phialides are subulate, i.e., widest at the base, narrowing to a point like an owl. The slimy conidia are macro- as well as micro type (Dube, 2009). The *F. solani* produces the micro-conidia on mono-phialides of long conidiogenous cells. They are oval shaped, having 0–1 septa. The macro-conidia are bigger in size and straight with slight curvature. They possess 3–4 well developed transverse septa but majority have 4 septa. Rarely they are 5 septate with pointed apical cell with barely notched basal cell (Summerell et al., 2003). The size of macro conidia ranges from 13–15 x 3–4 μm to 27–29 x 4–5 μm and the size of micro conidia range from 3–4 x 1–2 μm to 9–10 x 1–3 μm (Ravi and Reddi, 2012).

21.1.2.3 ECONOMIC IMPORTANCE

Among various diseases (Chen and Chen, 1990), dieback, incited by *F. solani* (Devi et al., 2012) is one of the important diseases in India. It is responsible for huge crop loss, since it infects tender pluckable tea shoots. Its severity depends on the reaction of grown tea cultivars and certain abiotic factors. Some of the cultivars such as TV 19, TV 23, TV25, TV 26, TV 29, etc., are reported to be more susceptible and it is distributed in entire tea growing regions of India.

21.1.2.4 SYMPTOMS

The pathogen first infects leaf petioles and then infection progresses upward and downward, resulting in chlorosis followed by shoot mortality, hence the named 'die back.'

21.1.2.5 DISEASE CYCLE

During off-season, the pathogen survives on crop and debris. On the onset of favorable weather conditions, the conidia get germinated and initiate the infection. From infected plants, the conidia are transmitted through air currents and deposited on healthy plant. After deposition there they cause new infection. Under unfavorable climatic conditions, it produces chlamydospores to maintain the cycle.

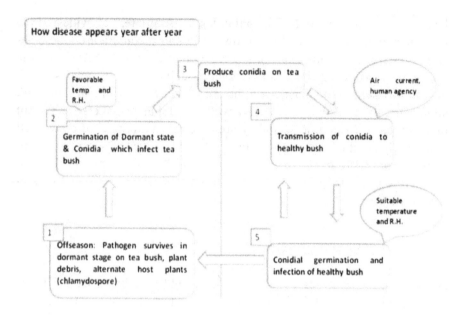

21.1.2.6 MANAGEMENT

Frequent monitoring of the crop removal and destroying the infected shoots followed by need-based application of copper oxychloride (0.25%) or copper hydroxide (0.25%) or hexaconazole (0.1%) or combi formulation of carbendazim 12% plus mancozeb 63% are the effective strategies to keep the pathogen under check (Kumhar et al., 2015).

21.1.3 GREY BLIGHT OF TEA

Pestalotiopsis theae, causing grey blight is yet another economically important pathogen that has been reported from all major tea growing countries of the world (Muraleedharan and Chen, 1997) and also as an endophyte (Worapong et al., 2003). In southern India, the grey blight of tea caused by this pathogen has resulted in 17% crop loss (Joshi et al., 2009) and 10–20% yield loss in Japan (Horikawa, 1986). Five species of *Pestalotiopsis* have been recorded from tea (Agnihothrudu, 1964), although *P. longiseta* (Speg.) H.T. Sun and R.B. Cao and *P. theae* are considered to be the major species causing grey blight (Joshi et al., 2009).

Economically Important Diseases of Tea (Camellia sp.) 443

21.1.3.1 THE PATHOGEN CURRENT NAME

Pseudopestalotiopsis theae (Sawada) Maharachch., K. D. Hyde and Crous, 2014

Synonymy:

Pestalotiatheae Sawada, 1915

*Pestalotiatheae*var. *minor* Steyaert, 1948 *Pestalotiatheae* (Sawada) var. *theae*, 1915 *Pestalotiopsis theae* (Sawada) Steyaert, 1949

Pestalotiopsis theae var. *minor* (Steyaert) Steyaert, 1949

Pestalotiopsis theae (Sawada) Steyaert var. *theae*, 1949

Teleomorph: *Pestalosphaeria* (Amphisphaeriaceae, Xylariales)

21.1.3.2 TAXONOMIC POSITION

Amphisphaeriaceae, Xylariales, Xylariomycetidae, Sordariomycetes, Pezizo-mycotina, Ascomycota, Fungi.

21.1.3.3 DISTRIBUTION

It is distributed in the entire tea growing areas of NE India, South India, and West Bengal including Dooars, Terai, and Darjeeling.

21.1.3.4 SUSCEPTIBLE CULTIVARS

Generally, it attacks a wide range of tea varieties/clones.

21.1.3.5 SYMPTOMS

This disease mainly infects maintenance foliage (mature leaf). Disease development takes place from middle portion of the leaves or from leaf margin. These spots are light to dark brown with grayish center on upper surface of leaves. These are more or less circular to oval, which are marked with concentric rings. Black and slightly bigger pustules are produced in concentric rings on the upper surface of the leaves (Sarmah, 1960).

21.1.3.6 PREDISPOSING FACTOR

The tea bushes suffering from abiotic stresses like sun scorching, imbalanced fertilizer application, hail damage, water stress, hard plucking, etc., are more prone to its attack apart from severe infestation of sap sucking insect pests like tea mosquito, red spider mite, and jassids.

21.1.3.7 SEASONAL OCCURRENCE

The grey blight disease incidence caused by *Pestalotiopsis longiseta* occurs during late June and reaches its highest peak incidence in late July (Oh et al., 2005).

21.1.3.8 PATHOGEN AND DISEASE CYCLE

The genus is widely distributed and has nearly 50 species. Its conidia are spindle or clavate shaped. The conidia are five celled, three central cells are dark colored while upper and lower most cells are hyaline. The upper most cells bear 2–3 cellular appendages, known as setullae, which may be branched in some species. The lower most cell bears a short hyaline pedicel. This genus causes gray blight of tea, *Pestalotiopsis theae* and mango leaf spot, *Pestalotiopsis mangiferae* (Dube, 2009).

Its infection takes place most probably by means of asexual conidia or fragmented spores (Espinoza et al., 2008). These conidia could survive during harsh weather conditions and may cause primary infection during favorable weather conditions. The plants subjected to abiotic stress are more susceptible to infection of this pathogen (Elliott et al., 2004; Keith et al., 2006). The source of the primary infection could be the wild tea plantations (Keith et al., 2006), flowers (Pandey, 1990), crop debris, soil, contaminated nursery tools (McQuilken and Hopkins, 2004), splashed water droplets (Hopkins and McQuilken, 1997; Elliott et al., 2004) and also air borne spores (Xu et al., 1999).

21.1.3.9 MANAGEMENT

It is better to collect and destroy/burn the diseased leaves present in the field. The disease can be managed by adopting recommended package of

Economically Important Diseases of Tea (Camellia sp.) 445

practices. The fungicides are the readily available means for its management however; their application should be done very wisely. The *Trichoderma* spp. and *Bacillus subtilis* are found effective against this disease. Oh et al. (2005) reported that *Bacillus subtilis* having the strongest antifungal activity against the grey blight pathogen. Kuberan et al. (2012) found that isolates of *Trichoderma* sp. showed variable degree of control of *Glomerella cingulata* and therefore such BCAs may be incorporated into integrated disease management practices for controlling brown blight disease of tea.

21.1.4 BROWN LEAF BLIGHT

21.1.4.1 THE PATHOGEN PROFILE

Colletotrichum camelliae Massee (Willis, 1899) Synonymy: *C. gloeosporioides* (Von Arx, 1957).

21.1.4.2 TAXONOMIC POSITION

Glomerellaceae, Incertaesedis, Hypocreomycetidae, Sordariomycetes, Pezizomycotina, Ascomycota, and Fungi.

The pathogen, Colletotrichum camelliae, was first described by Massee (Willis, 1899) from the living leaves of tea (*Camellia sinensis*) from Sri Lanka. It was synonymized as *C. gloeosporioides* (Von Arx, 1957). The pathogen's name as *C. camelliae* is still widely used in fungaria, websites, trade, and semi-popular literature as the causal agent of the brown blight disease of tea plantation (Weir et al., 2012; Hyde et al., 2009; Sosa de Castro et al., 2001). The *Glomerella cingulate* f. sp. *Camelliae* was proposed as the causal agent of disease on ornamental *Camellia saluenensis* hybrids (Dickens and Cook, 1989). Weir et al. (2012) reported that *G. cingulate* f. sp. *Camelliae* to belong to the *C. gloeosporioides* complex.

The genus *Colletotrichum* represents the anamorphic state of *Glomerella* (Phyllachoraceae, Phyllachorales). There are about 40 species of the genus causing anthracnose of many crops. Genus *Sphaceloma* and *Gloeosporium* are also known as 'anthracnose fungi,' which later on merged into genus *Colletotrichum* (Von Arx, 1957). There are 814 names in *Index Fungorum* (www.index fungorum.org; accessed 07 March 2016). It produces sub-cuticular, sub-epidermal, or epidermal acervuli with a characteristic peripheral ring of black, long, and stiff setae. The asexual spores (conidia) are hyaline, one-celled, cylindrical/falcate (sickle-shaped). The conidiophores (conidiogenous cells) are cylindrical in shape and termed as phialides (monophialic). This fungal pathogen causes anthacnose of jute, citrus, jowar, and red rot of sugarcane. Similar genus, i.e., *Gloeosporium* differs from *Colletotrichum* only in lacking setae.

The hyphae are hyaline initially, gradually it become dark colored. The conidia of the pathogen remain in viscid fluid, which swell under moist conditions and rupture the epidermis and liberates the conidial mass in the atmosphere.

It produces flat with an entire edge, aerial mycelium white, cottony, sparse colony on PDA. The colony is reverse white at first, then grey to black at the center. The hyaline, septate conidiophores develop directly on aerial mycelium. The conidiogenous cells are hyaline, cylindrical, and 16–42 x 1.5–4.5 μm in size. The conidia are hyaline, smooth-walled, cylindrical with obtuse ends, sometimes narrowed at the center or towards the base. The size of conidia is 9–25 x 3.5–7.5 μm. The appressoria are irregularly shaped, clavate, crenate, lobed, brown to dark brown, solitary, branched, catenate, with age sometimes complex chlamydospore like structures develop (size: 6.5–13.5 x 5.0–10.5).

21.1.4.3   SYMPTOMS

The brown blight is also an important foliage disease of tea that mainly infects older leaves. The brightening starts from leaf margin and progresses inwards. The edges of leaf spot well defined and more often marked with a delicate concentric zonation. The color of leaf spot remains yellowish to chocolate brown in the beginning, which gradually changes to brown. The browning starts from center and progresses outwards. The small fructification occurs on both sides of the affected leaves.

21.1.4.4   DISEASE CYCLE

Dissemination of conidia takes place through raindrop splashes and insects. The conidia seem pinkish colored when they are in mass, however hyaline individually, oval to oblong with rounded tip, aseptate, and measuring about 12–16 x 4–6 μ. Tea cultivars of large leaf size are most susceptible to brown blight than small and medium-sized leaves (Chen and Chen, 1982). It infects the older leaves, however occasionally it may also infect the branches and fruits.

Economically Important Diseases of Tea (Camellia sp.)

The pathogen overwinters on diseased leaves on the plant itself or shredded leaves. The high temp and high humidity are favorable for its infection. The production and dissemination of conidia are directly correlated with rainfall and high temp up to 35°C. The incubation period varies with tea varieties (Chen and Chen, 1982). In some regions of China, disease appears in the month of April. Rapidly develop during rainy season (May and June) and then declines during dry period (July). Within a period of 10 days, with average daily temperature of 28°C, total rainfall > 40 mm and RH > 80% may cause disease epidemics outbreak.

It may survive as a saprophyte for longer periods, as well. The humidity and moisture (>95%) are favorable for infection and disease development. Dastur (1921) reported that chili die back disease is closely dependent on heavy and prolonged dew deposits after rains. Secondary infection takes place through wind-borne conidia (Mehrotra, 1996).

21.1.4.5 MANAGEMENT

Same as mentioned above for grey blight disease management.

21.1.5 BROWN ROOT ROT DISEASE

21.1.5.1 THE PATHOGEN

Fomesnoxius Corner 1932

Phellinus noxius (Corner) G. Cunn. 1965

Phellinus orientalis Bondartseva and S. Herrera, Mikol.: 478 (1980)

Phellinidiumnoxium (Corner) Bondartseva and S. Herrera, 1992

21.1.5.2 TAXONOMIC STATUS

Hymenochaetaceae, Hymenochaetales, Incertaesedis, Agaricomycetes, Agaricomycotina, Basidiomycota, and Fungi.

21.1.5.3 HISTORICAL BACKGROUND AND GLOBAL DISTRIBUTION

The pathogen of brown root disease was first reported from Singapore as *Fomesnoxius* (Corner, 1932), later on, it was re-classified as *Phellinus noxius*

(Cunningham, 1965). Corner speculated that, this fungus was the cause of brown root rot of rubber trees and tea bushes. This fungus has a very wide host range and able to infect over 200 plant species of about 59 different families. Although, the host plants are mainly the woody plants, it can infect herbaceous plants, as well. It is widely distributed in tropical countries of Southeast Asia, Africa, Oceania, Central America, Caribbean (Pegler and Waterston, 1968), tropical Hainan Island of China (Tai, 1979), subtropical island of Okinawa in Japan (Abe et al., 1995) and both tropical and subtropical districts in Taiwan (Ann et al., 1999; Chang and Yang, 1998).

21.1.5.4 PATHOGEN'S ECOLOGY

The pathogen can flourish well at wide temperature regimes. Its optimum growth can be achieved at 30°C and completely seized at 8°C and hence restricted to tropical and subtropical regions of the world. The fungus prefers acidic condition and is capable of growing at a pH as low as 3.5 and its growth is completely seized at a pH of 7.5 (Ann et al., 1999).

21.1.5.5 HOST RANGE AND THEIR DISTRIBUTION

The pathogen is distributed in the African, Asian, Australian, Oceanian, Central American, and Caribbean countries. The pathogen has a marvelously wide host range and infecting hundreds of plants of monocotyledons and dicotyledonous which includes mahogany, teak, rubber, oil palm, tea, coffee, and cacao as well as a variety of fruit, nut, and ornamental trees. The updated list of host plants and their geographical distribution is available with the USDA – ARS systematic botany and mycology laboratory (https://nt.ars-grin.gov).

21.1.5.6 SYMPTOMS

The younger plants up to the age of 3 years, particularly grown in sandy soil, are comparatively more susceptible to this disease. Such plants ultimately lead to mortality within few months of infection. The affected plants start dying suddenly. The most characteristic symptoms of this disease is that the soil particles and pebbles remained attach with infected root through mycelial crust. This crust is so intact that it cannot be removed by several washing and even rubbing. Irregular, hard brown lines or reticulations are

Economically Important Diseases of Tea (Camellia sp.) 449

formed on infected wood of roots. Infected wood portion resembles honey comb structure.

21.1.5.7 SOURCE OF INOCULUM

The remains of a shade tree stump or a dead tea stump are the sources of infection. The secondary infection takes place through contact of infected roots or its parts.

21.1.5.8 MANAGEMENT

If young tea bush is infected, immediate uprooting of infected bush along with few bushes from the surrounding area and dispose them off properly, because such apparently healthy looking adjoining bushes may have inocula of this pathogen. The left over cut root portions of tea should be treated with copper fungicide @ 50 g/l of water. Remove all the roots of other trees from close vicinity and maintain proper sanitation. Grow green crop such as *Crotalaria anagyroides* in such area at least for 1–2 years and then start infilling. For any reasons, if uprooting could not be done immediately, then isolate the diseased area by digging trench (90 cm deep x 30 cm wide) around it.

If disease is found in old tea plantation, then mark the diseased area by digging a trench or uprooting two rows of healthy tea around it and then uproot.

The most efficient method of destroying the residual inocula is flooding the field (Chang, 1996), and the most practical way is to fumigate the infested soil with ammonia generated from urea amended in soil under alkaline conditions (Ann and Ko, 1994; Chang and Chang, 1999).

Application of *Trichoderma* spp. is effective in controlling the brown root rot of tea which can be applied into planting pits at the time of new planting. After planting it can also be applied with well-decomposed manure also (Barthakur and Dutta, 2011).

21.1.6 BRANCH CANKER

Pathogen: *Poriahypo brunnea* Petch (1916).

Current name: *Rigidoporus hypobrunneus* (Petch) Corner (1987)

21.1.6.1 TAXONOMIC POSITION

Polyporaceae, Polyporales, Incertaesedis, Agaricomycetes, Agaricomyco-tina, Basidiomycota, and Fungi. The branch canker is caused by *Poriahypo brunnea* Petch, was described by Tunstall and Sarma (1947) as a commonly occurring disease associated with the decay of dead woody tissues of tea in North East India. The pathogen is typical wound parasites that enter into tea plants normally through heavy pruning cuts. The fructification appearing as grey corky encrustations on the collar and branches are considered as an external symptoms (Sarmah, 1960).

21.1.6.2 MANAGEMENT

Careful removal of all dead woods during rejuvenation pruning (RP), medium pruning (MP) and light pruning (LP) is one of the means of its management. And immediately after pruning operation, apply *Trichoderma* spp. on such fresh pruned cuts that can take care of fresh infection.

It is reported that branch canker disease caused by *Macrophoma* sp. in south India, could be controlled by different ways. Among the different fungicides, tebuconazole was found to be the most effective in controlling the pathogen. Likewise, native bio-control agents such as *Bacillus* spp. *Pseudomonas* spp. and *Trichoderma* spp. were highly efficient in managing the pathogen (Mareeswaran et al., 2015).

21.1.7 CHARCOAL STUMP ROT PATHOGEN: USTULINAZONATA (LEV.) SACC.

Current name: *Kretzschmariazonata* (Lev.) P. M. D. Martin, 1976

Synonyms:

*Hypoxylondeustum*f. *Madagascariense* (Henn.) Hendr., 1948

*Hypoxylondeustum*f. *Madagascariense* (Henn.) Hendr.

Sphaeriazonata Lev., 1845

Ustulina vulgaris f. *madagascariensis* Henn., 1908

Ustulinazonata (Lev.) Sacc., 1882

Economically Important Diseases of Tea (Camellia sp.) 451

21.1.7.1 TAXONOMIC POSITION

Xylariaceae, Xylariales, Xylariomycetidae, Sordariomycetes, Pezizomycotina, Ascomycota, and Fungi.

The fundamental predisposing factor of this disease is lightning. This disease may cause sudden death of the infected plants. It occurs in India and Srilanka and has large number of host plants.

21.1.7.2 SYMPTOMS

The pathogen is soil borne and cause infection through roots. The air borne conidia may also cause infection on older branches through the wounds. The 'W' fan shaped patches of the mycelium on the wood below the bark is found. Such wood shows the presence of typical transverse double black line on primary root. On the collar region the pathogen produces black, hard, and effuse fructification which is wavy on the surface (Satyanarayana, 1980), in later stage, these fructification turn black and brittle and hence named as "Charcoal stump rot" (Sarmah, 1960).

Whitish or brownish-white fan-shaped mycelia is present on the wood of infected roots. The pathogen produces peculiar lines as drawn with a faulty nib of ink pen, which draws two parallel lines simultaneously. Fructification frequently found at the collar region of diseased tea bushes. The fructifications are somewhat ovoid in shape. Initially, they are thin, plate-shaped and grayish-white in color and dotted with black spots. A total of 24 records of this genus are available as on 10th March, 2016 (http://www.indexfungorum.org/names/Names.asp).

21.1.7.3 MANAGEMENT

Field sanitation may help in disease management to a great extent. Periodic monitoring of shade trees for the disease appearance may help in decision making for appropriate control measures. Soil application of *Trichoderma* spp. is the most effective and long-term management strategy for this disease.

21.1.8 RED RUST

The pathogen
Cephaleuros parasiticus, C. mycoides.

21.1.8.1 SYSTEMIC POSITION

Empire: Eukaryota, Kingdom: Plantae, Subkingdom: Viridiplantae, Infraking-dom: Chlorophyta, Phylum: Chlorophyta, Subphylum: Chlorophytina, Class: Ulvophyceae, Order: Trentepohliales Family: Trentepohliaceae, Genus: *Cephaleuros.*

21.1.8.2 THE PATHOGEN AND ITS HISTORICAL BACKGROUND

The red rust disease of tea caused by *Cephaleuros* spp. (Trentepohliaceae, Chlorophyta) was noticed for the first time way back in 1880 from Nowgong district, Assam, as blight of this crop. It was then termed as "white blight." Reason being, the leaves of infected bushes showed variegation, and in a few cases, the infected leaves were quite whitish. Later on, its occurrence was noticed in entire Assam Valley, Cachar, Sylhet, Chittagong, Dooars (Jalpaiguri), Darjeeling, and Terai region. The genus *Cephaleuros* is a pathogen of tea and other plants. Mann and Hutchinson (1907) identified the cause of red rust disease of tea as *C. virescens*. In India, way back in 1879 Cunningham reported presence of pathogenic alga as *Mycoidea parasitica* on tea plantation in Assam. He further recognized it as *C. virescens* Kunze in the year 1897. The incidence of this disease leads to adverse effect on both yield and quality (Ramya and Ponmurugan, 2012). The pathogen is aerophilic, filamentous green algae. It is aerophilic and terrestrial, and requires a film of water to complete its lifecycle. Three species of this alga have been reported from India, i.e., *C. parasiticus* Karsten, *C. solutes* Karsten and *C. virescens* Kunze. The first species *C. parasiticus* Karsten, is pathogenic on *Camellia sinensis* (Petch, 1923; Ponmurugan et al., 2010; Ramya et al., 2013), the second one *C. solutes* Karsten is pathogenic on *Pyrus* sp. (Chowdary and Jose, 1979) and the last one, i.e., *C. virescens* Kunze is infecting various plants (Cunningham, 1897; Mann and Hutchinson, 1904; Saxena, 1961; Panikkar et al., 1989; Gokhale and Shaikh, 2012).

21.1.8.3 GEOGRAPHICAL DISTRIBUTION

The disease is widely distributed in the zones of North-East India, Srilanka, and Bangladesh (Sana, 1989).

21.1.8.4 DISEASE INCIDENCE

In India, majority of the clones and seed stocks are susceptible to this disease. The changing climatic conditions have been playing a major role in aggravating the disease severity (Ramya et al., 2013). In Bangladesh, about 8–25% tea estates are affected by this disease (Sana, 1989).

21.1.8.5 SYMPTOMS

This disease infects the young stems and its symptoms normally apparent on the leaves. The leaf variegation with yellow patches is developed on the infected branches, which mimics the discoloration caused by mutation that are more common at higher elevation. The alga produces brick red or orange colored fructifications, which appear during the month of April to July on stems up to about two years of age. The patches are oval to oblong shaped. However, during August to March, the lesions appear purplish and there is no fructification. Severely affected stems of week plant show dieback in patches on the infected stems. The tissue necrosis on the stem appeared in patches.

The diseased patches (spots) on older leaves are usually circular and 5–7 mm in diameter. The patches can be long and oval, running along the entire length of the midrib (Anonymous, 2003).

Red rust disease symptoms: Leaf spot, leaf variegation and fructification on branch

454   *Diseases of Fruits and Vegetable Crops*

21.1.8.6   DISEASE CYCLE

The infection takes place during the fruiting period by means of spores, which are dispersed by wind or rain (Anonymous, 2003). The sporangia or thallus fragments with sporangia are deposited on susceptible host tissues. Infection occurs and symptoms develop under moist conditions when motile zoospoores are released from the sporangia, penetrate the host cuticle, and generate disc-like algal thalli with thread-like algal filaments. The infection reduces the photosynthetic area of leaves. It also leads to the defoliation, twig dieback and tissue necrosis. The pathogens reproduce and survive in spots on leaves or stems and in fallen plant host debris.

21.1.8.7   ALTERNATE HOST PLANTS

Species of *Cephaleuros* can also infect some other economically important tropical trees and shrubs such as *Piper methysticum, Piper nigrum, Magnolia grandiflora, Coffea arabica*, oil palm *Elaeisguineensis, Perseaamericana, Vanilla planifolia, Mangifera indica, Artocarpus altilis, Psidium guajava, Cocos nucifera, Theobroma cacao*, as well as *Citrus* spp (Nelson, 2008).

Acacia lenticularis, Adenanther apavonina, Albizzia amara, A. chinensis, A, lebbek, A. lucida, A. odoratissima, A. procera, A. Sumatrana, Crotalaria anagyroides, C. brownei, Dalbergia assamica, Derris robusta, Desmodium-gyroides, Gliricidiasepium, Indigoferadosua, I. teysmanii, Melia azedarach, Millettia dura, Parkiajavanica, Priotropiscytisoides, Tephrosia candida and *T. vogelii* (Sarmah, 1960).

Approximately 166 plant species have been found as its host in Assam, Mysore, Bihar, Tamil Nadu, North East, U.P., Kerala, Varanasi, Shillong, Darjeeling area of India (Jose and Chowdary, 1980).

21.1.8.8   MANAGEMENT

Adoption of proper cultural practices like proper drainage, optimum soil fertility, proper soil pH, adequate shade status and proper aeration help in disease management (Venkataramani, 1983; Keith et al., 2006) to the great extent. Pruning of severely affected sections is beneficial. The disease in young tea can be controlled by spraying copper oxychloride/copper hydroxide during cropping season (mid-April to mid-July). The first two sprays should be given at fortnightly interval and then the subsequent sprays

Economically Important Diseases of Tea (Camellia sp.) 455

should be done at monthly interval. Young shade trees/green crops in nurseries should be treated with copper oxychloride to avoid fresh infection. The rehabilitary spray of 2% MOP and urea in the severely affected sections take care of this disease (Huq et al., 2010). The *Streptomyces sannanensis* and *Streptomyces griseus* were found to be better performer than bacterial and fungal antagonists (Ramya et al., 2013); however, no satisfactory control was recorded due to the long and indefinite life cycle of the pathogen (Ponmurugan et al., 2011).

KEYWORDS

- **induce systemic resistance**
- **peroxidase**
- **phenylalanine ammonia-lyase**
- **plant protection code**
- **polyphenol oxidase**
- **rejuvenation pruning**

REFERENCES

Abe, Y., Kobayashi, T., Onuki, M., Hattori, T., & Tsurumachi, M., (1995). Brown root rot of trees caused by *Phellinus noxius* in windbreaks on Ishigaki island, Japan–incidence of disease, pathogen and artificial inoculation. *Annals of the Phytopathological Society of Japan, 61*, 425–433.

Agnihothrudu, V., (1964). A world list of fungi reported on tea (*Camellia* spp.). *Journal of Madras University, 34*, 155–271.

Albersheim, P., Jones, T. M., & English, P. D., (1969). Biochemistry of the cell wall in relation to infective processes. *Annual Review of Phytopathology, 7*, 171–194.

Ann, P. J., & Ko, W. H., (1994). Studies on ecology of brown root rot of fruit trees caused by *Phellinus noxius* and disease control. *Abstracts of Plant Pathology Bulletin, Taiwan, 3*, 69.

Ann, P. J., Lee, H. L., & Huang, T. C., (1999a). Brown root rot of 10 species of fruit trees caused by *Phellinus noxius* in Taiwan. *Plant Disease, 83*, 746–750.

Ann, P. J., Lee, H. L., & Tsai, J. N., (1999b). Survey of brown root disease of fruit and ornamental trees caused by *Phellinus noxius* in Taiwan. *Plant Pathology Bulletin, Taiwan, 8*, 51–60.

Anonymous, (2003). Protection of tea from red rust disease in the low country. Tea Research Institute of Srilanka, *Advisory circular, Serial No. 13/03*, pp. 1, 2.

Anonymous, (2015). *Crushed: Tea price record, The Telegraph, Siliguri* (p. 1). Reporter: Roopak Goswami and Abhranila Das.

456   *Diseases of Fruits and Vegetable Crops*

Arx, J. A. V., (1957). Die Arten der Gattung *Colletotrichum* Cda. *Phytopathologische Zeitschrift, 29*, 413–468.

Balasubramanian, S., Parathiraj, S., & Haridas, P., (2006). Effect of vermi-compost based *Trichoderma* (Vermiderma) on the recovery of pruned bushes and on the control of certain disease in tea (*Camellia sinensis* (L) O. Kuntze.). *Journal of Plantation Crops, 34*(3), 524–528.

Butler, E. J., & Bisby, G. R., (1931). *Fungi of India*, 93.

Chang, T. T., & Chang, R. J., (1999). Generation of volatile ammonia from urea fungicidal to *Phellinus noxius*in infested wood in soil under controlled conditions. *Plant Pathology, 48*, 337–344.

Chang, T. T., & Yang, W. W., (1998). *Phellinus noxius* in Taiwan: Distribution, host plants and the pH and texture of the rhizosphere soils of infected hosts. *Mycological Research, 102*, 1085–1088.

Chang, T. T., (1996). Survival of *Phellinus noxius* in soil and in the roots of dead host plants. *Phytopathology, 86*, 272–276.

Chen, T. M., & Chen, S. F., (1982). Diseases of tea and their control in the People's Republic of China. *Plant Disease*, 961–965.

Chen, Z. M., & Chen, X., (1990). *The Diagnosis of Tea Diseases and Their Control (Chinese)*. Shanghai, Shanghai Scientific and Technical Publishers, China.

Chowdary, Y. B. K., & Jose, G., (1979). Biology of *Cephaleuros* Kunze in nature. *Phykos, 18*, 1–9.

Corner, E. J. H., (1932). The identification of the brown root fungus. *Gard. Bull. Straits Settl., 5*, 317–350.

Cunningham, D. D., (1897). On certain diseases of fungal and algal origin affecting economic plants in India. *Sci. Mem. Med. Off. Army India, 10*, 95–130.

Cunningham, G. H., (1965). Polyporaceae of New Zealand. *N.Z. Dep. Sci. Indust. Res. Bull., 164*, 221–222.

Dastur, J. F., (1921). Dieback of chili in Bihar. *Mem. Deptt. Agr. India Bot., 11*, 129–144.

De Weille, G. A., (1959). Blister blight control in connection with climatic and weather conditions. *Archives of Tea cultivation, 20*, 1–116.

Devi, B. L., Thoudam, R., & Dutta, B. K., (2012). Control of leaf dieback disease of tea (*Camellia sinensis*) caused by *Fusarium solani. National Journal of Life Science, 9*(1), 55–58.

Dickens, J. S. W., & Cook, R. T. A., (1989). Glomerellacingulata on Camellia. *Plant Pathology, 38*, 75–85.

Dube, H. C., (2009). *An Introduction to Fungi* (3rd edn., p. 572). Publisher: Vikas Publishing House Pvt. Ltd., Delhi – 110 092.

Dutta, P., Sarmah, S. R., Begum, R., Phukan, I., Tanti, A. J., Kalita, J. N., Debnath, S., & Barthakur, B. K., (2005). 34thTocklai Conference. *Proc.*, 137–145.

Elliott, M. L., Broschat, T. K., Uchida, J. Y., & Simone, G. W., (2004). *Diseases and Disorders of Ornamental Palms*. American Phytopathological Society, St. Paul.

Espinoza, J. G., Briceno, E. X., Keith, L. M., & Latorre, B. A., (2008). Canker and twig dieback of blueberry caused by *Pestalotiopsis* spp. and a *Truncatella* sp. in Chile. *Plant Disease, 92*, 1407–1414.

Gokhale, M. V., & Shaikh, S. S., (2012). Host range of a parasitic alga *Cephaleurosvirescens* Kunz. ex Fri. from Maharashtra state. *India Plant Sciences Feed, 2*, 1–4.

Gulati, A., Gulati, A., Ravindranath, S. D., & Chakrabarty, D. N., (1993). Economic yield losses caused by Exobasidiumvexans in tea plantation. *Indian Phytopathology, 46*, 155–159.

Economically Important Diseases of Tea (Camellia sp.) 457

Hopkins, K. E., & McQuilken, M. P., (1997). *Pestalotiopsis on Nursery Stock*. in HDC Project News No 39. Horticultural Development Council, East Malling.

Horikawa, T., (1986). Yield loss of new tea shoots due to grey blight caused by *Pestalotialongiseta Spegazzini*. Bull Shizuoka Tea Exp Stn 12:1–8. Agnihothrudu, V. "A world list of fungi reported on tea (*Camellia* spp.)." *J. Madras Univ., B., 34*, 155–271, 1964.

Huq, M., Ali, M., & Islam, M. S., (2010). Efficacy of muriate of potash and foliar spray with fungtcides to control red rust disease (*Cephaleurousparasiticus*) of tea. *Bangladesh Journal of Agricultural Research, 35*(2), 273–277.

Hyde, K. D., Cai, L., Cannon, P. F., Crouch, J. A., Crous, P. W., et al., (2009). *Colletotrichum*–names in current use. *Fungal Diversity, 39*, 147–182.

Jose, G., & Chowdary, Y. B. K., (1980). Studies on the host range of the endophytic alga, *Cephaleuros* Kunze in India. *Revista De Biologia Tropical, 28*(2), 297–304.

Joshi, S. D., Sanjay, R., Baby, U. I., & Mandal, A. K. A., (2009). Molecular characterization of *Pestalotiopsis*spp. associated with tea (*Camellia sinensis*) in southern India using RAPD and ISSR markers. *Indian Journal of Biotechnology, 8*(4), 377–383.

Keith, L. M., Velasquez, M. E., & Zee, F. T., (2006). Identification and characterization of *Pestalotiopsis* spp. causing scab disease of guava, *Psidium guajava* in Hawaii. *Plant Disease, 90*, 16–23.

Keith, L., Ko, W. H., & Sato, D. M., (2006). Identification guide for diseases of tea (*Camellia sinensis*). *Plant Disease*, 1–4.

Kishor, C. K., Azariah, B., Mitali, B., Priyanka, B., & Tanima, D., (2015). Biological and chemical control of *Fusarium solani*, causing dieback disease of tea *Camellia sinensis* (L): An *in vitro* study. *International Journal of Current Microbiology and Applied Sciences, 4*(8), 955–963.

Maharachchikumbura, S. S. N., Hyde, K. D., Groenewald, J. Z., Xu, J., & Crous, P. W., (2014). Pestalotiopsis revisited. *Studies in Mycology, 79*, 121–186.

Mann, H. H., & Hutchinson, C. M., (1904). Red rust, a serious blight of the tea plant. *Indian Tea Association Bulletin, 4*, 1–26.

Mareeswaran, J., Nepolean, P., Jayanthi, R., Premkumar, Samuel, A. R., & Radhakrishnan, B., (2015). *In vitro* studies on branch canker pathogen (*Macrophoma*sp.) infecting tea. *Journal of Plant Pathology and Microbiology, 6*, 284. doi: 10.4172/21577471.1000284.

McQuilken, M. P., & Hopkins, K. E., (2004). Biology and integrated control of *Pestalotiopsis* on container-grown ericaceous crops. *Pest Management Science, 60*, 135–142.

McRae, W., (1910). *Bulletin of Agricultural Research Institute*, Pusa. No. 18.

Mehrotra, R. S., (1996). *Plant Pathology* (p. 771). Tata McGraw Hill Publishing Company Limited, New Delhi, India.

Muraleedharan, N., & Baby, U. I., (2007). Tea disease: ecology and control. In: Pimentel, D., (ed.), *Encyclopedia of Pest Management* (Vol. 2, pp. 668–671). [Electronic resource] CRC Press, Boca Raton.

Muraleedharan, N., & Chen, Z. M., (1999). Pests and diseases of tea and their management. *Journal of Plantation Crops, 25*, 15–43.

Nelson, S. C., (2008). *Cephaleuros Species, the Plant Parasitic Green Algae*. PD-43, 1–6. www.ctahr.hawaii.edu/freepubs (Accessed on 18 November 2019).

Oh, S. O., Kim, G. H., Lim, K. M., Hur, J. S., & Koh, Y. J., (2005). Disease progress of gray blight on tea plant and selection of a biocontrol agent from phylloplanes of the plant. *Research in Plant Diseases, 11*(2), 162–166.

Ordish, G., (1952). *Untaken Harvest* (p. 171). Constable et. Comp. ltd. London.

458 *Diseases of Fruits and Vegetable Crops*

Pandey, R. R., (1990). Mycoflora associated with floral parts of guava (*Psidium guajava* L.). *Acta Bot Sin.*, *18*, 59–63.

Panikkar, W. V. N., Ampili, P., & Chauhan, V. D., (1989). Observations on *Cephaleuros virescens* Kunze from Kerala. *Indian Journal of Economic and Taxonomic Botany*, *13*, 67–70.

Pegler, D. N., & Waterston, J. M., (1968). *Phellinus Noxius. No. 195 in: Descriptions of Pathogenic Fungi and Bacteria.* Commonw. Mycol. Inst., Kew, England.

Petch, T., (1923). *The Diseases of the Tea Bush* (p. 220). MacMillan & Co., London.

Phukan, I., Sarmah, S. R., Begum, R., Phukan, R., Dutta, P., Debnath, S., & Barthakur, B. K., (2006). *National Symposium on Microbial Diversity and plant Health* (p. 68). Department of Plant Pathology, B.C.K.V. West Bengal. Abstract.

Ponmurugan, P., Elango, V., Marimuthu, S., Chaudhuri, T. C., Saravanan, D., Gnanamangai, B. M., & Manjukarunambika, K., (2011). "Evaluation of actinomycetes isolated from southern Indian tea." *Journal of Plantation Crops*, *39*, 239–243.

Ponmurugan, P., Saravanan, D., & Ramya, M., (2010). Culture and biochemical analysis of a tea algal pathogen, *Cephaleurosparasiticus. Journal of Phycology*, *46*, 1017–1023.

Premkumar, R., (2001). *Report of the Plantpathology Division* (pp. 32, 33). Annual Report of UPASI TeaRes Foundation.

Premkumar, R., (2002). *Report of the Plantpathology Division* (pp. 35, 36). Annual Report of UPASI TeaRes Foundation.

Premkumar, R., (2003). *Report of the Plantpathology Division* (pp. 38, 39). Annual Report of UPASI TeaRes Foundation.

Rajalakshmi, N., & Ramarethinam, S., (2000). The role of *Exobasidium vexans* Massee in flavonoids synthesis by *Cammelia assamica* Shneider. *Indian Journal of Plant Protection*, *28*, 1–8.

Ramya, M., & Ponmurugan, P., (2012). Host pathogen interaction studies between susceptible and tolerant tea clones in relation to red rust disease. *International Journal on Algae*, *14*(4), 380–389. doi: 10.1615/InterJAlgae.v14.i4.80.

Ramya, M., Ponmurugan, P., & Saravanan, D., (2013). Management of *Cephaleurosparasiticus* Karst. (Trentepholiales: Trentepohliaceae), an algal pathogen of tea plant, *Camellia sinensis* (L) (O. Kuntze). *Crop Protection*, *44*, 66–74. doi: 10.1016/j.cropro.2012.10.023.

Ravi, C. M., & Reddi, K. M., (2012). Studies on cultural, morphological variability in isolates of *Fusarium solani* (Mart.) Sacc., incitant of dry root-rot of citrus. *Current Biotica*, *6*(2), 152–162.

Sana, D. L., (1989). *Tea Science* (pp. 224–226). Ashrafia BoiGhar, Dhaka.

Saravanakumar, D., Vijayakumar, C., Kumar, N., & Samiyappan, R., (2007). PGPR-induced defense responses in the tea plant against blister blight disease. *Crop Protection*, *26*(4), 556–565.

Sarmah, K. C., (1960). *Diseases of Tea and Associated Crops in North East India* (p. 68). Memorandum No. 26, Indian Tea Association, Tocklai Experimental Station, Jorhat, Assam.

Sarmah, S. R., Dutta, P., Begum, R., Tanti, A. J., Phukan, I., Debnath, S., & Barthakur, B. K., (2005). Int. Symp. on Innovation in Tea Science and Sustainable Development in Tea Industry. *Proc. China Tea Science.*

Saxena, P. N., (1961). *Algae of India. 1. Chaetophorales* (Vol. 57, pp. 1–59). Bulletin of National Botanical Garden.

Sosa de Castro, N. T., Cabrera, D. A. M. G., & Alvarez, R. E., (2001). *Primera Información de Colletotrichum Camelliae Comopatógeno de Camellia Japonica, en Corrientes.* Retrieved from: www.unne.edu.ar/cyt/2001/5-Agrarias/A-056.pdf (Accessed on 18 November 2019).

Subba Rao, M. K., (1946). *Blister Blight of Tea in South India* (p. 14). UPASI Coonoor Paper, No. 4.

Tai, F. L., (1979). *Sylloge Fungorum Sinicorum* (Vol. 19, pp. 78–92). Science Press, Academia Sinica, Peking, China.

Tubbs, F. R., (1947). Spraying and dusting in the control of blister blight of tea. Tea Q.

Venkata, R. C. S., (1964). Plant protection problems in tea blister blight control in Southern India. *Plant Protection Bulletin, New Delhi, 16*, 1–12.

Venkataramani, K. S., (1983). *Stem and Root Diseases of Tea* (p. 108). Tamil Nadu Tea Development Corporation Ltd., Coonoor, Tamilnadu, India.

Weir, B. S., Johnston, P. R., & Damm, U., (2012). The *Colletotrichum gloeosporioides* species complex. *Studies in Mycology, 73*, 115–180.

Willis, J. C., (1899). DCLII – Tea and coffee diseases. *Bulletin of Miscellaneous Information Royal Botanical Gardens Kew, 1899*, 89–94.

Worapong, J., Inthararaungsom, S., Stroble, G. A., & Hess, W. M., (2003). A new record of Pestalotiopsistheae, existing as an endophyte on Cinnamomum iners in Thailand. *Mycotaxon, 88*, 365–372.

Xu, L., Kusakari, S., Hosomi, A., Toyoda, H., & Ouchi, A., (1999). Post-harvest disease of grape caused by *Pestalotiopsis* species, Japan. *Annals of Phytopathological Society, 65*, 305–311.

CHAPTER 22

Symptomatology and Etiology of Alternariose in Root, Fruits, and Leafy Vegetables

UDIT NARAIN,[1] ALKA KUSHWAHA,[2*] RAJENDRA PRASAD,[1] and VED RATAN[1]

[1]*Department of Plant Pathology, C.S. Azad University of Agriculture and Technology, Kanpur, Uttar Pradesh, India*

[2]*Department of Botany, D.A-V, College, Kanpur, Uttar Pradesh, India*

[*]*Corresponding author. E-mail: alkakushwaha17march@gmail.com*

22.1 INTRODUCTION

Vegetables are the edible plants that store up reserved food in roots, stems, leaves, and fruits and which are eaten cooked or raw as a *salad*. The vegetables rank next to the cereals as the source of carbohydrate food. The nutritive value of vegetables is tremendous, because of the presence of indispensable mineral salts, and vitamins. India grows a large variety of vegetables which are affected by a number of *Alternaria* spp. from seedling to maturity stage in the field and in storage and transit and constitute common component of flora of vegetable seeds.

Miscellaneous types of symptoms are seen in diseases incited by *Alternaria* spp. as necrosis, chlorosis, scab, leaf-spot, ring spotting (zonations), leaf blight, blossom blight, blotch, defoliation, bud rot, fruit rot, root rot and also some other types of abnormalities in infected/affected vegetable crops (Walker, 1952; Narain and Kant, 2008). Most of the common names of Alternaria diseases are: early blight, stem, and fruit canker, purple blotch, target leaf spot, Alternaria leaf blight, black/brown leaf spot, and black rot (Neergaard, 1945; Ellis, 1971).

There are about one hundred types of vegetables grown in India and the world, which are utilized in different ways. Like other crops, they are also affected by several diseases of different origins in which Alternaria diseases are one of them. Because of their occurrence from seedling to maturity stage and storage, they reduce the quantity and quality of their produce.

Various types of root, fruits, and leafy vegetables have been included in this chapter which is affected by *Alternaria* species. The symptoms produced by *Alternaria* spp. (symptomatology) and characteristic morphological features (etiology) of causative species have been given in greater detail. Where there is the involvement of more than one or two species with the causation of Alternaria disease, a very simple (easy) and feasible key has been framed and given for ready and correct identification of species (Narain et al., 2016).

Forty-two vegetables grouped into roots (six), fruits (24) including 19 cucurbits, leguminous pods (7) and 5 leafy vegetables were found to suffer from eleven species of *Alternaria* in which the most dominant was *Alternaria alternata*, a cosmopolitan in distribution on 27 hosts followed by *A. cucumerina* on 16 cucurbits, *A. tenuissima* on 13 hosts, one on each host *A. brassicae, A. brassicicola, A. radicina, A. cyamopsidis,* and *A. solani* on five and two each *A. dauci* and *A. raphani* affected in all 42 vegetables (Table 22.1).

Description of Altarnariose has been given in grouping the vegetables taken up for discussion.

TABLE 22.1 Association of *Alternaria* spp. with Vegetable Crops

Sl. No.	Vegetables	Botanical Name	Family	Alternaria spp.
I	**Root Vegetables**			
1.	Radish	*Raphanus sativus*	Brassicaceae	*A. alternata, A. raphani*
2.	Turnip	*Brassica rapa*	Brassicaceae	*A. raphani*
3.	Knol knol	*Brassica olracea* var. *gongylodes*	Brassicaceae	*A. brassicae, A. brassicicola*
4.	Sweet potato	*Impomoea batatas*	Convolvulaceae	*A. solani, A. alternata*
5.	Carrot	*Daucus carota*	Apiaceae	*A. dauci, A. radicina*
6.	Ratalu	*Dioscorea* sp.	Dioscoraceae	*A. alternata*
II	**Fruit Vegetables**			
7.	Tomato	*Lycopersicon esculentus*	Solanaceae	*A. alternata, A. solani*
8.	Brinjal	*Solanum melogena*	Solanaceae	*A. alternata, A. melogenae, A. solani*

Symptomatology and Etiology of Alternariose in Root 463

TABLE 22.1 *(Continued)*

Sl. No.	Vegetables	Botanical Name	Family	Alternaria spp.
9.	Chili	*Capsicum annuum*	Solanaceae	*A. alternata,* *A. solani*
10.	Pepper	*Capsicum annuum*	Solanaceae	*A. alternata,* *A. solani*
11.	Okra (Lady's finger)	*Abelmoschus esculentus*	Malvaceae	*A. tenuissima*
	Cucurbits			
12.	Round gourd	*Cucurbita vulgaris* var. *fitulosus*	Cucurbitacea	*Alternaria cucumerina*
13.	Scarlet gourd	*Coccinia indica*	Cucurbitacea	*A. cucumerina,* *A. alternata*
14.	Muskmelon	*Cucumis melo*	Cucurbitacea	*A. cucumerina*
15.	Watermelon	*Citrullus lanatus*	Cucurbitacea	*A. cucumerina,* *A. alternata*
16.	Phoot	*C. melo* var. *momordica*	Cucurbitacea	*A. cucumerina,* *A. tenuissima*
17.	Long gourd	*C. melo* var. *utilissimus*	Cucurbitacea	*A. cucumerina,* *A. tenuissima*
18.	Cucumber	*Cucumis sativus*	Cucurbitacea	*A. cucumerina,* *A. alternata*
19.	Winter squash	*Cucurbita maxima*	Cucurbitacea	*A. cucumerina,* *A. alternata*
20.	Summer squash	*Cucurbita pepo*	Cucurbitacea	*A. cucumerina*
21.	Bottle gourd	*Ligenaria siceraria*	Cucurbitacea	*A. cucumerina,* *A. alternata*
22.	Ridge gourd	*Luffa acutangula*	Cucurbitacea	*A. cucumerina,* *A. alternata*
23.	Bitter gourd	*Momordica charantia*	Cucurbitacea	*A. cucumerina,* *A. tenuissima,* *A. alternata*
24.	Small bitter gourd	*Momordica dioica*	Cucurbitacea	*A. cucumerina,* *A. alternata*
25.	Pointed gourd	*Trichosanthes dioica*	Cucurbitacea	*A. cucumerina,* *A. alternata*
26.	Sponge gourd	*Luffa cylindrica*	Cucurbitacea	*A. alternata*
27.	Snake gourd	*Trichosanthes anguina*	Cucurbitacea	*Alternaria tenuissima*
28.	Cucurbit	*Cucurbita vulgaris*	Cucurbitacea	*A. alternata*
29.	Ash gourd	*Benincasa hispda*	Cucurbitacea	*Alternaria cucumerina*

TABLE 22.1 *(Continued)*

Sl. No.	Vegetables	Botanical Name	Family	Alternaria spp.
30.	Pumpkin	*Cucurbita moschata*	Cucurbitacea	*A. cucumerina*
	Legumes and Pods			
31.	Guar	*Cyamopsis tetragonaloba*	Fabaceae	*Alternaria cyamopsidis, A. tenuissima*
32.	Cowpea (Lobia)	*Vigna unguiculata*	Fabaceae Fabaceae	*A. tenuissima, A. alternata*
33.	French bean	*Phaseolus vulgaris*	Fabaceae	*A. alternata*
34.	Sem	*Lablab purpurea*	Fabaceae	*A. tenuissima, A. alternata*
35.	Moth bean	*Vigna acontifolia*	Fabaceae	*A. alternata*
36.	Winged bean	*Psophocarpus tetragonolobus*	Fabaceae	*A. alternata*
37.	Pea	*Pisum sativum*	Fabaceae	*A. alternata A. tenuissima*
III	**Leafy Vegetables**			
38.	Lettuce	*Lactuca sativa*	Asteraceae	*Alternaria tenuissima*
39.	Spinach	*Spinacea oleracea*	Chenopodiaceae	*A. tenuissima*
40.	Bathuwa	*Chenopodium album*	Chenopodiaceae	*A. tenuissima*
41.	Sova	*Anethum sowa*	Umbelliferae (Apiaceae)	*A. alternata, A. dauci*
42.	Chaulai	*Amaranthus blitum*	Amaranthaceae	*A. tenuissima*

22.1.1 RADISH

Radish (*Raphanus sativus* L. Family Brassicaceae) is attacked by two important species of *Alternaria viz., A. raphani* (Atkinson, 1950) and *A. alternata* (Suhag et al., 1985).

The symptoms produced by *A. raphani* appear as brown to dark brown patches starting from margin of leaves, which develop inward in irregular fashion. The effected portions (areas) soon turn blighted and wither (Neergaard, 1945; Narain and Saksena, 1975; Narain et al., 1982; Sangwan et al., 2002). On seed pods of radish, the causative fungus produces circular black spots, up to 4 mm in diameter as "Black pod blotch" of radish (Ellis, 1971). *A. raphani* also occurs on *Matthiola incana* and other Brassicaceous hosts.

Symptomatology and Etiology of Alternariose in Root 465

The symptoms of Alternaria leaf spot of radish caused by *Alternaria* alternata, initially appear as minute dark brown lesions which start from margin or center of leaves. At the later stage of disease development, a number of spots coalesce to from the irregular dark brown patches (Suhag et al., 1985). The fungus also infects some other plants of crucifers like broccoli (Chand et al., 2007), rapeseed, and mustard (Khalid et al., 2004).

Both the pathogens are seed borne in nature (McLean, 1947; Thakur et al., 1981). In a report, *A. brassicae* is known to infect radish also (Mitter and Tandon, 1930).

22.1.2 KNOL KHOL

The Alternaria disease in this cole crop, knol khol (*Brassica oleracea* var. *gongylodes*) may be attributed to two species of *Alternaria viz., A. brassicicola* and *A. brassicae*. The symptoms overlap to some extent and both the pathogens may attack simultaneously but there are some points of distinction.

The leaf spot disease due to *Alternaria brassicicola* form dark brown to almost black, circular spots on leaves which may vary from 4–10 mm in size. In case of severe infection, several spots/lesions coalesce to cover larger areas. The spots are more often surrounded by concentric zonations and darker in color because of propose sporulation of the causative fungus. More severe occurrence of disease may be noticed in seed crop of the knol knol. Severely infected leaves lastly dry up and may fall prematurely (Chupp, 1925; Chupp and Sherf, 1960).

A. brassicae also occurs on this cole crop. The spot incited by *A. brassicae* has much in common with *A. brassicicola*. It trends to remain smaller in diameter and also lighter in color (Anonymous, 1969). *A. brassicae* and *A. brassicicola* are seed borne and they may cause shriveling and low germination. Seedlings growing from infected seeds may be attacked in the hypocotyls region (Atkinson, 1950).

22.1.3 TURNIP

In the case of turnip (*Brassica rapa*), when foliage is infected, the roots also get affected which result in the reduction of number and size of root. The lesions are nearly circular, often zonate and show various shades of brown or black coloration. The disease is caused by *Alternata raphani*, which also

466   *Diseases of Fruits and Vegetable Crops*

causes leaf spots on the plants kept for seeds (Atkinson, 1950). A rot of turnip in storage has been described by Chupp (1935).

A. raphani is also known to produce black spots on seed pods of radish crop as "Black pod blotch" (Walker, 1952). Narain and Saksena (1975) have reported *A. raphani* on turnip to cause the leaf spot disease in India.

The species of *Alternaria* which occur on Brassicae have been dealt by Wiltshire (1947).

22.1.3.1 ALTERNARIA BRASSICAE (BERK.) SACC. (MICHELIA, 2: 129, 1880)

Mycelium immersed, hyphae branched, septate, hyaline, smooth, 4–8µm thick; *Conidiophores* arising in groups of 2–10 or more from the hyphae, emerging through stomata, usually simple, erect, straight or flexuous, frequently geniculate, more or less cylindrical but often slightly swollen at the base, septate, mid-pale grayish olive, smooth, up to 170µm long, 6–11µm thick, bearing one to several small but distinct conidial scars; *Conidia* solitary or occasionally in chains of up to 4, acropleurogenous arising through small pores in the conidiophore wall, straight or slightly curved, obclavate, rostrate, with 6–19 (usually 11–15) transverse and 0–8 longitudinal or oblique septa, pale or very pale olive or grayish olive, smooth or infrequently, very inconspicuously warted, 75–350µm long and usually 20–30µm (sometimes up to 40µm) thick in the broadest part, *Beak* about 1/3 to 1/2 the length of the conidium and 5–9µ m thick.

22.1.3.2 ALTERNARIA BRASSICICOLA (SCHW.) WLTSHIRE. (MYCOL. PAP., 20: 8, 1947)

Colonies amphigenous, effuse, dark olivaceous brown to dark blackish brown, velvety, *Mycelium* immersed, hyphae branched, septate, hyaline at first, later brown or olivaceous brown, inter-, and intracellular, smooth, 1.5–7.5µm thick; *Conidiophores* arising singly or in groups of 2–12 or more, emerging through stomata, usually simple, erect or ascending, straight or curved, occasionally geniculate, more or less cylindrical but often slightly swollen at the base, septate, pale to mid olivaceous brown, smooth, up to 70 µm long, 5–8 µm thick; *Conidia* mostly in chains of up to 20 or more, sometimes branched, acropleurogenous, arising through small pores in the conidiophore wall, straight, nearly cylindrical, usually tapering slightly towards the apex or

Symptomatology and Etiology of Alternariose in Root

obclavate, the basal cell rounded, the beak usually almost non-existent, the apical cell being more or less rectangular or resembling a truncated cone, occasionally better developed but then always short and thick, with 1–11, mostly less than 6 transverse septa and usually few but up to 6 longitudinal septa, often slightly constricted at the septa, pale to dark olivaceous brown, smooth or becoming slightly warted with age, 18–130μm long, 8–20μm thick in the broadest part.

22.1.3.3   ALTERNARIA RAPHANI GROVES AND SKOLKO, (CAN. J. RES., SECT. C, 22: 227, 1944)

Conidiophores simple or occasionally branched, septate, olivaceous brown up to 150μm long, 3–7μm thick, sometimes swollen slightly at the tip and usually with a single conidial scar; *Conidia* commonly in chains of 2–3, straight or slightly curved, obclavate or ellipsoidal, generally with a short beak, mid to dark golden brown or olivaceous brown, smooth or sometimes minutely verruculose, with 3–7 transverse and often a number of longitudinal or oblique septa; constricted at the septa, 50–130 (70) μm long, 14–30 (22) μm thick in the broadest part; *Chlamydospores* formed abundantly in culture, sometimes in chains, at first 1-celled, round, finally many-celled and irregular, brown; conidiophores often develop from them.

22.1.3.4   KEY TO ALTERNARIA SPP. PARASITIC ON BRASSICACEOUS HOSTS

I.   Conidia solitary or occasionally in chains up to four, obclavate, rostrate, tapering gradually into thick cylindrical beak. *A. brassicae.*

II. Conidia in long chains consisting of twenty or even more in a chain:

    a.   Conidia usually cylindrical, basal cell rounded and apical cell more or less rectangular and beak usually almost non-existent. *A. brassicicola.*

    b.   Conidia usually polymorphic, often with short conical or cylindrical short beaks. *A. alternate.*

III. Conidia 3–4 in chain, straight or curved, obclavate, generally with short beak, chlamydospores formed abundantly in culture. *A. raphani.*

468   *Diseases of Fruits and Vegetable Crops*

22.1.4   *CHILI*

Chili (*Capsicum annuum* L.) is used as condiment in every house hold of India. It is preferred by both vegetarian and non-vegetarian masses to increase the palatability and taste in cooked food and vegetables. Leaf spot and fruit rot due to *Alternaria alternata* (Fr.) Keissler causes considerable quantitative and qualitative losses to the crop and has been reported quite early from India (Dutt, 1937).

22.1.4.1   *SYMPTOMS*

The symptoms of disease are observed from seedling to maturity stage of crop and even early blight symptoms are observed in seedling stage (Narain and Bhale, 2000). Necrosis of tender twigs also occurs from tip downward (Edward and Shrivastava, 1954). Characteristics symptoms of the disease are pronounced on the leaves and particularly on fruits (Sreekantiah et al., 1973).

1. **On Leaves:** Initially, small scattered, brown necrotic spots that gradually enlarge in size are formed. The mature spots are irregular to circular in shape and up to 8–10 mm in diameter. The spots are accompanied by narrow chlorotic margins. Finally, the spots coalesce, which result in the weathering and shredding of leaves. Due to humid conditions prevailing at ground level, lower leaves are affected first and the infection spreads to upper leaves later on (Narain and Bhale, 2000; Edward and Shrivastava, 1954; Sujatha et al., 1993).

2. **On Fruits:** In the beginning, small, blackish brown, circular to elongated water-soaked depressed lesions are formed on the pericarp of fruits, which lead the rotting of fruits in later stage (Sreekantiah et al., 1973). Sometimes the characteristic lesions are not seen on green fruits and fruit rot is visible on maturity stage (Sujatha et al., 1993; Uma, 1981).

In the form of minute, grayish green lesions are found at the blossom end, which later turn darker and become wrinkled with age and lead the blighting and extensive rotting of the receptable. Longitudinal cracks on the pericarp can also be seen which often coalesce and lastly cover the appreciable area of the fruits with abundant sporulation of fungus (Sultana et al., 1992; Khodke and Gahukar, 1993).

Apart from the symptoms on leaves, twigs, fruits, *A. alternata* also incites internal infection of fruits as "internal mold" during post-harvest with handling, transit, and storage (Adisa, 1985; Leyendecker, 1954; Mathur and Agnihotri, 1961; Sultana et al., 1992). A fruit rot of chili (*Capsicum annuum* L.) caused by *Alternaria solani* (Ell. and Mart.) Jones and Grout also have been reported from India (Subramaniam, 1954).

The disease is considered to be seed-borne in nature (Sultana et al., 1988; Bhale et al., 1998).

A similar type of symptoms of leaf spot and fruit rot has been observed in bell pepper also (Quebral and Shurtleff, 1965; Bremer, 1945; Alam et al., 1981).

22.1.5 TOMATO

Tomato (*Lycopersicon esculentum* Mill., Family Solanaceae) is a major vegetable crop and is grown and used widely. Due to the introduction of hybrids, it has emerged as an important major crop. Tomato suffers greatly from the following two serious diseases caused by *Alternaria* spp.

The species of *Alternaria* spp. *viz.*, *A. alternata*, and *A. solani* are known to be associated with tomato causing leaf spot with stem canker and early blight, respectively (Ciccarone, 1953; Rama Krishnan et al., 1971).

22.1.5.1 FRUIT ROT

Elliot (1917) reported for the first time the black spot or black rot of tomato caused by *Alternaria alternata*. On stems, dark brown to black cankers with concentric zonation can be seen (Nene, 1998). Fruit rot of symptoms of tomato caused by the pathogen is characterized as slightly wrinkled, dark brown lesions from minute pinheads to areas extending completely across the fruits (Ciccarone, 1953). Later on, velvety mats of conidia of *A. alternata*, the causal organism of the disease, are more frequently noticed on lesions (Warner, 1936; Srivastava and Tandon, 1966).

The pathogen forms grey lesions on the leaves and yellow-brown to black areas on the fruits. Generally, the spots on leaf are small, circular, scattered, and dark brown in color.

470   *Diseases of Fruits and Vegetable Crops*

22.1.5.2   EARLY BLIGHT

It is caused by *Alternaria solani* which was reported very early by Rands (1917). It attacks the main season in plains. Dark brown spots with concentric rings are produced on lower leaves, which spread on upper leaves with the progress of disease. In humid weather, the spots increase in size and coalesce forming big necrotic patches, which result in pre-mature defoliation of affected plants. On stems and branches lesions are produced which turn angular light brown to dark brown in color, sometimes girdling them and showing concentric rings (Nene, 1998).

McWhorter (1927) says that *A. solani* is one of the commonest causes of seedling blight or damping-off of tomatoes, causing dark lesions on rootlets. It also often girdles the large seedlings just below the collar. Stems, fruits, and leaves are affected. This is recorded from most parts of the world.

22.1.6   BRINJAL (EGG PLANT)

Alternata solani Sorauer occurs on potato, tomato, brinjal (eggplant), and chili (not so frequently) and also some other plants belonging to the family Solanaceae. The fungus causes early blight symptoms on potato and tomato affecting all parts above the ground.

22.1.6.1   LEAF SPOT AND FRUIT ROT

Alternaria leaf spots are characterized by brown and irregular spots produced on the leaves. Several spots usually coalesce to form large necrotic patches. Characteristically concentric rings are usually seen on older spots due to growth of *Alternaria solani*.

Fruits are also affected by the same fungus and large necrotic sunken spots develop on the fruits which may turn yellow and drop prematurely (Kapoor and Hingorani, 1958).

22.1.6.2   LEAF SPOT AND FRUIT ROT OF BRINJAL

This disease is caused by *Alternaria melogenae* and was reported for the first time by Rangaswami and Sambandam in 1960 from Tamil Nadu, India.

Symptomatology and Etiology of Alternariose in Root 471

The leaf spot infection first appears as a light brown speck about 2 mm, which when fully developed attains about 5 mm, circular, light brown and bound by a dark brown peripheral ring. The disease appears mainly on mature leaves, forming up to twenty spots or less on each leaf. Shot-hole symptoms also appear on older spots.

On fully mature fruits, the lesions first appear as brown, circular flat spots. Fully developed lesions are in the form of large scab, circular or elliptical, 2 to 4 cm broad. The affected skin is leathery, the flesh below hard and discolored, dirty brown to a depth of 1 cm. It was also observed that the same fungus caused fruit rot of chili and red pepper (*Capsicum annuum* L.) in the field (Rangasami and Sambandam, 1960).

22.1.7 PEPPER

The fungus which incites the early blight of potato and tomato, *Alternaria solani* (Ell. and Mart.) Jones and Grout, has been reported on pepper (*Capsicum annuum*) to cause the leaf spot. In general, however, the disease is not considered important as that of potato, tomato, and chili. The leaf spots are dark, leathery, and up to 10 mm in diameter. When the spots are numerous, old leaves die prematuraly and drop (Walker, 1952). The fruits also get infected (Bremer, 1945).

Another species of *Alternaria, A. alternata* incites spots on pepper fruits. It also causes blossom end-rot and fruit rot and forms profuse sporulation of the fungus on affected tissues (Bremen, 1945; Quebral and Shurtleff, 1965; Alam et al., 1981).

22.1.7.1 ALTERNARIA MELONGINAE RANGASWAMI AND SAMBANDAM (MYCOLOGIA, 52:517–520, 1960)

Mycelium both intercellular and intracellular, hyaline to light brown, 3–6µm in diameter; *Conidiophores* amphigenous, arising in fascicles of 2–3 through stomata and epidermis, septate, light brown, straight or geniculate, 65–91 x 5–7 µm; *Conidia* in chains of 2–5, brown, obclavate, muriform with 4–7 cross and 0–5 vertical septa and constriction at septa, smooth-surfaced, 28–84 x 7–17.5 µm, beaked; *Beaks* light brown and 10.5 to 52.5 µm in length.

The fungus appears to be distinct in certain morphological characters and in producing profuse soluble purple pigment in synthetic as well as complex agar and liquid media. *A. melongenae* sprulates profusely in all the media tested, even after repeated sub-culturing for over 15 months (Rangaswami and Sabandham, 1960).

22.1.7.2 *ALTERNAIRA SOLANI SORAUER, Z. PFLKRANKHA, 6: 6, 1986. (MACROSPORIUM SOLANI ELLIS AND MARTIN, AM. NAT., 16: 1003, 1982)*

Conidiophores arising singly or in small groups, straight or flexuous, septate, rather pale brown or olivaceous brown, up to 110 µm long, 6–10 µm thick; *Conidia* usually solitary, straight or slightly flexuous, obclavate or with the body of the conidium oblong or ellipsoidal tapering to a beak which is commonly of the same length as or rather longer than the body, pale or mid pale golden or olivaceous brown, smooth, overall length usually 150–300 µm, 15–19 µm thick in the broadest part, with 9–11 transverse and 0 or a few longitudinal or oblique septa; *Beak* flexuous, pale, sometimes branched, 2.5–5 µm thick tapering gradually.

The key to *Alternaria* spp. parasitic on Solanaceous hosts are given below:

I. Conidia Formed in Chains:

a. Conidial chain longer, conidia usually polymorphic, often with short, conical or cylindrical beaks. *A. alternate.*

b. Conidial chain shorter (3–6), conidia obclavate, rostrate, beaks often slightly swollen. *A. tenuissima.*

II. Conidia Solitary or Seldom Formation of a Secondary Conidium:

a. Beak flexuous and sometimes branched, culture on PDA chromogenic. *A. solani.*

b. Beak long, unbranched, tapering gradually, culture on PDA non-chromogenic. *A. melogenae.*

22.1.8 *DIOSCOREA*

22.1.8.1 *LEAF SPOT/BLIGHT*

The disease symptoms appear on *Dioscorea floribunda, D. composita* and *D. bulbifera* as light brown, small circular spots on the upper surface of the leaves. The spots increase in size in the later stage and coalesce forming dark brown to black patches of typical blight symptoms. The spots are up

Symptomatology and Etiology of Alternariose in Root 473

to 1.5 cm in diameter. The dead tissues appear firm and dry. Under humid conditions, brown to black masses of spores of the fungus may be visible on the surface of lesions. The disease is caused by a cosmopolitan fungus, *Alternaria alternata* (Solankure and Rao, 1972).

22.1.9 CARROT

Carrot (*Daucus carot* L.) is one of the popular and commonly consumed vegetables. It is grown all over India. It is taken raw as well as in cooked form. Carrot juice is a rich source of carotene. Carrot suffers from many foliar and root diseases (Walker, 1952). Two species of *Alternaria viz., A. radicina* and *A. dauci* are responsible to cause "Black rot" and "Leaf blight" diseases in carrot crop, respectively.

22.1.9.1 BLACK ROT

The black rot of carrot caused by *Alternaria radicina* was reported by Meire et al., as early as in 1922. Black rot is the disease that affects the foliage as well as the fleshy roots. It is worldwide in distribution.

In the case of this disease, the primary lesions on older plants begin at the base of the petioles where dark, usually shallow lesions spread into the crown and sides of the roots. Secondary lesions develop below the ground and are often coincident with cracks and spilt. On the stems, the fungus causes black sunken lesions of irregular to circular outline. The decayed tissues become grayish black to jet black due to presence of the masses of spores. The disease is known to cause the damage both in field and storage (Meir et al., 1922; Lauritzen, 1926; Neergaard, 1945) and is both seed and soil borne (Uppal et al., 1949; Scott and Wenham, 1972; Lambat et al., 1985).

The infection of *A. radicina* occurs apart from carrot on celery, dill, and parsnip also (Ellis, 1971) and the most characteristics morphological features of fungus have been described by Ellis and Holliday (1972).

22.1.9.2 ALTERNARIA RADICINA MEIER, DRECHSLER AND EDDY (PHYTOPATHOLOGY, 12: 157–166, 1922)

Colonies growing as dark blackish-brown to black in color; *Conidiophores* arising usually singly from hyphae, simple or occasionally branched, straight

474   *Diseases of Fruits and Vegetable Crops*

or flexuous, cylindrical, septate, pale to mid-brown or olivaceous brown, smooth up to 200 µm long, 3–9 µm thick, with 1 or several conidial scars; *Conidia* solitary or in chains of 2 or rarely 3, and are typically dark olive-brown to natal brown, highly variable in shape, often ellipsoidal, obclavate or obpyriform with 1 or several longitudinal or oblique septa, sometimes constricted at septa, 27–57 µm long, 9–27 µm thick in the broadest part.

22.1.9.3 ALTERNARIA BLIGHT

Alternaria leaf spot/blight of carrot was first described in Germany in 1855, where it was reported causing sporadic posses in several Northern European countries (Kuhn, 1855).

Small dark brown to black colored irregular spots are usually formed on the leaves and petioles of infected plants. Characteristically the number of spots gradually increases, and when the infected are severe, whole top may be killed. Reduction of leaf surface prevents the full development of the root.

Standberg (1987) reported that the disease may reduce carrot yield significantly due to loss of photosynthetic tissues or may cause complete damage of a crop when collapse of the crop canopy prevents the mechanical harvest.

The disease is caused by *A. dauci* and its characteristic features have been described by David (1988) and Pryor and Standberg (2001). Meire et al. (1922) and Lauritzen (1926) reported that the disease was mistakenly attributed to the fungus, *Sporidesmum exitiosum* (Kuhn) *v. dauci* Kuhn (Syn. *Alternaria dauci*, the causal agent of Alternaria leaf blight). Mohanty (1961), Roy (1969), and Singh et al. (1975) recorded the disease from India. The disease is seed borne in nature (Scott and Wenham, 1972; Puttoo, 1973; Lambat et al., 1985).

Although the fungus causes leaf blight of carrot but has been recorded on other umbelliferous plants (Ellis, 1971). The comparative study of two carrot disease has been made by Hooker (1944).

22.1.9.4 ALTERNARIA DAUCI (KUHN) GROVES AND SKOLKO (CAN. J. RES., SECT. C., 22: 222, 1944): (SPORIDESMIUM EXITIOSUM KUHN VAR. DAUCI KUHN, HEDWIGIA, 1: 91, 1855)

Conidiophores arising singly or in small groups, straight or flexuous, sometimes geniculate, septate, pale or mid pale olivaceous brown or brown, up to

Symptomatology and Etiology of Alternariose in Root 475

80μm long, 6–10μm thick; *Conidia* usually solitary, occasionally in chains of 2, straight or curved, obclavate, rostrate with beak up to 3 times the length of the body of the spore, at first pale olivaceous brown, often becoming brown with age, smooth, overall 100–450μm long, 16–25 (20) μm thick in the broadest part, with 7–11 transverse and 1 to several longitudinal or oblique septa; *Beaks* often once branched, flexuous, hyaline or pale, 5–7μm tick at base tapering to 1–3μm.

The key to *Alternaria* species parasitic on carrot are given below:

Conidia solitary (Acatenate):

a. Conidia without beak (unbeaked). *A. radicina.*
b. Conidia with long filiform, often branched beaks. *A. dauci.*

22.1.10   SWEET POTATO

The Alternaria leaf spot of sweet potato (*Ipomoea batatas*) occurs on the old leaves at the center of the hill. Spots are irregular in shape with well-defined margins, black, and shiny on the upper side, dull brown to black on the lower side (Walker, 1952). The necrotic tissue cracks and becomes ragged.

A foot rot attributed to the potato early blight organisms, *Alternaria solani* (Ell. and Mart.) Jones and Grout, was described in Delaware and Texas (U.S.A.) by Taubenhaus (1925).

The leaf spot disease with other diseases of sweet potato has been described by Harter and Weimer (1929) and Weber and West (1930).

22.1.11   LADY'S FINGER

Lady's finger (okra) or *Bhindi* (*Abelmoschus esculentus* (L.) Moench. is attacked by *Alternaria tenuissima* which was first reported by Mehrotra and Narain (1969) from Allahabad, U.P. The causative fungus produces dark brown spots which are scattered throughout the lamina. In a later stage of disease development, the spots coalesce together to form larger irregular areas. Severely infected leaves ultimately dry up and may fall prematurely. The infected seeds of *Bhindi* exhibit a species of *Alternaria,* which was un-named (Gupta and Shukla, 1955).

22.1.12 ALTERNARIA DISEASES OF CUCURBITS

The members of the family Cucurbitaceae are the major source of vegetables in India and constitute an important part of the daily diet. These vegetables are mostly grown throughout the year in India in all the seasons of *Rabi, Kharif, and Zaid.* Family Cucurbitaceae is represented by 90 genera and 750 species, which are especially used for their edible fruits (Chakravarty, 1959). Some of the important cultivated plants are *Citrullus lanatus* (watermelon) *Cucumis melo* (musk melon), *Cucumis sativus* (cucumber), *Lagenaria siceraria* (bottle gourd), *Cucurbita maxima* (winter squash), *Momordica charantia* (bitter gourd), *Trichosanthes dioica* (pointed gourd), *Coccinia indica* (scarlet gourd), *Citrullus vulgaris* var. *fistulosus* (round gourd), *Cucurbita pepo* (winter squash), *Trichosanthes anguina* (snake gourd) and *Luffa acutangula* (ridge gourd).

Alternaria cucuerina is the major pathogen to cause the leaf spot and blight in cucurbits but apart from this, two more species, *A. alternata* and *A. teniussima* are also involved to cause two types of diseases in cucurbits i.e., foliar blight (leaf spot) and fruit rot. The causative fungus was originally described as *Macrosporium cucumerinum* by Ellis and Everhart (Brisley, 1923), which was changed to *Alternaria cucumerina* by Elliot in 1917.

Some of the important diseases of Cucurbitaceous crops caused by *Alternaria* species reported from India are given in Table 22.2.

In general, the leaf blight begins as small tan spots which enlarge as roughly circular areas, often coalescing to involve most of leaf. The spots later become dark in color with age often showing concentric ridges. Rotting and decay of fruits may be caused in storage and transit. Watermelon and musk melon are the great sufferers. Fruit rot of pointed gourd (*Trichosanthes dioica*) was reported from Sabour (Bihar) by Sahu Kritagyan and Singh (1980) due to *A. alternata*. The same fungus was also found capable to cause fruit rot of *Momondica dioica* in Jabalpur (Agrawal, 1961) and snake gourd by Singh (1974) from Bangalore due to *A. tenuissima*. Seed borne nature of *Alternaria alternata* was established by Sohi and Mahally (1977) in *Citrullus vulgaris* and of *A. cucumerina* by Khandelwal and Prasada (1970) in watermelon.

The casual organism of blight of watermelon was identified in India as *Alternaria cucumerina* (E. and E.) Elliot by Khandelwal and Prasada (1970) and also Ibrahim et al. (1975) identified its causal organism as *A. cucumerina* from Egypt.

TABLE 22.2 *Alternaria* Species Associated with Cucurbits in India

Sl. No.	Disease	Host	Causal Organism	References
1.	Leaf blight	Water melon	*Alternaria alternata*	Narian et al. (1985)
2.	Leaf spot	Watermelon	*Alternaria cucumerina*	Singh et al. (1975)
3.	Fruit rot	Watermelon	*Alternaria cucumerina*	Mathur and Shekhawat (1992)
4.	Fruit rot	Bottle gourd	*Alternaria alternata*	Singh and Chohan (1980)
5.	Fruit rot	Bottle gourd	*Alternaria cucumerina*	Narain and Srivastava (2000)
6.	Leaf blight	Bottle gourd	*Alternaria cucumerina*	Chahal et al. (1970)
7.	Leaf spot	Cucumber	*Alternaria alternata*	Laxminarayanan and Reddy (1976)
8.	Leaf blight	Cucumber	*Alternaria cucumerina*	Chahal et al. (1970)
9.	Fruit rot	Cucumber	*Alternaria tenuissima*	Narain and Srivastava (2000)
10.	Leaf blight	Pumpkin	*Alternaria cucumerina*	Chahal et al. (1970)
11.	Leaf blight	Scarlet gourd	*Alternaria cucumerina*	Narain and Prasad (1981)
12.	Leaf blight	Bitter gourd	*Alternaria cucumerina*	Ahmad and Narain (2000)
13.	Fruit rot	Bitter gourd	*Alternaria cucumerina*	Narain and Srivastava (2000)
14.	Fruit rot	Round gourd	*Alternaria alternata*	Narain and Srivastava (2000)
15.	Fruit rot	Pointed gourd	*Alternaria cucumerina*	Narain and Srivastava (2000)
16.	Fruit rot	Sponge gourd	*Alternaria tenuissima*	Narain and Srivastava (2000)
17.	Fruit rot	Bitter gourd	*Alternaria tenuissima*	Sharma and Bhargava (1977)
18.	Fruit rot	Pointed gourd	*Alternaria alternata*	Sahu Kritigyan and Singh (1980)
19.	Fruit rot	Snake gourd	*Alternaria tenuissima*	Singh (1974)
20.	Leaf spot	Long gourd	*Alternaria cucumerina*	Narain et al. (2003)
21.	Leaf spot	Phoot	*Alternaria cucumerina*	Narain et al. (2003)

478   *Diseases of Fruits and Vegetable Crops*

Later on, the leaf blight, leaf spot, veinal necrosis and fruit rot of cucurbits caused by *A. cucumerina, A. alternata* and *A. tenuisima* were reported from Punjab (Chahal et al., 1970), Rajasthan (Khandelwal and Prasada, 1970), Andhra Pradesh (Laxminarayanan and Reddy, 1976) and Uttar Pradesh (Narain and Prasad, 1981; Ahmad and Narain, 2000; Narain and Srivastava, 2000; Narain et al., 2003; Narain and Kant, 2008).

The diagnostic characters of all the three *Alternaria* species associated with different Cucurbitaceous hosts are described in subsections.

22.1.12.1 ALTERNARIA ALTERNATA (FR.) KEISSLER (BEIH. BOT. ZBL., 29: 434, 1912)

Colonies moderately fast growing, usually black or olivaceous black, sporulation abundant; *Mycelium* septate, branched, olive buff, 2.5–9.5 μm in width, sometimes swollen (12.5–23.5 μm) to from chain of chlamydospores; *Conidiohpores* arising singly or in groups, usually simple, septate, straight or variously curved, geniculate, pale to mid olivaceous brown, smooth, and 24.5–68.5 μm in length and 3.5–7.25 μm in width; *Conidia* formed in long and often branched chains, obclavate, obpyriform ovoid or ellipsoidal, dark brown, smooth or verruculose, 8.5–53.50 x 5.30–21.50 μm in size with 3–10 cross and 0–8 longitudinal septa, *Beaks* often short, conical or cylindric but never equaling length of conidium, usually lighter in color, 1.0–45.5 μm in length and 1.5–6.0 μm in width and 0–3 septate.

22.1.12.2 ALTERNARIA TENUISSIUMA (KUNZE EX PERS.) WLTSHIRE (TRANS. BR. MYCOL. SOC., 18: 157, 1933)

Conidiophores solitary or in groups, simple or branched, straight or flexuous, more or less cylindrical, septate, pale or mid pale brown, smooth, with 1 or several conidial scares, up to 115μm long, 4–6μ thick; *Conidia* solitary or in short chains (3–6), straight or curved, obclavate or with the body of the conidium ellipsoidal tapering gradually to the beak, usually shorter, sometimes tapered to a point but more frequently swollen at the apex where there may be several scars, pale to mid clear golden brown, usually smooth, sometimes minutely verruculose, generally with 4–7 transverse and several longitudinal or oblique septa, slightly or not constricted at the septa, overall length 22–95(54) μm, 8–19(13.8) μm thick in the broadest part; *Beak* 2–4μm thick, swollen apex, 4–5μm wide.

Symptomatology and Etiology of Alternariose in Root

22.1.12.3 *ALTERNARIA CUCUMERINA (ELLIS AND EVERH.) ELLIOT (AM. J. BOT., 4: 472, 1917)*

Colonies amphigenous, *Conidiophores* arising singly or in small groups, erect, straight or flexuous, sometimes geniculate, cylindrical, septate, pale to mid-brown, up to 110μm long, 6–10μm thick, usually with several well developed conidial scars; *Conidia* solitary or occasionally in chains of 2, obclavate, rostrate, the beak longer, often much longer than the body of the spore, pale to golden brown, smooth to verruculose, overall 130–220 (180) μm long, 15–20μm thick in the broadest part; body with 6–9 transverse and several, sometimes many longitudinal and oblique septa; *Beak* pale brown, septate, not branched, 4–5μm thick at the base rapidly narrowing to 1–2.5μm.

The comparative morphological differences of all the three species of *Alternaria* parasitic on cucurbits are given in Table 22.3.

TABLE 22.3 Comparative Differentiation in Three *Alternaria* spp. Occurring on Cucurbitaceous Hots

Morphological Characters	A. Alternata	A. Tenuissima	A. Cucurmerina
Conidiophores	Olive brown, erect, simple, septate	Dark brown, erect simple or branched, septate	Dark brown, erect, simple, septate
Length	24.5–68.5 μm	19.5 0–115.0 μm	42.5–110.0 μm
Width	3.50–7.25 μm	4.0–6.0 μm	6.0–9.0 μm
Conidia			
Conidia in chain	Up to 20 or even more	3–6	Single (rarely 2)
Shape	Polymorphic	Conical to oval	Oblavate rostrate
Color	Dark brown	Dark olive buff	Olive brown
Cross septa	3–10	4–7	6–9
Long. septa	0–8	0–8	2–12
Length	8.50–53.50 μm	22.0–95.0 μm	54.3–80.2 μm
Width	5.30–21.50 μm	8.00–9.00 μm	15.0–20.0 μm
Beak			
Length	1.0–45.5 μm	3.50–70.0 μm	25.2–75.0 μm
Width	1.5–6.0 μm	3.50–5.50 μm	1.0–2.5 μm
Cross septa	0–3	0–3	0–5

480   *Diseases of Fruits and Vegetable Crops*

Key to *Alternaria* spp. occurring on Cucurbitaceous hosts

- Conidia formed in long and often branched chains consisting of up to 20 or even more.
- Conidia usually polymorphic, often with short conical or cylindrical beaks. *A. alternate*.
- Conidia formed in short chains (3–6).
- Conidia conical to oval with many cross and longitudinal septa, sometimes swollen terminally and may be of up to the same length of spore body. *A. tenuissima*.
- Conidia formed singly or rarely up to two.
- Conidia obclavate, reostrate, comparatively larger with many cross and longitudinal septa and with long filiform beaks. *A. cucumerina*.

22.1.13   *CLUSTERBEAN OR GUAR*

Clusterbean or guar [*Cyamopsis tetragonoloba* (L.) Taub.], a hardy, salt tolerant leguminous crop is grown for feed, fodder, vegetable, and gum production. *Alternaria cyamopsidis* causing leaf spot of guar is responsible for considerable losses and reducing the production per unit area and also affecting the quantity and quality of gum (Lodha, 2000). The disease manifests itself in the form of small, light colored spots on leaves varying from 2–10 mm in diameter. Concentric rings are also formed in the necrotic tissues. In the humid weather, the affected areas coalesce to form patches involving a major portion of the leaf blade. In severe attack, leaves shrivel and fall down.

The disease caused by *A. cyamopsidis* was first reported by Rangaswami and Venkata Rao (1957) from Coimbatore (Tamil Nadu). According to Bhatnagar and Jain (1987), the disease caused up to 43% loss in seed yield of cluster bean in Rajasthan when relative humidity (RH) was 80% and temperature ranged from 25–30°C. Singh and Prasada (1973) studied the physiology and control of *A. cyamopsidis*, the incitant of blight disease of guar. *Alternaria tenuissima* is also known to cause the similar type of disease of cluster bean (Chand and Verma, 1968). Earlier in 1935, Narasimhan recorded an unidentified species of *Alternaria* on guar in Mysore.

Alternaria cyamopsidis Rangaswami and Rao (*Indian Phytopath.*, 10: 18–25, 1957):

Hyphae both inter and intracellular; *Conidiophores* aphigenous, arising in groups (2–8) through stomata, grayish olive, geniculate with prominent scars; *Conidia* in chains up to four, usually straight, obclavate light grayish

Symptomatology and Etiology of Alternariose in Root 481

olive, smooth walled, constricted at septa, muriform with 6–11 cross and 0 to 5 vertical or oblique septa, 62–146µm in length and 12–19µm in width; *Beak* lighter in color than conidia, measuring 50–140µm in length.

22.1.14 MOTH BEAN

Moth bean [*Vigna acontifolia* (Jacq.) Marechel] is a high source of protein, delicious taste of soaked grains, high forage product, overall drought resistant, makes a more loveable legume to the masses of India. The crop suffers from an Alternaria leaf spot disease.

22.1.14.1 ALTERNARIA LEAF SPOT

It is a severe disease of moth bean caused by *Alternria alternata* (Fr.) Keissler, which was reported from Jobner (Rajasthan) in 1979 (Toor, 1980). The disease initially appears as small, water-soaked specks, later enlarging and becoming reddish often coalesce to produce large areas of dead tissues.

Alternaria alternata and *A. longissima* have been found to be associated with moth bean seeds externally/internally (Sharma, 1986).

22.1.15 OTHER PULSES USED AS VEGETABLES

Lobia (*Vigna unguiculata*), Bakla (*Vicia faba*), Sem (*Lablab purpurea*) and pea (*Pisum sativum*) are the pulse crops which are used as vegetables in cooked form and are attacked by *Alternaria tenuissima*. The infection of *Alternaria alternata* has also been noticed to cause leaf spot/blight. Although the disease is of minor importance but in their seed borne nature in which the seeds get infected by these two species cause rotting and discoloration of seeds and sometimes their poor germination, thus causing the losses. The disease in known in Lobia (Shukla et al., 1977), bean (Shukla and Bhargava, 1976), sem (Rao, 1966), pea (Agarwal, 1985; Mehrotra and Narain, 1969).

Key to *Alternaria* spp. parasitic on Leguminous host (pulses)

1. Conidia in long chains usually polymorphic, often with short conical or cylindrical short beak. *A. alternata.*
2. Chain short with 3–6 spores, beak up to the length of spore body with terminal swellings. *A. tinuissima.*

3. Conidia in chain (up to 4), straight, obslavate with longer filiform beaks. *A. cyamopsidis.*

22.1.16 LEAFY VEGETABLES

Leafy vegetable like *Chenopodium album* (Bathuwa), Spinach (*Spinacea oleracea*), Indian dill (Sowa), Chaulai (*Amaranthus blitium*) and lettuce (*Lactuca sativa*) play a significant role in human diet providing the rich source of vitamin 'A.' Lettuce is utilized as shallot. All these leafy vegetable are attacked by species of *Alternaria viz., A. alternata* and *A. tenussima* (Mehrotra and Narain, 1969) expect lettuce which suffers from infection of *A. lactucae* (Rao, 1966); *A. sonchi* (Rao, 1963; Ellis, 1971) and *A. tenuissima* also (Mehrotra and Narain, 1969). In general, disease initially appears as brown spots on the margin of leaves, which develop in ward to produce irregular lesions. Scattered light brown spots appear on leaves and *Lactuca sativa* whereas in the *Chenopodium album,* the spots are dark brown with concentric rings (Mehrotra and Narain, 1969).

22.1.17 INDIAN DILL

India dill [*Anethum sowa* (Roxb.) et Flem.] is the oldest cultivated medicinal plant and leafy vegetable in India. Its young aromatic foliage is used in culinary, whereas fruits have carminative properties.

Indian dill or sowa is attacked by *A. alternata.* The small brown spots first appear on leaves which increase in size and also coalescing leading the death of leaves and whole crown. The inflorescence also becomes infected. Linear dark brown lesions appear on petioles and sometimes also on branches to girdle them. Profuse aporulation of the fungus becomes visible in infected portion. Seed borne nature of the causative fungus was established by Swarup and Mathur (1972).

Infected seeds can transmit *A. dauci* and *A. alternata* into new carrot and Indian dill production areas, respectively (Scott and Wenham, 1972).

Dill (*Anethum sowa*) is also known to be infected by *Alternaria radicina* (Ellis, 1971) and details of symptom appearance and etiology of the disease have already been discussed with Black rot of carrot (*A. radicina*).

Most of *Alternaria* survives on dead plant material. They can survive off of plant debris both on the surfaces of soil or below ground level. They can survive in plant seeds. These fungi produce spores (conidia) that are spread by wind or rain. Assuming the weather conditions are viable, secondary spread (cycle) can occur. Once the pathogen has infected its host, leaves will develop an array of dark spots. Eventually, as the plant weakens, they will fall and survive on plant debris for next season. Considering the source of primary inoculum, mode of survival, and secondary spread of pathogens, the management strategy can be adopted accordingly.

KEYWORDS

- *Alternaria cucumerina*
- *Alternaria* leaf spot
- Indian dill
- leafy vegetables
- moth bean
- symptomatology

REFERENCES

Adisa, V. A., (1985). Micro-organisms associated with spoilage of *Capsicum annuum* and *Capscicum frutescens* in Nigeria. *Fitopath. Braz., 10,* 427–432.

Agarwal, D. K., (1985). New host records for *Alternaria* spp. from India. *Indian Phytopath., 38,* 392–393.

Agarwal, G. P., (1961). Fungi causing plant diseases at Jabalpur (M. P.). III. *J. Indian bot. Soc., 40,* 404–408.

Ahmad, S., & Narain, U., (2000). A new host record of *Alternaria cucumerina* on bottle gourd. *Indian Phytopath., 53,* 234.

Alam, K. B., Bakr, A., & Ahmad, H. U., (1981). Fruit rot of paper. *FAO Plant Prot. Bull., 29,* 28–29.

Anonymous, (1969). *CMI Distribution Map of Alternaria Brassicicola,* CMI, Kew, England.

Atkinson, R. G., (1950). Studies on the parasitism and variation of *Alternata raphani. Can. J. Res. Sect. C., 28,* 288–317.

Bhale, U., Bhale, M. S., & Khare, M. N., (1998). Anternariose of chili and its management. *Nat. Seminar, J.N.K.V.V., Jabalpur,* (M.P.).

Bhatnagar, S. K., & Jain, J. P., (1987). Studies on *Alternaria* blight of guar in Rajasthan. *Indian J. Mycol. Pl. Pathol., 17,* 108.

Bremer, H., (1945). On pod spots in peper. *Phytopathology, 35,* 283–287.

Brisley, H. R., (1923). Studies on blight of cucurbits caused by *Macrosporium cucurmerinum* E. &. E. *Phytopathology, 13,* 199–204.

Chahal, D. S., Chohan, J. S., & Sidhu, G. S., (1970). Alternaria leaf spot of cucurbits in Punjab. *Indian Phytopath., 23*, 580–581.

Chakravarty, N. L., (1959). Monograph on Indian Cucurbitaceae (taxonomy and distribution). *Botanical Survey of India, 17*(1), 6, 7.

Chand, G., Narain, U., Kumar, M., & Verma, S., (2007). Symptomatology, etiology and ecofriendly management of Alternaria leaf spots and blight of broccoli. In: Shahid, A., & Narain, U., (eds.), *Ecofriendly Management of Plant Disease* (pp. 461–472). Daya Publishing House, New Delhi.

Chand, J. N., & Verma, P. S., (1968). Occurrence of new *Alternaria* leaf spot of cluster bean (*Cyamopsis tetragonoloba*) in India. *Plant Dis. Reptr., 52*, 145–147.

Chupp, C., & Sherf, A. F., (1960). Crucifer diseases. In: *Vegetable Diseases and Their Control* (Vol. 8, pp. 237–288). Ronald Press Company, New York, Chapter.

Chupp, C., (1925). *Alternaria* leaf spot and blight of crucifers In: *Manual of Vegetable Garden Diseases* (pp. 146–150). The McMillan Co., New York.

Chupp, C., (1935). Macrosporium and Colletotrichum rots of turnip roots. *Phytopathology, 25*, 269–274.

Ciccarone, A., (1953). Phytopathological notes. I. Economic importance of a basal rot of tomato fruits in some areas of Compania. *Boll. Staz. Pat. Veg. Roma Ser., 3*(9), 57–90.

David, J. C., (1988). *Alternaria Dauci No. 951. CMI Description of Pathogenic Fungi and Bacteria.* Commonwealth Mycological Institute, Kew, England.

Dutt, K. M., (1937). *Alternaria* species on chili in India. *Curr. Sci., 6*, 96–97.

Edward, J. C., & Shrivastava, K. N., (1954). Die back of chili caused by *Alternaria* sp. *Curr. Sci., 23*, 301.

Elliot, J. A., (1917). Taxonomic characters of the genus *Alternaria* and *Macrosporium. Amer. J. Bot., 4*, 439–476.

Ellis, M. B., & Holliday, P., (1972). *Alternaria Radicina No. 346, CMI Description of Pathogenic Fungi and Bacteria,* Commonwealth Mycological Institute, Kew, England.

Ellis, M. B., (1971). *Dematiaceous Hyphomycetes* (pp. 411–427). Commonwealth Mycological Institute, Kew, Surrey, England.

Gupta, S. K., & Shukla, T. N., (1955). Mycoflora Kanpurensis-I. *Kanpur Agric. Coll. J., 14*, 66–72.

Harter, L. L., & Weimer, J. L., (1929). A mographic study of sweet potato diseases and their control. *U. S. Dept. Agr. Tech. Bull., 99*, 1929.

Hooker, W. J., (1944). Comparative studies of two carrot leaf diseases. *Phytopathology, 34*, 606–612.

Ibrahim, A. N., Abdul-Hak, T. M., Fadl, F. A., & Mahrous, M. M., (1975). Certain morphological and pathological characters of *Alternaria cucumerina,* cause of water melon leaf spot disease in Egypt. *Acta Phytopathologica Academiae Scientirum Hungaricae, 10*(3/4), 301–307.

Kapoor, J. N., & Hingorani, M. K., (1958). Alternaria leaf spot and fruit rot of brinjal. *Indian J. Agric. Sci., 28*, 109–114.

Khalid, A., Akram, M., Narain, U., & Srivastava, M., (2004). Characterization of *Alternaria* spp. associated with Brassicacaous vegetables. *Farm Sci. J., 13*, 195–196.

Khandelwal, G. L., & Prasada, R., (1970). Taxonomy, physiology and control of Alternaria leaf spot of water melon (*Citrullus lanatus*). *Indian Phytopath., 23*, 81–84.

Khodake, S. W., & Gahukar, K. B., (1993). Fruit rot of chili caused by *Alternaria alternata* (Fr.) Keissler in Maharashtra. *PKV-Res. J., 17*, 206–207.

Kuhn, J., (1855). Uber das Vervallen des Rapses and die Krankheit der Mohrenblatter. *Hedwigia, 1*, 86–92.

Lambat, A. K., Ram, N., Agarwal, P. C., Khetarpal, R. K., Usha, D., Kaur, P., Majumdar, A., Varshney, J. L., Mukewar, P. M., & Indra, R., (1985). Pathogenic fungi intercepted in imported seeds and planting materials during 1982. *Indian Phytopath., 38*, 109–111.

Lauritzen, J. I., (1926). The relation of black rot in the storage of carrots. *J. Agric. Res., 38*, 1025–1041.

Laxminarayanan, P., & Reddy, S. M., (1976). Post-harvest disease of some Cucurbitaceous vegetables from Andhra Pradesh. *Indian Phytopath., 29*, 57–59.

Leyendeker, P. G., (1954). Fungi associated with internal contamination of sundried chili in Mexico. *Bull. Torrey Bot. Soc., 81*, 400–404.

Lodha, S., (2000). Diseases of legumes. In: Trivedi, P. C., (ed.), *Plant Diseases* (pp. 147–148). Pointer Publication, Jaipur.

Mathur, K., & Shekhawat, K. S., (1992). Fruit rot of water melon. *J. Mycol. Pl. Pathol., 22*, 80.

Mathur, R. L., & Agnihotri, J. P., (1961). Internal moulds of chili caused by *Alternaria tenuis* Auct. *Indian Phytopath., 14*, 104–105.

McLean, D. M., (1947). *Alternaria* blight and seed infection, a cause of low germination in certain radish seed crops. *J. Agric. Res., 75*, 71–79.

McWhorter, F. P., (1927). The early blight diseases of tomato. *Virginia Truck Exp. Stn. Bull., 59*, 547–566.

Mehrotra, B. S., & Narain, U., (1969). Studies on the genus *Alternaria* I. Some new records and a new species. *Indian Phytopath. Soc. Bull., 5*, 1–7.

Meier, F. C., Drechsler, C., & Eddy, E. D., (1922). Black rot of carrots caused by *Alternaria radicina n. sp. Phytopathology, 12*, 157–168.

Mitter, J. H., & Tandon, R. N., (1930). Fungus flora of Allahabad. *J. Indian Bot. Soc., 9*, 190–198.

Mohanty, N. N., (1961). Alternaria blight of carrot. *Proc. Indian Sci. Congr. Part III, 256*–257.

Narain, A., Swain, N. C., Sahoo, K. S., Das, S. K., & Shukla, V. D., (1985). A new leaf blight and fruit rot of water melon. *Indian Phytopath., 38*, 149–150.

Narain, U., & Bhale, U., (2000). Alternaria leaf spot and fruit rot of chili. In: Narain, U., Kumar, K., & Srivastava, M., (eds.), *Advances in Plant Disease Management* (pp. 163–173). Advance Pub. Concept, New Delhi.

Narain, U., & Kant, S., (2008). Diversity of species and parasitism in the genus *Alternaria*. *Appl. Bot. Abst., 28*, 272–281.

Narain, U., & Prasad, R., (1981). Alternaria leaf spot of *Kundru* from India. *Plant Sci., 13*, 96.

Narain, U., & Saksena, H. K., (1975). A new leaf spot of turnip. *Indian Phytopath., 28*, 98–100.

Narain, U., & Srivastava, M., (2000). Detection and diagnosis of *Alternaria* spp. associated with fruits of Cucurbitaceous vegetables. *Proc. Indian Phytopath. Soc. Golden Jubilee, 2*, 718–719.

Narain, U., Singh, J., & Koul, A. K., (1982). Leaf spot of candytuft caused by *Alternaria rephani. Nat. Acad. Sci. Letters, 5*, 13.

Narain, U., Srivastava, M., & Rani, P., (2003). A new record of *Alternaria cucumerina* on Cucurbitaceous hosts. *Farm Sci. J., 12*(1), 80–81.

Narasimhan, M. J., (1935). Report of the work done in mycological section for the year 1933–1934. *Adm. Rept. Agric. Dept. Mysore (1933–34),* pp. 19–22.

Narian, U., Kant, S., & Chand, G., (2016). Characterization of species and parasitism in the genus *Alternaria*. In: Chand, G., & Kumar, S., (eds.), *Crop Diseases and Their Management: Integrated Approaches* (pp. 385–404). Apple Academic Press, Waretown, USA.

Neergaard, P., (1945). *Danish Species of Alternaria and Stemphyrium: Taxonomy, Parasitism, Economical Significance* (p. 560). Humphry Millford, Oxford Univ. Press, Landon.

Nene, Y. L., (1998). *Tomato Disorders: Identification Handbook* (pp. 2, 3, 10, 11). Tech. Bull. No. 1. P.D.S. Enterprises, Ashok Path, Pune, India.

Pryor, B. M., & Strandberg, J. O., (2001). Alternaria leaf blight of carrot. In: Davis, R. M., & Raid, R. N. (eds.), *Compendium of Umbelliferous Crop Diseases.* American Phytopathological Society, St. Paul, Mn.

Puttoo, B. L., (1973). Seed mycoflora of carrot. *Indian J. Mycol. Pl. Pathol., 3*, 201–202.

Quebral, F. C., & Shurtleff, M. C., (1965). Alternaria rot-a serious disease of bell pepper in Illinois. *Phytopathology, 55*(10), 1051–1085(Abstr.).

Ramakrishnan, L., Kamalnathan, & Krishnamurthy, C. S., (1971). Studies on *Alternaria* leaf spot of tomato. *Madras Agric. J., 58*, 275–280.

Rands, R. D., (1917). Early blight of tomato and related plants. *Wisconi. Agri. Exp. Sta. Res. Bull., 41*, 48.

Rangaswami, G., & Sambandam, G. N., (1960). *Alternaria melengenae* causing leaf spot and fruit scab of eggplant and fruit rot of chili. *Mycologia, 52*, 517–520.

Rangaswami, G., & Venkata, R., A., (1957). Alternaria blight of cluster beans. *Indian Phytopath., 10*, 18–25.

Rao, V. G., (1963). Some new host records of *Alternaria* species from India. *Mycopath. et Mycol. Appl., 19*, 181–183.

Rao, V. G., (1966). An account of the market and storage diseases of fruits and vegetables in Bombay, Maharasthra. *Mycopath. et Mycol. Appl., 28*, 165–176.

Roy, A. K., (1969). Studies on leaf blight of carrot caused by *Alternaria dauci. Indian Phytopath., 22*, 105–109.

Sahu, K. S. P., & Singh, S. P., (1980). Fruit rot of pointed gourd (*Trichosanthes dioica*) in Bihar. *Indian Phytopath., 33*, 308–309.

Sangwan, M. S., Mehta, N., & Gandhi, S. K., (2002). Some pathological studies on *Alternaria raphani* causing leaf and pod blight of radish. *J. Mycol. Pl. Pathol., 32*, 125–126.

Scott, D. J., & Wenham, H. T., (1972). Occurrence of two seed borne pathogens. *Alternaria radicina* and *Alternaria dauci* on imported seeds in New Zealand. *N. Z. J. Agric. Res., 16*, 247–250.

Sharma, N., & Bhargava, K. S., (1977). Fruit rot bitter gourd. *Indian Phytopath., 30*, 557–558.

Sharma, U. N., (1986). Studies on seed mycoflora of moth bean [*Vigna aconitifolia* (Jacq.) Marchel]. *M.Sc. (Ag.) Thesis, Sukhadia Univ., Udaipur*, p. 42.

Shukla, D. N., & Bhargava, S. N., (1976). Some pathogenic fungi from pulse and oilseed crops. *Proc. Nat. Acad. Sci. India, 46*(B), 530–531.

Shukla, P., Lal, B., Singh, R. P., & Singh, P. N., (1977). A new *Alternaria* leaf spot of cowpea. *Indian J. Mycol. Pl. Pathol., 7*, 159–160.

Singh, C. S., Kanaujia, R. S., & Chaudhary, R. L., (1975). Parasitic fungi of Faizabad. *Alternaria. Botanique, 6*, 191–194.

Singh, R. S., & Chohan, J. S., (1980). Fungal fruit rot of bottle gourd in North India. *Indian Phytopath., 33*, 598–599.

Singh, S. D., & Prasada, R., (1973). Studies on physiology and control of *Alternaria cyamopsidis*, the incintant of blight disease of guar. *Indian J. Mycol. Pl. Pathol., 3*, 33–39.

Singh, S. J., (1974). A fruit rot of snake gourd caused by *Alternaria tenuissima. Indian Phytopath., 27*, 384–385.

Sohi, H. S., & Mahally, M. N., (1977). Studies on mycoflora of water melon seeds. *India J. Mycol. Pl. Pathol., 7*, 25–28.

Solankure, R. T., & Rao, V. G., (1972). Alternaria leaf spot of taro from India. *Indian Phytopath.*, *25*, 457.

Sreekantiah, R. R., Nagaraj, K. S., & Ramchandra Rao, T. N., (1973). A virulent strain of *Alternaria alternata* causing leaf and fruit of chili. *Indian Phytopath.*, *26*, 600–6003.

Srivastava, M. P., & Tandon, R. N., (1966). Post-harvest diseases of tomato in India. *Mycopath. Mycol. Appl.*, *29*, 254–264.

Standberg, J. O., (1987). Detection of *Alternaria dauci*. *Phytopathology*, *77*, 1008–1012.

Starter, L. L., & Weimer, J. L., (1929). A monographic study of sweet potato diseases and their control. *U. S. Dept. Agr. Tech. Bull.*, *99*, 1929.

Subramaniam, C. L., (1954). A fruit rot of chili (*Capscium annuum* L.) caused by *Alternaria solani* (Ell. & Mart.) Jones & Crout. *Madras Agric. J.*, *41*, 96–101.

Suhag, L. S., Singh, R. M., & Malik, Y. S., (1985). Epidemiology of pod and leaf blight of radish caused by *Alternaria alternata*. *Indian Phytopath.*, *38*, 148–149.

Sujatha Bai, E., Seetharaman, K., & Shivaprakasam, (1993). Alternaria fruit rot disease of chili, a serious malady in Tamil Nadu. *Indian Phytopath.*, *46*, 338(Abst.).

Sultana, N., Khan, S. A. J., & Khanzada, A. K., (1988). Studies on seed-borne mycoflora of chili and control of fruit rot disease. *Pak. J. Sci. & Ind. Res.*, *31*, 365–368.

Sultana, N., Khanzada, A. K., & Alam, M., (1992). A new cause of fruit rot of chilies in Pakistana. *Pak. J. Sci. & Ind. Res.*, *35*, 461–462.

Swarup, J., & Mathur, R. S., (1972). Seed mycoflora of some Umbelliferous spices. *Indian Phytopath.*, *25*, 125.

Taubenhaus, J. J., (1925). A new foot rot of sweet potato. *Phytopathology*, *15*, 238–240.

Thakur, V., Sharma, A. D., & Munjal, R. L., (1981). Seed mycoflora of *Raphanus sativus*, its pathogenicity and control in India. *Indian J. Mycol. Pl. Pathol.*, *11*, 161–163.

Toor, M. S., (1980). Studies on Alternaria leaf spot disease of Moth bean [*Vigna aconitifolia* (Jacq.) Marechel]. *MSc (Ag.) Thesis* (p. 42). Sukhadia Univ., Udaipur.

Uma, N. U., (1981). Four post harvest diseases of the Nigerian red pepper (*Capsicum annuum* L.) *Plant Dis.*, *65*, 915–916.

Uppal, B. N., Patel, M. K., & Kamat, M. N., (1949). Fungi of Bombay, Supplement I. *Indian Phytopath.*, *2*, 142–155.

Walker, J. C., (1952). Disease of crucifers. In: *Diseases of Vegetable Crops* (pp. 150–152). McGraw Hill Book Co., New York, Chapter 6.

Warner, E. E., (1936). Black rot of tomato caused by *Alternaria tomato*. *Fla. Agric. Exp. Sta. Bull.*, *332*, 54.

Weber, G. F., & West, E., (1930). Diseases of sweet potatoes in Florida. *Fla. Agr. Expt. Sta. Bull.*, *212*, 1930.

Weimer, J. L., (1924). *Alternaria* leaf spot and brown rot of cauliflower. *J. Agric. Res.*, *29*, 421–441.

Wltshire, S. P., (1947). Species of *Alternata* on Brassicae. *Imp. Mycol. Instt.*, *Mycol. Paper*, *20*, 1947.

CHAPTER 23

Micronutrients Deficiency in Vegetable Crops and Their Management

SHWETA SHAMBHAVI, RAKESH KUMAR, RAJKISHORE KUMAR, and MAHENDRA SINGH

Department of Soil Science and Agricultural Chemistry,
Bihar Agricultural University, Sabour, Bhagalpur, Bihar, India,
E-mail: rbinnu@gmail.com

23.1 INTRODUCTION

Mineral nutrients play a significant role in plant growth and development. However, a large number of elements are required for the growth of plants. Of these nutrients, only a few are required in large amounts for agricultural production. Deficiencies of the remaining elements which are required in lesser amounts are most frequently related to specialized crops or certain types of soils. But as cropping system becomes more intensive, changes in soil management practices frequently alter micronutrients availability, and depletion of nutrients not added in fertilizers becomes more rapid. As demand for higher yields increases and the plants requirement for major elements is more efficiently met, other nutrients are more likely to become limiting. Micronutrients have also been called minor or trace elements, indicating that their concentrations in plant tissues are minor or in trace amounts relative to the macronutrients (Mortvedt, 2000). The essential micronutrients for field crops are B, Cu, Fe, Mn, Mo, and Zn. The accumulation of these micronutrients by plants generally follows the order of Mn>Fe>Zn>B>Cu>Mo. However, this order may change among plant species and growth conditions. Other mineral nutrients at low concentrations considered essential to growth of some plants are Ni and Co. Plant roots require certain conditions to obtain these nutrients from the soil. First, the soil must be sufficiently moist to allow the roots to take up and transport the nutrients (Yin et al., 2009). Sometimes correcting improper watering strategies eliminate nutrient deficiency symptoms. Second, the pH of the soil must be within a certain range for nutrients to be release-able from the soil particles (Fernández, and Hoeft, 2009). Third, the temperature of the soil must fall within a certain range for nutrient uptake to occur (Pregitzer and King, 2005). The optimum range of temperature, pH, and moisture is different for different species of plants. Thus, nutrients may be physically present in the soil, but not available to plants. Further, micronutrient content in soils depends on the nature of parent material, type of minerals, soil type and climatic factors. Knowledge of soil pH, texture, and history can be very useful for predicting what nutrients may become deficient (McCauley, 2011).

Micronutrient deficiencies in crop plants are widespread because of (i) increased micronutrient demands from intensive cropping practices and adoption of high yielding cultivars which may have higher micronutrient demand, (ii) enhanced production of crops on marginal soils that contain low levels of essential nutrients, (iii) increased use of high analysis fertilizers with low amounts of micronutrient contaminations, (iv) decreased use of animal manures, composts, and crop residues; and (v) use of soils that are inherently low in micronutrient reserves.

Micronutrient Content in Indian Soils (ppm)

Micronutrient	Total	Available
Zinc	21.5–88.7	0.2–9.2
Molybdenum	0.6–11.6	Traces – 75
Copper	1.8–960	Traces – 16.8
Manganese	92–11500	0.7–85.90
Iron	4000–270,000	Trace – 982
Boron	7–630	Traces – 12.2

23.1.1 ROLE OF THE MICRONUTRIENTS

Micronutrients play many complex roles in plant nutrition. While most of the micronutrients participate in the functioning of a number of enzyme systems, there is considerable variation in the specific functions of the various micronutrients in plant and microbial growth processes. For example, copper, iron, and molybdenum are capable of acting as electron carriers in the enzyme systems that bring about oxidation reduction reactions in plants. Such reactions are essential steps in photosynthesis and many metabolic processes. Zinc and manganese function in many plant enzyme systems as bridges to connect the enzyme with the substrate upon which it is meant to act (Chatzistathis, 2014) (Table 23.1).

TABLE 23.1 Functions of Several Micronutrients in Plants

Micronutrient	Function in Higher Plants
Zinc	Present in several dehydrogenate, proteinase, and peptidase enzymes; promotes growth hormones and starch formation; promotes seed maturation and production.
Iron	Present in several peroxides, catalase, and cytochrome oxidase enzymes; found in ferredoxin, which participates in chlorophyll formation.
Copper	Present in laccase and several other oxidase enzymes; important in photosynthesis, protein, and carbohydrate metabolism, and probably nitrogen fixation.
Manganese	Activates decarboxylase, dehydrogenase, and oxidase enzymes; important in photosynthesis, nitrogen metabolism and nitrogen assimilation.
Nickel	Essential for urease, hydrogenases, and methyl reductase; needed for grain filling, seed viability, iron absorption, and urea and ureide metabolism (to avoid toxic levels of these nitrogen fixation products in legumes).
Boron	Activates certain dehydrogenase enzyme, facilitates sugar translocation and synthesis of nucleic acids and plant hormones; essential for cell division and development.
Molybdenum	Present in nitrogenase (nitrogen fixation) and nitrate reductase enzymes; essential for nitrogen fixation and nitrogen assimilation.
Cobalt	Essential for nitrogen fixation; found in vitamin B_{12}.

Molybdenum and manganese are essential for certain nitrogen transformations in microorganisms as well as in plants. Molybdenum and iron are components of the enzyme nitrogenize (Hille, 1999), which is essential for the processes of symbiotic and non-symbiotic nitrogen fixation. Molybdenum is also present in the enzyme nitrate reductase, which is responsible for the reduction of nitrates in soils and plants.

Nickel has only recently been added to the list of elements shown to be essential to plants. It is essential for the function of several enzymes (Ragsdale, 2009), including urease, the enzyme that breaks down urea into ammonia and carbon dioxide. Nickel deficient plants accumulate toxic levels of urea in their leaves and seeds. Copper is involved in both photosynthesis and respiration, and in the use of iron. It also stimulates lignifications of cell walls. The roles of boron have yet to be clearly defined, but boron appears to be involved with cell division, water uptake, and in the synthesis of proteins and nucleic acids. Manganese seems to be essential for photosynthesis, respiration, and nitrogen metabolism.

The role of chlorine is still somewhat obscure; however, it is known to influence photosynthesis and root growth. Cobalt is essential for the symbiotic fixation of nitrogen (Weisany et al., 2013). In addition, legumes, and some other plants have a cobalt requirement independent of nitrogen fixation, although the amount required is small compared to that for the nitrogen fixation process.

23.1.2 VISUAL DEFICIENCY SYMPTOMS

Visual deficiency symptoms are generally characteristic enough to permit easy identification of the deficiency of a nutrient, as these appear on particular plant parts at specific growth stage(s). Appearance of deficiency and toxicity symptoms is related with mobility of nutrient in plants.

Element	Mobility in Plants	Form Absorbed
Chlorine	Mobile	Cl^-
Zinc	Low mobility	Zn^{2+}
Molybdenum	Moderately mobility	MoO_4^{2-}
Copper	Relatively immobile	Cu^+, Cu^{2+}
Manganese	Relatively immobile	Mn^{2+}, Mn^{4+}
Iron	Relatively immobile	Fe^{2+}, Fe^{3+}
Boron	Relatively immobile	BO_3

In case the symptoms are less characteristics, their presence needs to be confirmed through soil and plant analysis as some conditions other than the nutrient deficiency may be responsible for such a situation. It should be remembered that nutrient deficiencies or toxicities can resemble non-nutritional disorders such as disease or herbicide damage. The location of the symptoms of nutrient deficiencies on plant depends on the extent and rate of mobility of nutrients from the old to new emerging leaves (Vitosh et al., 1994).

Micronutrients Deficiency in Vegetable Crops

23.1.3 DIAGNOSING NUTRIENT DEFICIENCY AND TOXICITY SYMPTOMS IN VEGETABLE CROPS

23.1.3.1 MANGANESE

Manganese has shown to be an essential nutrient for healthy plant growth by McHargue in 1922. Plant roots have a Mn absorption mechanism that provides sufficient Mn for healthy growth in most soils. Mn concentration in the earth's crust averages 10, 000 ppm and is found mostly in Fe-Mn rocks. Total Mn in soils generally ranges between 20 to 3000 ppm and averages about 600 ppm. In some acid soils, however, solution Mn concentration may reach very high levels. Many plants will then absorb more Mn than they require internally.

1. **Manganese Deficiency Symptoms:** Chloroplasts are the most sensitive of cell organelles to Mn deficiency (Mengel and Kirkby, 2001). As a result, a common symptom of Mn deficiency is interveinal chlorosis in young leaves. Manganese deficient crops become yellowish to olive-green in color. However, unlike with Fe deficiency, there is no sharp distinction between veins and interveinal areas, but rather a more diffuse chlorotic effect. Manganese deficiency in soybeans, dry edible beans, snap beans, sugar beets, celery, cucumbers, and cabbage often causes marked yellowing between the leaf veins while the veins themselves remain dark green. Iron deficiency mostly occurs on new growth while Mn deficiency occurs over the entire plant. In sugar beets and potatoes, chlorosis begins in the younger leaves. Later, grey and black freckling may develop along the veins. Potatoes also show reduced leaf size (Allen et al., 2012). Manganese-deficient onions are olive-green and the leaves may appear wilted. Manganese deficiency is sometimes confused with nitrogen deficiency and for its differentiation nitrogen tissue test is performed. Nitrate N tissue test easily determines which nutrient is deficient. Manganese-deficient plants usually have higher nitrate-nitrogen because of the lack of enzymes required for the conversion of nitrate to protein N.

2. **Correcting Manganese Deficiency:** Manganese deficiency in crops can be prevented by band application of manganese fertilizer to the soil, spraying it on the foliage or making the soil more acidic. For micronutrient metals, the foliar application is often the most effective

way to correct low or deficient levels (Johnson, 2016). Steam or chemical fumigation also corrects it temporarily. Generally, when manganese is deficient, manganese sulfate or manganous oxide is mixed with the fertilizer and applied in a band near the seed. Commercial manganese sulfate has 26 to 28% manganese (Mn); manganous oxide usually has 41 to 68% manganese. Studies have shown that manganous oxide should be finely ground to be effective. Granular manganous oxide (8 mesh) are largely ineffective. Manganous oxide powders (200 and 325 mesh) are less effective than manganese sulfate but acceptable. These materials do not blend well with other fertilizer materials as segregation problems occur because of differences in particle sizes. However, the use of a sticker such as liquid fertilizer has made it possible to use these finely ground materials in the bulk blending process. Broadcast application of manganese is not recommended because of high fixation in the soil. Residual carryover of available manganese fertilizer is usually low. Therefore, manganese must be applied every year on a deficient soil. Foliar applications of manganese are recommended when:

1. Fertilizer is not applied in a band near the seed;
2. Deficiency symptoms appear on the foliage; and
3. Regular fungicide and insecticide sprays are applied.

The recommended rate is 1 to 2.5 kg of manganese per ha in 100 liters of water, using the 1 kg rate if plants are small and the 2.5 kg rate if plants are medium to large. Spray grades of the manganese carriers are recommended to prevent nozzle plugging. Acidifying the soil with materials such as sulfur and aluminum sulfate can correct manganese deficiency. Acid-forming nitrogen and phosphorus fertilizers promote the release of fixed soil manganese, especially if banded near the plant. Some of the benefits accredited to band placement of fertilizer may be due to the release of fixed soil manganese.

3. **Manganese Toxicity:** Excessive manganese is a problem in extremely acid soils (<pH 5.0). A toxic manganese situation may also develop in plants if excessive soil and/or foliar applications are used. Liming soils to the desired pH range for the crop will usually prevent any manganese toxicity. In the early stages of plant growth, manganese toxicity symptoms may be similar to deficiency symptoms. Mn toxicity symptoms are generally characterized by blackish-brown or red spots on older leaves and an uneven distribution of chlorophyll, causing chlorosis and necrotic lesions on leaves (Havlin et al., 1999).

Micronutrients Deficiency in Vegetable Crops

In potatoes, the symptoms are chlorosis and black specks on the stems and undersides of the leaves, followed by death of the lower leaves. The following crops are sensitive to excess manganese: alfalfa, cabbage, cauliflower, clover, dry edible beans, potatoes, sugar beets and tomatoes. Plant tissue analysis is helpful in diagnosing manganese status. Values below 20 ppm are usually considered deficient. Readings of 30 to 200 ppm are normal, and those above 300 ppm are considered excessive or toxic.

23.1.3.2 ZINC

Zinc is essential for plant growth because it controls the synthesis of Indole acetic acid, which dramatically regulates plant growth. Zinc is also active in many enzymatic reactions and is necessary for chlorophyll synthesis and carbohydrate formation. Because zinc is not readily translocated within the plant, deficiency symptoms first appear on younger leaves. Soils associated with zinc deficiency are usually neutral to alkaline in reaction. The more alkaline the soil, the greater the need for zinc. Deficiency is particularly noticeable on crops growing where calcareous subsoils have been exposed by land leveling or erosion, or where subsoil is mixed with topsoil, such as after tiling and spoil-bank leveling. Crops on poorly drained organic soils show a deficiency probably because of restricted root growth. Field and vegetable crops often show differences in response to zinc fertilizer. Dry edible beans, corn, onions, sorghum, snap beans, spinach, and sweet corn are the most responsive crops. High soil phosphorus levels have been known to induce zinc deficiency, especially in responsive crops. For years, the cause of this interaction was suspected to be the formation of an insoluble zinc phosphate, which reduced the concentration of zinc in the soil solution to deficiency levels. Zinc phosphate has since been shown to be soluble in soil and is an acceptable source of zinc when finely ground. High levels of phosphorus in plants have been shown to restrict zinc movement within the plant, resulting in accumulation in the roots and deficiency in the tops. Therefore, large applications of phosphorus fertilizer may contribute to zinc deficiency in zinc-responsive crops.

1. **Zinc Deficiency Symptoms:** It's hard to tell the difference between zinc deficiency and other trace element or micronutrient deficiencies by looking at the plant because they all have similar symptoms. The main difference is that chlorosis due to zinc deficiency begins on

the lower leaves, while chlorosis due to a shortage of iron, manganese or molybdenum begins on the upper leaves. Zinc deficiency inhibits both vegetative and reproductive growth. The deficiency results in shortened internodes, downward curling of leaflets, and chlorosis. Under severe deficiency, oozing of cell contents as brown fluid is seen from the leaves in tomato (Jeyakumar and Balamohan, 2018). Zinc-deficient crops at the initial stage become light green in color. When the deficiency is severe, the area between the leaf veins becomes pale green and then yellow near the tips and outer edges. In the early stages of deficiency, the leaves become deformed, dwarfed, and crumpled. In later stages, they look like sunscald leaves. On zinc-deficient plants, the terminal blossoms set pods that drop off, delaying maturity. Areas of the leaf near the stalk may develop a general white to yellow discoloration. In case of severe deficiency, the plants have shortened internodes, and the lower leaves show a reddish or yellowish streak about one-third of the way from the leaf margin. Plants growing in dark sandy or organic soils usually show brown or purple nodal tissues when the stalk is split. This is particularly noticeable in the lower nodes. In chili, the entire foliage of the crop will be reduced in size with interveinal chlorosis, when zinc is in short supply, and the plant will present a stunted growth (Jeyakumar and Balamohan, 2018). Deficiency in onions shows up as stunting, with marked twisting and bending of yellow-striped tops. In potatoes, early symptoms are similar to leaf roll. The plants are generally more rigid than normal, with smaller than normal leaves and shorter upper internodes.

2. **Correcting Zinc Deficiency:** Several zinc compounds can be used to correct zinc deficiency. Zinc sulfate, zinc oxide, zinc chloride, zinc sulfide, and zinc carbonate are common inorganic salts. Organic compounds such as zinc chelates (zinc EDTA and zinc NTA) are about five times more effective than inorganic salts with equivalent amounts of zinc. Organic carriers, however, have a lower zinc concentration, ranging from 9 to 14%. The zinc concentration of zinc sulfate ranges from 25 to 36%, and that of zinc oxide, 70 to 80%. In field tests, granular zinc oxide was not as effective as the powdered formulation. The test also showed that mixing the zinc carrier with the fertilizer was more effective than incorporating the carrier in the granule. In the case of soil application, zinc must be applied near the seed at planting time for its higher effectiveness. Mixing zinc with a phosphate fertilizer, such as 6-24-24, is acceptable while, seed

Micronutrients Deficiency in Vegetable Crops

treatment with zinc oxide is not recommended. If a zinc deficiency problem is diagnosed after emergence, foliar spray with 0.5 to 1 kg of zinc per ha (from 2 to 5 kg of zinc sulfate) is advisable. Foliar application of zinc sulfate @ 500 g and urea @ 100 g dissolved in 100 lit of water along with 100 ml soap solution is to be given two to three times at an interval of 20 days depending upon the extent of severity (Jeyakumar and Balamohan, 2018). Foliar application has to be supplemented with soil application of zinc sulfate @ 8 kg/acre is given basally during the last plowing of the field. Response to spray applications is usually obvious within 10 days. For plants with waxy leaves, such as onions, a wetting agent in the water may be needed to obtain good foliage cover. If foliage sprays are used, they should be applied when plants are small to obtain best results.

3. **Soil and Plant Tissue Tests for Zinc:** Plant tissue tests can help diagnose a need for zinc. Tissues containing less than 20 ppm of zinc are often deficient. Values of 30 to 100 ppm are normal; values above 300 ppm may be considered excessive or toxic.

4. **Zinc Toxicity:** Excessive soil zinc levels may occur on extremely acid soils (< pH 5.0) or in areas where zinc-enriched municipal sewage sludge or industrial waste has been added to cropland as a soil amendment. High levels of available soil zinc that result in 100 to 300 ppm zinc in crown leaf tissue seldom result in zinc toxicity in corn, which is highly zinc tolerant. However, if the soil levels result in 40 to 50 ppm or more of zinc in the leaf tissue of some varieties of dry edible beans, toxicity may occur because dry edible beans are a zinc sensitive crop (Miyazawa et al., 2001). Vegetable crops are generally sensitive to high zinc levels, while grasses usually tolerate high levels of available soil zinc.

23.1.3.3 COPPER

Copper is believed to be the first metal employed by man. It was first reported by Sommers, Lipman, and MacKinney in 1931. It is an essential constituent of all living materials. Cu concentration in the earth's crust averages about 50 to 70 ppm while in the soil, it ranges from 1 to 40 and averages about 9 ppm. Normal Cu concentration in plant tissues ranges from 5 to 20 ppm. Copper is essential for plant growth and the activation of many enzymes. A copper deficiency interferes with protein synthesis and causes a buildup of soluble nitrogen compounds. Normal plants contain 8 to 20 ppm copper;

deficient plants usually contain less than 6 ppm. Without copper, all crops fail to grow. Peaty soils, which have low ash content, are generally the only soils deficient in copper. If the problem does appear on mineral soils, it will most likely be on acid soils that have been heavily cropped but well fertilized with N, P, and K. Copper applied to soil is not easily leached; nor is it extensively used by the crop. Consequently, no further copper fertilization is needed on organic soils if a total of 25 kg per ha has been applied for low responsive crops and 45 kg per ha for highly responsive crops.

1. **Copper Deficiency Symptoms:** Copper deficiency in many plants shows up as wilting or lack of turgor and development of a bluish-green tint before leaf tips become chlorotic and die. Carrot roots, wheat grain and onion bulbs show poor pigmentation. Alfalfa, lettuce, oats, onion, spinach, Sudan grass, table beet and wheat are the most responsive crops on organic soils. Both vegetative and reproductive growths are reduced. Wilting of terminal shoots occurs, which is followed by frequent death. Leaf color is often faded due to the reduction of carotene and other pigments. The foliage may, however, show the burning of the margins or chlorosis or rosetting, and multiple bud formation and gummosis may also occur. The chlorotic leaves subsequently become bronzed and later into brown with development of necrosis at the margins and blackening of veins (Jeyakumar and Balamohan, 2018). In crops, young leaves wither and show marginal chlorosis (yellowing grey) of the tips. It is known as Reclamation White Tip disease. Heads are dwarfed, distorted, and tips tend to be chlorotic.

2. **Correcting Copper Deficiency:** Rates of copper commonly used in highly responsive crops are 3.5 to 7 kg per ha, depending on the soil test level. Soil application of $CuSO_4$ @ 5–10 kg/ha and Cu-EDTA @1–2 kg/ha is recommended. Foliar application of $CuSO_4$ or Cu-chelates @ 0.1% solutions is also recommended. These rates should be doubled on fields that have never received copper.

3. **Copper Toxicity:** Excessive soil copper levels have not been a problem in crop production. However, the potential for copper toxicity does exist because copper is applied annually for some vegetables, either as a soil amendment or a component of some fungicides. Copper toxicity often results in plant stunting, a bluish tint to leaf color, and leaf cupping followed by chlorosis or necrosis. When the copper concentration exceeds 150 ppm in mature leaf tissue, toxicity may occur. Cumulative copper applications of 110

Micronutrients Deficiency in Vegetable Crops

kg per ha have reduced cucumber and snap bean yields on sandy soils. Copper is tightly adsorbed by most soils and will not leach. Therefore, once a copper toxicity problem develops, it may be very difficult, if not impossible, to alleviate it. The improved crop performance of grafted plants was attributed to their strong capacity to inhibit Cu accumulation in the aerial parts and to maintain a better plant nutritional status (Rouphael et al., 2008).

23.1.3.4 IRON

Iron represents essential elements which are usually available in oxidized form but must be reduced, often using metabolic energy, before they can be utilized. Iron, after oxygen, silicon, and aluminum is the fourth most abundant where it concentration is 4.7%. In soils, the concentration of Fe may range from 0.5 to 3.6%, a concentration which in comparison with other plant nutrients such as N and P is extraordinarily high. Iron is a constituent of many organic compounds in plants. It is essential for synthesizing chlorophyll, which gives plants their green color. It functions in a plant in many ways but a lack of Fe in the growing medium most often evidenced by a yellowing of leaves commonly referred to as iron-chlorosis. Iron-chlorosis affects the production and well being of many kinds of plants including field crops, nursery stocks, large, and small fruits, as well as forage and turf plants. Iron deficiency can be induced by high levels of manganese. High iron levels can also cause manganese deficiency. Copper deficiency can be responsible for promoting an iron deficiency (Krämer, 2012).

1. **Iron Deficiency Symptoms:** Deficiency symptoms are marked and show up first in terminal leaves as a light yellowing. The symptoms are very similar to those of manganese deficiency. A lack of iron in field and vegetable crops is not common in soils with pH below 7.0 (Walworth, 2012; Shalau, 2010). Iron deficiency is common in the areas, where the soils contain considerable amount of sodium and calcium. Iron deficiency in many woody plants appears when they are grown in soils low in organic matter and high in pH.
2. **Correcting Iron Deficiency:** Soil treatments usually require applications of iron chelates at a rate equivalent to 0.5 to 1 kg of iron per ha. Soil applications are effective if soils are acid or neutral in reaction. Under alkaline soil conditions, foliage sprays are recommended. Iron chelates are normally recommended for foliar spray,

500   *Diseases of Fruits and Vegetable Crops*

although more expensive than iron sulfate, but it persists for longer period of time. Sometimes the best cure for Fe deficiency is to grow varieties that are not sensitive to Fe deficiency. For instance, some soybean varieties are more sensitive to Fe deficiency than others. To help prevent an iron problem, avoid using excessive amounts of lime or phosphate. Apply chemicals or fertilizers to increase the soil acidity and add organic matter.

3. **Iron Toxicity:** Injury due to high soil iron concentrations is not common under neutral or high pH soil conditions. Toxic situations occur primarily on acid soils ($<$ pH 5.0) and where excess soluble iron salts have been applied as foliar sprays or soil amendments. The first symptoms of iron toxicity are necrotic spots on the leaves. Some iron-rich, low pH, low manganese soils create an environment in which an interaction between the iron and manganese in the soil reduces manganese uptake by plants. The symptoms observed on the plants are of manganese deficiency, but the low plant uptake of manganese is caused by excessive available iron in the soil (Heenan and Campbell, 1982). The addition of iron chelates or manganese chelates, which rapidly convert to the iron form under these soil conditions, aggravates the situation by increasing the amount of available iron and without solving the manganese deficiency problem.

23.1.3.5 BORON

Boron primarily regulates the carbohydrate metabolism in plants. It is essential for protein synthesis, seed, and cell wall formation, germination of pollen grains and growth of pollen tubes. Boron is also associated with sugar translocation. Boron requirements vary greatly from crop to crop. Rates required for responsive crops such as alfalfa, celery, sugar beets, and table beets can cause serious damage to small grains, beans, peas, and cucumbers. Boron deficiency may occur under a wide range of soil conditions. Alkaline soils have reduced uptake of boron due to high pH. Leached soils may be boron deficient because of low boron reserves. The soil types most frequently deficient in boron are sandy soils, organic soils, and some fine-textured lakebed soils. Boron deficiency frequently develops during drought periods when soil moisture is inadequate for maximum growth.

1. **Boron Deficiency Symptoms:** Gupta (1979) reported that the deficient and toxic levels of boron are associated with plant disorders

and/or reductions in the yield of crops. In cauliflower, turnip, radish, cabbage, and other root crops, its deficiency commonly causes brown heart. Boron deficiency in crops causes a breakdown of the growing tip tissue or a shortening of the terminal growth (Anonymous, 2012). This may appear as rosetting. Internal tissues of beets, turnips, and rutabagas show breakdown and corky, dark discoloration. Boron deficiency and leafhopper damage in alfalfa are often confused. Boron deficiency shows up as a yellowish to reddish-yellow discoloration of the upper leaves, short nodes, and few flowers. Growing tips of alfalfa may die, with regrowth coming after a new shoot is initiated at a lower axis. Leafhopper damage shows up as a V-shaped yellowing of the affected leaves and may appear on any or all parts of the plant; the growing tip is usually normal, and the plant may support abundant flowers. When the soil is dry, and plant growth is retarded, both boron deficiency and leafhopper injury often occur in the same field. Deficiency in cauliflower shows up as a darkening of the head and is associated with hollow and darkened stems. Hollow stem can also be caused by adverse weather conditions. Boron deficiency usually appears in small spots and may spread until the entire head is discolored. In sugar beets, the first symptoms are white, netted chapping of upper blade surfaces or wilting of tops. Later, if the deficiency becomes severe, transverse (crosswise) cracking of petioles develops, the growing point dies and the heart of the root rots. In celery, the first symptoms are brownish mottling along the margins of the bud leaves and brittle stems with brown stripes along the ribs. Later, crosswise cracks appear on the stems.

2. **Correcting Boron Deficiency:** Alfalfa, cauliflower, celery, table beets and turnips are the most responsive crops for boron deficiency. The boron recommendations for soil applications are 1.5 to 3.5 kg for highly responsive crops and 0.5 to 1 kg per ha for medium responsive crops. Occasionally, certain deficient soils may require up to 6 ha of boron per ha for cauliflower and table beets. The suggested rate for foliar application is 0.3 kg of boron per ha in 100 liters of water for highly responsive crops and 0.1 kg for low to medium responsive crops. The boron carrier most frequently used in fertilizer is sodium borate, which ranges from 10 to 20% boron. "Solubor" is a trade name for a sodium borate that is 20.5% boron. This compound is commonly used in foliar sprays or in liquid fertilizers. Because boron is fairly mobile in soils, several methods of application can be used. Boron may be mixed with regular N-P-K fertilizer, applied

502   *Diseases of Fruits and Vegetable Crops*

separately on the soil, sprayed on the plant, top dressed (for alfalfa) or side dressed (for row crops). Be sure to mix completely when boron is combined with other fertilizers. Segregation due to particle size differences is often a problem. Be careful when banding fertilizers containing boron near the seed or plants. Too much boron near the seed or plant may be toxic to young plants or germinating seeds.

3. **Boron Toxicity:** Toxicity symptoms typically show first on older leaf tips and edges as a yellowing, spotting or drying of leaf tissues. It is also evident from the previous work of Valmis and Ulrich (1971) that the supply of boron affects the distribution of boron in various plant parts. Boron toxicity is usually limited to situations where boron-containing fertilizers are used at planting time on highly sensitive crops such as dry edible beans, corn, grass, and small grains. Toxicity to crops has also occurred when sensitive crops were planted where fertilizers containing boron had been used earlier in the season. Similar problems may occur where sensitive vegetable crops are planted with high rates of boron in the starter fertilizer. Unlike copper, zinc, and manganese, boron is rapidly leached out of the soil or fixed in the soil so there is little potential for toxic carryover from year to year. Some wastewaters used for irrigation may have high boron levels. Boron toxicity is characterized by yellowing of the leaf tips, interveinal chlorosis and progressive scorching of the leaf margins. High levels of calcium may increase the boron tolerance of plants. Average boron concentrations in mature leaf tissues can be used to estimate plant boron status as follows: deficient—less than 15 ppm; sufficient—20 to 100 ppm; and excessive or toxic—over 200 ppm.

23.1.3.6 *MOLYBDENUM*

Molybdenum functions largely in the enzyme systems of nitrogen fixation and nitrate reduction. Plants that cannot fix adequate N or incorporate nitrate into their metabolic system because of inadequate molybdenum may become nitrogen deficient (Sardesai, 1993). The usual carriers of molybdenum are sodium or ammonium molybdate. These salts contain about 40% of the element. Molybdenum is required in very small amounts. Normal tissues usually contain between 0.8 and 5ppm; some plants may contain up to 15 ppm. Deficient plants usually contain less than 0.5 ppm. Certain nonresponsive crops such as grass and corn may contain as little

Micronutrients Deficiency in Vegetable Crops 503

as 0.1 ppm. The responsive crops are clover, cauliflower, broccoli, lettuce, onions, spinach, and table beets.

1. **Molybdenum Deficiency Symptoms:** Molybdenum deficiency in clover shows up as a general yellow to greenish yellow foliage color, stunting, and lack of vigor. The symptoms are similar to those caused by nitrogen starvation. Early stages of the deficiency in cauliflower and broccoli appear as a marginal scorching, rolling or curling upward, and withering and crinkling of the leaves. In later growth stages, the deficiency shows up as "whiptail," especially in the younger leaves. The leaf blade is often very narrow or non-existent. Older leaves show crinkling and marked yellow mottling between the veins. In onions, molybdenum deficiency shows up as dying leaf tips. Below the dead tip, the leaf shows 1 or 2 inches of wilting and flabby formation. As the deficiency progresses, the wilting and dying advance down the leaves. In severe cases, the plant dies (Chipman et al., 1970).

2. **Correcting Molybdenum Deficiency:** Molybdenum deficiency can be corrected by seed treatment and/or foliar applications. For seed treatment, dissolve 1 kg of the molybdenum compound in 3 tablespoons of water and mix with sufficient seed to plant one ha. Using excess water can cause the chemical to penetrate and injure the seed embryo. Mix the seed thoroughly and let dry. It is advisable to use a suitable fungicide dust to help dry the seed. For foliar sprays, apply 2 to 3 kg of the compound per ha. Use wetting agents in the spray when applying the solution to cauliflower or onions. For some cauliflower varieties, repeated applications at two-week intervals are beneficial. Soil acidity has a marked influence on the need for molybdenum— the greater the acidity, the greater the need for molybdenum. Liming from pH 4.9 to pH 6.7 increased the molybdenum concentration of cauliflower fivefold. Liming severely deficient soils, however, will not completely correct the deficiency.

3. **Molybdenum Toxicity:** Plants appear quite tolerant of high soil molybdenum concentrations. There is no record of molybdenum toxicity under field conditions. High levels of Mo in forages can induce Cu deficiencies in animals (Kubota, 1975; Vlek and Lindsay, 1977). This disorder, referred to as molybdenosis, occasionally results in death. In greenhouse studies, tomato leaves turned golden-yellow and cauliflower seedlings turned purple. Animals fed foliage high in molybdenum may need supplemental copper to counteract the molybdenum.

23.1.4 MICRONUTRIENT FERTILIZERS

There are several ways to add secondary and micronutrients to nitrogen, phosphorus, and potassium (N-P-K) fertilizers. They may be incorporated into granulated fertilizers during the granulation process so that each granule of fertilizer contains an equal amount of all nutrients. They may be blended with N-P-K fertilizers at a bulk blending plant. If the particle size of secondary and micronutrients is greatly different from the size of particles containing the primary nutrients, a sticker may be needed to prevent particle size separation. Separation can lead to segregation of particle sizes and non-uniform application. These nutrients may also be added to liquid or suspension fertilizers. Chelated secondary and micronutrient formulations of these nutrients are generally preferred to non-chelated materials for mixing with liquid fertilizer because a larger amount of the nutrient can be added before precipitation occurs. The amount of secondary or micronutrients required in mixed fertilizers depends on the application rate (Chatterjee and Dube, 2012).

23.1.5 FOLIAR APPLICATION OF MICRONUTRIENTS

Nutrients can be absorbed through plant leaves. In some situations, foliar-applied micronutrients are more readily available to the plant than soil-applied micronutrients, but foliar applications do not provide continuous nutrition as do soil applications. Foliar spray programs may be used to supplement soil applications of fertilizer or to correct deficiencies that develop in midseason (Hosier and Bradley, 1999).

When spray equipment is available, secondary, and micronutrient needs of plants may be met with a good spray program (Mueller and Diaz, 2011). Use low rates for young plants and higher rates when plants develop dense foliage. Micronutrient chelates are generally no more effective than water-soluble inorganic sources when foliar applied. Chelates, however, are more compatible when mixed with other spray materials. For a preventive spray program, spray the crop about four weeks after emergence or transplanting. Because many micronutrients are not readily translocated within the plant, a second spray will be needed two weeks later to cover the new foliage. When a known nutrient deficiency develops, spray the crop with the appropriate nutrient at the recommended rate every 10 days until the deficiency is corrected. Complete coverage of the foliage is important, especially for iron. Adding a wetting agent to the spray solution will improve the coverage and

Micronutrients Deficiency in Vegetable Crops

may increase absorption, especially in crops with waxy surfaces, such as cauliflower and onions. Micronutrients may be mixed with most fungicides and insecticides. However, some combinations are incompatible and may injure crops. When in doubt, spray only a limited acreage until compatibility is established. Any injury will usually appear within 48 hours. In developing a spray program, remember that some fungicides and insecticides contain copper, manganese or zinc. The amounts of micronutrients present in these materials may or may not be sufficient to correct a deficiency but should be considered when determining a spray program.

KEYWORDS

- **boron toxicity**
- **copper toxicity**
- **iron toxicity**
- **micronutrients**
- **molybdenum toxicity**
- **zinc toxicity**

REFERENCES

Allen, J. Williams, J. Olson, TL, Tufts, A., Oyala, P., & Lee, W., (2012). New Light Shined on Photosynthesis. University of Arizona. http://www.newswise.com/articles/new-light-shined-on-photosynthesis (Accessed on 18 November 2019).

Allen, V. B., & Pilbeam, D. J., (2007). *Handbook of Plant Nutrition*. CRC Press. ISBN 978-0-8247-5904-9.

Anonymous, (2012). *Boron Deficiency Symptoms*. Agronomy note. http://www.borax.com/docs/agronomy-notes/borondeficiencysymptoms-final-feb2012.pdf?sfvrsn=2 (Accessed on 18 November 2019).

Chatterjee, C., & Dube, B. K., (2012). *Fruit and Vegetable Diseases: Volume 1 of the Series Disease Management of Fruits and Vegetables* (pp. 145–188).

Chatzistathis, T., (2014). *Micronutrient Deficiency in Soils & Plants*, 934–938.

Chipman, E. W., Mackay, D. C., Gupta, U. C., & Cannon, H. B., (1970). Response of cauliflower cultivars to Molybdenum deficiency. *Can. J. Plilt Sci., 50*, 163–167.

Fernández, F. G., & Hoeft, R. G., (2012). Managing Soil pH and Crop Nutrients. In: *Illinois Agronomy Handbook* (pp. 91–112). University of Illinois.

Gupta, U. C., (1979). Boron nutrition of crops: Advances in agronomy. *Am. Soc. Agron., 31*, 273–307.

Havlin, J. L., Beaton, J. D., Tisdale, S. L., & Nelson, W. L., (1999). *Soil Fertility and Fertilizers* (6th edn., 499). Upper Saddle River, N. J. Prentice-Hall, Inc.

Heenan, D. P., & Campbell, L. C., (1982). Manganese and iron interactions on their uptake and distribution in soybean (Glycine max (L.) Merr.). *Plant and Soil, 70*, 317–326.

Hille, R. Reètey, J., Bartlewski-Hof, U., Reichenbecher, W., & Schink, B., (1999). Mechanistic aspects of molybdenum-containing enzymes. *FEMS Microbiology Reviews, 22*, 489–501.

Hosier, S., & Bradley, L., (1999). *Guide to Symptoms of Plant Nutrient Deficiencies.* University of Arizona, Extension Cooperative, Az1106.

Jeyakumar, P., & Balamohan, T. N., (2018). *Micronutrients for Horticultural Crops.* http://agritech.tnau.ac.in/agriculture/PDF/Micronutrients%20for%20horticultural%20crops.pdf (Accessed on 18 November 2019).

Johnson, G., (2016). *Correcting Nutrient Deficiencies in Vegetable Crops.* https://extension.udel.edu/weeklycropupdate/?p=9302 (Accessed on 18 November 2019).

Krämer, U., (2012). *"Nutrient and Toxin All at Once: How Plants Absorb the Perfect Quantity of Minerals".* http://esciencenews.com/articles/2012/04/13/nutrient.and.toxin.all.once (Accessed on 18 November 2019).

Kubota, J., (1975). The poisoned cattle of willow creek. *Soil Conservation, 40*(9), 18–21.

Mário, M., Sôniam, N. G., Maria, J. S. Y., Edson, L. O., & Marcos, Y. K., (2002). Absorption and toxicity of copper and zinc in bean plants cultivated in soil treated with chicken manure. *Water, Air, and Soil Pollution, 138*, 211–222.

Marschner, P., (2012). *Marschner's Mineral Nutrition of Higher Plants* (3rd ed. ed.). Amsterdam: Elsevier/Academic Press. ISBN 9780123849052.

McCauley, A., (2011). *Plant Nutrient Functions and Deficiency and Toxicity Symptoms.* Nutrient Management. Montana State University Extension.

Mengel, K., & Kirkby, E. A., (2001). *Principles of Plant Nutrition* (p. 849). The Netherlands. Kluwer Academic Publishers.

Mortvedt, J. J., (2000). Bioavailability of micronutrients. In: Sumner, M. E., (ed.), *Handbook of Soil Science.* (pp. D71–D88). CRC Press, Boca Raton, FL.

Mueller, N D., & Diaz, D. A. R., (2011). *Micronutrients as Starter and Foliar Application for Corn and Soybean* (Vol. 27). North Central Extension -Industry Soil Fertility Conference.

Pregitzer, K. S., & King, J. S., (2005). Effects of soil temperature on nutrient uptake. *In Nutrient Acquisition by Plants, 181*, 277–310.

Ragsdale, S. W., (2009). Nickel-based enzyme systems. *The Journal of Biological Chemistry, 284*(28), 18571–18575.

Rouphael, Y., Cardarelli, M., Elvira, R., & Giuseppe, C., (2008). Grafting of cucumber as a means to minimize copper toxicity. *Environmental and Experimental Botany, 63*, 49–58.

Sardesai, V. M., (1993). Molybdenum: An essential trace element. *Nutr. Clin. Pract., 8*(6), 277–281.

Shalau, J., (2010). *Laboratories Conducting Soil, Plant, Feed or Water Testing.* Publication AZ1111, College of Agriculture and Life Science, University of Arizona.

Valmis, J., & A. Ulrich, (1971). Boron nutrition in the growth and sugar content of sugar beets. *J. Am. Soc. Sugarbeet Technol., 16*, 428–439.

Vitosh, M. L., Warncke, D. D., & Lucas, R. E., (1994). *Secondary and Micronutrients for Vegetables and Field Crops* (p. E-486). Michigan State University Extension.

Vlek, P. L. G., & Lindsay, W. L., (1977). In: Chappell, W. R., & Petersen, K. K., (eds.), *"Molybdenum in the Environment"* (Vol. 2, pp. 619–650). Dekker, New York.

Walworth, J. L., (2012). *Soil Sampling and Publication.* AZ1412, College of Agriculture and Life Science, University of Arizona.

Weisany, W., Raei, Y., & Allahverdipoor, K. H., (2013). Role of some of mineral nutrients in biological nitrogen fixation. *Bull. Env. Pharmacol. Life Sci., 2*(4), 77–84.

Yina, C., Xueyong, P., & Ke, C., (2009). The effects of water, nutrient availability and their interaction on the growth, morphology and physiology of two poplar species. *Environmental and Experimental Botany, 67,* 196–203.

Index

A

Abiotic
  factors, 30, 217, 441
  stresses, 177, 444
Abscisic acid, 119
Acervuli, 50, 79, 90, 97, 137, 280, 384, 412, 445
Actinomycetes, 150, 221
Aeciospores, 25, 246
Aeonium canariese, 414
Aerial
  blackleg, 148
  infection, 102
  sprays, 407
  tubers, 163
Aerosols, 148, 199, 369
Aesculin hydrolysis, 315
Agaricomycetes, 447, 450
Agrimycin, 138, 212, 336
Agrobacterium tumefaciens, 35
Airborne
  conidia, 70
  spore, 419
Albedo tissue, 58
Aldicarb, 342, 424
Algae, 96, 97, 122, 125, 436, 452
  cells, 117
  disease, 94, 117, 141, 142
    algal leaf spots, 117
  leaf, 92, 93, 96, 107
  matrix, 141
Algicides, 142
Alkaline soils, 230, 239, 326, 397, 499
Alomae
  bobone complex, 222
  virus disease, 223
Alternaria, 329, 330
  alternata, 108, 115, 116, 462, 465, 468, 469, 473, 476, 477, 481
  cucumerina, 379, 381, 463, 476, 477, 483
  disease, 109, 122, 461, 462, 465

leaf spot, 70, 72, 310, 311, 320, 383, 407, 408, 465, 470, 474, 475, 481, 483
  disease cycle, 383
  disease distribution, 383
  disease occurrence, 383
  favorable conditions, 70
  management, 70
  spread mode
  survival mode, 70
  symptoms, 70, 383
  melogenae, 470
  solani, 153, 154, 187, 311, 469–471, 475
  species, 462, 475, 476, 478
Alternate host, 24, 25, 398, 399, 407, 436, 438
Ammonium
  molybdate, 502
  sulfate, 163, 326
Amorphophallus, 214, 216
  mosaic virus, 214
  paeonifolius, 214
Ampelocissus species, 72
Ampelovirus, 15, 17
Angiospermic parasites, 125
Angular leaf spot, 379, 385, 391, 392, 400, 407
Anthracnose, 68, 69, 72, 79, 80, 89–91, 112, 113, 116, 127, 128, 132, 275, 279–281, 353, 355, 359, 363, 379, 384, 390, 391, 394, 398, 399, 407, 412, 413, 445
  birds eye disease, 68
    etiology, 69
    favorable conditions, 69
    management, 69
    spread mode, 69
    survival mode, 69
    symptoms, 68
  biology, 384
  diagnostic symptoms, 384
  disease cycle, 384
  disease distribution, 384

510   *Index*

disease occurrence, 384
molecular characterization, 384
Antibacterial properties, 268
Antibiotics, 71, 428
Antifungal compounds, 109
Antisporulant activity, 234
Apex, 20, 67, 80, 83, 84, 130, 218, 235, 244, 250, 309, 466, 478
Aphid, 51, 54, 167–169, 223, 319, 320, 341, 371–375, 395, 420–422
  vector, 167, 291, 421
    population, 371, 372, 375
Apiaceae, 324, 331, 462, 464
Apothecia, 328, 329, 410
Apothecium, 410
Apple mosaic
  disease, 37
  virus, 37, 38
Araceae, 215, 216
Armillaria, 29, 30
  mellea, 30
  root rot, 29
  species, 29
  tabescens, 29
Arthritis, 268
Ascocarps, 50
Ascochyta blight, 249, 253
  causal organism, 250
  disease cycle, 250
  epidemiology, 250
  management, 251, 252
  symptoms, 249
Ascopspores, 20, 21, 47, 50, 69, 76, 101, 232, 250, 251, 288, 328, 387
Ascus, 23, 101, 288, 387
Asiatic citrus psyllids, 42
Asparagines, 160
Aspergillus
  flavus, 116
  niger, 89, 115, 116, 221
Aureofungin, 27, 31, 138, 140
Axillary shoots, 315
Azadirachta indica, 242, 264, 426
Azoxystrobin, 68, 109, 111, 114

B

Bacilliform virus, 223
Bacillus

  polymyxa, 150
  solanacearum, 149
Bacteria ooze, 405
Bacterial disease, 12–15, 33–36, 41–46, 85, 137–140, 147–151, 197–201, 212, 217, 276–279, 308, 324–326, 336, 337, 343, 353, 366–370, 379, 380, 385, 398, 401, 402, 405, 428
  angular leaf spot, 276
    causal organism, 277
    disease cycle, 277
    management, 277
    symptoms, 276
  bacterial canker, 201, 366
    cause and conditions, 201, 366
    management, 201, 366
    symptoms, 201, 366
  bacterial heart rot, 14
    management, 14
    survival and spread, 14
    symptoms, 14
  bacterial leaf blight, 212
  bacterial leaf spot, 279
    causal organism, 279
    management, 279
    symptoms, 279
  bacterial speck, 200
    cause and conditions, 200
    management, 200
    symptoms, 200
  bacterial spot, 198, 368
    causes and conditions, 199, 369
    management, 199, 369
    symptoms, 198, 368
  bacterial stem and peduncle canker, 367
    disease development conditions, 367
    management, 367
    symptoms, 367
  bacterial wilt, 85, 277, 369
    causal organism, 85, 278
    disease cycle, 278
    disease development conditions, 370
    favorable environment, 85
    management, 86, 278, 370
    survival, 85
    symptoms, 85, 277, 370
  black leg and soft rot, 148
    bacterial wilt, 149

Index 511

brown rot, 149
management, 149
potato ring rot, 151
potato scab, 150
black pit, 45
management, 45
symptoms, 45
black spot, 137
disease development favorable conditions, 138
management, 138
symptoms, 138
brown spot, 402–405
cells, 44, 138, 140, 344, 406
citrus (bacterial) blast, 44
management, 45
occurrence, 44
symptoms, 44
citrus canker, 43
management, 44
spread, 44
symptoms, 43
citrus greening, 41
diseases spread, 42
hosts, 42
management, 42
symptoms, 41
citrus variegated chlorosis, 46
management, 46
symptoms, 46
crown gall, 35
causal organism, 35
disease cycle, 35
epidemiology, 35
introduction, 35
management, 36
symptoms, 35
disease cycle, 404
fire blight, 33
causal organism, 34
disease cycle, 34
epidemiology, 34
introduction, 33
management, 34
symptoms, 33
management, 405
mango bacterial canker disease, 139
disease development conditions, 140

management, 140
symptoms, 139
marbling, 12
management, 13
survival and spread, 13
symptoms, 13
pink disease, 13
management, 14
survival and spread, 13
symptoms, 13
soft rot, 217, 336, 392, 400
syringae seedling blight and leaf spot, 367
disease development conditions, 368
management, 368
symptoms, 368
tomatoes bacterial wilt, 197
cause and conditions, 198
management, 198
symptoms, 197
wilt, 86, 149, 150, 197, 198, 261, 262, 277, 278, 314, 315, 320, 353, 370, 379, 402, 404
Bactericides, 406
Bacterium, 33–35, 42, 44, 46, 71, 85, 151, 152, 175, 198–200, 262, 277, 279, 325, 337, 344, 369, 370, 402–404
Bamoviruses, 204, 376
Basidiomycota, 447, 450
Basidiospores, 25, 30, 244, 245
Beet curly top virus (BCTV), 372, 373, 376
Begomoviruses, 202
Beta-amino butyric acid (BABA), 202, 208, 366, 376
Bio-agents, 158, 161, 162, 194, 241, 354, 355
Bio-control agents (BCAs), 61, 89, 101, 112, 192, 225, 241, 263, 269, 270, 306, 445, 450
Biofumigation, 267
Bioinoculants, 98
Biological factors, 328
Biomass, 230
Biotechnological tools, 122
Biotic
constraints, 65
factors, 30
stresses, 155, 177
Bitunicate walls, 20

512 *Index*

Black
 banded, 135, 142
 encrustation, 158
 fibers, 10
 lesions, 33, 100, 200, 280, 311, 412
 mildew, 133
 spot, 47, 48, 68, 78, 80, 127, 128, 137, 142,
  160, 308, 384, 390, 451, 464, 466, 469
Blackening, 160, 164, 169, 216, 337, 340,
 344, 498
Blackleg disease, 148
Bleaching, 150, 203, 337, 410
Blossoms, 19, 33, 34, 48, 132, 133, 196,
 199, 362, 369, 496
Blotches, 32, 57, 80, 96, 119, 120, 132, 389
Blue mold, 60
Bordeaux, 45, 46, 70, 82, 97, 100, 101, 103,
 130, 135, 136, 138, 142, 263, 265, 266,
 308
Boric acid, 119, 149, 151, 158, 163, 170
Boron
 deficiency, 9, 335, 340, 348, 501
 toxicity, 502, 505
Botryodiplodia, 80, 98, 110, 116, 212, 221,
 436
 theobromae, 80, 98, 110, 221, 436
Brassicaceae, 335, 343, 462, 464
Brassicol paint, 27, 29
Breeding, 177, 220, 223, 225, 233, 349
Brinjal diseases, 303
 alternaria leaf spots, 310
  casual organism, 311
  disease cycle, 311
  disease management, 311
  epidemiology, 311
  symptoms, 310
 bacterial wilt, 314
  causal organism, 314
  disease cycle, 315
  disease management, 315
  epidemiology, 315
  symptoms, 314
 brinjal little leaf, 315
  causal organism, 316
  disease cycle, 316
  disease management, 316
  epidemiology, 316
  symptoms, 316

damping off, 304
 causal organisms, 304
 disease cycle, 305
 disease management, 305
 epidemiology, 305
 post-emergence damping-off, 304
fruit rot, 319
 disease management, 319
leaf spot, 308
 casual organism, 308
 disease cycle and epidemiology, 309
 disease management, 309
 symptom, 308
mosaic, 319
 disease management, 319
phomopsis blight, 306
 causal organism, 307
 disease cycle, 307
 disease management, 307
 epidemiology, 307
 symptoms, 306
root knot disease, 317
 casual organism, 317
 disease cycle, 318
 disease management, 318
 epidemiology, 318
 symptoms, 317
verticillium wilt, 311
 causal organism, 312
 disease cycle, 313
 disease management, 314
 epidemiology, 313
 symptoms, 312
Brown
 citrus aphid, 53, 54
 gummy lesions, 46
 lesions, 25, 91, 153, 155, 161, 219, 267,
  276, 280, 365, 403, 416, 465, 469, 482
 liquid ooze, 148
 rot, 48, 61, 62, 149, 197, 198, 370
 spot bacteria, 404
Browning, 7, 114, 149, 158, 161, 197, 201,
 314, 366, 370, 446
Budding, 36, 37, 49, 56, 57, 312
Budwood, 46, 53, 55, 56, 57
Bulbous hyphae, 235
Burkholderia solanacearum, 149, 198, 369,
 370

C

Calcium oxalate crystals, 216
Callophyllum inophyllum, 264
Candidatus liberobacter asiaticus, 42
Canopy, 41, 48, 51, 61, 90, 91, 93, 94, 100, 108, 110, 118, 119, 136, 141, 149, 309, 410, 474
Captan, 21, 22, 32, 100, 213, 219, 241, 281, 417
Carbendazim, 6, 8, 11, 21, 22, 24, 27, 32, 70, 89, 98, 100, 101, 111, 112, 114, 116, 126, 128, 130, 137, 192, 194, 214, 234, 237, 241, 249, 252, 270, 281, 289, 326, 329, 354, 409, 411, 413, 414, 418, 419, 442
Carbofuran, 169, 174, 175, 216, 320, 332, 341, 342, 420, 424, 426
Carbohydrate, 95, 167, 461, 491, 495, 500
Carbon dioxide, 12, 491
Cardiovascular diseases, 268
Carrot, 163, 205, 323–335, 382, 473–475, 482
  bacterial diseases, 324
    bacterial leaf blight, 324
    bacterial soft rot, 325
    carrot scab, 326
  fungal diseases, 327
    alternaria leaf blight, 329
    Cercospora leaf blight, 330
    powdery mildew, 331
    Sclerotinia rot, 327
  nematodes, 332
    root knot nematodes, 333
  physiological disorders, 334
    cavity spot, 335
    heat canker, 334
    splitting, 335
  viral diseases, 331
    yellows, 332
Cayenne, 5, 7, 13, 14
Cell
  division, 119
  expansion, 119
  wall, 171, 305, 347, 438, 500
Cephaleuros virescens, 92, 96, 117, 141
Cercospora, 308–310, 327, 330, 358, 362, 380, 393, 407
  leaf spot, 358, 388, 419
  melongenae, 309

Cereals, 171, 191, 197, 248, 264, 326, 329, 333, 345, 364, 419, 461
Chalara paradoxa, 6, 11
Charcoal
  rot, 153, 160, 161, 390, 400
  stump rot, 436, 451
Chaubatia paste, 27, 29
Chenopodiaceae, 168, 464
Chitosanase activity, 269
Chlamydospores, 4, 6, 28, 101, 111, 161, 239, 286, 338, 415, 441, 467, 478
Chlorine, 59, 360, 367, 407, 492
Chlorophyll, 176, 421, 491, 494, 495, 499
Chlorosis, 46, 54, 171, 172, 204, 218, 287, 295, 312, 337, 367, 397, 403, 415, 423, 425, 441, 461, 493–496, 498, 499, 502
Chlorothalonil, 21, 98, 281, 283, 310, 409
Chlorotic
  areas, 46
  halo, 139, 368, 409
  lesions, 308, 341, 344, 348, 358
  margins, 468
  mottle, 291, 374
  spots, 169, 308, 344
  streak, 215
Circulifer tenellus, 56, 372, 423
Citrus
  canker, 43, 44
  clonal protection program, 53
  diseases, 43
  exocortis viroid (CEVd), 57, 58, 62
  groves, 53
  growing areas, 53, 55
  orchards, 46, 49, 56
  plants, 51, 53
  psorosis virus, 53
  scab diseases, 50
  sinensis, 212
  species, 54
  stubborn, 55, 56
  trees, 46, 52–55, 57
  tristeza virus (CTV), 53–55, 62
  variegation chlorosis, 46
  varieties, 52, 54
Clavate vesicle, 76
Cleistothecia, 23, 66, 195, 231, 288, 331, 357, 387
Clover yellow vein virus (CYVV), 275, 293

514 *Index*

Club root, 345, 347, 349

Clusters, 102, 127, 135, 244, 309

Coalesce, 20, 77, 79, 82, 90, 96, 110, 127, 134, 136–138, 141, 151, 162, 165, 170, 185, 264, 266, 270, 276, 279, 280, 310–312, 329, 330, 359, 365, 368, 389, 403, 411, 414, 416, 465, 468, 470, 472, 475, 480, 481

Collar
  region, 28, 213, 268, 305, 354, 418, 419, 451
  rot, 27, 28, 102, 185, 212, 213, 399, 437

Colletotrichum, 79, 89, 97, 101, 110, 112, 115, 116, 127, 266, 275, 280, 355, 359, 381, 384, 437, 445, 446
  capsici, 101, 270, 355
  gloeosporioides, 79, 97, 110, 112, 115, 116
  leaf spot, 266
    causal organism, 266
    management, 266
    symptoms, 266
  zingiberi, 266

Colocasia, 205, 211, 215–222, 224, 225, 227
  bacterial diseases, 217
  fields, 218
  fungal diseases, 218
    brown leaf spot, 221
    colocasia phytopthora blight, 219
    pythium root rot, 218
    sclerotium rot, 222
    storage rot, 221
  nematode, 224
    root-knot nematode, 224
  physiological disorder, 224
    loliloli, 225
    metsubre, 224
  tissue, 221
  viral disease, 222
    alomae and bobone virus disease complex, 222
    dasheen mosaic virus (DSMV) disease, 223

Colonization, 278, 328, 428

Concentric rings, 70, 77, 154, 161, 169, 185, 249, 250, 307, 309, 310, 338, 345, 358, 359, 372, 383, 408, 411, 414, 443, 470, 480, 482

Conidia, 6, 20, 21, 23, 32, 50, 66, 69, 70, 76–80, 83, 90, 98, 101, 108, 114, 126, 132, 155, 221, 231–235, 250, 280, 281, 288, 289, 307, 309, 311, 313, 330, 384, 387, 412, 414, 440, 441, 444–447, 451, 466, 467, 469, 471, 472, 474, 475, 478–482

Conidiophores, 20, 21, 23, 32, 66, 76–80, 83, 90, 126, 135, 195, 231, 239, 288, 307, 309, 330, 357, 384, 440, 446, 466, 467, 471–474, 478–480

Copper
  bactericides, 325
  fungicides, 142
  oxychloride, 22, 29, 77–83, 85, 97, 100, 103, 109–111, 114, 118, 128, 130, 134, 135, 139, 140, 142, 184, 188, 198, 252, 262, 263, 281, 287, 330, 331, 370, 419, 439, 442, 454, 455
  sulfate, 103, 114
  toxicity, 498, 499, 505

Correcting
  boron deficiency, 501
  molybdenum deficiency, 503

Corrugated fiber board (CFB), 116, 122

Cortex, 88, 197, 239, 267, 315, 333–335, 370

Corticium salmonicolor, 82, 103

Corynebacterium flaccumfaciens, 402, 428

Cottony
  growth, 153, 319, 388
  leak, 284, 388
  mycelial growth, 26, 388

Cotyledons, 250, 277, 278, 337, 343, 355, 368, 389, 405, 412

Crinkling, 132, 168, 503

Crop
  debris, 70, 76, 78, 79, 81, 83, 84, 159, 162, 164, 195, 196, 213, 236, 239, 245–247, 281, 311, 325, 330, 338, 345, 358, 360, 362, 389, 390, 407–414, 416, 418, 419, 423, 444
  plants, 17, 312, 317, 440, 490
  rotation, 150, 151, 155, 158–160, 162–165, 171, 173, 174, 187, 191, 199, 201, 205, 213, 216, 219, 220, 222, 236, 251, 264, 276, 279, 281, 307, 314, 315, 318, 325, 326, 329, 330, 333, 337, 338,

Index 515

340, 344, 345, 347, 348, 365, 366, 369, 397, 405, 408–411, 413, 414, 416–419, 421, 424, 426
straw, 162
stress, 164
Crown gall, 35, 36, 38
Crucifers, 56, 188, 290, 339, 340, 344, 356, 465
Cucumber mosaic virus (CMV), 275, 289, 290, 292, 300, 371, 372, 374, 376, 379, 380, 382
Cuticle, 21, 32, 48, 58, 454
Cyst
nematode, 173, 423, 426
population, 173
Cytokinins, 131

D

Dark
brown lesions, 365
pycnidia, 32
Dasheen mosaic virus (DSMV), 215, 223, 225
Dead host cells, 50
Deblossoming, 131
Defoliation, 19, 20, 26, 45, 52, 56, 67, 70, 71, 77, 85, 91, 129, 134, 164, 184, 196, 212, 290, 310–312, 358, 362, 365, 372, 375, 409, 411, 414, 415, 454, 461, 470
Dehydration, 323
Denematization, 267
Die back
phase, 89
symptoms, 103
Difenoconazole, 109, 111, 114, 248, 270
Discoloration, 7, 30, 31, 33, 61, 66, 80, 85, 88, 93, 97, 129, 139, 149, 163, 164, 166, 175, 188, 191, 197, 198, 238, 262, 276, 312, 337, 356, 367, 370, 372, 404, 417, 428, 453, 481, 496, 501
Disease
cycle, 35, 38, 220, 232, 253, 307, 311, 365, 385, 405, 436
dynamics, 121, 122
management, 96, 147, 153, 177, 225, 338, 344, 346, 398, 401, 419, 428, 445, 447, 451, 454
prone areas, 90, 413

resistant cultivars, 225, 411, 416
Disorders, 122, 206, 268, 334, 379, 426, 492, 500
catface, 207, 208
tomatoes blossom-end rot, 206, 207
Distortion, 50, 91, 132, 202, 203, 214, 215, 223, 238, 289–292, 341, 374, 375
Dithanon, 21, 22
Dodine, 21, 22
Domestic quarantine, 159
Downy
blight, 114, 115
mildew, 67, 68, 72, 231, 234–237, 281–283, 345, 348, 349, 379, 384, 386, 387, 398
favorable conditions, 68
management, 68
pathogen, 67
spread and survival, 68
symptoms, 67
Drought stress, 207, 331
Dry rot, 153, 158, 160, 161, 216, 265, 365, 383
causal organism, 265
management, 265
symptoms, 265

E

Early blight, 153, 155, 185, 303, 311, 461
Eggplant, 163, 188, 262, 284, 303, 308–310, 312–314, 316, 317, 321, 356, 358, 361, 371, 372, 375, 470
Electrophoresis, 58
Elephant foot yam, 211–215, 225
Endophyte, 266, 442
Endophytic bacteria (EB), 263, 270
Endosperm, 204, 376
Epidemics, 27, 44, 102, 232, 253, 328, 329, 447
Epidemiology, 102, 103, 109, 132, 328, 387
Epidermal
cell walls, 32
layers, 311
Epidermis, 48, 58, 91, 126, 155, 165, 235, 267, 335, 359, 367, 410, 446, 471
Epinasty, 57, 373
Eradication, 43, 169, 171, 277, 317, 320, 338, 398

516   *Index*

Erosion, 5, 335, 407, 495
Erwinia
  amylovora, 33, 34
  ananas, 12
  chrysanthemi, 14, 17
Erysiphe
  cichoracearum, 287, 288, 379, 381, 386
  heraclei, 331
  pisi, 231
  polygoni, 230, 231
  trifolii, 231
Etiology, 103, 316, 462, 482
Exocortis, 57, 58, 62

F

Fertilization, 45, 93, 164, 192, 195, 207,
  245, 292, 296, 299, 305, 358, 498
Fertilizers, 149, 155, 163, 198, 240, 287,
  295, 298, 326, 336, 370, 427, 489, 490,
  494, 500–502, 504
Fibrosin bodies, 23, 288
Fire blight, 33, 34, 38
Flagella, 35, 220, 277, 278, 314
Flavescence, 171, 172
Floral
  calyx, 196, 362
  cavity, 7
Fludioxonil, 59, 61, 409
Fluorescent light, 102
Flyspeck, 31, 32
Foliar
  application, 15, 252, 314, 422, 439, 493,
    497, 498, 501
  blight, 110, 310, 476
  diseases, 23, 68, 94, 349
  spray, 70, 110, 118, 169, 234, 249, 252,
    265, 320, 341, 346, 397, 398, 497,
    499–501, 503, 504
  wilting, 163
Food and Agriculture Organization (FAO),
  229, 253
Formaldehyde, 27, 86
Fructification, 141, 446, 450, 451, 453
Fruit
  blight, 108, 109
  canker, 44, 91–94, 201, 366, 461
  cracking, 118, 119, 122
  deformation, 20

  rot, 8, 9, 12, 28, 80, 81, 86, 89–91, 93,
    96, 98–100, 115, 116, 127, 283, 284,
    306, 319, 355, 356, 359, 360, 379, 388,
    461, 468, 469, 471, 476, 478
  spot, 82, 83, 92, 93
Fruitlet, 7, 9, 10, 12, 13, 21, 24
Fuhgicides, 234
Fungal/fungal-like diseases, 108, 152, 155,
  182, 337, 340, 428
  algal leaf spot, 92, 96
    causal organism, 96
    disease cycle, 97
    disease symptoms, 92
    etiology, 96
    favorable conditions, 93, 97
    management, 97
    survival and spread, 92
    symptoms, 96
  alternaria leaf spot, 77, 408
    causal organism, 77
    favorable environment, 77
    management, 78
    survival, 77
    symptoms, 77
  angular leaf spot, 409
  anthracnose fruit rot, 359
    causal agent, 359
    symptoms, 359
  anthracnose, 79, 112, 127, 279, 412
    causal organism, 79, 280
    disease development conditions, 128
    epidemiology, 280
    favorable environment, 80
    management, 80, 128, 281
    spread mode, 280
    survival, 80
    symptoms, 79, 127, 280
  apple scab, 19
    disease cycle, 21
    epidemiology, 21
    management, 21
    symptoms, 19
  armillaria root rot, 29
    causal organism, 29
    disease cycle, 30
    epidemiology, 30
    introduction, 29
    management, 30

Index

symptoms, 29
aschochyta leaf and pod spot, 411
ashy stem blight, 414
base (butt) rot, 6
   management, 6
   survival and spread, 6
   symptoms, 6
black banded, 134
   management, 135
   symptoms, 135
black canker, 78
   causal organism, 78
   favorable environment, 78
   management, 79
   survival, 79
   symptoms, 78
black scurf, 155
causal organism, 238
cercospora leaf spot, 358
   causal agent, 358
   disease development conditions, 358
   management, 358
   symptoms, 358
charcoal rot, 160
choanephora blight, 360
citrus black spot, 47
   management, 47
   spread, 47
   symptoms, 47
citrus gummosis, 48
   management, 49
   spread, 49
   symptoms, 48
citrus powdery mildew, 51
   management, 52
   spread, 52
   symptoms, 51
citrus scab, 50
   management, 50
   symptoms, 50
collar rot, 213
colletotrichum root rot, 163
cylindrocladium leaf spot, 76
   causal organism, 76
   favorable environment, 76
   management, 76
   survival, 76
   symptoms, 76

damping off and root rot, 353
   disease development conditions, 354
   management, 354
   symptom, 354
damping-off of vegetables, 192
   causes and conditions, 193
   management, 194
   symptoms, 192
die back, 97, 129
   causal organism, 98
   disease development favorable conditions, 129
   epidemiology, 98
   etiology, 98
   management, 98, 130
   spread and survival mode, 98
   symptoms, 97, 129
die-back and fruit rot, 355
   causal organism, 355
   disease development conditions, 355
   management, 356
   symptoms, 355
dieback and anthracnose fruit rot, 89
   disease symptoms, 89
   favorable conditions, 91
   management, 91
   pathogen, 90
   survival and spread, 90
disease cycle and epidemiology, 239
downy blight, 114
downy mildew, 281
   causal organism, 282
   epidemiology, 282
   management, 283
   survival and spread mode, 282
   symptoms, 281
early blight, 154
fruit and vine rot, 285
   causal organism, 286
   disease cycle, 286
   epidemiology, 286
   management, 287
   symptoms, 286
fruit canker, 91
   disease symptoms, 91
   favorable conditions, 92
   survival and spread, 92
fruit rot, 12, 80, 98, 116, 283

518   *Index*

causal organism, 80, 99, 284
epidemiology, 99, 284
etiology, 99
favorable environment, 81
management, 12, 81, 100, 284
survival and spread mode, 12, 99, 284
symptoms, 12, 80, 98, 283
fruit spot, 82
causal organism, 83
favorable environment, 83
management, 83
survival, 83
symptoms, 82
fruitlet core rot (green eye), 7
management, 7
survival and spread, 7
symptoms, 7
fungal fruit rots, 359
symptoms, 359
fungal wilt, 188
cause and conditions, 191
fusarium wilt, 188
management, 191
symptoms, 188
verticillium wilt, 189
fusariosis, 7
management, 8
survival and spread, 8
symptoms, 8
Fusarium wilt, 161, 188, 191, 237, 238, 242, 312, 356
causal agent, 356
disease development conditions, 356
gray leaf spot, 361
disease development conditions, 361
management, 361
symptoms, 361
green fruit rot, 8
management, 9
survival and spread, 8
symptoms, 8
grey blight, 137
management, 137
symptoms, 137
growth, 81, 126, 135, 154, 192, 194, 195, 282, 307, 357, 360, 410, 418
guava rust, 91
inter fruitlet corking, 9

late blight, 135, 153
management, 136
symptoms, 135
leaf spot, 100, 110, 362
causal organism, 101
cause and condition, 362
epidemiology, 101
etiology, 101
management, 101, 364
symptoms, 100, 362
leathery pocket, 9
management, 10, 11
survival and spread, 10
symptoms, 9, 10
mango malformation, 130
factors, 131
management, 131
symptoms, 130
pathogens, 110
penetration, 21
phoma blight, 134
management, 134
symptoms, 134
phomopsis blight, 365
disease development conditions, 365
management, 365
symptoms, 365
phytophthora heart (top) rot, 4
management, 4
survival and spread, 4
symptoms, 4
phytophthora root rot, 5, 27
causal organism, 28
disease cycle, 28
disease management, 28
epidemiology, 28
introduction, 27
management, 5
survival and spread, 5
symptoms, 5, 28
phytophthora soft rot, 101
causal organism, 102
epidemiology, 102
etiology, 102
management, 102
symptoms, 101
pink disease, 81, 103, 136
causal organism, 82, 103

Index 519

favorable environment, 82
management, 82, 103, 136
survival, 82
symptoms, 81, 103, 136
pink rot, 164
potato wart, 159
powdery mildew, 194
cause and conditions, 195
management, 195
symptoms, 194
powdery mildew, 22, 125, 287, 357, 413
causal agent, 357
causal organism, 23, 287
disease cycle, 23
disease development conditions, 126, 357
epidemiology, 23, 288
introduction, 22
management, 23, 126, 289, 358
survival and spread mode, 288
symptoms, 22, 126, 287, 357
powdery scab, 162
purple blotch, 84
causal organism, 84
favorable environment, 84
management, 84
survival, 84
symptoms, 84
rhizopus rot, 359
disease development conditions, 360, 361
management, 360, 361
symptoms, 359, 360
rust, 24, 407
causal organism, 25
disease cycle, 25
epidemiology, 25
introduction, 24
management, 25
symptoms, 24
scab, 131
disease development favorable conditions, 132
management, 132
symptoms, 132
septoria leaf spot of vegetables, 196
cause and condition, 196
management, 197

symptoms, 196
silver scurf, 165
sooty blotch and fly speck, 31
causal organism, 32
disease cycle, 32
epidemiology, 32
introduction, 31
management, 32
symptoms, 32
sooty mold, 51, 133
disease development favorable conditions, 133
management, 51, 133
symptoms, 51, 133
southern stem blight, 30
disease cycle, 31
disease management, 31
epidemiology, 31
introduction, 30
symptoms, 30
spores, 99, 166, 309, 313, 346, 365, 389, 411, 412
styler end rot, 93
disease symptoms, 93
favorable conditions, 93
management, 93
survival and spread, 93
tomato early blight, 184
cause and condition, 187
management, 187
symptoms, 185
tomato late blight, 182
cause and conditions, 184
management, 184
symptoms, 182
tuber rot, 213
twig blight, 110
water blister, 10
white leaf spot, 11
management, 11
survival and spread, 11
symptoms, 11
white mold, 409
white root rot/Hite root rot, 26
causal organism, 26
disease cycle and epidemiology, 27
introduction, 26
management, 27

symptoms, 26
wilt, 87, 111
economic importance, 87
favorable conditions, 88
management, 88
symptoms, 88

G

Gall index, 334
Gelatinous matrix, 174, 425
Genus potyvirus, 215
Geotrichumcitri-aurantii, 58
Germination, 49, 59, 67, 90, 126, 155, 192, 204, 232, 233, 236, 241, 244–246, 251, 252, 270, 277, 281, 283, 288, 304, 305, 333, 340, 341, 345, 346, 348, 354, 385, 408, 412, 413, 417, 465, 481, 500
Gibberellins, 119
Gliocladium roseum, 242, 252
Golden
mosaic virus, 419, 420
nematode, 173
Grafting, 36, 37, 54–56, 215, 371, 421, 422
Graft-transmissible diseases, 53
Gram-negative, 34, 35, 85, 199, 314
Green
canopy, 42
manure, 121, 151, 158, 160, 163, 173, 333
mold, 59, 60
parasitic alga, 96
pods, 401
scarf, 96
spores, 59
tip stage, 21
Greenhouse, 54, 97, 208, 310, 250, 313, 315, 318, 367, 372, 374, 376, 386, 388, 396, 503
Greening, 35, 41, 42, 56, 176, 177
Guava, 87–89, 91, 92, 94
Gummosis, 48, 49, 102, 128, 129, 498
Gummy stem blight, 390, 391, 398, 400
Guttation salt injury, 295
Gymnosporangium, 24, 25
Gypsum, 89, 151, 326, 397

H

Halo blight, 401–405
Harbor viruliferous aphids, 167

Harvest tubers, 149
Haustoria, 23, 67, 101, 235, 244, 288, 387
Heart rot, 6, 9
Helicotylenchus dihystera, 88, 118
Herbaceous plants, 35, 57, 448
Hexaconazole, 21, 112, 234, 241, 248, 249, 439, 442
Hollow
heart, 177, 299
stem, 348, 349
Honeydew, 15, 51, 133, 276, 295, 391
Horticultural, 26, 34, 51
Host
cells, 57, 326, 346
penetration, 195, 357, 387
plants, 42, 194, 195, 313, 354, 357, 396, 408, 410, 419, 425, 436, 438, 448, 451
range, 26, 29, 30, 150, 160, 168, 205, 223, 290, 292, 313, 316, 372, 373, 380, 391, 416, 418, 448
Hyaline, 23, 50, 67, 69, 76, 78, 79, 82–84, 90, 235, 244, 245, 264, 280, 282, 286–288, 307, 309, 312, 384, 387, 440, 444–446, 466, 471, 475
Hybridization, 122
Hybrids, 160, 469
Hydathodes, 217, 405
Hydrogen
peroxide, 33
sulfide, 315
Hyphae, 4, 26, 32, 67, 83, 102, 220, 221, 235, 239, 245, 282, 286, 288, 307, 312, 415, 440, 446, 466, 473, 480
Hypocotyls, 304, 389, 404, 412, 415, 416, 465

I

Imazalil, 59, 60, 61
Immunosorbent assay, 214
In vitro
antagonism, 263
conditions, 252
Indole production, 315
Induce systemic resistance (ISR), 439, 455
Inflorescence, 7–9, 65, 87, 91, 98, 99, 108, 128, 131, 338, 482
Inoculation, 102, 158, 169, 215, 233, 240, 242, 248, 290, 420, 425

Index 521

Insecticide, 16, 43, 98, 131, 167, 241, 278, 291, 371, 373, 374, 505
Integrated crop management (ICM), 107, 122
Intercropping, 220, 420
Internodes, 56, 130, 172, 238, 286, 290, 311, 316, 496
Iron
  deficiency symptoms, 499
  toxicity, 500, 505
Irrigation
  management, 406
  water, 36, 85, 148, 150, 174, 187, 200, 309, 325, 336, 347, 356, 358, 367, 391, 396

K

Karathane, 24, 127, 289, 414
Kholrabi, 343, 349
Knol Khol, 323, 343–346, 348, 349, 351
  bacterial diseases, 343
    black rot, 343
  fungal diseases, 344
    alternaria leaf spot, 345
    club root, 346
    damping off, 345
    downy mildew, 348
  physiological disorder, 348
    hollow stem, 348
Konjac mosaic virus (KoMV), 214, 225

L

Lanceolate, 20
Leaf
  blight, 108–110, 142, 219, 295, 324, 327, 330, 461, 474, 476, 478
  epinasty, 58
  lamina, 137, 214, 215, 312, 341, 389
  lesions, 20, 182, 279, 280, 420
  petiole, 44, 45
  spot
    causal organism, 270
    management, 270
    symptoms, 269
  variegation, 453
Leafhoppers, 56, 171, 172, 293, 316, 332, 372, 373, 501
Leaflets, 166, 168, 171, 172, 188, 201, 238, 325, 356, 366, 496

Leafy vegetables, 462, 483
Legumes, 197, 364, 412, 421, 428, 491, 492
Lenticels, 150, 151, 158, 160, 164
Lesions, 20, 43, 45, 132, 139, 155, 165, 182, 184, 250, 307, 310, 326, 338, 340, 358, 359, 365, 368, 389, 403, 410, 412
Leveilulla taurica, 195, 357
Light pruning (LP), 450
Litchi, 108, 110–114, 116–118, 120–122
Loliloli corm, 225
Low temperature injury, 295

M

Macrophomin aphaseolina, 153, 380
Macular degeneration, 95
Maize, 121, 158, 160, 198, 240, 370
Malady, 119, 130, 427
Malathion, 317
Malformation, 130, 131, 203, 207, 224, 346, 420
Malling Merton 106 (MM.106), 27
Mancozeb, 8, 9, 21, 22, 26, 77–81, 83, 85, 91, 94, 103, 154, 155, 162, 220, 222, 248, 270, 281, 283, 287, 331, 339, 340, 408, 413, 419
Manganese, 380, 382, 397, 490–494
Mangifera indica, 117, 137
Manuring, 89, 141, 158, 401, 419
Maroon spots, 325
Mealybug, 5, 15, 16, 51, 133, 223
Medium pruning (MP), 450
Meloidogyne incognita, 118, 215, 224, 267, 318, 380, 382, 423
Meristem, 223, 235, 420
Mesocarp tissue, 93
Methyl bromide, 417
Methylesterase, 55
Microcracks, 116
Microdiplodia itchi, 110
Micrometers, 425
Micronutrients, 489, 490, 504, 505
Microorganisms, 181, 262, 491
Microsclerotia, 312, 313
Mild strains, 36, 55
Moisture stress, 163, 174, 401
Molecular testing, 34
Molybdenosis, 503

Molybdenum, 275, 380, 382, 397, 490, 491, 496, 502, 503, 505
  deficiency, 295, 503
  toxicity, 503, 505
Monocots, 56
Monocrotophos, 169, 170, 214, 215, 341, 342, 420–422
Moth bean, 481, 483
Motile
  bacterium, 35
  zoospores, 4, 49, 286, 340, 346
Mulching, 48, 77–79, 81, 83, 84, 119, 287, 298, 329
Multilobate appressoria, 66
Mummification, 69
Mushroom phase, 30
Mushy rot, 148, 337
Mycelia, 23, 28, 30, 31, 59, 61, 82, 98, 99, 101, 113, 194, 284, 287, 313, 328, 346, 354, 359, 360, 386, 410, 440, 451
Mycelial
  extrusions, 29
  growth, 32, 135, 163, 213, 221, 222, 235, 241, 327
  masses, 102
  mat, 30
Mycelium, 23, 27, 30, 48, 50, 59, 60, 66–70, 76–80, 82–84, 92, 98, 99, 114, 116, 155, 160, 161, 195, 235, 239, 242, 244, 245, 280, 282, 284, 286–288, 304, 305, 307, 327, 328, 331, 338–340, 357, 359, 384, 385, 387, 414, 417, 418, 446, 451, 466, 471, 478
  septate, 78, 80, 478
Mycoplasma, 171, 175, 331, 332, 341, 342
Mycoplasmal diseases, 171
  management, 172
  marginal flavescence, 172
  purple top roll, 171
  witch broom, 172
Myzus persicae, 167–169, 214, 223, 291, 319, 420

N

Necator stage, 82
Necroses, 154
Necrosis, 25, 54, 67, 97, 108, 152, 161, 166–170, 172, 204, 222, 223, 290, 292, 295, 355, 367, 375, 380, 397, 403, 415, 420, 421, 453, 454, 461, 478, 498
Necrotic
  areas, 71, 90, 91, 128, 168, 170, 290, 368, 375, 408
  flecks, 330
  lesions, 43, 90, 168, 420, 494
  patches, 470
  spots, 79, 138, 397, 468, 500
Neem bark methanol extract (NBM), 233, 253
Nematicide application, 118
Nematode diseases, 118, 205, 215, 293, 379
  lesion nematode, 216
  nematode life cycle, 294
  non-parasitic diseases, 295
  root-knot nematode, 215
    symptoms, 294
  sting nematodes, 294
    management, 294
    symptoms, 294
  potato cyst nematodes, 173
    management, 173
  root knot nematode, 174
    management, 174
  tomatoes root-knot, 205
    cause and conditions, 205
    management, 205
    symptoms, 205
Nitrogen, 380, 382, 396
  fertility, 149
  fertilizer, 176, 208, 330, 334, 335, 410
  fixation, 491, 492, 502
  starvation, 503
Non-host crops, 164, 171, 173, 199, 200, 205, 236, 240, 358, 369, 405, 411, 420
Non-parasitic disease, 175, 177, 300
  greening, 176
  hollow heart, 176
  potato black heart, 175
Nucleic acid, 54
Nutritional disorders, 380, 396

O

Oak root fungus disease, 29
Oblique septa, 466, 467, 472, 474, 475, 478, 479, 481
Oidiopsis taurica, 195, 357

Index 523

Oidium
  farinosum, 23
  lycopersici, 195, 357
Ontogeny, 335
Oogonia, 286, 305
Oomyceteous, 218, 219
Oospores, 28, 68, 114, 164, 194, 220,
  234–237, 282, 284, 305, 339, 340, 354,
  385, 388, 416, 417
Orchard hygiene, 114, 116
Organic
  debris, 31, 100
  growers, 234, 274
  manures, 121, 263
  material, 100
  matter, 3, 31, 158, 230, 284, 345, 388,
    415, 417, 419, 499, 500
  soils, 495, 496, 498, 500
Ostioles, 20, 78, 250, 264
Oxycarboxin, 82, 248
Oxydemeton methyl, 204
Ozone injury, 426, 428, 429

P

Paecilomyces lilacinus, 216, 318, 319, 334,
  426
Pathogen
  multiplication, 34
  population, 109
  profile, 440
Pathogenicity, 238, 326
Pea downy mildew, 234
  causal organism, 235
  disease cycle, 236
  epidemiology, 236
  management, 236
    chemical, 237
    cultural, 236
    host resistance, 237
  symptoms, 234
Pedicel, 19, 339
Peduncles, 358, 361
Penicillium
  citrinum, 89
  digitatum, 59
  funiculosum, 7, 9, 10
  italicum, 60
Pentalonia nigronervosa, 214, 215, 223

Pepper
  mild mottle virus (PMMV), 375, 376
  mottle virus (PepMoV), 374, 375
Perennation, 172, 250
Perennial plants, 195, 357
Pericarp, 114, 119, 468
Periderm, 162, 166
Peridermal layers, 176
Peripheral ring, 445, 471
Perithecia, 76, 101, 250, 387
Peronospora pisi, 230, 235
Peroxidase (PO), 439, 455
Peroxyacetic acid, 33
Pestalotiopsis mangiferae, 137, 444
Pesticides, 147, 248, 439
Petioles, 19, 24, 44, 57, 138, 139, 153,
  168, 188, 195, 220, 222, 223, 231, 243,
  276, 280, 290, 308, 316, 325, 327, 329,
  356–358, 361, 372, 389, 392, 441, 473,
  474, 482, 501
Phenamiphos, 342, 424, 426
Phenylalanine ammonia-lyase (PAL), 114,
  439, 455
Phloem
  cells, 316
  translocation, 166
Phoma
  blight, 134, 142
  glomerata, 134
Phomopsis, 61, 303, 304, 307, 365, 437
  annonacearum, 78
Phosphamidon, 169, 320, 342
Photosynthesis, 51, 137, 348, 490–492
Phyllody, 341, 342, 419, 422
Phylloplane bacteria, 428
Phyllosticta leaf spot, 264
  causal organism, 264
  management, 265
  symptoms, 264
Physiological disorder, 41, 103, 118, 181,
  224, 225, 295–300, 334, 335, 348, 426
  blossom end rot, 297, 298
  fruit cracking, 118
  hollow heart, 299
  light belly color, 299
  pointed gourd unfruitfulness, 295, 296
  sunburn, 119
Phytopathogens, 266, 440

524   *Index*

Phytophthora, 4–6, 8, 27, 28, 48, 49, 61, 86, 96, 101–103, 153, 184, 193, 212, 218, 219, 275, 284, 286, 304, 319, 354, 359, 360, 381
   cinnamomi, 8
   erythroseptica, 153
   infections, 49
   omnivore, 27
   palmivora, 84, 86, 102
   soft rot, 96, 103
   species, 49
Phytoplasma, 275, 293, 316, 324
Phytoplasmal diseases, 289–293, 315
   aster yellows, 293
      management, 293
      symptoms, 293
Phytoplasmas, 316
Phytotoxic reaction, 252
Pigeon pea cyst nematode, 426
Pink
   disease, 14, 17, 86, 136, 142
   encrustation, 103
Plant
   cell, 36, 57, 293
   debris, 4, 21, 70, 71, 76, 77, 79, 92, 93, 161, 187, 200, 201, 204, 222, 231, 232, 244, 252, 277, 280, 288, 289, 305, 307, 309, 313, 318, 327, 337, 339, 340, 343–346, 348, 355, 356, 358, 361, 365, 366, 376, 383, 403, 404, 413, 414, 482, 483
   density, 408
   genotypes, 34, 232
   growth-promoting (PGP), 263, 270
   heart, 16
   hormones, 36, 491
   pathogenic species, 309
   pathologists, 177
   protection code (PPC), 439, 455
   rhizospheres, 111
   rot disease, 218
Plantation, 5, 17, 27, 129, 435, 445, 449, 452
Plasmodium, 163, 347
Pod
   infection, 235, 250
   lesions, 403, 411
   surface, 235, 405
   symptoms, 402, 403

Podding stage, 232
Pod-fill stage, 402
Podosphaera leucotricha, 23
Pointed gourd, 274, 285, 296–298, 380–382, 476
Polar flagella, 314
Pollination, 296, 297, 299, 376
Polyembryonic cultivars, 139
Polyethylene, 287, 425
Polyphenol oxidase (PPO), 114, 439, 455
Post-harvest, 108, 114, 116, 122, 128
   decay, 61, 360, 367
   diseases, 58
      blue mold of citrus, 60
      brown rot, 61, 62
      green mold of citrus, 59
      phompsis end rot of citrus, 61
      sour rot of citrus, 58, 59
   fruit decay, 116
   lesions, 47
   stages, 61
Potassium, 221, 380, 382, 397
Potato
   mottle virus, 167
   tuber yield, 166
   virus X (PVX), 167–169, 177
   virus Y (PVY), 168, 169, 177, 319, 374
Potential
   catastrophic losses, 53
   damage, 108
Potyvirus, 373
Powdery
   causal organism, 231
   disease cycle, 23, 231
   epidemiology, 23, 231
   favorable conditions, 66
   growth, 67, 194, 357, 386
   infection, 232
   management, 66, 232, 233
   mass, 162, 231, 232, 331, 414
   mildew, 22, 23, 51, 52, 65, 66, 126, 194, 195, 230–233, 275, 287–289, 297, 327, 331, 353, 357, 358, 379, 386, 394, 398, 399
   pathogen, 66
   scab, 163
   spores, 52, 71
   spread and survival mode, 66

Index

symptoms, 65, 231
Premature
 chlorosis, 26
 yellowing, 20, 173
Propagule, 214, 305, 328, 418
Propiconazole, 234, 248, 270, 439
Proteinase, 55, 491
Proteolytic enzymes, 305
Pseudomonas
 chlamydosporia, 206
 fluorescens, 112, 150, 213, 225, 334, 439
 solanacearum, 148, 149, 198, 314, 315, 369, 370
 viticola, 71, 72
Pseudoperonospora cubensis, 282, 379, 385
Pseudostem, 261, 262, 265, 268, 269
Psidium
 cattleianum, 89
 galgava L., 87
Psorosis, 52, 53
Pustules, 25, 50, 61, 117, 136, 162, 185, 243, 325, 339, 407, 408, 443
Pycnidia, 47, 81, 98, 110, 135, 249, 250, 264, 307, 365, 411, 414
Pycnidium, 78, 264, 307
Pycniospores, 245, 414
Pythium, 5, 193, 218, 252, 269, 275, 284, 304, 305, 345, 354, 379, 381, 407, 415–417, 437
 aphanidermatum, 218, 263, 269, 284, 305, 345, 379, 381
 arrhenomanes, 5
 graminicolum, 269
 oligandrum, 252
 root rot, 218, 407, 415–417
 sylvaticum, 417

R

Radish, 151, 163, 173, 188, 326, 333, 335–342, 349, 356, 382, 464–466, 501
 bacterial diseases, 336
  bacterial soft rot, 337
  black rot, 337
  root rot of radish, 336
 fungal diseases, 337
  alternaria blight, 338
  black root, 340
  white rust, 339

mosaic virus, 341
nematode, 342
 stunt nematodes, 342
phyllody, 341, 349
physiological disorders, 343
viral diseases, 341
 radish mosaic, 341
 radish phyllody, 341
Ralstonia solanacearum, 149, 198, 262, 263, 369, 370
Rejuvenation, 450, 455
 pruning (RP), 450, 455
Relative humidity (RH), 21, 23, 32, 38, 51, 68, 72, 76, 77, 80, 81, 84, 92, 98, 100, 101, 114, 116, 119, 126, 128–130, 140, 141, 149, 184, 195, 220, 246, 251, 295, 309, 326, 338, 355, 357, 367, 386, 391, 408, 413, 414, 447, 480
Reniform nematode, 423, 425
Rhizoctonia, 153, 158, 194, 219, 284, 304, 345, 353, 354, 381, 407, 415, 418, 436
 root rot, 417
 solani, 153, 158, 219, 304, 345
Rhizome, 233, 261–263, 265–270
 fly, 262
 rot, 263, 268
  causal organism, 269
  management, 269
  symptoms, 268
  temperatures, 263
  treatment, 263
Rhizomorphs, 29, 30
Rhizopus, 99, 100, 212, 221
 artocarpus, 99
 oryzae, 99
 stolonifera, 99
Rhizosphere, 112, 118
Ribosomes, 316
Rigidity, 15
Ring
 nematode, 118
 rot bacteria, 152
Root
 debris, 27, 204
 elongation, 333
 grafts, 37
 knot, 173, 174, 294, 317, 318, 320, 332–334, 349, 379, 396, 398

disease, 320
galls, 423
nematode, 173, 175, 205, 216, 224, 266, 317, 332, 334, 423, 424
population, 174
symptoms, 215
morphology, 333
Rootstocks, 27, 31, 35, 53–58, 89
Rotylenchulus reniformis, 118, 380, 425, 429
Rust, 24, 25, 71, 72, 91, 96, 117, 141, 169, 243, 245–249, 337, 404, 407, 408, 412, 437, 452
causal organism, 25, 243
disease cycle, 25, 72, 245
epidemiology, 245
management, 72, 247
chemical control, 248
cultural practices, 247
host resistance, 247
symptoms, 71, 243
Rutabagas, 501

S

Saline soils, 75, 230
Sanitation, 36, 44, 49, 52, 88, 93, 97, 109, 138, 140, 155, 159, 167, 169, 170, 173, 184, 191, 195, 197, 200, 220, 222, 223, 236, 240, 247, 277, 307, 318, 345, 358, 360, 364, 367, 376, 399, 406, 408, 418, 449, 451
practices, 36, 97, 152, 155, 159, 162, 173, 184, 195, 197, 200, 222, 345, 358, 364, 376
Sanitization, 59, 60
Saprophytes, 99, 221, 417
Scab
disease, 147, 151, 163
fungi, 50
fungus, 132
lesions, 20, 151, 326
stromata, 50
symptoms, 19, 150, 151
Sclerotia, 27, 30, 31, 158, 160, 163, 194, 213, 222, 312, 327–329, 338, 354, 410, 414, 417, 418
Sclerotinia, 327, 329, 344, 407, 415
Sclerotium rolfsii, 31, 212, 221, 222, 304

Secondary
infection cycle, 21
inoculums, 71
molds, 206
organisms, 4, 347
root system, 416
soft-rot bacteria, 183
Seed treatment, 158, 192, 241, 248, 252, 266, 318, 325, 344, 405, 406, 413, 503
Septate, 26, 76–80, 82, 83, 98, 101, 239, 244, 280, 288, 307, 309, 338, 384, 440, 441, 446, 466, 467, 471, 472, 474, 478, 479
Septoria
leaf spot, 310, 390
lycopersici, 196, 362
Single polar flagellum, 85, 199, 369
Sodium
hypochlorite, 32, 58
solution, 169
ortho-phenylphenate (SOPP), 59, 60, 62
Soft rot, 10, 67, 96, 98–103, 147–151, 217, 262, 265, 268, 276, 325–327, 344, 360, 367, 380, 388, 389, 392, 416
causal organism, 263
management, 263
symptoms, 67, 262
Soil
aeration, 31
amendment, 162, 240
borne
conidia, 101
diseases, 417
fertility, 225, 401, 416, 454
fumigants, 417
fumigation, 164, 314
inoculums, 102
moisture, 27, 31, 119, 120, 160, 194, 207, 284, 286, 298, 304, 305, 318, 340, 342, 347, 354, 356, 383, 410, 415, 416, 418, 428, 500
particles, 27, 194, 340, 354, 388, 424, 448, 490
temperature, 28, 111, 149, 160, 161, 176, 206, 239, 240, 294, 342, 347, 417, 418, 423–425
Solanaceous
crops, 150, 160, 163, 167, 187, 262, 304, 311, 356, 376

Index

527

cultivars, 168
plants, 374
weeds, 310, 361
Solarization, 27, 86, 89, 94, 194, 201, 206, 219, 263, 267, 294, 314, 318, 329, 334, 354, 366, 424, 425
Sooty
blotch, 31, 32, 133
mold, 51, 133, 142
Sphaerotheca fuliginea, 275, 287, 288, 386
Spiroplasma diseases, 55
citrus stubborn disease, 55
management, 56
spread, 56
symptoms, 56
Sporangia, 4, 67, 68, 84, 101, 102, 117, 194, 219, 220, 235, 236, 282–284, 286, 304, 305, 339, 354, 359, 360, 385, 386, 416, 417, 454
Sporangiophores, 67, 114, 220, 235, 282
Sporangium, 84, 99, 102, 304
Spores, 8, 187, 195, 196, 250, 307, 311, 346, 357, 358, 362
Sporulation, 20, 83, 84, 86, 220, 234, 235, 283, 309, 361, 412, 413, 465, 469, 471, 478
Spray program, 310, 358, 504, 505
Squash mosaic virus (SqMV), 275, 289, 300
Stomata, 44, 67, 117, 199, 235, 277, 282, 308, 330, 339, 369, 384, 405, 466, 471, 480
Streptocycline, 140, 150, 198, 337, 370
Streptomyces
griseoviridis, 61
griseus, 455
sannanensis, 455
scabies, 150, 151, 326
Streptomycin, 15, 71, 199, 369, 398, 406
sulfate, 15, 138, 140
Stromata, 83, 90, 280, 330
Stubborn disease, 56
Styler end rot, 94
Subglobose, 244, 264, 288
Subtropical
areas, 103, 158
regions, 47, 136, 153, 198, 324, 448
Sunken lesions, 69, 113, 160, 412, 473
Sunscald, 196, 275, 295, 353, 362, 426, 427, 429, 496

Surface
drip irrigation, 329
lesions, 150
Symptomatology, 462, 483
Syringae seedling blight, 368

T

Taxonomic position, 436, 440
Tebuconazole, 234, 450
Teliospores, 25, 72, 244–246, 407
Terminal leaves, 7, 9, 152, 499
Tetracycline
antibiotics, 57
solution, 317
Thiabendazole (TBZ), 59–62, 252
Thiophanate methyl, 21, 22, 27, 98, 109, 110, 126, 128, 289
Thrips, 16, 17, 170, 203, 208
Tobacco
mosaic virus (TMV), 204, 208, 375, 376
ring spot virus (TRSV), 275, 292
Tomato
leaf curl virus (ToLCV), 202, 208
mosaic virus (ToMV), 204, 208, 375, 376
ring spot virus (TmRSV), 275, 292
spotted wilt virus (TSWV), 16, 17, 169, 170, 203, 208
tobacco mosaic virus (TMV), 203
tomato
spotted wilt virus (TSWV), 203
yellow leaf curl virus (TYLCV), 202
virus diseases, 202
yellow leaf curl virus (TYLCV), 202, 208
Tospoviruses, 16, 17, 169, 203
Toxicity, 494, 497, 498, 502
Transfer DNA (T-DNA), 36
Transplanting, 89, 199, 204, 206, 306, 317, 320, 368, 369, 504
Transplants, 196, 199–201, 308, 309, 311, 314, 346, 348, 362, 365, 366, 369, 371, 376
Tree
canopy, 50, 51, 62, 108
ventilation, 118
Trichoderma
harzianum, 89, 112, 158, 161, 162, 206, 213, 216, 225, 233, 242, 243, 263, 334, 399

528   *Index*

koningii, 252
viride, 89, 101, 243, 269, 418, 439
Tridemorph, 24, 82, 126, 127, 234, 248, 289, 408
Tristeza virus, 53–55, 62
Tropical
  areas, 61, 166
  legume, 401
  storms, 42
True potato seed (TPS), 172, 177
Tuber, 148–155, 158–172, 174–176, 198, 212–218, 333, 370
  crop improvement, 214
  periderm, 151, 165
  rot, 160, 163, 212
  sprouting, 171
  surface, 158, 164, 165
  symptoms, 170
Turgidity, 188, 327, 356
Twig dieback, 45, 454

U

Umbelliferae, 324, 464
Uncinula necator, 65
Uredospores, 72

V

Vascular
  bundles, 88, 149, 197, 344, 370, 404, 405
  parenchyma cells, 333
  region, 197, 370
  ring, 149, 152
  system, 152, 188, 201, 238, 239, 276, 277, 312, 313, 344, 366, 420, 424
  tissue, 35, 129, 139, 188, 201, 217, 312, 333, 344, 356, 366, 404, 405
Vector, 14, 42, 43, 47, 53, 54, 167–170, 214, 215, 290, 291, 293, 332, 371, 375, 420, 421, 423
  activity, 34
  aphid, 167
  populations, 203, 316
  reservoirs, 374
Vegetative
  malformation, 130
  propagation, 17, 42, 46
  propagules, 214
  shoots, 34

Vermicompost, 121, 233
Verticillium, 189–191, 303, 312, 313, 320, 380, 382
  wilt, 189–191, 303, 313, 320, 380
Viral diseases, 15–17, 36, 52–55, 166–171, 214, 215, 268, 290, 291, 319, 332, 341, 349, 371, 379, 398, 419, 428
  amorphophallus mosaic, 214
  apple mosaic virus, 36
    causal organism, 37
    disease cycle, 37
    introduction, 36
    management, 37
    symptoms, 36
    transmission, 37
  beet curly top, 372
    disease development conditions, 373
    management, 373
    symptoms, 372
  chili veinal mottle, 371
    disease development conditions, 371
    management, 371
    symptoms, 371
  cucumber mosaic, 371
    disease development conditions, 372
    management, 372
    symptoms, 372
  dasheen mosaic virus (DSMV), 215
  mealybug wilt disease, 15
    management, 16
    survival and spread, 15
    symptoms, 15
  pepper mottle, 374
    disease development conditions, 374
    management, 375
    symptoms, 374
  pepper yellow mosaic, 373
    disease development conditions, 373
    management, 374
    symptoms, 373
  potato leaf roll, 166
    management, 167
  potato mosaic, 167
    crinkle of potato, 169
    management, 169
    mild or latent mosaic, 167
    rugose mosaic, 168
    vein banding severe mosaic, 168

Index 529

potato spindle tuber, 170
management, 171
potato stem necrosis, 169
management, 170
psorosis, 52
management, 53
symptoms, 52
squash mosaic virus (SQMV), 291, 292
tobamo viruses, 375
disease development conditions, 375
management, 376
symptoms, 375
tristeza disease complex, 53
management, 55
spread, 54
symptoms, 54
yellow spot, 16
management, 17
survival and spread, 16
symptoms, 16
Viroid, 57, 58, 170, 171
exocortis, 57
management, 58
spread, 57
symptoms, 57

W

Water
blister, 10
conductivity, 49
management, 149, 325, 336, 337, 345, 413
scarcity, 335
Waterlogging, 251, 263
Watermelon mosaic virus
1 (WMV-1), 275, 289, 291, 292
2 (WMV-2), 275, 289, 292
Wax treatments, 60

Weed
control, 51
species, 168, 292, 404
Wet rot, 183, 360
White
blight, 452
leaf spot, 11, 17
mildew, 153
powdery spots, 413

X

Xanthomonas
amorphophalli, 212
axonopodis, 405
campestris, 139, 199, 217, 275, 279, 324,
337, 343, 368, 369, 380, 382, 402, 428
Xiphenema inequale, 118
Xylem, 46, 111, 149, 152, 239, 278, 286,
312, 313, 344

Y

Yellow mosaic, 275, 289, 292, 293, 300,
373, 374, 419, 421, 429

Z

Zinc, 89, 380, 382, 397, 490, 491, 495–497
sulfate, 89, 496
toxicity, 497, 505
Zingiberaceae, 268
Zoospoores, 454
Zoosporangium, 305, 347
Zoospores, 67, 68, 84, 93, 102, 117, 163,
220, 235, 282–284, 286, 305, 339, 340,
347, 385, 388, 416, 417
Zucchini yellow mosaic virus (ZYMV),
275, 289, 292, 300